Valorization of Agro-Industrial Byproducts

This book covers sustainable approaches for industrial transformation pertaining to the valorization of agro-industrial byproducts. Divided into four sections, it starts with information about the agro/food industry and its byproducts, including their characterization, followed by different green technologies (principle, process strategies, and extraction of bioactive compounds) applied for the management of agro-industrial byproducts. It further explains biotechnological interventions involved in the value addition of these byproducts. Various regulatory and environmental concerns related to by-product management along with the biorefinery concept and future strategies are provided as well.

Features:

- Provides extensive coverage of agro-industrial byproducts and their environmental impact.
- Details the production of value-added products from agro-industrial waste.
- Describes environmental legislations and future strategies.
- Presents multidisciplinary approaches from fundamental to applied and addresses biorefineries and the circular economy.
- Includes innovative approaches and future strategies for management of agro-industrial waste.

This book is aimed at researchers, graduate students, and professionals in food science/food engineering, bioprocessing/biofuels/bioproducts/biochemicals and agriculture, bioeconomy, food waste processing, post-harvest processing, and waste management.

Valorization of Agro-Industrial Byproducts

Sustainable Approaches for Industrial Transformation

Edited by
Anil Kumar Anal and Parmjit S. Panesar

CRC Press
Taylor & Francis Group
Boca Raton London New York

CRC Press is an imprint of the
Taylor & Francis Group, an **informa** business

First edition published 2023
by CRC Press
6000 Broken Sound Parkway NW, Suite 300, Boca Raton, FL 33487-2742

and by CRC Press
4 Park Square, Milton Park, Abingdon, Oxon, OX14 4RN

CRC Press is an imprint of Taylor & Francis Group, LLC

ISBN: 9780367646554 (hbk)
ISBN: 9780367646578 (pbk)
ISBN: 9781003125679 (ebk)

DOI: 10.1201/9781003125679

Typeset in Times
by Deanta Global Publishing Services, Chennai, India

Contents

About the Editors

Prof. Anil Kumar Anal is the Professor of Food Engineering and Bioprocess Technology and the Food Innovation, Nutrition, and Health programme at the Department of Food Agriculture and Bioresources in the Asian Institute of Technology (AIT), Thailand. He obtained his PhD from AIT in Food Engineering and Bioprocess Technology (FEBT) in 2003. He has experience as a researcher and academic at Otago University and Massey University of New Zealand. His primary research focuses on food safety, food and nutrition security, food bioprocessing and preservation, and bioeconomy. At the same time, Prof. Anal specializes in the valorization of food waste and agro-residues for high value, antimicrobial resistance and risk analysis, and delivery systems for enhancing the bioavailability and bioaccessibility of bioactive compounds. He has invented a patented encapsulation system that has been commercialized to encapsulate probiotics and bio nutrients during processing. He has authored more than 150 refereed international journal articles, 50 book chapters, 1 authored book, 9 edited books, and several international conference proceedings. He has been invited as keynote speaker and presenter at various food, biotechnology, agro-industrial processing, and veterinary as well as life sciences conferences and workshops organized by national, regional, and international agencies. Prof. Anal has served as an advisory member, associate editor, and member of editorial boards of numerous regional and international peer-reviewed journal publications. He has also conducted innovative research and product development funded by influential donor agencies, including the Newton Fund, ADB, European Union, World Bank, Food and Agriculture Organization, various government agencies, CTCN, USAID, and various food and biotech industries.

Parmjit S. Panesar is Professor and Dean (planning and development) at the Sant Longowal Institute of Engineering and Technology (established by the government of India), Longowal, Punjab, India. Prof. Panesar has also been Dean (research and consultancy), and Head of the Department of Food Engineering and Technology, Sant Longowal Institute of Engineering & Technology (SLIET) Longowal, India. His research is focused on the area of value addition of food industry byproducts, the green extraction of bio actives, food enzymes, prebiotics (lactulose, Galactooligosaccharide GOS), and bio pigments, among others. He also has more than 5000 citations and a h-index of 37 (as per Google Scholar).

In 2005, he was awarded a BOYSCAST (Better Opportunities for Young Scientists in Chosen Areas of Science and Technology) fellowship by the Department of Science and Technology, Government of India, to carry out advanced research at Chembiotech Laboratories, University of Birmingham Research Park, UK. His postdoctoral research focused on the development of immobilized cell technology for the production of lactic acid from whey. Additionally, during his stay in the UK he was involved in US and Australian sponsored projects. In 2016, he was also offered a visiting professorship at the Tehran University of Medical Sciences, Tehran, Iran.

Prof. Panesar has successfully completed seven research projects funded by DBT, CSIR, MHRD, AICTE, New Delhi and two projects are in progress. He has published more than 150 international/national scientific papers, 50 book reviews in peer-reviewed journals, 32 book chapters, and has authored/edited 8 books. He has guided 11 PhD students and 8 students are under progress. He is a member of the editorial advisory boards of national/international journals including *International Journal of Biological Macromolecules*, *Journal of Food Science & Technology*, and *Carbohydrate Polymer Technologies & Applications* and currently serves as a member of the national level scientific panel on water and beverages (alcoholic/non-alcoholic) constituted by the Food Safety and Standards Authority of India.

In recognition of his work, Prof. Panesar was granted a Fellow Award 2018 by the Biotech Research Society of India and Fellow Award 2019 by the National Academy of Dairy Science India. He also served as the national vice president of the Association of Food Scientists and Technologists of India in 2019. In 2020, he was selected for the prestigious award of the Indian National Science Academy Teachers. Recently, he has also been listed in the most coveted World Ranking of Top 2% Scientists published by Stanford University, USA.

Prof. Panesar is also a life member of the International Forum on Industrial Bioprocesses, the Biotechnology Research Society of India, the Association of Microbiologists of India, the Association of Food Scientists and Technologists, and the Indian Science Congress Association. He has also visited several countries like the UK, USA, Canada, Switzerland, New Zealand, Australia, Germany, France, Singapore, Malaysia, China, Iran, and Thailand.

Contributors

Shilpi Ahluwalia
Dr. S.S. Bhatnagar University Institute of
    Chemical Engineering & Technology
Punjab University
Chandigarh, Punjab, India

Anil Kumar Anal
Department of Food, Agriculture and
    Bioresources
Asian Institute of Technology
Pathumthani, Thailand

Tuan Pham Anh
Institute of Agricultural Engineering and Post-
    Harvest Technology
Hanoi, Vietnam

Camella Anne Arellano
Department of Food, Agriculture and
    Bioresources
School of Environment, Resources and
    Development
Asian Institute of Technology
Pathumthani, Thailand

Nuntarat Boonlao
Department of Food Agriculture and
    Bioresources
School of Environment, Resources and
    Development
Asian Institute of Technology
Klong Luang, Thailand

Chaichawin Chavapradit
Department of Food Agriculture and
    Bioresources
School of Environment, Resources and
    Development
Asian Institute of Technology
Klong Luang, Thailand

Meera Christopher
CSIR- National Institute for Interdisciplinary
    Science and Technology (CSIR-NIIST)
Thiruvananthapuram, Kerala, India

and

Academy of Scientific and Innovative Research
    (AcSIR)
Ghaziabad, Uttar Pradesh, India

Son Chu-Ky
School of Biotechnology and Food Technology
Hanoi University of Science and Technology
Hanoi, Vietnam

and

International Joint Laboratory
Tropical Bioresources and Biotechnology
Université Bourgogne AgroSup Dijon
    (France) – Hanoi University of Science and
    Technology (Vietnam)

Angelica Corpuz
Department of Food, Agriculture and
    Bioresources
School of Environment, Resources and
    Development
Asian Institute of Technology
Pathumthani, Thailand

and

Department of Chemical Engineering
College of Engineering and Architecture
Cagayan State University
Cagayan, Philippines

Dipak Das
Department of Food Engineering and
    Technology
Sant Longowal Institute of Engineering and
    Technology
Longowal, Punjab, India

Valan Rebinro Gnanaraj
CSIR- National Institute for Interdisciplinary
    Science and Technology (CSIR-NIIST)
Thiruvananthapuram, Kerala, India

Anushuya Guragain
Department of Food Agriculture and
    Bioresources
School of Environment, Resources and
    Development
Asian Institute of Technology
Klong Luang, Thailand

Sushma Gurumayum
Department of Basic Engineering and Applied
    Sciences
College of Agricultural Engineering and
    Post-harvest Technology
Ranipool, Sikkim, India

Gunjan K. Katoch
Department of Food Technology and Nutrition
Lovely Professional University
Phagwara, Punjab, India

Navneet Kaur
Department of Food Processing Technology
Sri Guru Granth Sahib World University
Fatehgarh Sahib, Punjab, India

Rupinder Kaur
Department of Food Engineering and
    Technology
Sant Longowal Institute of Engineering and
    Technology
Longowal, Punjab, India

Sawinder Kaur
Department of Food Technology and Nutrition
Lovely Professional University
Phagwara, Punjab, India

Sushil Koirala
Department of Food Agriculture and
    Bioresources
School of Environment, Resources and
    Development
Asian Institute of Technology
Klong Luang, Thailand

Prajeesh Kooloth-Valappil
CSIR- National Institute for Interdisciplinary
    Science and Technology (CSIR-NIIST)
Thiruvananthapuram, Kerala, India

and

Academy of Scientific and Innovative Research
    (AcSIR)
Ghaziabad, Uttar Pradesh, India

Anjineyulu Kothakota
Agro-Processing and Technology Division
CSIR-National Institute for Interdisciplinary
    Science and Technology
Trivandrum, Kerala, India

Thanh-Ha Le
School of Biotechnology and Food Technology
Hanoi University of Science and Technology
Hanoi, Vietnam

Hong-Nga Luong
School of Biotechnology and Food Technology
Hanoi University of Science and Technology
Hanoi, Vietnam

Thatchajaree Mala
Department of Food Agriculture and
    Bioresources
School of Environment, Resources and
    Development
Asian Institute of Technology
Klong Luang, Thailand

Reshma M. Mathew
CSIR- National Institute for Interdisciplinary
    Science and Technology (CSIR-NIIST)
Thiruvananthapuram, Kerala, India

and

Academy of Scientific and Innovative Research
    (AcSIR)
Ghaziabad, Uttar Pradesh, India

Akanksha Mishra
National Institute of Industrial Engineering
    (NITIE)
Mumbai, India

Neegam Nain
Department of Food Technology and Nutrition
Lovely Professional University
Phagwara, Punjab, India

Chinh-Nghia Nguyen
School of Biotechnology and Food Technology
Hanoi University of Science and Technology
Hanoi, Vietnam

Loc Thai Nguyen
Department of Food, Agriculture and
    Bioresources
School of Environment, Resources and
    Development
Asian Institute of Technology
Pathumthani, Thailand

Tien-Cuong Nguyen
School of Biotechnology and Food Technology
Hanoi University of Science and Technology
Hanoi, Vietnam

Tien-Thanh Nguyen
School of Biotechnology and Food Technology
Hanoi University of Science and Technology
Hanoi, Vietnam

Neeraj Pandey
National Institute of Industrial Engineering
    (NITIE)
Mumbai, India

Gaurav Panesar
Department of Food Engineering and
    Technology
Sant Longowal Institute of Engineering and
    Technology
Longowal, Punjab, India

Parmjit S. Panesar
Department of Food Engineering and
    Technology
Sant Longowal Institute of Engineering and
    Technology
Longowal, Punjab, India

Suwan Panjanapongchai
Department of Food Agriculture and
    Bioresources
School of Environment, Resources and
    Development
Asian Institute of Technology
Klong Luang, Thailand

Divyani Panwar
Department of Food Engineering and
    Technology
Sant Longowal Institute of Engineering and
    Technology
Longowal, Punjab, India

Tuan-Anh Pham
School of Biotechnology and Food Technology
Hanoi University of Science and Technology
Hanoi, Vietnam

Quyet-Tien Phi
Institute of Biotechnology
Vietnam Academy of Science and Technology
Hanoi, Vietnam

Velayudhan Pillai-Prasannakumari Adarsh
CSIR- National Institute for Interdisciplinary
    Science and Technology (CSIR-NIIST)
Thiruvananthapuram, Kerala, India

and

Academy of Scientific and Innovative Research
    (AcSIR)
Ghaziabad, Uttar Pradesh, India

Anoop Puthiyamadam
CSIR- National Institute for Interdisciplinary
    Science and Technology (CSIR-NIIST)
Thiruvananthapuram, Kerala, India

and

Academy of Scientific and Innovative Research
    (AcSIR)
Ghaziabad, Uttar Pradesh, India

Ngoc Tung Quach
Institute of Biotechnology
Vietnam Academy of Science and Technology
Hanoi, Vietnam

Le-Ha Quan
School of Biotechnology and Food Technology
Hanoi University of Science and Technology
Hanoi, Vietnam

Prasad Rasane
Department of Food Technology and Nutrition
Lovely Professional University
Phagwara, Punjab, India

Sujitta Raungrusmee
Department of Home Economics
Kasetsart University
Bangkok, Thailand

Anuradha Saini
Department of Food Engineering and
    Technology
Sant Longowal Institute of Engineering and
    Technology
Longowal, Punjab, India

Meena Sankar
CSIR- National Institute for Interdisciplinary
    Science and Technology (CSIR-NIIST)
Thiruvananthapuram, Kerala, India

and

Academy of Scientific and Innovative Research
    (AcSIR)
Ghaziabad, Uttar Pradesh, India

Anjali Shrestha
Department of Food Agriculture and
    Bioresources
School of Environment, Resources and
    Development
Asian Institute of Technology
Klong Luang, Thailand

Gisha Singla
Department of Food Engineering and
    Technology
Sant Longowal Institute of Engineering and
    Technology
Longowal, Punjab, India

AthiraRaj Sreeja-Raju
CSIR- National Institute for Interdisciplinary
    Science and Technology (CSIR-NIIST)
Thiruvananthapuram, Kerala, India

and

Academy of Scientific and Innovative Research
    (AcSIR)
Ghaziabad, Uttar Pradesh, India

Rajeev K. Sukumaran
CSIR- National Institute for Interdisciplinary
    Science and Technology (CSIR-NIIST)
Thiruvananthapuram, Kerala, India

and

Academy of Scientific and Innovative Research
    (AcSIR)
Ghaziabad, Uttar Pradesh, India

Lai Phuong Phuong Thao
Institute of Natural Products Chemistry
Vietnam Academy of Science and Technology
Hanoi, Vietnam

Phan Thi Phuong Thao
Faculty of Food Science and Technology
Vietnam National University of Agriculture
    (VNUA)
Hanoi, Vietnam

Sarina Pradhan Thapa
Department of Food Agriculture and
    Bioresources
School of Environment, Resources and
    Development
Asian Institute of Technology
Klong Luang, Thailand

Tran Thi Thu Hang
Faculty of Food Science and Technology
Vietnam National University of Agriculture
    (VNUA)
Hanoi, Vietnam

Yakindra Timilsena
Melbourne Reliable Consulting Group
Melbourne, Australia

Kim-Anh To
School of Biotechnology and Food Technology
Hanoi University of Science and Technology
Hanoi, Vietnam

Tran Quoc Toan
Institute of Natural Products Chemistry
Vietnam Academy of Science and Technology
Hanoi, Vietnam

Muhammad Umar
Department of Food Agriculture and
    Bioresources
School of Environment, Resources and
    Development
Asian Institute of Technology
Klong Luang, Thailand

Nguyen-Thanh Vu
Food Industries Research Institute
Hanoi, Vietnam

Thu-Trang Vu
School of Biotechnology and Food Technology
Hanoi University of Science and Technology
Hanoi, Vietnam

1 Agro-Industrial Waste as Wealth
Principles, Biorefinery, and Bioeconomy

Anil Kumar Anal, Parmjit S. Panesar, and Rupinder Kaur

CONTENTS

1.1 INTRODUCTION

Agro-processing is a collection of technological and economic processes used to preserve and handle agricultural products to make them usable as food, feed, fibre, fuel, or industrial raw material. While the expansion of food-processing industrialization has resulted in a wide variety of processed food products, created jobs, and generated economic benefits, it has also resulted in a massive amount of biodegradable waste/byproducts that pose a threat to the environment if not properly disposed of (Prajapati et al., 2021). Globally, the food and agricultural industries generate enormous amounts of waste each year, posing a severe disposal challenge. Not only does waste creation represent an economic loss, but it also entails additional costs for waste management. Strict environmental regulations have made a big contribution to the global adoption of sustainable waste management strategies. Over the past several decades, there has been a growing trend towards the value addition of agro-industrial waste via the use of green and biotechnological approaches to develop a variety of value-added products with uses in a variety of areas. Utilizing these procedures to valorize agro-industrial leftovers will provide an alternative substrate and aid in reducing environmental contamination and the manufacture of value-added goods (Thapa et al., 2021). The production of value-added goods from these wastes can also contribute to some cost reduction.

Agricultural waste represents the waste generated during various agricultural practices, whereas agro-industrial waste results from the industrial processing of agricultural produce (Portugal-Pereira et al., 2015). With burgeoning global growth in the population and advancements in industrial sectors, several tonnes of residues are generated annually, accounting for approximately 50 million tonnes (Zhao et al., 2012). The sudden upsurge in the food processing sector has led to the ample generation of residues and waste streams, accounting for about 38% among all the wastes

DOI: 10.1201/9781003125679-1

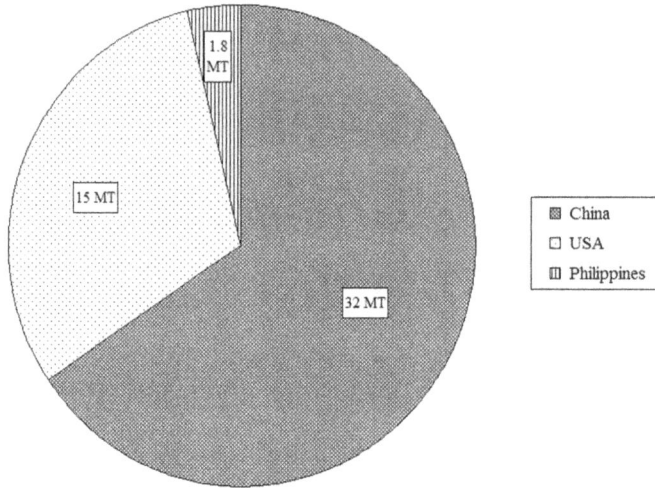

FIGURE 1.1   Waste generated by countries worldwide (in million tonnes).

generated from various industries. According to the Food and Agriculture Organization's statistics, about 1.3 billion tonnes of food are wasted worldwide, from processing to reaching consumers. The fruit and vegetable processing sector contributes about 44% of the waste, followed by roots and tubers (20%) and cereal processing (19%) (Nora et al., 2017). The leading players contributing globally to the highest waste generation are China, followed by the USA and the Philippines (Figure 1.1).

Generally, most of the waste generated is either burned or disposed of in an uncontrolled manner or released in landfills, thereby leading to serious environmental hazards and biological and economic loss. Hence, sustainable management of these residues is important, and strategies have to be developed to effectively handle these surplus residues (Pattanaik et al., 2019). Thus, an integrated approach, known as biorefinery, has been devised to manage these agro-industrial residues effectively. In addition, a new term, "circular economy", has also come into existence to exploit the various lignocellulosic biomasses (Campos et al., 2020). The biocircular economy is a concept based on the biorefinery technique that facilitates reducing, reusing, and recycling residues to obtain high-valued components (Velis, 2015). Therefore, agro-industrial residues have been a growing area of research worldwide in the past few years due to various constituents that make these residues suitable substrates for producing various high value-added products (Beltrán-Ramírez et al., 2019).

Currently, the food industry is concentrating its efforts on waste management and recycling through valorization or the exploitation of byproducts and the development of new economically viable value-added goods. Waste valorization is an intriguing new idea that provides various waste management options for disposal or landfill. Valorization enables us to investigate the possibilities of recycling nutrients in the manufacturing of primary goods, highlighting possible benefits. Historically, waste was utilized in various ways, including as animal feed, fertilizer, and disposal. However, its usage has been restricted due to legal constraints, ecological concerns, and financial constraints. As a result, emphasis is being placed on efficient, cost-effective, and environmentally friendly methods of waste disposal (Mala et al., 2021). Food industry waste often contains dietary fibres, proteins and peptides, lipids, fatty acids, and phenolic chemicals, depending on the product's nature. For instance, waste from the meat and poultry sector contains proteins and lipids, whereas waste from the fruit and vegetable processing industry and cereal manufacturing industry contains phenolic compounds and dietary fibres. The isolation of these bioactive chemicals is critical for their commercialization as nutraceuticals and medicinal products (Paritosh et al., 2017).

1.2 PHYSICO-CHEMICAL CHARACTERISTICS OF AGRO-INDUSTRIAL RESIDUES

Today, waste from the agro-industry is of major concern globally since it is generated in huge quantities and is a source of pollution. Generally, organic waste comes from (i) agricultural or forestry and (ii) industrial activities. Waste originating from agriculture or forestry comprises straw, manure, crop residues, etc., whereas waste from industrial processing includes bagasse, peels, seeds, molasses, and much more (Venglovsky et al., 2006).

These residues are rich in lignocellulosic materials, such as hemicellulose, cellulose, and lignin (Figure 1.2), responsible for the cell wall's structural conformation. Typically, the main lignocellulosic materials' main unit comprises 40–50% cellulose, 20–30% hemicellulose, and 10–25% lignin (Iqbal et al., 2011).

(i) Hemicellulose: Hemicellulose is the second-most abundant heterogeneous polymer that constitutes repeated units of pentoses and hexoses, such as glucomannan, glucuronoxylan, and other polysaccharides (Wyman et al., 2005). The backbone of the polymer is linked with $\beta(1-4)$ bonds and depicts a molecular weight of <30 kDa. Hemicellulose and cellulose are covalently bonded on the micro-fibril structure of cellulose (Vercoe et al., 2005).

(ii) Cellulose: Cellulose $(C_6H_{10}O_5)_n$ comprises (1,4) D-glucopyranose units linked together by $\beta(1-4)$ with a molecular weight of about 100 kDa (Himmel et al., 2007). Although in its native state, intermolecular hydrogen bonding is present in the cellulose molecules, these can also form intramolecular hydrogen bonds and intermolecular bonds that increase its rigidity and are therefore responsible for its mechanical strength (Taherzadeh and Karimi, 2008).

(iii) Lignin: Lignin is the most complex heterogeneous polymer composed of phenyl-propane units, methoxy groups, and non-carbohydrate poly-phenolic substance, linked together with ether bonds responsible for binding the components of the cell wall. This lignocellulose acts as the glue between the cellulose and hemicellulose complexes (Hamelinck et al., 2005).

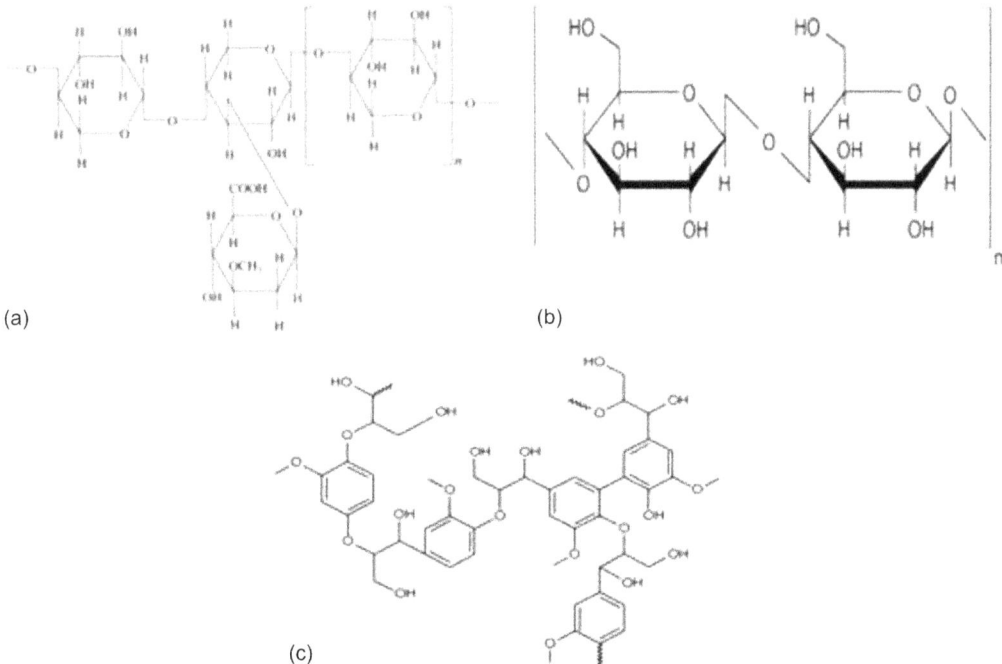

(a)

(b)

(c)

FIGURE 1.2 Structures of lignocellulosic materials (a) hemicellulose, (b) cellulose, and (c) lignin.

Although this lignocellulosic biomass is present in every agro-industrial residue or plant material, the concentration varies with the source and genetic variability (Bertero da la Puente and Sedran, 2012).

1.3   STRATEGIES FOR VALORIZATION OF AGRO-INDUSTRIAL RESIDUES

Valorization of agro-industrial residues is one of the world's greatest challenges, as millions of tonnes of waste are generated annually. Nevertheless, for any integrated recovery or sustainable management of these residues, it is crucial to understand the sources of the waste throughout the cycle from supply to production to the consumer's plate and evaluate the future of these processes (Ronzon and Piotrowski, 2017; Camia et al., 2018). For instance, the cereal industry produces husk as waste, the juice industry generates pulp and peels as waste, and large amounts of coffee pulp are obtained as waste from the coffee industry (Sadh et al., 2018). In addition, in the case of the fruit processing industry, the quantity of waste generated depends upon the location and the method of harvest; for mangoes and citrus fruits, the percentage of waste obtained is 30–50%, 40% for pomegranate, and 20% in the case of bananas (Parfitt et al., 2010). On the other hand, the scale of the industry also has a huge impact on the amount of waste generated. For example, waste generated in small-scale industries during processing would be less than those from large-scale industries.

On the contrary, waste disposal is apriority in big companies as they pay to dispose of it in landfills. The situation is different in developing and developed countries. In developing countries like India, the average transport cost was estimated to be $11–15 per ton per trip, accounting for $300 million of the total landfill cost (FICCI, 2010). Apart from this, considering the environmental impact of landfills on greenhouse gas emissions, it is vital to develop innovative strategies to efficiently utilize the waste (Roggeveen, 2010).

From a biotechnological point of view, these lignocellulosic materials, because of their high nutritional composition, pose prime importance as potential substrates for the production of high-value components of industrial importance, such as enzymes, biofuels, bio-pigments, and much more (Panesar and Marwaha, 2014) by microbial fermentation (Figure 1.3). In recent years, green biotechnology has been in the spotlight of various researchers and is being widely explored to generate various products (Asgher et al., 2013; Panesar et al., 2016).

Apart from this, these byproducts represent a rich source of valuable components, also known as bioactive substances with nutritional value, such as phenolic compounds, pectin, protein, fibres, vitamins, and antioxidants that have a beneficial impact on human health (Tadmor et al., 2010).

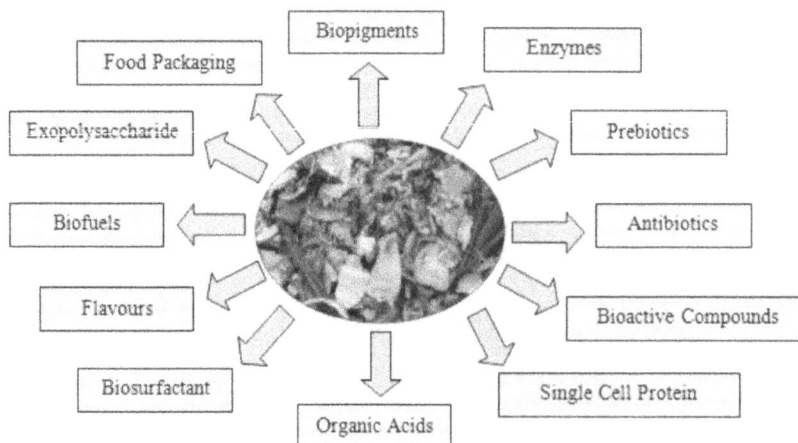

FIGURE 1.3   Valorization of agro-industrial residues to high-valued products.

Extraction of these compounds from the natural matrices has been carried out by both conventional (soxhlet) and non-conventional techniques (microwave extraction, supercritical fluid extraction, ultrasound treatment) to date. However, it is a difficult as well as an important task to commercialize these products. Nonetheless, biotechnological tools can be explored in the future to extract these valuable compounds to fulfil human demands (Dey et al., 2016; Campos et al., 2020).

In this regard, projects have been carried out by both India and Europe, namely "New advances in the integrated management of food processing waste in India and Europe-Namaste and Healthgrain Forum" ("Project NAMASTE An Epic Diversity", 2015; "Healthgrain EU Project", 2017). This project intends to develop new products using these extracted bioactive ingredients and ensure their therapeutic properties (Banerjee et al., 2018).

The conversion of lignocellulosic biomass into value-added products requires certain pre-treatment steps wherein these residues are subjected to various treatments (physical, chemical, or a combination of methods) to improve the yield of the product (Hassan et al., 2019).

1.4 PRE-TREATMENT OF AGRO-INDUSTRIAL RESIDUES

Pre-treatment is an important step during the valorization process, in which depolymerization of the complex carbohydrate molecules to simple fermentable sugars is carried out (Krishnan et al., 2010). This is done in two ways: (i) removal of the hemicellulose to leave lignin as well as cellulose together (cellulignin) and (ii) removing the lignin and leaving the cellulose and hemicellulose together (holocarbohydrate). These can be done with different techniques, such as chemical, physical, biological, or a combination of these methods, to facilitate the breakdown of complex sugars into simpler ones for enhanced fermentation and yield (Ravindran and Jaiswal, 2016).

Any effective pre-treatment should fulfil the following criteria: reduction in the loss of carbohydrates, release of maximum sugars, minimize the generation of inhibitors, and be cost-effective (Hamzeh et al., 2013). Examples of the pre-treatment of agro-industrial residues are listed in Table 1.1.

1.4.1 ENZYMATIC HYDROLYSIS

Enzymatic hydrolysis is a cost-effective and efficient process that cleaves polysaccharides into sugar monomers; this is carried out by filamentous fungi, especially *Trichroderma* and *Aspergillus* (García-Cubero et al., 2010). Various factors, such as pH, time, enzyme, and temperature, have a huge impact on the efficiency of the process. It can be carried out in two ways: one method is known as separate hydrolysis and fermentation, in which enzymatic saccharification and fermentation are carried out separately.

The other method is simultaneous saccharification and fermentation. In contrast to the former, this process is widely used since both hexose(s) and pentose(s) can be co-fermented in a single step using genetically engineered microbes, indicating this method to be feasible and economical (Wyman et al., 2005).

The filamentous fungi secrete a cellulase enzyme cocktail, including endo β-1,4-glucanases, exoglucanase, and β-glycosidase that break down the various cellulosic bonds (Xue et al., 2017). Endo β-1,4-glucanases cleave the β-1,4 intramolecular bonds present in the low crystal regions of the cellulose fibre, leading to cellobiose production with a generation of free radicals. Exoglucanase is responsible for the polymer chain reduction, and β-glycosidase hydrolyses cellobiose, thereby forming sugar monomers (Canilha et al., 2012).

1.4.2 CHEMICAL PRE-TREATMENT

Chemical treatment is one of the most widely used techniques for breaking down cellulose and hemicellulose polymers from the lignocellulosic biomass. This treatment can be carried out in two ways:

TABLE 1.1

Pre-treatment of Various Agro-Industrial Residues

Techniques	Chemicals	Agro-industrial Waste	End Product	Reference(s)
Chemical	Sulphuric acid	Sugarcane bagasse, rice hulls, peanut shells, cassava stalks	Ethanol	Martin et al. (2007)
	Sulphuric acid	Sugarcane molasses	Inulinase	Sguarezi et al. (2009)
	Hydrochloric acid	Molasses and olive mill wastewater	Bioethanol	Sarris et al. (2014)
	Acid-base	Sugarcane bagasse	Lipase	Pinotti et al. (2017)
	Dilute sulphuric acid	Grapevine trimmings, olive tree trimmings, almond shell	Bioethanol	Nitsos et al. (2018)
	Sulphuric acid	Sugarcane bagasse, soybean husks, empty fruit bunches of oil palm	Cellulase/xylanase enzyme complex	Karp et al. (2020)
	Sodium hydroxide	Apricot industrial waste	Ethanol	Zoubiri et al. (2020)
	Sulphuric acid	Wheat straw, rice straw, corn cobs, jute stick	Cellulase cocktail	Jain et al. (2020)
	Dilute acid	Wheat straw	Bioethanol	Olofsson et al. (2010)
	Sulphuric acid	Micro-algae biomass with raisin extract	Bioethanol	Tsolcha et al. (2021)
	Sulphuric acid	Tender coconut husk	Biodegradable plastics	De et al. (2021)
Alkali peroxide delignification	Sodium hydroxide+hydrogen peroxide			
Ionic liquids	1-Ethyl-3-methylimidazolium chloride			
Neoteric solvent	Quaternary ammonium salt with choline chloride and oxalic acid dehydrate			
Thermochemical	(Sulphuric acid+ sodium hydroxide)	Sunflower straw, olive stones	Bioethanol	Antonopoulou et al. (2019)
Enzymatic	Celluclast 1.5 L(cellulase) and Novozyme 188 (cellobiase)	Grapevine trimmings, olive tree trimmings, almond shell	Bioethanol	Nitsos et al. (2018)
	Cellic CTec 2			
Mechanical	Ultrasound and microwave	Swine hair, chicken feathers	Keratinase	Czapela et al. (2020)
	Thermal-torrefaction	Tomato peels, virgin olive husk	—	Brachi et al. (2018)
Hydrothermal	Dilute sulphuric acid	Grapevine trimmings, olive tree trimmings, almond shell	Bioethanol	Nitsos et al. (2018)
		Corn stover	Bioethanol	Chen and Fu (2016)
Steam	Saturated steam	Grapevine trimmings, olive tree trimmings, almond shell	Bioethanol	Nitsos et al. (2018)

acid hydrolysis and alkali hydrolysis. Acid hydrolysis is generally carried out with concentrated hydrochloric or sulphuric acid and reactions at low temperatures. The major limitation of this process is the corrosion caused by the higher concentration (30–70%) of acids (Wyman, 1999). On the contrary, dilute acid treatment overcomes toxicity, acid recovery, and corrosion (Sun and Chen, 2007).

During alkali-based treatment, bases such as sodium hydroxide or ammonium hydroxide are used to hydrolyze the carbohydrates in the lignocellulose biomass. This method causes various structural variations, such as diminution of the lignin barrier, swelling of the cellulose, decrystallization, and solvation of hemicellulose and cellulose, respectively, resulting in the disruption of complex polysaccharides and release of sugar monomers (Ibrahim et al., 2011).

1.4.3 BIOLOGICAL PRE-TREATMENT

This technique involves using several microbes, especially wood-degrading ones, such as white rot fungi, brown rot fungi, and bacteria, to modify the structure and delignification of the lignocellulosic materials (Iqbal and Asgher, 2013). As compared with the other treatments, this method is effective, environmentally friendly, and economical. Despite this, it is a very slow process, and it requires controlled environmental parameters for growth and a large space for treatment. In addition, the microbes can not only consume lignin but also attack hemicellulose or cellulose. Hence, these drawbacks limit their use (Eggeman and Elander, 2005).

Although various methods have been widely explored, and each has its own merits and demerits, amongst all the treatment methods, acid hydrolysis is the traditional and most widely used method to date as it is efficient in the removal of pentose sugar from the biomass.

1.5 CONCLUSIONS AND FUTURE PERSPECTIVES

Recent urbanization and rapid growth in the industrialization sector contribute to the annual generation of ample quantities of waste. The food processing industry is one of the largest contributors to agro-industrial waste, as in 2012, approximately 17 million tonnes of waste was generated. This situation will worsen in the future, especially in developing countries, due to the lack of proper infrastructure and disposal facilities. Despite being rich in various nutritional components, these are underutilized and often disposed of, leading to serious environmental issues. Nevertheless, in recent years, global perception about agro-industrial residues has changed with a view to sustainable management of the various residues. Hence, biotechnological strategies have been developed to recharge this lignocellulosic biomass into highly valued products of both nutritional and commercial importance. Although various developed nations have created opportunities for the utilization of agro-industrial residues in industries, commercial production of various products from the residues is at a bottleneck in developing countries. In the future, advancements in biotechnological tools would aid the discovery of novel enzymes and microbes that would further help in the generation of cost-effective strategies for waste utilization. An integrated valorization approach would reduce the negative impact on the ecological system and generate an additional source of income in the global market.

REFERENCES

Antonopoulou, G., Kampranis, A., Ntaikou, I. et al. 2019. Enhancement of liquid and gaseous biofuels production from agro-industrial residues after thermochemical and enzymatic pretreatment. *Frontiers in Sustainable Food Systems* 3: 92.

Asgher, M., Ahmad, Z., and Iqbal, H.M.N. 2013. Alkali and enzymatic delignification of sugarcane bagasse to expose cellulose polymers for saccharification and bio-ethanol production. *Industrial Crops and Products* 44: 488–495.

Banerjee, S., Ranganathan, V., Patti, A. et al. 2018. Valorisation of pineapple wastes for food and therapeutic applications. *Trends in Food Science and Technology* 82: 60–70.

Beltrán-Ramírez, F., Orona-Tamayo, D., Cornejo-Corona, I. et al. 2019. Agro-industrial waste revalorization: The growing biorefinery. *Biomass for Bioenergy-Recent Trends and Future Challenges*. https://doi.org /10.5772/intechopen.83569.

Bertero, M., de la Puente, G., and Sedran, U. 2012. Fuels from bio-oils: Bio-oil production from different residual sources, characterization and thermal conditioning. *Fuel* 95: 263–271.

Brachi, P., Miccio, F., Ruoppolo, G. et al. 2018. Pressurized steam torrefaction of wet agro-industrial residues. *Chemical Engineering Transactions* 65: 49–54.

Camia, A., Robert, N., Jonsson, R. et al. 2018. Biomass production, supply, uses and flows in the European Union. *JRC Science for Policy Report*. Available online: https://ec.europa.eu/jrc/en/publication/eur-sci-entific-and-technical-research-eports/biomass-production-supply-uses-and-flows-european-union-first -results-integrated-assessment (accessed on 10 January 2020).

Campos, D.A., Gómez-García, R., Vilas-Boas, A.A. et al. 2020. Management of fruit industrial by-products— A case study on circular economy approach. *Molecules* 25(2): 320.

Canilha, L., Chandel, A.K., Milessi, T.S.S. et al. 2012. Bioconversion of sugarcane biomass into ethanol: An overview about composition, pretreatment methods, detoxification of hydrolysates, enzymatic sacchari-fication, and ethanol fermentation. *Journal of Biomedicine and Biotechnology* 2012: 1–15.

Chen, H., and Fu, X. 2016. Industrial technologies for bioethanol production from lignocellulosic biomass. *Renewable and Sustainable Energy Reviews* 57: 468–478.

Czapela, F.F., Kubeneck, S., Preczeski, K.P. et al. 2020. Reactional ultrasonic systems and microwave irra-diation for pretreatment of agro-industrial waste to increase enzymatic activity. *Bioresources and Bioprocessing* 7(1): 1–13.

De, D., Sai, M.S.N., Aniya, V. et al. 2021. Strategic biorefinery platform for green valorization of agro-indus-trial residues: A sustainable approach towards biodegradable plastics. *Journal of Cleaner Production* 290: 125184.

Dey, T.B., Chakraborty, S., Jain, K.K. et al. 2016. Antioxidant phenolics and their microbial production by submerged and solid state fermentation process: A review. *Trends in Food Science and Technology* 53: 60–74.

Eggeman, T., and Elander, R.T. 2005. Process and economic analysis of pretreatment technologies. *Bioresource Technology* 96(18): 2019–2025.

FICCI. 2010. Bottlenecks in Indian food processing industry. Available online: http://ficci.in/sedocument /20073/Food-Processing-bottlenecks-study.pdf.

García-Cubero, M.T., Coca, M., Bolado, S. et al. 2010. Chemical oxidation with ozone as pre-treatment of lignocellulosic materials for bioethanol production. *Chemical Engineering Transactions* 21: 1273–1278.

Hamelinck, C.N., Van Hooijdonk, G., and Faaij, A.P. 2005. Ethanol from lignocellulosic biomass: Techno-economic performance in short-, middle-and long-term. *Biomass and Bioenergy* 28(4): 384–410.

Hamzeh, Y., Ashori, A., Khorasani, Z. et al. 2013. Pre-extraction of hemicelluloses from bagasse fibers: Effects of dry-strength additives on paper properties. *Industrial Crops and Products* 43: 365–371.

Hassan, S.S., Williams, G.A., and Jaiswal, A.K. 2019. Lignocellulosic biorefineries in Europe: Current state and prospects. *Trends in Biotechnology* 37(3): 231–234.

Heathgrain EU Project. 2017. Retrieved January 1, 2017, from https://www.healthgrain.org/eu_project/about. The definition of whole grain.

Himmel, M.E., Ding, S.Y., Johnson, D.K. et al. 2007. Biomass recalcitrance: Engineering plants and enzymes for biofuels production. *Science* 315(5813): 804–807.

Ibrahim, M.M., El-Zawawy, W.K., Abdel-Fattah, Y.R. et al. 2011. Comparison of alkaline pulping with steam explosion for glucose production from rice straw. *Carbohydrate Polymers* 83(2): 720–726.

Iqbal, H.M.N., Ahmed, I., Zia, M.A. et al.2011. Purification and characterization of the kinetic parameters of cellulase produced from wheat straw by *Trichoderma viride* under SSF and its detergent compatibility. *Advances in Bioscience and Biotechnology* 2(3): 149–156.

Iqbal, H.M.N., and Asgher, M. 2013. Characterization and decolorization applicability of xerogel matrix immobilized manganese peroxidase produced from Trametes versicolor IBL-04. *Protein and Peptide Letters* 5(5): 591–600.

Jain, L., Kurmi, A.K., Kumar, A. et al. 2020. Exploring the flexibility of cellulase cocktail obtained from mutant UV-8 of *Talaromyces verruculosus* IIPC 324 in depolymerising multiple agro-industrial ligno-cellulosic feedstocks. *International Journal of Biological Macromolecules* 154: 538–544.

Karp, S.G., Osipov, D.O., Semenova, M.V. et al. 2020. Effect of novel *Penicillium verruculosum* enzyme preparations on the saccharification of acid-and alkali-pretreated agro-industrial residues. *Agronomy* 10(9): 1348.

Krishnan, C., Sousa, L.D.C., Jin, M. et al. 2010. Alkali-based AFEX pretreatment for the conversion of sugarcane bagasse and cane leaf residues to ethanol. *Biotechnology and Bioengineering* 107(3): 441–450.

Mala, T., Sadiq, M.B., and Anal, A.K. 2021. Comparative extraction of bromelain and bioactive peptides from pineapple byproducts by ultrasonic-and microwave-assisted extractions. *Journal of Food Process Engineering* 44(6): e13709.

Martin, C., Alriksson, B., Sjöde, A. et al. 2007. Dilute sulfuric acid pretreatment of agricultural and agro-industrial residues for ethanol production. *Applied Biochemistry and Biotechnology* 136–140(1–12): 339–352.

Nitsos, C., Matsakas, L., Triantafyllidis, K. et al. 2018. Investigation of different pretreatment methods of Mediterranean-type ecosystem agricultural residues: Characterisation of pretreatment products, high-solids enzymatic hydrolysis and bioethanol production. *Biofuels* 9(5): 545–558.

Nora, S.M.S., Ashutosh, S., and Vijaya, R. 2017. Potential utilization of fruit and vegetable wastes for food through drying or extraction techniques. *Novel Techniques in Nutrition and Food Science* 2017: 1.

Olofsson, K., Palmqvist, B., and Lidén, G. 2010. Improving simultaneous saccharification and co-fermentation of pretreated wheat straw using both enzyme and substrate feeding. *Biotechnology for Biofuels* 3(1): 1–9.

Panesar, P.S., Kaur, R., Singla, G. et al.2016. Bio-processing of agro-industrial wastes for the production of food grade enzymes: Progress and prospects. *Applied Food Biotechnology* 3(4): 208–227.

Panesar, P.S., and Marwaha, S.S. 2014. Value addition of agro-industrial waste and residues. In: *Biotechnology in Agriculture and Food Processing: Opportunities and Challenges* (Eds. Parmjit S.Panesar, Satwinder S.Marwaha) (pp 557–607). CRC Press, Boca Raton, FL.

Parfitt, J., Barthel, M., and Macnaughton, S. 2010. Food waste within food supply chains: Quantification and potential for change to 2050. *Philosophical Transactions of the Royal Society of London Series B Biological Sciences* 365(1554): 3065–3081.

Paritosh, K., Kushwaha, S.K., Yadav, M. et al. 2017. Food waste to energy: An overview of sustainable approaches for food waste management and nutrient recycling. *BioMed Research International* 2017: 2370927.

Pattanaik, L., Pattnaik, F., Saxena, D.K. et al. 2019. Biofuels from agricultural wastes. In: (Eds. A. Basile and F. Dalena) pp. 103–142. Elsevier, Amsterdam

Pinotti, L.M., Lacerda, J.X., Oliveira, M.M. et al. 2017. Production of lipolytic enzymes using agro-industrial residues. *Chemical Engineering Transactions* 56: 1897–1902.

Portugal-Pereira, J., Soria, R., Rathmann, R. et al. 2015. Agricultural and agro-industrial residues-to-energy: Techno-economic and environmental assessment in Brazil. *Biomass and Bioenergy* 81: 521–533.

Prajapati, S., Koirala, S., and Anal, A.K. 2021. Bioutilization of chicken feather waste by newly isolated keratinolytic bacteria and conversion into protein hydrolysates with improved functionalities. *Applied Biochemistry and Biotechnology* 193(8): 2497–2515.

Project Namaste an Epic Diversity. 2015. Retrieved January 1, 2017, from http://www.pdpu.ac.in/downloads /namaste-1.pdf.

Ravindran, R., and Jaiswal, A.K. 2016. Exploitation of food industry waste for high-value products. *Trends in Biotechnology* 34(1): 58–69.

Roggeveen, K. 2010. *Tomato Journey from Farm to Fruit Shop: Greenhouse Gas Emissions and Cultural Analysis*. University of Wollongong, Wollongong.

Ronzon, T., and Piotrowski, S. 2017. Are primary agricultural residues promising feedstock for the European bioeconomy?*Industrial Biotechnology* 13(3): 113–127.

Sadh, P.K., Duhan, S., and Duhan, J.S. 2018. Agro-industrial wastes and their utilization using solid state fermentation: A review. *Bioresources and Bioprocessing* 5(1): 1–15.

Sarris, D., Matsakas, L., Aggelis, G. et al. 2014. Aerated vs non-aerated conversions of molasses and olive mill wastewaters blends into bioethanol by *Saccharomyces cerevisiae* under non-aseptic conditions. *Industrial Crops and Products* 56: 83–93.

Sguarezi, C., Longo, C., Ceni, G. et al. 2009. Inulinase production by agro-industrial residues: Optimization of pretreatment of substrates and production medium. *Food and Bioprocess Technology* 2(4): 409–414.

Sun, F., and Chen, H. 2007. Evaluation of enzymatic hydrolysis of wheat straw pretreated by atmospheric glycerol autocatalysis. *Journal of Chemical Technology and Biotechnology: International Research in Process, Environmental and Clean Technology* 82(11): 1039–1044.

Tadmor, Y., Burger, J., Yaakov, I. et al. 2010. Genetics of flavonoid, carotenoid, and chlorophyll pigments in melon fruit rinds. *Journal of Agricultural and Food Chemistry* 58(19): 10722–10728.

Taherzadeh, M.J., and Karimi, K. 2008. Pretreatment of lignocellulosic wastes to improve ethanol and biogas production: A review. *International Journal of Molecular Sciences* 9(9): 1621–1651.

Thapa, S.S., Shrestha, S., Sadiq, M.B. et al. 2021. Xylitol production from sugarcane bagasse Through ultrasound-assisted alkaline pretreatment and enzymatic hydrolysis followed by fermentation. *Sugar Tech*, https://doi.org/10.1007/s12355-021-01066-3.

Tsolcha, O.N., Patrinou, V., Economou, C.N. et al. 2021. Utilization of biomass derived from cyanobacteria-based agro-industrial wastewater treatment and raisin residue extract for bioethanol production. *Water* 13(4): 486.

Velis, C.A. 2015. Circular economy and global secondary material supply chains. *Waste Management and Research* 33(5): 389–391.

Venglovsky, J., Martinez, J., and Placha, I. 2006. Hygienic and ecological risks connected with utilization of animal manures and biosolids in agriculture. *Livestock Science* 102(3): 197–203.

Vercoe, D.Q., Stack, K.R., Blackman, A.J. et al. 2005. A study of the interactions leading to wood pitch deposition. In: *59th Appita Annual Conference and Exhibition* held at Auckland, New Zealand from 16-19 May, 2005, pp. 123–130.

Wyman, C.E. 1999. Biomass ethanol: Technical progress, opportunities, and commercial challenges. *Annual Review of Energy and the Environment* 24(1): 189–226.

Wyman, C.E., Dale, B.E., Elander, R.T. et al. 2005. Coordinated development of leading biomass pretreatment technologies. *Bioresource Technology* 96(18): 1959–1966.

Xue, D., Lin, D., Gong, C. et al. 2017. Expression of a bifunctional cellulase with exoglucanase and endoglucanase activities to enhance the hydrolysis ability of cellulase from a marine Aspergillus niger. *Process Biochemistry* 52: 115–122.

Zhao, X., Zhang, L., and Liu, D. 2012. Biomass recalcitrance. Part I: The chemical compositions and physical structures affecting the enzymatic hydrolysis of lignocellulose. *Biofuels, Bioproducts and Biorefining* 6(4): 465–482.

Zoubiri, F.Z., Rihani, R., and Bentahar, F. 2020. Golden section algorithm to optimise the chemical pretreatment of agro-industrial waste for sugars extraction. *Fuel* 266: 117028.

2 Sources, Composition, and Characterization of Agro-Industrial Byproducts

Dipak Das, Parmjit S. Panesar,
Gaurav Panesar, and Yakindra Timilsena

CONTENTS

2.1 INTRODUCTION

Discarding or burning agro-industrial byproducts can generate enormous quantities of toxic compounds that cause air and water pollution. However, high moisture byproducts generated from agro-based industries are challenging to manage. The composition and characteristics of numerous byproducts possess enormous potential to produce valuable products for human consumption as well as animal feed (Mirzaei-Aghsaghali and Maheri-Sis, 2008). Many byproducts like oil meals, beet molasses, corn distillers, and brewers grain are well recognized and have been used productively for many years. There are several other byproducts, but their use is inconsistent because of ambiguity regarding obtainability, palatability, and presence of hazardous toxic compounds (Denek and Can, 2006). Byproducts formed in the agro-industry are known as an inexpensive source of highly valued elements. Current technologies recycle these elements in the food processing chain as useful ingredients to increase the functionality and nutritional quality of various products. Furthermore, converting agro-industrial waste into valuable products using novel technologies can reduce the environmental impact of their dumping and allow the sustainability of high-quality ingredients in the food processing chain (Grumezescu and Holban, 2018).

DOI: 10.1201/9781003125679-2

Many agro-industrial byproducts like peels, seeds, skins, pericarp, and pomace are generated from fruits and vegetables such as apples, grapes, guava, lychee, carrot, and potato in processing industries (Figure 2.1). In addition, other byproducts like whey from dairy plants, molasses from sugarcane-processing plants, bran from barley and rice plants, and bran and wheatgerm from wheat-processing plants are also produced in huge quantities (Kumar et al., 2017). These wastes are considered a potential source of bioactive phenolic compounds that can act as antioxidants, enzymatic inhibitors, and thickening agents, for example, pectin, fibre, and other free-radical scavenging compounds (García et al., 2009; Llobera and Canellas, 2007). As a source of biologically active compounds, intensive research on agro-industrial byproducts is ongoing to treat and prevent carcinogenic human diseases. There is a high potential to develop therapeutic compounds from the bioactive substances of agro-industrial byproducts, as they interact with protein chains and biomolecules to provide beneficial results (Ajikumar et al., 2008). Consumers are showing more interest in natural bioactive compounds due to their useful effects in improving health and reducing the risk of disease.

Agro-industrial byproducts can be used as an ingredient or in processing; for example, they can be dried or used in other innovative technologies to collect useful bioactive substances that could provide health benefits in addition to nutritional value. These bioactive substances are known to provide antimicrobial, antioxidant, and anti-inflammatory activity, which can further be used to formulate various nutraceutical, pharmaceutical, and food products (Simitzis and Deligeorgis, 2011). Furthermore, these byproducts are enriched with various health-improving biologically active substances such as dietary fibre, phenolic compounds, flavonoids, and anthocyanins (Abbasi-Parizad et al., 2021). Currently, various agro-industrial byproducts are pre-treated or processed to feed ruminants, which would otherwise be thrown away in landfills. Feedstuffs produced from byproducts nourish ruminants and maintain healthy growth and outcomes in manufacturing high-quality edible meat and dairy products. The popularity of agro-industrial byproducts as feedstuffs is increasing day by day as these are easily available at reasonable prices compared with other products (Simitzis and Deligeorgis, 2018). Many agro-industrial byproducts constitute procyanidins, stilbenes, cinnamic acids, and other polyphenolic compounds and can be used as food additives, antioxidants, and antimicrobial compounds. In addition, these can be used in several biotechnological applications (Kumar et al., 2017; Sette et al., 2020).

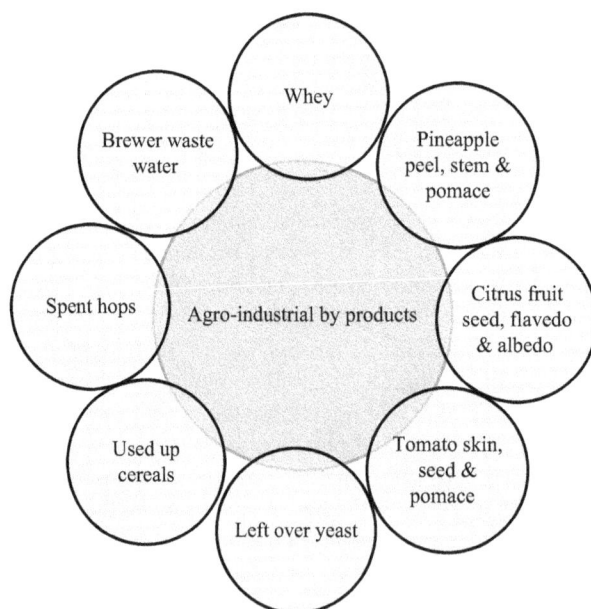

FIGURE 2.1　Various byproducts obtained from agro-industries.

Agricultural byproducts are produced in different stages of the food-handling chain, including harvesting agricultural products, manufacturing, processing operations, and distributions. Approximately 42% of food byproducts are generated by households, 39% from food processing sectors, 5% is generated during distribution; around 14% of food byproducts are lost from food-service organizations (Kumar et al., 2017).

Agro-industrial byproducts contain some harmful organic materials that may have a huge impact on environmental pollution and may cause a human health hazard if they are not treated properly. Seeing the increasing production of dairy products, it has been estimated that about 230 million tonnes of dairy byproducts, especially whey and whey permeates, will be produced globally by 2023 (Rama et al., 2019). These dairy byproducts can cause huge environmental problems when they are not discarded properly, as these are rich in organic matter like mineral salts and lactose.

Wastewater generated from the brewing industry should undergo several primary and secondary treatments before being discarded into rivers or land, as it is rich in suspended solids, fatty acids, sugars, soluble starch, and ethanol. However, the high concentration of organic matter in brewery wastewater makes it useful in the production of microalgae, hydroponic plant cultivation, and other valuable product development (Karlović et al., 2020). On the other hand, the fruit and vegetable processing industry produces approximately 38% of total agro-industrial byproducts (peel, seed, core, pomace, damaged parts, etc.) both in solid and liquid form. Hence, a large quantity of fruit and vegetable byproducts is globally added to formulate food products as an ingredient by multiple food-processing organizations. These byproducts also contain significant quantities of organic substances that may cause environmental pollution as well as health hazards (Panwar et al., 2021a). Therefore, to decrease the effect of these byproducts and improve environmental problems, it is necessary to develop some approaches to reusing these resources for valuable purposes. These strategies may either use the fruit and vegetable byproducts directly as livestock feed or, through further treatments (coagulation and extraction), convert them into valuable food products.

2.2 BYPRODUCTS FROM THE DAIRY INDUSTRY

In the dairy industry, numerous byproducts are generated during the production of milk products like butter, cheese, paneer, flavoured milk, Yakult, and yoghurt. Whey is the largest by-product generated by the dairy industry (Kaur et al., 2009; Panesar et al., 2007). The production of cheese yields whey as a by-product of about 90% of the total volume of milk used in the processing chain. Out of this, approximately 70% of the whey is used to prepare various products, and the rest (30%) is used in animal feed preparation, fertilizer, or discarded into rivers or the sea (Khezri et al., 2016). The cheese preparation flow diagram of milk generating whey as a by-product is described in Figure 2.2. Based on acidity, whey is classified into three groups: sweet whey, medium acid whey, and acid whey. Generally, sweet whey is produced when cheese and casein are processed with rennet, while acid whey is produced during the processing of acid cheese and fresh acid casein, for example, ricotta or cottage cheese. The major components present in sweet whey are 46.0–52.0 g/l (w/v) lactose, 6.0–10.0 g/l (w/v) protein, and 2.0 g/l (w/v) lactate, whereas those present in acid whey were 44.0–46.0 g/l (w/v) lactose, 6.0–8.0 g/l (w/v) protein, and 6.4 g/l (w/v) lactate. The use of rennet for whey processing tends to retain mineral components like phosphorus and calcium with the curd. However, when acid is used to process casein, some of the phosphorus and calcium components are transferred to the whey. Medium acid whey is generally produced during the preparation of paneer. The composition of whey varies depending on the milk source (cow, buffalo, or any other livestock) and milking seasons (summer, rainy, or winter). Whey produced from cow milk after draining the curd contains 5.5–7% total solids, 4.1–5.2% lactose, 0.3–1% protein, and 0.01–0.4% fat, whereas whey produced from buffalo milk after draining the curd contains 5.2–5.9% total solids, 4.1–4.7% lactose, 0.4–0.6% protein, and 0.01–0.2% fat (Zall, 1992). For a long time, whey has been discarded without any treatment, which causes environmental pollution because of its high protein content and other organic matter.

```
┌─────────────────────────────────────────┐
│                Raw milk                  │
└─────────────────────────────────────────┘
                    ⬇
┌─────────────────────────────────────────┐
│              Pasteurization              │
└─────────────────────────────────────────┘
                    ⬇
┌─────────────────────────────────────────┐
│               Cheese milk                │
└─────────────────────────────────────────┘
                    ⬇
┌─────────────────────────────────────────┐
│    Cheese processing with starter culture │
└─────────────────────────────────────────┘
                    ⬇
┌─────────────────────────────────────────┐
│              Whey drainage               │
└─────────────────────────────────────────┘
                    ⬇
┌─────────────────────────────────────────┐
│                 Brining                  │
└─────────────────────────────────────────┘
                    ⬇
┌─────────────────────────────────────────┐
│                 Cheese                   │
└─────────────────────────────────────────┘
```

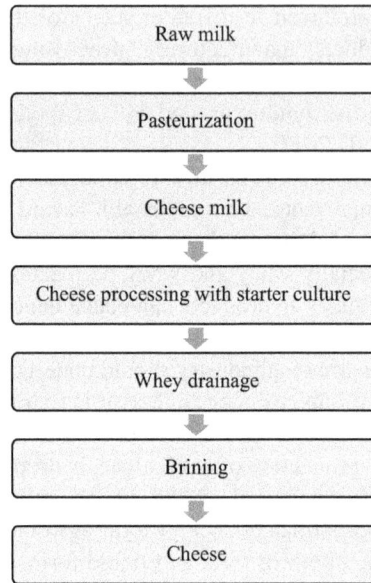

FIGURE 2.2   Flow diagram of the cheese-making process generating whey as a by-product (source: Atamer et al., 2013).

Cheese whey is considered one of the highly valuable byproducts of the dairy industry due to its value-added compounds like protein, minerals, and sugar. These compounds of cheese whey can be used as an ingredient to prepare functional and dietetic food. Generally, 9 kg of whey and 2.3 l of whey permeate are produced during the preparation of 1 kg of cheese (Romaní et al., 2018). Pre-treatments and handling processes like pasteurization, centrifugation, heat treatment, microbial culture used, screening operation, and addition of chemical compounds to improve colour can significantly affect the composition of the whey.

The characteristics of whey mainly include colour, flavour, acidity, and surface tension. Whey is generally greenish and semi-transparent in nature. The greenish colour of whey is attributed to the presence of riboflavin. Riboflavin is water-soluble and less sensitive to heat but highly sensitive to light and ionizing irradiation. When placed under light and radiation, the green colour of whey fades. Flavours like the sweetness, bitterness, umami, sourness, and saltiness of whey are highly dependent on the concentration of whey. For example, skimmed milk with 20% whey has umami, whereas 40% whey is sweet and bitter, and 100% whey is salty and astringent (McGugan et al., 1979). However, under neutralized conditions, all flavour characteristics change significantly. Depending on the source of milk, the titratable acidity and pH of the whey may vary. For example, the acidity of cow's milk whey varies from 0.12 to 0.16%, and that of goat's milk whey varies from 0.13 to 0.15% (Borba et al., 2014). Surface tension is one of the important characteristics of whey. The surface tension of whey increases with an increase in the total dry extract and decreases with an increase in temperature. The variations in the surface tension of whey are attributed to the variation in their lipid content.

The characteristic constituents of whey, such as proteins, lactose, vitamins, and minerals, make it useful for developing many valuable food products. There are whey-based products in a wide range of food applications, such as confectionery, ready-to-drink beverages, bakery products, dietetic foods, and pharmaceutical and nutraceutical products (Khezri et al., 2016). Whey protein concentrate (WPC) and whey protein isolate (WPI) are the most widely produced industrial whey products and are used in a variety of food product development. Depending on the whey's functional characteristics such as solubility, foaming capacity, gelation kinetics, emulsion properties, water- and lipid-binding capacity, WPC and WPI are used in many foods as raw materials or as functional ingredients. These whey products have the potential to improve the appearance, texture, structure, and rheological characteristics of foods. Whey undergoes various physical, chemical, and biotechnological treatments before it is used as an ingredient for product development (Panesar and Kennedy, 2012; Prazeres et al., 2012). Physicochemical methods such as membrane separation, crystallization, and chromatographic techniques are used to recover and prepare WPC and WPI. It has been observed that whey protein, in combination with methoxyl pectin, can be used as a fat replacer as well as texturizing agents in low-fat yoghurt (Krzeminski et al., 2014). As the whey is rich in protein, it can be used to prepare high protein beverages after eliminating astringent flavour (Nagar and Nagal, 2013). Due to its high nutritive value and emulsifying characteristics, it can be incorporated into the production of functional beverages and sports drinks. After removing protein, a large amount of lactose remains in whey, and it can be used for the production of bioethanol and β-galactosidase using yeast fermentation (Panesar et al., 2014). Many lactose-based functional compounds such as lactone, gluconic acids, lactobionic acids, and lactosucrose are produced (Khezri et al., 2016). In addition, whey lactose is used as an additive in the confectionery industry to extend the shelf life and increase the weight of products without changing the original characteristics. Protein isolated from whey can also be used to prepare edible films or coatings.

2.3 BYPRODUCTS FROM THE BREWING INDUSTRY

The brewing industry is considered one of the largest producers of agricultural byproducts. Water, grains, yeast, and hops are used as the main raw materials in the brewing industry. The major byproducts produced by the brewing industry are wastewater, used-up grains, hops, yeast, and germ (Karlović et al., 2020). A flow diagram of the brewing process is shown in Figure 2.3. Used up grains and yeasts are rich in protein and can be further used as animal feed. Most of the brewing byproducts are easily available at a low cost, are rich in nutrients, and are highly valuable compounds. They can also be used to develop new products in food, nutraceuticals, and the biotechnological field.

2.3.1 Brewer Wastewater

The preparation of brewing products like beer involves the blending and fermentation of different cereals such as barley, corn, and sorghum using yeasts and requires a high volume of water. The preparation of one litre of beer takes around ten litres of water, from which around seven litres of wastewater is generated (Simate et al., 2011). In the brewing industry, a large quantity of water is used as a direct ingredient and as part of the main process such as steam generation, cleaning floors, and washing brewing equipment. With the increasing demand for the reuse of water, discharging brewer wastewater into the environment has become a vital problem. The main concern is the water-soluble starch, sugars, fatty acids, ethanol, phosphate, nitrate, nitrite, and other organic and inorganic matter in brewing wastewater these compounds have a significant effect on the aquatic environment (Kovoor et al., 2012). Brewing wastewater contains about 1096.4–8926.08 mg/l of total chemical oxygen demand (COD) and 1178.64–5847.74 mg/l of soluble COD. Whereas the biological oxygen demand (BOD) value of brewing wastewater ranged from 1609 to 3980 mg/l. Other characteristics like total dissolved solids varied from 2020 to 5940 mg/l, total dry matter 1900–8000 mg/l, acidity of 3–12 pH, and temperature of 18–40°C (Mielcarek et al., 2013). It has been found that the characteristics of brewing water do not match the discharge standards (Enitan et al., 2015). Therefore, to reduce the negative impact of brewing wastewater and meet the standard limit, it is subjected to some primary and secondary treatments. In the primary treatments, large particulate matter is removed by sieving, particle size reduction, flow balancing, sedimentation, and filtration operations, whereas in secondary treatments, chemical and biological compounds are removed from wastewater through coagulation, flocculation, precipitation, adsorption, disinfection, and microbial treatment.

```
                              ┌─────────────────────────────┐
                              │   Raw material (grains)     │
                              └─────────────────────────────┘
                                            ▼
                              ┌─────────────────────────────┐
                              │         Steeping            │
                              └─────────────────────────────┘                    ┌──────────────────┐
                                            ▼ ◄──────────────────────────────────│   Malting loss   │
  ┌──────────────────┐        ┌─────────────────────────────┐                    └──────────────────┘
  │     Malting      │◄───────│        Germination          │
  └──────────────────┘        └─────────────────────────────┘
                                            ▼
                              ┌─────────────────────────────┐
                              │          Kilning            │
                              └─────────────────────────────┘
                                            ▼
                              ┌─────────────────────────────┐
                              │          Milling            │
                              └─────────────────────────────┘
                                            ▼
                              ┌─────────────────────────────┐                    ┌──────────────────┐
                              │          Mashing            │◄───────────────────│      Water       │
                              └─────────────────────────────┘                    └──────────────────┘
                                            ▼
                              ┌─────────────────────────────┐                    ┌──────────────────┐
                              │          Filtering          │───────────────────►│  Used up grains  │
                              └─────────────────────────────┘                    └──────────────────┘
  ┌──────────────────┐                      ▼
  │       Hops       │───────►┌─────────────────────────────┐
  └──────────────────┘        │         Wort boiling        │
                              └─────────────────────────────┘
                                            ▼
                              ┌─────────────────────────────┐                    ┌──────────────────┐
                              │        Clarification        │───────────────────►│    Spent hops    │
                              └─────────────────────────────┘                    └──────────────────┘
                                            ▼
                              ┌─────────────────────────────┐
                              │          Cooling            │
                              └─────────────────────────────┘
  ┌──────────────────┐                      ▼
  │     Malting      │───────►┌─────────────────────────────┐                    ┌──────────────────┐
  └──────────────────┘        │        Fermentation         │───────────────────►│  Leftover yeast  │
                              └─────────────────────────────┘                    └──────────────────┘
                                            ▼
                              ┌─────────────────────────────┐
                              │        Conditioning         │
                              └─────────────────────────────┘
                                            ▼
                              ┌─────────────────────────────┐
                              │          Packaging          │
                              └─────────────────────────────┘
```

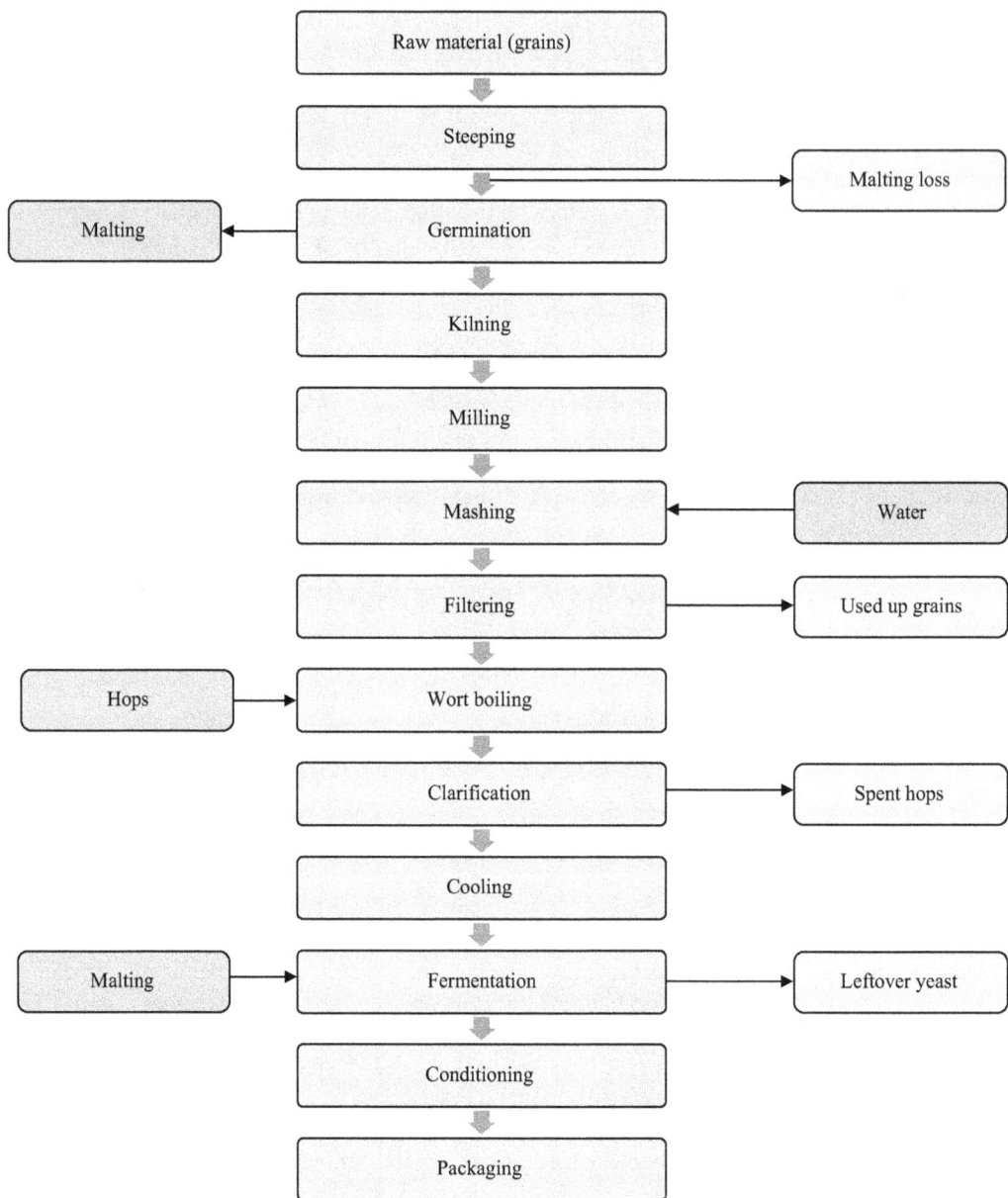

FIGURE 2.3   Flow diagram of the brewing process (source: Pascari et al., 2018).

2.3.2 USED-UP GRAINS

Used-up grains or spent grains are the most widely generated by-product of the brewing industry. Nearly 30% (w/w) of used-up grain is generated during the malting operation; it is considered one of the brewing industry's most widely available and low-cost byproducts. Brewing byproducts generally include husks, endosperm residue, and other soluble and insoluble substrates produced during the mashing of cereals such as corn, barley, sorghum, wheat, and rice. The composition (Table 2.1) of used-up grains may vary based on the quality of the cereals, time of harvest, germination conditions during malting, and the quality of other raw materials (Robertson et al., 2010). Used-up grain from the brewing industry is prone to microbial spoilage as it is rich in nutrients such as

TABLE 2.1

Chemical Characteristics of Byproducts Obtained from the Brewing Industry

Composition	Used-Up Grains (%, w/v)	Leftover Yeast (%, w/v)	Spent Hops (%, w/v)
Crude protein	18–29.6	48–50	22–23
Crude fat	6–19.6	1.0	4.5
Crude fibre	12–18.1	3.0	23–26
Carbohydrate	–	36–46	6–6.5
Ash	6–6.5	6–6.5	46.0
Dry matter	21–29.2	–	–
Moisture	69.8–79	–	–

Sources: Hardwick, 1995; Mussatto, 2009; Muthusamy, 2014.

carbohydrates and protein. Therefore, these used-up grains need to be stabilized immediately after production to prevent microbial spoilage. The moisture content of used-up grains varies from 70% to 80%. Hence, to retain the quality, prevent microbial spoilage, and increase the shelf life of used-up grains, it is necessary to remove the moisture content with different drying processes. The final moisture content of used-up grains should not be greater than 10% (Karlović et al., 2020). Drying reduces the risk of spoilage and decreases the volume for the storage and transportation of used-up grains. Besides drying, the shelf life of used-up grains can also be improved with some chemical compounds such as lactic acid, potassium sorbate, acetic acid, formic acid, etc. as preservatives (Kuentzel and Sonnenberg, 1997).

2.3.3 LEFTOVER YEAST

Leftover yeast is one of the biggest byproducts generated by the brewing industry. It has acquired lots of attention because of the high amount of protein, carbohydrate, and other chemical compounds in it. Leftover yeasts are generally deposited in the bottommost and topmost part of the brewing tank in the case of cold fermentation and warm fermentation, respectively. Though yeasts can be simply collected after settling down or floating in a brewing tank, some may need to be separated using a centrifuge or filtration unit (Kunze, 2014). Carbon is the most copious element found in yeast and makes up approximately 50% of dried leftover yeast. Besides this, some other elements such as oxygen, nitrogen, hydrogen, and phosphorus are present in yeast. The composition of yeast varies based on functional conditions and the attained growth phase (Mussatto, 2009). The chemical composition of leftover yeast generated from the brewing industry is described in Table 2.1. The amino acid composition of a yeast cell indicates that leucine, lysine, and tyrosine are the most abundant bound amino acids in yeast. However, some other bound amino acids like arginine, cysteine, and glycine are also present in the yeast cell. Niacin is the most abundant amino acid in the yeast cell; pantothenic acid, biotin, riboflavin, folic acid, and thiamine are also present. Nearly 5–10% mineral content is found in yeast. Phosphorous and potassium are two mineral elements mostly present in yeast. In addition, copper, cobalt, manganese, and selenium are also present (Huige, 2006). Beta-glucan and yeast extract (a combination of protein, peptides, and other soluble compounds) are two industrially important compounds that can be separated from yeast. Beta-glucan is used as a thickener, emulsifier, foamer, water binding and oil binding agent, and the yeast extract is used as a flavour enhancer in many foods like snacks, soup, sauces, and canned food.

2.3.4 SPENT HOPS

Hops is agricultural produce specifically used in the manufacturing of brewing products. Hops is available in many forms, such as a cone, powder, pellets, and extract. The chemical composition of spent hops generated from the brewing industry is described in Table 2.1. It contains some bitter compounds such as humulones, lupulones, and ethereal oils, which provide bitterness and an aroma to beer. The addition of hops increases the stability and antimicrobial activity of beer. This may be due to the presence of antimicrobial compounds in hops. About 15% of hops is used in beer, whereas the remaining 85% end up as spent materials (Mussatto, 2009). Lupulones are the most discarded compounds in spent hops, as they are not soluble at the pH conditions of wort and are not isomerized during the heat treatment of hops. On the other hand, phenolic constituents such as gallic acids, ferulic acids, caffeic acids, catechins, and flavones, etc., of hops are also removed during wort heating.

Hops are made up of a large quantity of protein, crude fibre, and nitrogen-free extract. The crude fibre of hops mostly contains glucose and xylose. In addition, other sugar compounds such as rhamnose, mannose, glucose, and arabinose are also part of the crude fibre content of hops. Around 46% of the carbohydrate compounds of hops is made up of arabinose, rhamnose, galactose, pectic sugar, and uronic acids (Oosterveld et al., 2002). When hops is used in brewing as a powder, extract, or pellet form, a major portion of these is removed in the trub. The trub produced in the brewing process is generally of two types, i.e., hot trub and cold trub. The former consists of 40–70% protein, 7–15% bitter compounds, and 20–30% mineral elements, whereas the latter consists of approximately 50% protein, 15–25% phenolic compounds, and 20–30% carbohydrate substances (Esslinger and Narziss, 2005). From an industrial point of view, hops can be used to develop many food products as it is rich in flavour compounds, polysaccharides, and organic acids. These compounds can be efficiently extracted from hops by oxidation and hydrolysis. Hop acids have high industrial importance among the other hop extracts, as they can act as a natural antibacterial agent. They can be used to control the bacterial reaction during ethanol fermentation and can be used as a substitute antibiotic in the fermentation process (Rückle and Senn, 2006).

Wastewater produced in the brewing industry contains high quantities of carbon and additional nutrients that are required for the growth of microalgae. Hence, brewing wastewater can be useful for the preparation of microalgae. It is possible to produce biofuels like hydrogen (applicable as an energy source) by using brewing wastewater during the production of microalgae (Maintinguer et al., 2017). Spent grains produced from the brewing industry have various applications in biotechnological operations, livestock feed preparation, and biofuel production. Beer trub obtained during the brewing process is rich in nutrients and other chemical compounds. Hence, it can be used as a medium for growing beneficial micro-organisms. In the biotechnological field, used-up grains can also be used to produce artificial sweeteners and lactic acid. Sweeteners like xylitol and lactic acids are generally produced by hydrolyzing and fermenting used-up grains. Lactic acids obtained from spent cereal fermentation have wide applications in multiple fields such as food product development, pharmaceuticals, nutraceuticals, and chemical and textile industries (Carvalheiro et al., 2007). Leftover yeast generated in the brewing process has a high potential for the production of livestock feed and nutrient supplements as it is rich in protein, minerals, and vitamins. The composition of extract prepared from leftover yeast includes amino acids, peptides, nucleotides, and other soluble compounds. Due to the presence of these compounds, more attention has been given to leftover yeast by food industries because it can be used as a raw material for developing food products like canned foods, snack foods, soups, sauces, etc. (Thammakiti et al., 2004).

2.4 BYPRODUCTS FROM THE FRUIT AND VEGETABLE PROCESSING INDUSTRY

Parts of plants such as the leaves, seeds, peels, pomace, husk, hulls, stems, stalks, etc., are the main waste generated from the fruit and vegetable processing industry (Figure 2.4). These parts of plants are taken as waste or used to develop some valuable products after the removal and utilization of the consumable parts (Saini et al., 2019). According to a study, approximately 20–60% of fruits and vegetable processing waste is found in solid and liquid form, of which around 42% comes from

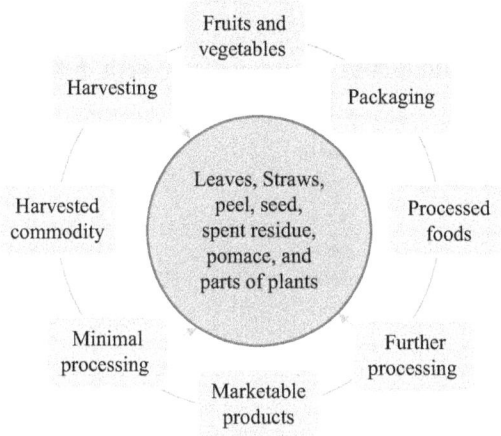

FIGURE 2.4 Byproducts generated from the fruit and vegetable processing industry.

household sources, 38% is from the fruit and vegetable processing industry, and the rest (20%) is generated from other sources (Wadhwa and Bakshi, 2013). Hence, large amounts of vegetable and fruit waste worldwide come from various food processing industries. Many countries like India, China, the Philippines, and the United States of America, in combination, produced approximately 55 million tonnes of fruit and vegetable processing waste per year. This waste is mainly produced during unit operations like sorting, grading, peeling, halving, packaging, and even during the distribution of products to consumers. A study by Malik et al. (2019) reported that onion peel, waste from the sugar beet process, and brewery waste of approximately 3.5, 5, and 1 million tonnes, respectively, are generated every year.

Discarding the fruit and vegetable processing waste in water streams and land may lead to environmental pollution and, subsequently, multiple health problems (Chaboud and Daviron, 2017). This is why it is necessary to develop technical approaches to convert this waste into valuable products and minimize their harmful effect on the environment. This waste is rich in highly utilizable organic substances and bioactive compounds such as antioxidants, phytochemicals, and other biologically active components and can be used as ingredients in many food and pharmaceutical products (Sharma et al., 2016). These byproducts can also be utilized to produce animal feeds and human foods if appropriate precautions have been taken during the processing. These waste byproducts can be used to develop supplement foods as they contain significant amounts of vitamins and mineral compounds. Different products such as edible and essential oils, edible films, enzymes, vinegar, pigments, antioxidant derivatives, single-cell protein, biofuel, and many others can be produced from fruit and vegetable processing byproducts (Balasundram et al., 2006; Singla et al., 202). Extracts of biologically active compounds from these byproducts have been reported to have anticarcinogenic, anti-heart-disease, and anti-tumour activity (Spatafora and Tringali, 2012).

2.4.1 Pineapple Processing Byproducts

In fresh-cut, minimally processed food, pineapple is considered one of the major fruits globally, especially in Brazil. Pineapple processing byproducts are generally produced in the form of the core, stem, peel, and crown. Approximately 49% of the whole fruit is a by-product, consisting of peels and crowns. Of these byproducts, 57% are composed of peels that remain unused after pineapple processing. These byproducts are rich in vitamins, minerals, and various polyphenolic compounds (de Ancos et al., 2016). The concentration of bioactive compounds present in different portions of the pineapple varies significantly. For instance, bromelain, a polymeric enzyme, is highly concentrated in the leaves, stems, and peel, whereas citric acid and ferulic acids are found in the leaves,

TABLE 2.2

Composition of Major Bioactive Compounds Present in Pineapple Byproducts

Compounds	Parts of Pineapple	Content	Reference
Bromelain (µm/ml)	Crown	113.79	Amini et al. (2016)
	Peel	13.16	
	Core	24.13	
	Stem	23.33	
Ferulic acid (mg/100 g)	Core	17.57–21.43	Li et al. (2014)
Dietary fibre (% w/w)	Core	99.8	Prakongpan et al. (2002)
Starch (% w/w)	Stem	97.9	Ashwar et al. (2016)

and the precursor of vitamin C is mostly found in the core (Salve and Ray, 2020). Polyphenolic compounds like coumaric acids, cinnamic acids, ferulic acids, syringic acids, and salicylic acids of pineapple processing byproducts have many applications in the pharmaceutical and food sectors as these can act as potent antioxidants and antimicrobial agents. The composition of major bioactive compounds present in various portions of the pineapple is shown in Table 2.2. The moisture content of the pineapple leaves and crown varies from 9–9.3% and 6.75–6.85%, respectively, on a dry weight basis (Brito et al., 2020). Although, variation in the total moisture content of pineapple byproducts is highly dependent on the drying conditions, like time, temperature, and the portion of the byproducts. The protein and lipid content of the pineapple peel varies from 3.83% to 4.17% and 1.27% to 1.33%, respectively (Martínez et al., 2012), whereas those of the pineapple crown varies from 6.93% to 8.25% and 1.78% to 1.94%, respectively (Brito et al., 2020).

Pineapple byproducts, especially pineapple pomace, are rich in both soluble and insoluble dietary fibre in a balanced quantity. Hence, pineapple byproducts are suitable for producing low calorie, fibre-rich foods. Ferulic acid is one of the most valuable hydroxyl cinnamic acids and is mainly present in pineapple peel and leaves. Ferulic acid is generally present in plant cells in the ester linkage with polysaccharides and ether. Citric acid is one of the most commonly used organic acids, and the demand for this acid is approximately 6.0×10^5 tonnes each year. It is mostly used in beverages, dairy, and medicinal plants to improve the taste and acidify their respective products. It has been noted that citric acid produced from pineapple waste using *Aspergillus foetidus* gives a higher yield compared with that produced from other agro-industrial waste like apple pomace and wheat bran (Nakthong et al., 2017).

2.4.2 CITRUS FRUITS BYPRODUCTS

Citrus fruits are one of the largest traded fruits throughout the world. Approximately 14% of the total fruit crops cultivated in India are citrus fruit, and with this, India is the world's sixth-largest citrus fruit producer. Citrus fruits are rich in ascorbic acids, polyphenolic compounds, flavour components, and other nutritionally important bioactive compounds (Jandrić et al., 2017; Panwar et al., 2021b). Citrus fruit varieties such as mandarins, sweet orange, bitter orange, lemons, limes, grapefruit, and pomelos are popular throughout the world; mandarin, sweet orange, lime, and lemon are mostly cultivated in different parts of India. About 7.8 million tonnes of citrus fruit processing waste is produced in India, whereas its worldwide production is about 119.7 million tonnes (Rezzadori et al., 2012). Out of the total waste generated by citrus fruit processing plants, a vast quantity of this waste is burned or discarded into water streams. This may increase the pollution levels in the environment, reduce the level of dissolved oxygen in water, and degrade the soil quality (Wadhwa and Bakshi, 2013). Citrus fruit processing byproducts are mainly composed of peel, albedo, and core. The peels of citrus fruits are composed of a cuticle on the outer part covered by a layer known as flavedo. The flavedo mainly consists of oil pockets containing flavour compounds, and these have a high market value. The albedo of citrus fruits is a spongy layer and consists of 20% pectin. While the centre portion of citrus fruits is a spongy white layer and looks the same as albedo, it is known as the core of the fruit. Approximately 50% of citrus fruits generate waste residues like peel, seeds, and pulp. These citrus fruit byproducts contain about 80% moisture (Rezzadori et al., 2012).

The chemical and phytochemical compositions of citrus fruits depend on many factors such as variety, maturity level, and atmospheric conditions during growth. During the growing stage of citrus fruits, various phytochemical compounds such as polyphenols, flavonoids, terpenoids, and acetophenones are produced. Afterwards, these compounds accumulate in different parts of the fruit (Sharma et al., 2017). The chemical composition of citrus fruits is described in Table 2.3. Usually, citrus fruits consist of a significant quantity of lipids, including linolenic, linoleic, oleic, stearic, and palmitic acids. Sucrose, fructose, and glucose are the main sugars present in citrus fruit, whereas cellulose and pectin are the major carbohydrate compounds, and citric and maleic acids are the main acids produced by citrus fruit processing. Carotenes and xanthophyll are the main colour components present in citrus fruits (Simitzis and Deligeorgis, 2018).

2.4.3 Tomato Processing Byproducts

Tomato comes in second in terms of fruit consumption throughout the world. It makes up approximately 19% of the total fruit consumed. Its appealing taste, colour, and diversity make the tomato a valuable fruit worldwide. A variety of products like ketchup, puree, paste, sauces, and canned tomatoes are produced on a large scale. Besides these main products, a number of byproducts such as tomato peels, seeds, pomace, and vascular tissues are generated in a significant quantity from tomato processing plants. The major compounds present in tomato processing byproducts are described in Table 2.3.

Tomato processing waste includes 14–20% protein, 4%–5% solvent extract, about 22% fibre comprising both lignin and cellulose, 40–60% saccharides, especially sugars, and 5–10% pectin. Sometimes, these tomato processing byproducts can be harmful if they have been discarded without proper treatment, but these are rich in many valuable bioactive substances (Szabo et al., 2018). Tomato pomace constitutes approximately 2%–3% of total weight after it has been processed. It is used to produce 0.8 to 1.3 million tonnes of renewable feedstock per year. In addition to animal feedstock, tomato pomace can also be used to develop human food, as it comprises a high quantity of protein (approximately 10–20% w/w). Tomato peel is a by-product of the tomato processing industry, and it is rich in carotenoids (lycopene, β-carotene, and lutein) and total polyphenols (quercetin, rutin, and isochlorogenic acids). The total carotenoid and total polyphenol content of tomato peel range from 0.60 to 5.31 mg/100 g and 37 to 155 mg/100 g, respectively (Szabo et al., 2019). It has been found that tomato peel has the potential to act as both antioxidant and antimicrobial agents. The antioxidant activity (DPPH) of tomato peel ranges from 120 to 255 μmol TE/100 g, and it may be correlated with the synergistic relations between the active components

---

TABLE 2.3
Composition of the Major Bioactive Compounds Present in Tomato and Citrus Byproducts

Byproducts	Parts of Fruit	Total Polyphenol (mg GAE/g dm)	Total Flavonoids (mg QE/g dm)	Carotenoids (mg/g dm)	Antioxidant Activity (μmoles Fe2+/g dm)	Reference(s)
Citrus	Peel	141.3–153.9	47.87–50.53	–	645.27–667.09	Wang et al., 2008;
	Pulp	53.6–61.8	1.09–1.21	–	441–472.4	Goulas and Manganaris, 2012
Tomato	Roots	21.98–24.5	10.0–18.0	0.35–0.36	–	Silva-Beltrán
	Leaves	83.35–125.5	33.03–61.96	1.19–1.25	–	et al., 2015

present in tomato peel. Tomato peel with a minimum inhibitory concentration of 10 to 2.5 mg/ml shows antimicrobial activity against gram-positive and gram-negative bacteria. The antimicrobial activity of tomato peel is mainly correlated with its carotenoid content (Silva-Beltrán et al., 2015). Tomato seeds are another valuable by-product generated from tomato processing plants. They are rich in total phenolic content and generally vary from 1.97 to 2 mg Gallic acid equivalent (GAE)/100 g (Choe et al., 2021).

Byproducts produced from various fruit and vegetables such as pineapple, citrus fruit, and tomato processing plants are sustainable, easily available, and the most abundant plant residue that can be used in multipurpose applications. The presence of several bioactive components and flavour compounds and the chemical composition of pineapple peel is suitable for wine production. Wine from pineapple peel is generally produced by using *S. cerevisiae* in fermentation. Chemical compounds like vinegar, which have antioxidant activity, can be produced from pineapple peel. *Acetobacter* bacteria is commonly used in the fermentation of pineapple peel for producing vinegar. The vinegar obtained from pineapple peel gives an antioxidant activity of 2077 mg ascorbate equivalents (AE)/100 ml (Rabiu et al., 2018). Bromelain is a proteolytic enzyme that can be extracted from the pineapple stem and has multiple applications in the food industry. Citrus fruit processing byproducts such as albedo, flavedo, and seeds have many food, phytochemical, and pharmaceutical applications. Essential oils obtained from citrus byproducts can be used as flavour compounds in various food products and beverages (Anagnostopoulou et al., 2006). After removing the essential oil, citrus residues can be used to produce dietary fibre. One study suggested that these residues are rich sources of cellulytic fibre (Chau and Huang, 2003). A large amount of by-product residues, around 24,000 kg/ha is generated during the harvesting of tomatoes. In addition, industrially processed tomato by-product residues, including immature, defective, and damaged tomatoes, are produced in significant quantities. These residues are rich in several soluble and insoluble fibres, such as pectin, lignin, cellulose, and hemicellulose, which makes them suitable as an ingredient for animal feed and organic fertilizer development (Fritsch et al., 2017).

2.5 BYPRODUCTS FROM THE CEREAL PROCESSING INDUSTRY

Cereals belong to the grass family of *Poaceae*, also recognized as *Gramineae*, and the cultivation of cereals started thousands of years ago. A variety of cereals such as rice, wheat, barley, maize, millet, oat, rye, and sorghum are the most valuable cereals globally (Hill and Li, 2016). Approximately 2 billion tonnes of cereal is produced per year, and it is considered the most significant food source for humans. Though, during processing, handling, storage, and transportation, approximately 30% of it is wasted (FAO, 2011). Numerous byproducts like germ, bran, hull, husk, and the aleurone layer are produced from cereal processing industries. The steps in the cereal milling process and by-product generation are described in Figure 2.5. These byproducts generated from milling industries are rich in fibre, vitamins, minerals, bioactive compounds, polyphenolic compounds, etc. The quality of cereal processing byproducts mainly depends on the types of milling process used, such as dry or wet milling. Hence, these byproducts can also be used for animal feed production and various food and non-food products depending on the milling process. Some of the important cereal byproducts are discussed below.

2.5.1 WHEAT BRAN AND GERM

Wheat milling creates byproducts such as bran, germ, wheat middling, wheat red dog, etc., from approximately 25–30% of total wheat. These byproducts are mainly used as animal feed, as they are a rich source of amino acids, phosphorous, and energy (Huang et al., 2014). The major byproducts obtained from wheat milling are wheat bran and germ. Byproducts are classified into different types, such as coarse bran, fine bran, wheat middlings, and red dog, based on endosperm content and particle size of the wheat bran. Wheat bran constitutes approximately 38% non-starch polysaccharide,

```
┌─────────────────────────────┐
│        Cereal grains        │
└─────────────────────────────┘
              ⬇
┌─────────────────────────────┐
│     Cleaning and grading    │
└─────────────────────────────┘
              ⬇
┌─────────────────────────────┐
│       Pitting (optional)    │
└─────────────────────────────┘
              ⬇
┌─────────────────────────────┐
│    Addition of oil or water │
└─────────────────────────────┘
              ⬇
┌─────────────────────────────┐
│   Conditioning or tempering │
└─────────────────────────────┘
              ⬇
┌─────────────────────────────┐
│           Drying            │
└─────────────────────────────┘
              ⬇
┌─────────────────────────────┐        ┌──────────────────────────────┐
│           Milling           │───────▶│  Germ, bran, hull and husk   │
└─────────────────────────────┘        └──────────────────────────────┘
              ⬇
┌─────────────────────────────┐
│           Sieving           │
└─────────────────────────────┘
              ⬇
┌─────────────────────────────┐
│ Dehulled whole seed or dehulled │
│     and split products      │
└─────────────────────────────┘
```

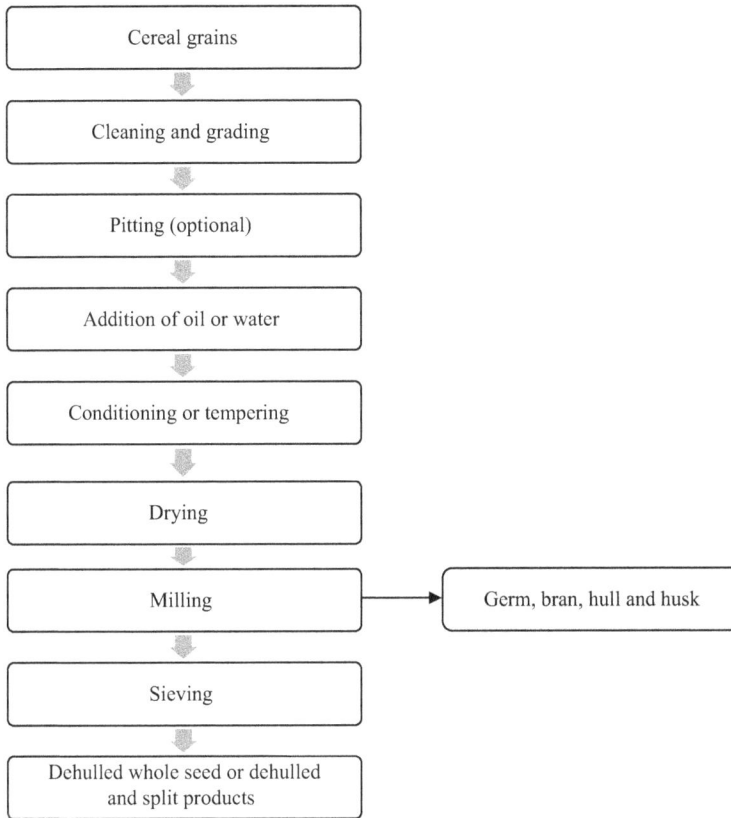

FIGURE 2.5 The cereal milling process and by-product generation (source: Wood and Malcolmson, 2011).

19% starch, 18% protein, and 6% lignin (Papageorgiou and Skendi, 2018). Polysaccharide compounds such as arabinoxylans and the beta-glucan content of wheat bran are known for lowering the risk of many diseases like cardiovascular disease, colorectal cancer, diverticular disease, and type II diabetes (Poutanen et al., 2014). In addition to the major compounds of wheat bran, it also contains a significant amount of dietary fibre and bioactive compounds like polyphenols. Among the phenolic compounds, ferulic acid is rich in wheat bran. Other polyphenolic compounds like vanillic acid, trans-coumaric acid, p-hydroxybenzoic acid, and p-hydroxybenzaldehyde are also present in wheat bran. Wheat germ is another highly nutritious by-product obtained during the wheat milling operation. It is a rich source of vitamin E and vitamin B groups. In addition, dietary fibre, protein, and minerals are also found in wheat germ in a significant quantity (Anson et al., 2012). The concentration of most of the essential amino acids present in wheat germ is higher than those present in egg protein. Wheat germ is also rich in many unsaturated fatty acids like linoleic, α-linoleic, and oleic acids, along with some bioactive compounds like glutathione, octacosanol, sterols, and flavonoids (Zhu et al., 2006).

2.5.2 RICE BRAN AND GERM

Rice is recognized as one of the chief cereal grains and staple food consumed by most of the global population, particularly in Asian countries. During rice milling, approximately 70% of rice endosperm, 20% husk, 8% bran, and 2% germ are produced (Van Hoed et al., 2006). The composition of major rice processing byproducts is presented in Table 2.4. Currently, byproducts such as rice bran and germ obtained from rice milling are getting more attention due to their high content of

TABLE 2.4

Composition of Major Rice Processing Byproducts

Composition	Husk (%, w/v)	Bran (%, w/v)	Straw (%, w/v)
Crude protein	2.1–4.3	12.37–13.12	1.2–7.5
Crude fat	0.30–0.93	21.53–24.39	0.8–2.1
Crude fibre	30.0–53.4	11.29–13.20	33.5–68.9
Ash	13.2–24.4	10.53–11.08	12.2–21.4
Carbohydrate	22.4–35.3	29.23–33.54	39.1–47.3

Sources: Esa et al., 2013; Rosniyana et al., 2007.

polyphenolic compounds, fibres, minerals, and vitamins. These compounds play a significant role in lowering the risk of harmful diseases like cardiovascular disease and anti-atherogenic activity (Wilson et al., 2002). Rice bran is the biggest by-product obtained during rice milling and is mainly composed of the outer layer of the rice grain, i.e., the aleurone layer, with some parts of the endosperm and germ. Rice bran constitutes approximately 10.6–16.9% protein, 7.0–18.9% fibre, and 1.3–1.9% phosphorous (Esa et al., 2013). It is rich in lipid content but deteriorates rapidly during storage. Due to this, it may affect the final quality of the food and pharmaceutical products developed from rice bran. The storage quality of rice bran has been found to be affected by some enzymes like lipoxygenases and lipases (Dhaliwal et al., 1991). Polyphenolic compounds and antioxidants such as γ-oryzanol, tocotrienols, and tocopherols are also rich in rice bran, and these components have a significant role in the prevention of cardiovascular disease, free-radical scavenging, and some cancer (Devi and Arumughan, 2007). Rice germ is another important by-product generated during rice milling. It is the well-known reproductive part of rice that is responsible for the germination and development of new plants. It is rich in vitamin E and contains approximately five times more vitamin E than rice bran. Alpha-tocopherol is the main source of vitamin E component in rice germ; it is recognized as the most active form of vitamin E. On the other hand, γ-tocopherol is the main source of vitamin E found in rice germ. It is also rich in fibre, vitamin B groups, and gamma-amino-butyric acid. As it contains significant amounts of gamma-aminobutyric acid, rice germ is believed to have potential health benefits like lowering blood pressure, lowering blood glucose levels, and improving cognition (Esa et al., 2013).

2.5.3 RYE BRAN

Rye is known as an important dietary fibre source among cereals worldwide. A variety of cereal-based products are produced from the whole grain flour of rye. A by-product, i.e., rye bran, obtained during rye milling can also be used as a raw material in the formulation of different cereal-based food products (Nordlund et al., 2012). In addition to dietary fibre, rye bran also contains some bioactive substances like polyphenolic compounds, tocopherols, and phytosterols. Among the bioactive compounds of rye, steryl ferulates and alkylresorcinol content have been found to play an important role as antioxidants and in anticancer activity (Patel, 2012). The fermentation of rye bran with baker's yeast helps improve its nutritional and functional quality. After fermentation, rye bran showed an approximately 90% increase in phenolic compounds compared with unfermented bran. Ferulic acid content was also increased significantly in fermented rye bran compared with unfermented bran (Katina et al., 2007). The bioprocessing of rye bran with an enzyme and baker's yeast helps improve the overall quality of bread prepared. This may be due to the breakdown of the cell structure of rye bran during the bioprocess and consequently increase the solubility of several nutritional compounds like proteins, dietary fibres, phenolic compounds, and arabinoxylans (Nordlund et al., 2012).

2.6 BYPRODUCTS FROM THE MEAT AND POULTRY PROCESSING INDUSTRY

The meat and poultry processing industries generate a number of byproducts such as offal, bones, blood, skins, and feathers. Most of these byproducts are generated during the slaughtering process. Bovine and porcine processing plants generate approximately 56% and 48% of byproducts, respectively (Mullen et al., 2017). The steps in by-product generation during meat processing are described in Figure 2.6. These byproducts are generally discarded or disposed of in landfills. Although disposal of these byproducts is difficult as they contain a high quantity of liquid and are highly susceptible to oxidation and enzymatic degradation, they can cause serious environmental problems if not disposed of or used appropriately (Anal, 2017). However, these byproducts can be a huge source of both macro- and micro-nutrients, as red meat is rich in these nutrients with a high biological value. Offal produced from meat and poultry processing plants is mostly used as animal feed. However, it requires rapid processing as it may putrefy and produce a bad odour.

Byproducts of edible meat are rich in essential nutrients. As these byproducts are rich in amino acids, fatty acids, vitamins, minerals, and hormones, they can be used to develop many food and pharmaceutical products. Except for blood, other meat byproducts such as the kidney, spleen, lungs, and brain contain higher moisture than meat. Byproducts like the liver and kidneys contain a higher quantity of carbohydrates compared with meat (Devatkal et al., 2004). These are rich sources of riboflavin, vitamin B6, vitamin B12, and folate. The concentration of these vitamins present in meat byproducts is significantly higher than the recommended dietary allowances values. Minerals like iron, copper, manganese, phosphorous, and potassium are also present in meat byproducts in a high quantity. Meat byproducts constitute a large quantity of connective tissue, and that is why the amino acid composition of these byproducts differs from that of lean tissue. Hence, byproducts such as tripe, lungs, ears, and stomach comprise higher quantities of glycine, hydroxyproline, and proline and a lower quantity of tyrosine (Jayathilakan et al., 2012).

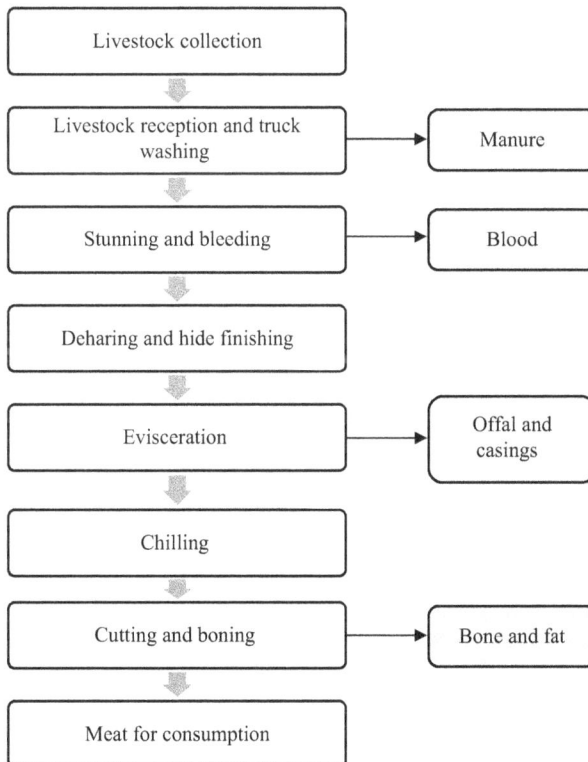

FIGURE 2.6 By-product formation during meat processing (source: Fornarelli et al., 2017).

Blood is one of the major byproducts of meat and poultry processing plants. It may cause environmental problems due to its high pollutant load. However, blood is also composed of many valuable components, like protein, that make it an important edible by-product. Blood obtained from healthy animals is considered sterile and is approved for use in food product formulation. With the increasing demand for foods with high protein content, the use of blood as a source of protein and bioactive peptides can be a good alternative (Anal, 2017). Feathers are highly produced byproducts generated from the poultry processing industry. Approximately 1.8 million tonnes of feathers are produced every year as animal processing byproducts. They are mainly composed of 90% protein, 8% water, and 1% lipid (Wang and Cao, 2012). The fermentation of feathers with a bacterium called *Bacillus licheniformis* can degrade and convert the feathers into feather lysate. It is also known as a digestible protein and can be used for feed production. Feathers are rich in keratin protein and are mainly composed of cysteine and serine.

2.7   CONCLUSIONS AND FUTURE PERSPECTIVES

The quantity of agro-industrial byproducts is increasing with the rising number of food industries. In addition, agricultural byproducts are also generated during harvesting and from household use. The use of these byproducts in animal feed and human food development will improve the economic status of farmers as well as reduce environmental problems, waste disposal, and management of waste. Furthermore, various treatments to improve these byproducts will facilitate sustainability and value addition in the agro-food chain. The development of healthy products with improved composition, absence of any harmful compounds, and presence of bioactive substances having health benefits, etc., were found to have high demand. As agro-industrial byproducts are rich in bioactive compounds such as dietary fibre, essential oils, phenolic acids, flavonoids, anthocyanins, etc. and remarkably contain twice the quantity of these compounds found in the edible parts of agricultural products, these can be used as suitable sources for the development of healthy food products. Agricultural byproducts have been found to be a source of many beneficial compounds that possess antioxidant and antimicrobial activity. Using novel techniques such as ultrasounds, supercritical fluid extraction, high pressure processing, and pulse electric fields, these beneficial compounds can be extracted at a high yield and used to prepare value-added products. There is a requirement for more advancement related to the use of agro-industrial byproducts from the food chain by maintaining food and feed safety.

REFERENCES

Abbasi-Parizad, P., De Nisi, P., Scaglia, B., Scarafoni, A., Pilu, S., and Adani, F. (2021). Recovery of phenolic compounds from agro-industrial by-products: Evaluating antiradical activities and immunomodulatory properties. *Food and Bioproducts Processing, 127*, 338–348.

Ajikumar, P. K., Tyo, K., Carlsen, S., Mucha, O., Phon, T. H., and Stephanopoulos, G. (2008). Terpenoids: Opportunities for biosynthesis of natural product drugs using engineered microorganisms. *Molecular Pharmaceutics, 5*(2), 167–190.

Amini, A., Masoumi-Moghaddam, S., and Morris, D. L. (2016). *Utility of Bromelain and N-Acetylcysteine in Treatment of Peritoneal Dissemination of Gastrointestinal Mucin-Producing Malignancies.* Springer.

Anagnostopoulou, M. A., Kefalas, P., Papageorgiou, V. P., Assimopoulou, A. N., and Boskou, D. (2006). Radical scavenging activity of various extracts and fractions of sweet orange peel (*Citrus sinensis*). *Food Chemistry, 94*(1), 19–25.

Anal, A. K. (Ed.). (2017). *Food Processing By-Products and Their Utilization.* John Wiley and Sons.

Anson, N. M., Hemery, Y. M., Bast, A., and Haenen, G. R. (2012). Optimizing the bioactive potential of wheat bran by processing. *Food and Function, 3*(4), 362–375.

Ashwar, B. A., Gani, A., Shah, A., Wani, I. A., and Masoodi, F. A. (2016). Preparation, health benefits and applications of resistant starch—A review. *Starch-Stärke, 68*(3–4), 287–301.

Atamer, Z., Samtlebe, M., Neve, H., Heller, K. J., and Hinrichs, J. (2013). Review: Elimination of bacteriophages in whey and whey products. *Frontiers in Microbiology, 4*, 191.

Balasundram, N., Sundram, K., and Samman, S. (2006). Phenolic compounds in plants and agri-industrial by-products: Antioxidant activity, occurrence, and potential uses. *Food Chemistry*, 99(1), 191–203.

Borba, K. K. S., Silva, F. A., Madruga, M. S., de Cássia Ramos do Egypto Queiroga, R., de Souza, E. L., and Magnani, M. (2014). The effect of storage on nutritional, textural and sensory characteristics of creamy ricotta made from whey as well as cow's milk and goat's milk. *International Journal of Food Science and Technology*, 49(5), 1279–1286.

Brito, T. B. N., Pereira, A. P. A., Pastore, G. M., Moreira, R. F. A., Ferreira, M. S. L., and Fai, A. E. C. (2020). Chemical composition and physicochemical characterization for cabbage and pineapple by-products flour valorization. *LWT*, 124, 109028.

Carvalheiro, F., Duarte, L. C., Medeiros, R., and Gírio, F. M. (2007). Xylitol production by *Debaryomyces hansenii* in brewery spent grain dilute-acid hydrolysate: Effect of supplementation. *Biotechnology Letters*, 29(12), 1887–1891.

Chaboud, G., and Daviron, B. (2017). Food losses and waste: Navigating the inconsistencies. *Global Food Security*, 12, 1–7.

Chau, C. F., and Huang, Y. L. (2003). Comparison of the chemical composition and physicochemical properties of different fibers prepared from the peel of *Citrus sinensis* L. Cv. Liucheng. *Journal of Agricultural and Food Chemistry*, 51(9), 2615–2618.

Choe, U., Sun, J., Bailoni, E., Chen, P., Li, Y., Gao, B., … and Yu, L. L. (2021). Chemical composition of tomato seed flours, and their radical scavenging, anti-inflammatory and gut microbiota modulating properties. *Molecules*, 26(5), 1478.

de Ancos, B., Sánchez-Moreno, C., and González-Aguilar, G. A. (2016). Pineapple composition and nutrition. In M. G. Lobo and R. E. Paull (Eds.), *Handbook of Pineapple Technology* (pp. 221–239).

Denek, N., and Can, A. (2006). Feeding value of wet tomato pomace ensiled with wheat straw and wheat grain for Awassi sheep. *Small Ruminant Research*, 65(3), 260–265.

Devatkal, S., Mendiratta, S. K., and Kondaiah, N. (2004). Quality characteristics of loaves from buffalo meat, liver and vegetables. *Meat Science*, 67(3), 377–383.

Devi, R. R., and Arumughan, C. (2007). Phytochemical characterization of defatted rice bran and optimization of a process for their extraction and enrichment. *Bioresource Technology*, 98(16), 3037–3043.

Dhaliwal, Y. S., Sekhon, K. S., and Nagi, H. P. S. (1991). Enzymatic activities and rheological properties of stored rice. *Cereal Chemistry*, 68(1), 18–21.

Enitan, A. M., Adeyemo, J., Kumari, S. K., Swalaha, F. M., and Bux, F. (2015). Characterization of brewery wastewater composition. *International Journal of Environmental and Ecological Engineering*. https://doi.org/https://doi.org/10.51415/10321/2992

Esa, N. M., Ling, T. B., and Peng, L. S. (2013). By-products of rice processing: An overview of health benefits and applications. *Rice Research: Open Access*. http://dx.doi.org/10.4172/jrr.1000107

Esslinger, H. M., and Narziss, L. (2005). *Beer*. Wiley-VCH Verlag GmbH and Co. KGaA.

FAO. (2011). *Global Food Losses and Food Waste - Extent, Causes and Prevention*. FAO.

Fornarelli, R., Bahri, P. A., and Moheimani, N. (2017). *Utilization of Microalgae to Purify Waste Streams and Production of Value Added Products*. Australian Meat Processor Corporation.

Fritsch, C., Staebler, A., Happel, A., Cubero Márquez, M. A., Aguiló-Aguayo, I., Abadias, M., … and Belotti, G. (2017). Processing, valorization and application of bio-waste derived compounds from potato, tomato, olive and cereals: A review. *Sustainability*, 9(8), 1492.

García, Y. D., Valles, B. S., and Lobo, A. P. (2009). Phenolic and antioxidant composition of by-products from the cider industry: Apple pomace. *Food Chemistry*, 117(4), 731–738.

Goulas, V., and Manganaris, G. A. (2012). Exploring the phytochemical content and the antioxidant potential of Citrus fruits grown in Cyprus. *Food Chemistry*, 131(1), 39–47.

Grumezescu, A. M., and Holban, A. M. (Eds.). (2018). *Food Quality: Balancing Health and Disease*, Vol. 13. Academic Press.

Hardwick, W. A. (1995). *Handbook of Brewing*. Marcel Decker.

Hill, C. B., and Li, C. (2016). Genetic architecture of flowering phenology in cereals and opportunities for crop improvement. *Frontiers in Plant Science*, 7, 1906.

Huang, Q., Shi, C. X., Su, Y. B., Liu, Z. Y., Li, D. F., Liu, L., … and Lai, C. H. (2014). Prediction of the digestible and metabolizable energy content of wheat milling by-products for growing pigs from chemical composition. *Animal Feed Science and Technology*, 196, 107–116.

Huige, N. J. (2006). Brewery by-products and effluents. In G. P. Fergus and G. S. Graham (Eds), *Handbook of Brewing* (pp. 670–729). CRC Press.

Jandrić, Z., Islam, M., Singh, D. K., and Cannavan, A. (2017). Authentication of Indian citrus fruit/fruit juices by untargeted and targeted metabolomics. *Food Control*, 72, 181–188.

Jayathilakan, K., Sultana, K., Radhakrishna, K., and Bawa, A. S. (2012). Utilization of byproducts and waste materials from meat, poultry and fish processing industries: A review. *Journal of Food Science and Technology, 49*(3), 278–293.

Karlović, A., Jurić, A., Ćorić, N., Habschied, K., Krstanović, V., and Mastanjević, K. (2020). By-Products in the malting and brewing industries—Re-usage possibilities. *Fermentation, 6*(3), 82.

Katina, K., Laitila, A., Juvonen, R., Liukkonen, K. H., Kariluoto, S., Piironen, V., … and Poutanen, K. (2007). Bran fermentation as a means to enhance technological properties and bioactivity of rye. *Food Microbiology, 24*(2), 175–186.

Kaur, G., Panesar, P. S., Bera, M. B., and Singh, B. (2009). Optimization of permeabilization process for lactose hydrolysis in whey using response surface methodology. *Journal of Food Process Engineering, 32*(3), 355–368.

Khezri, S., Seyedsaleh, M. M., Hasanpour, I., Dastras, M., and Dehghan, P. (2016). Whey: Characteristics, applications and health aspects. In: *3rd International Conference on Science and Engineering* (pp. 1–8). April, 2016. http://www.bipublication.com

Kovoor, P. P., Idris, M. R., Hassan, M. H., and Yahya, T. F. T. (2012). A study conducted on the impact of effluent waste from machining process on the environment by water analysis. *International Journal of Energy and Environmental Engineering, 3*(1), 1–12.

Krzeminski, A., Prell, K. A., Busch-Stockfisch, M., Weiss, J., and Hinrichs, J. (2014). Whey protein–pectin complexes as new texturising elements in fat-reduced yoghurt systems. *International Dairy Journal, 36*(2), 118–127.

Kuentzel, U., and Sonnenberg, H. (1997). Conservation of pressed brewers grain with potassium sorbate. *Monatsschrift fuer brauwissenschaft (Germany), 50*, 175–181

Kumar, K., Yadav, A. N., Kumar, V., Vyas, P., and Dhaliwal, H. S. (2017). Food waste: A potential bioresource for extraction of nutraceuticals and bioactive compounds. *Bioresources and Bioprocessing, 4*(1), 1–14.

Kunze, W. (2014). *Technology Brewing and Malting.* 5th Revised English edition. VLB Berlin.

Li, T., Shen, P., Liu, W., Liu, C., Liang, R., Yan, N., and Chen, J. (2014). Major polyphenolics in pineapple peels and their antioxidant interactions. *International Journal of Food Properties, 17*(8), 1805–1817.

Llobera, A., and Canellas, J. (2007). Dietary fibre content and antioxidant activity of Manto negro red grape (Vitis vinifera): Pomace and stem. *Food Chemistry, 101*(2), 659–666.

Maintinguer, S. I., Lazaro, C. Z., Pachiega, R., Varesche, M. B. A., Sequinel, R., and de Oliveira, J. E. (2017). Hydrogen bioproduction with Enterobacter sp. isolated from brewery wastewater. *International Journal of Hydrogen Energy, 42*(1), 152–160.

Malik, A., Erginkaya, Z., and Erten, H. (Eds.). (2019). *Health and Safety Aspects of Food Processing Technologies.* Springer Nature.

Martínez, R., Torres, P., Meneses, M. A., Figueroa, J. G., Pérez-Álvarez, J. A., and Viuda-Martos, M. (2012). Chemical, technological and in vitro antioxidant properties of mango, guava, pineapple and passion fruit dietary fibre concentrate. *Food Chemistry, 135*(3), 1520–1526.

McGugan, W. A., Emmons, D. B., and Armond, E. L. (1979). Influence of volatile and nonvolatile fractions on intensity of Cheddar cheese flavor. *Journal of Dairy Science, 62*(3), 398–403.

Mielcarek, A., Janczukowicz, W., Ostrowska, K., Jóźwiak, T., Kłodowska, I., Rodziewicz, J., and Zieliński, M. (2013). Biodegradability evaluation of wastewaters from malt and beer production. *Journal of the Institute of Brewing, 119*(4), 242–250.

Mirzaei-Aghsaghali, A., and Maheri-Sis, N. (2008). Nutritive value of some agro-industrial by-products for ruminants-A review. *World Journal of Zoology, 3*(2), 40–46.

Mullen, A. M., Álvarez, C., Zeugolis, D. I., Henchion, M., O'Neill, E., and Drummond, L. (2017). Alternative uses for co-products: Harnessing the potential of valuable compounds from meat processing chains. *Meat Science, 132*, 90–98.

Mussatto, S. I. (2009). Biotechnological potential of brewing industry by-products. In P. Singh nee' Nigam and A. Pandey (Eds.), *Biotechnology for Agro-Industrial Residues Utilisation* (pp. 313–326). Springer.

Muthusamy, N. (2014). Chemical composition of brewers spent grain. *International Journal Science and Environment Technology, 3*, 2109–2112.

Nagar, S., and Nagal, S. (2013). Whey: Composition, role in human health and its utilization in preparation of value added products. *International Journal of Food and Fermentation Technology, 3*(2), 93–100.

Nakthong, N., Wongsagonsup, R., and Amornsakchai, T. (2017). Characteristics and potential utilizations of starch from pineapple stem waste. *Industrial Crops and Products, 105*, 74–82.

Nordlund, E., Aura, A. M., Mattila, I., Kössö, T., Rouau, X., and Poutanen, K. (2012). Formation of phenolic microbial metabolites and short-chain fatty acids from rye, wheat, and oat bran and their fractions in the metabolical in vitro colon model. *Journal of Agricultural and Food Chemistry, 60*(33), 8134–8145.

Oosterveld, A., Voragen, A. G., and Schols, H. A. (2002). Characterization of hop pectins shows the presence of an arabinogalactan-protein. *Carbohydrate Polymers*, *49*(4), 407–413.

Panesar, P. S., Bera, M. B., and Kaur, S. (2014). Bioutilization of whey for ethanol production using yeast isolate. *International Journal of Food and Fermentation Technology*, *4*(2), 107.

Panesar, P. S., and Kennedy, J. F. (2012). Biotechnological approaches for the value addition of whey. *Critical Reviews in Biotechnology*, *32*(4), 327–348.

Panesar, P. S., Kennedy, J. F., Gandhi, D. N., and Bunko, K. (2007). Bioutilisation of whey for lactic acid production. *Food Chemistry*, *105*(1), 1–14.

Panwar, D., Panesar, P. S., and Chopra, H. K. (2021a). Recent trends on the valorization strategies for the management of citrus by-products. *Food Reviews International*, *37*(1), 91–120.

Panwar, D., Saini, A., Panesar, P. S., and Chopra, H. K. (2021b). Unraveling the scientific perspectives of citrus by-products utilization: Progress towards circular economy. *Trends in Food Science and Technology*, *111*, 549–562.

Papageorgiou, M., and Skendi, A. (2018). Introduction to cereal processing and by-products. In C. M. Galanakis *Sustainable Recovery and Reutilization of Cereal Processing By-Products* (pp. 1–25). Woodhead Publishing.

Pascari, X., Ramos, A. J., Marín, S., and Sanchís, V. (2018). Mycotoxins and beer. Impact of beer production process on mycotoxin contamination. A review. *Food Research International*, *103*, 121–129.

Patel, S. (2012). Cereal bran: The next super food with significant antioxidant and anticancer potential. *Mediterranean Journal of Nutrition and Metabolism*, *5*(2), 91–104.

Poutanen, K., Sozer, N., and Della Valle, G. (2014). How can technology help to deliver more of grain in cereal foods for a healthy diet? *Journal of Cereal Science*, *59*(3), 327–336.

Prakongpan, T., Nitithamyong, A., and Luangpituksa, P. (2002). Extraction and application of dietary fiber and cellulose from pineapple cores. *Journal of Food Science*, *67*(4), 1308–1313.

Prazeres, A. R., Carvalho, F., and Rivas, J. (2012). Cheese whey management: A review. *Journal of Environmental Management*, *110*, 48–68.

Rabiu, Z., Maigari, F. U., Lawan, U., and Mukhtar, Z. G. (2018). Pineapple waste utilization as a sustainable means of waste management. In Z. Zakaria *Sustainable Technologies for the Management of Agricultural Wastes* (pp. 143–154). Singapore: Springer

Rama, G. R., Kuhn, D., Beux, S., Maciel, M. J., and de Souza, C. F. V. (2019). Potential applications of dairy whey for the production of lactic acid bacteria cultures. *International Dairy Journal*, *98*, 25–37.

Rezzadori, K., Benedetti, S., and Amante, E. R. (2012). Proposals for the residues recovery: Orange waste as raw material for new products. *Food and Bioproducts Processing*, *90*(4), 606–614.

Robertson, J. A., I'Anson, K. J., Treimo, J., Faulds, C. B., Brocklehurst, T. F., Eijsink, V. G., and Waldron, K. W. (2010). Profiling brewers' spent grain for composition and microbial ecology at the site of production. *LWT - Food Science and Technology*, *43*(6), 890–896.

Romaní, A., Michelin, M., Domingues, L., and Teixeira, J. A. (2018). Valorization of wastes from agrofood and pulp and paper industries within the biorefinery concept: Southwestern Europe scenario. In T. Bhaskar, A. Pandey, S. V. Mohan, D. J. Lee and S. K. Khanal (Eds.), *Waste Biorefinery* (pp. 487–504). Elsevier.

Rosniyana, A., Hashifah, M. A., and Norin, S. S. (2007). The physico-chemical properties and nutritional composition of rice bran produced at different milling degrees of rice. *Journal of Tropical Agriculture and Food Science*, *35*(1), 99.

Routray, W., and Orsat, V. (2017). Plant by-products and food industry waste: A source of nutraceuticals and biopolymers. In A. M. Grumezescu, A.M and A. M. Holban (Eds.), *Food Bioconversion* (pp. 279–315). Academic Press.

Rückle, L., and Senn, T. (2006). Hop acids can efficiently replace antibiotics in ethanol production. *International Sugar Journal*, *108*(1287), 139–147.

Saini, A., Panesar, P. S., and Bera, M. B. (2019). Valorization of fruits and vegetables waste through green extraction of bioactive compounds and their nanoemulsions-based delivery system. *Bioresources and Bioprocessing*, *6*(1), 1–12.

Salve, R. R., and Ray, S. (2020). Comprehensive study of different extraction methods of extracting bioactive compounds from pineapple waste-A review. *The Pharma Innovation Journal*, *9*(6), 327–340.

Sette, P., Fernandez, A., Soria, J., Rodriguez, R., Salvatori, D., and Mazza, G. (2020). Integral valorization of fruit waste from wine and cider industries. *Journal of Cleaner Production*, *242*, 118486.

Sharma, K., Mahato, N., Cho, M. H., and Lee, Y. R. (2017). Converting citrus wastes into value-added products: Economic and environmently friendly approaches. *Nutrition*, *34*, 29–46.

Sharma, S. K., Bansal, S., Mangal, M., Dixit, A. K., Gupta, R. K., and Mangal, A. K. (2016). Utilization of food processing by-products as dietary, functional, and novel fiber: A review. *Critical Reviews in Food Science and Nutrition*, 56(10), 1647–1661.

Silva-Beltrán, Saul, Ruiz-Cruz, S., Cira-Chávez, L. A., Estrada-Alvarado, M. I., Ornelas-Paz, Jde J., López-Mata, M. A., … and Márquez-Ríos, E. (2015). Total phenolic, flavonoid, tomatine, and tomatidine contents and antioxidant and antimicrobial activities of extracts of tomato plant. *International Journal of Analytical Chemistry*, 2015, 284071.

Simate, G. S., Cluett, J., Iyuke, S. E., Musapatika, E. T., Ndlovu, S., Walubita, L. F., and Alvarez, A. E. (2011). The treatment of brewery wastewater for reuse: State of the art. *Desalination*, 273(2–3), 235–247.

Simitzis, P. E., and Deligeorgis, S. G. (2011). The effects of natural antioxidants dietary supplementation on the properties of farm animal products. *Animal Feed: Types, Nutrition and Safety*. Nova Science Publishers, Inc., 155–168.

Simitzis, P. E., and Deligeorgis, S. G. (2018). Agroindustrial by-products and animal products: A great alternative for improving food-quality characteristics and preserving human health. In A. M. Holban, and A. M. Grumezescu (Eds.), *Food Quality: Balancing Health and Disease* (pp. 253–290). Academic Press.

Singla, G., Panesar, P. S., Sangwan, R. S., and Krishania, M. (2021). Enzymatic processing of Citrus reticulata (Kinnow) pomace using naringinase and its valorization through preparation of nutritionally enriched pasta. *Journal of Food Science and Technology*, 58, 3853–3860.

Spatafora, C., and Tringali, C. (2012). Valorization of vegetable waste: Identification of bioactive compounds and their chemo-enzymatic optimization. *The Open Agriculture Journal*, 6(1), 9–15.

Szabo, K., Cătoi, A. F., and Vodnar, D. C. (2018). Bioactive compounds extracted from tomato processing by-products as a source of valuable nutrients. *Plant Foods for Human Nutrition*, 73(4), 268–277.

Szabo, K., Diaconeasa, Z., Cătoi, A. F., and Vodnar, D. C. (2019). Screening of ten tomato varieties processing waste for bioactive components and their related antioxidant and antimicrobial activities. *Antioxidants*, 8(8), 292.

Thammakiti, S., Suphantharika, M., Phaesuwan, T., and Verduyn, C. (2004). Preparation of spent brewer's yeast β -glucans for potential applications in the food industry. *International Journal of Food Science and Technology*, 39(1), 21–29.

Van Hoed, V., Depaemelaere, G., Ayala, J. V., Santiwattana, P., Verhé, R., and De Greyt, W. (2006). Influence of chemical refining on the major and minor components of rice brain oil. *Journal of the American Oil Chemists' Society*, 83(4), 315–321.

Wadhwa, M., and Bakshi, M. P. S. (2013). Utilization of fruit and vegetable wastes as livestock feed and as substrates for generation of other value-added products. *Rap Publication*, 4, 1–67.

Wang, Y. C., Chuang, Y. C., and Hsu, H. W. (2008). The flavonoid, carotenoid and pectin content in peels of citrus cultivated in Taiwan. *Food Chemistry*, 106(1), 277–284.

Wang, Y. X., and Cao, X. J. (2012). Extracting keratin from chicken feathers by using a hydrophobic ionic liquid. *Process Biochemistry*, 47(5), 896–899.

Wilson, T. A., Idreis, H. M., Taylor, C. M., and Nicolosi, R. J. (2002). Whole fat rice bran reduces the development of early aortic atherosclerosis in hypercholesterolemic hamsters compared with wheat bran. *Nutrition Research*, 22(11), 1319–1332.

Wood, J. A., and Malcolmson, L. J. (2011). Pulse milling technologies. In B. K. Tiwari, A. Gowen, and B. Mckenna (Eds.), *Pulse Foods: Processing, Quality and Nutraceutical Applications* (pp. 193–221). Academic Press

Zall, R. R. (1992). Sources and composition of whey and permeate. In J. G. Zadow (Ed.), *Whey and Lactose Processing* (pp. 1–72). Springer.

Zhu, K. X., Zhou, H. M., and Qian, H. F. (2006). Proteins extracted from defatted wheat germ: Nutritional and structural properties. *Cereal Chemistry Journal*, 83(1), 69–75.

3 Microwave-Assisted Extraction for the Valorization of Agro-Industrial Waste

Camella Anne Arellano, Angelica Corpuz,
and Loc Thai Nguyen

CONTENTS

DOI: 10.1201/9781003125679-3

3.1 MICROWAVE

3.1.1 MICROWAVE PROCESSING

In the past 20 years, there have been rapid advances in the applications of microwave (MW) technology in the food processing industry (Li et al., 2019; Zhang et al., 2017). MW heating is increasingly preferred over traditional heating methods due to its rapid and uniform heating, shorter exposure time, enhanced product quality, simplicity and safety of operation, and ease of maintenance (Nguyen et al., 2020).

MW radiation lies between infrared (IR) and radio frequencies within 300 MHz–300 GHz on the electromagnetic (EM) spectrum. While home MW ovens are operated at 2.45 GHz, industrial MW systems are usually designed at 900 MHz to balance cost and efficiency (Zhang et al., 2017). MW is considered a non-ionizing radiation due to its low energy per photon (~0.00001 eV), even at the highest frequency. Therefore, this technology primarily relies upon its thermal effects, although MW-induced nonthermal effects of certain materials have also been reportedly exploited (Li et al., 2019; Stuerga, 2012).

3.1.2 APPLICATIONS OF MICROWAVE TECHNOLOGY

MW technology has been applied in various fields due to its notable benefits. Some popular applications in the food industry include drying, heating or cooking, extraction, pasteurization, and preservation (Nguyen et al., 2020). The unique features of the microwave play a critical role in improving the quality and safety of food products (Zhang et al., 2017; Marić et al., 2018). MW technology may also be used to synthesize organic, inorganic, and organometallic compounds. Other applications include pre-treatment of environmental or biological samples for subsequent analysis or processing (Gomez et al., 2020). Some chemical processing techniques such as distillation, membrane separation, and adsorbent regeneration can also benefit from this technique (Li et al., 2019).

3.2 MICROWAVE-ASSISTED EXTRACTION

3.2.1 DEVELOPMENT AND MAIN PRINCIPLES

The increasing demand for sustainable production and consumption has called for the exploration and adoption of more efficient and environmentally friendly processes to facilitate a more extensive conversion of biomass or low-value agricultural byproducts into valuable ingredients (Benmoussa et al., 2018; Kaderides et al., 2019; Llompart et al., 2019; del Pilar Sánchez-Camargo et al., 2021). There is a range of high-value compounds that can be extracted from various biomass materials. Depending on the nature of biomass, intended product forms and quality, extraction techniques could be classified as (i) Soxhlet extraction using hot water, acidic solution, or organic chemical as a solvent with external heating and (ii) heat reflux extraction, pressing (maceration, percolation, digestion), decoction, sublimation, and steam distillation and hydrodistillation, among others. However, these conventional extraction (CE) techniques inadvertently suffer from a tedious and energy-intensive process, massive use of solvents, and utilization of hazardous solvents and thus, formation of toxic byproducts, long extraction times, and poor product quality and yields. The extraction performance using conventional methods may be improved by using it together with MW, ultrasonication, or pulsed electric fields; although stand-alone, integrated, or sequential use of these technologies have been demonstrated to be sufficiently efficient (Gomez et al., 2020; Li et al., 2019; Macedo et al., 2021; Su et al., 2019; del Pilar Sánchez-Camargo et al., 2021). Among these, microwave-assisted extraction (MAE) is one of the most widely used extraction techniques. The process has gained immense popularity as a preferred environmentally friendly and efficient technique for the extraction of bioactive compounds (Kaderides et al., 2019; Benmoussa et al., 2018; Gomez et al., 2020; Mirzadeh et al., 2021). The unique extraction mechanism (presented in a later section) helps speed up the process and enhance the extraction yield. Short exposure time and reduced processing temperature lead to better preservation of the extracted compounds and lower energy consumption (Benmoussa et al., 2018; Li et al., 2019). MAE is also more friendly to the environment since it allows the use of selected polar and non-organic solvents and minimizes the consumption of organic solvents.

The first application of MAE dates back to 1986 in Hungary, when it was used to extract secondary plant metabolites in a domestic oven (Ciriminna et al., 2016). In the 20th century, MAE was mainly applied in environmental analysis to extract organic and organometallic pollutants (Llompart et al., 2019). Through the years, its application has steadily expanded to the recovery of organic and high-value compounds from several types of organic matter, such as vegetables and fruits, wild plants, animal tissues, and wastes (Benmoussa et al., 2018; Kaderides et al., 2019; Llompart et al., 2019; Mirzadeh et al., 2021). In MAE, extraction occurs due to the rupture of cellular structures caused by EM waves and may be categorized based on solvent usage (Veggi et al., 2012). Originally, MW is combined with traditional solvent extraction (SE) to separate target chemicals (non-volatile compounds) from sample matrices in the solvent. This approach could greatly reduce extraction time and solvent usage while enhancing the extraction yield. For volatile compounds, solvent-free microwave extraction (SFME), which employs *in situ* water as the extraction solvent and a cooling system that condenses the volatilized components (e.g., fragrances, flavours), can be conveniently used. This configuration further reduces solvent consumption and processing time. Microwave hydro diffusion and gravity (MHG) extraction is a special configuration that avoids commonly used distillation and evaporation steps and may be used for non-volatile compounds. It exploits the *in situ* water of the sample and gravity for atmospheric or vacuum extraction of desired compounds from the plant matrices (Benmoussa et al., 2018; Gomez et al., 2020; Zhang et al., 2017; Llompart et al., 2019). MAE can also be categorized according to the mode of MW application and the sample containment configuration, as discussed in section 3.2.3.2. The application of a vacuum can be used to extract thermolabile and oxygen-sensitive components, whereas atmospheric pressure is sufficient to extract more stable compounds. The commercial applications of MAE are discussed in section 3.6.

3.2.2 MECHANISM OF MICROWAVE EXTRACTION

3.2.2.1 Volumetric Heating

Energy transfer is the primary feature of MW heating (Veggi et al., 2012). The capacity of a material to produce heat under MW irradiation depends on its dielectric properties that dictate its interactions with EM radiation. These properties are related to the amount and phases of water in foods, chemical make-up, and structure of food materials. Dielectric properties are made up of a dielectric constant (ε') and dielectric loss factor (ε''), which represent the ability to store and convert EM energy to heat, respectively. For the former, heat loss primarily is a result of ionic conduction and dipole rotation. These molecular-level interactions enable heat to be generated in the material's interior, leading to higher heat transfer rates within the entire process volume (Nguyen et al., 2020). The energy absorption within the solid or solvent causes the dissipation of the electric field penetrating it. In addition, the penetration depth varies inversely to the frequency and the corresponding dielectric properties. Thus, the penetration depth of the MW must be comparable to the thickness of the material to achieve the volumetric heating effect. Consequently, the proper selection of the working frequency and thickness of the sample matrix is crucial (Veggi et al., 2012).

In MAE, MW radiation induces molecular motion by the dual mechanism of the movement of ions and the rotation of the dipoles present in both the solvent and sample matrix (Li et al., 2019). In the first mechanism, the electrophoretic migration of charged species across the solution in response to the applied EM field generates friction and, in turn, homogeneous heat (Gomez et al., 2020). It was established that ionic solutes could be heated more efficiently than their polar covalent counterparts; hence, there is increasing interest in ionic liquids as an alternative solvent and/or additives in

MAE (Llompart et al., 2019; Krishnan et al., 2020). In the second mechanism, the rapidly oscillating electric field of MW induces the self-alignment of dipoles according to the alternating phases. Energy is dissipated as heat from the motion of these dipoles in the agitated medium. It is said that optimal heating can occur at optimum coupling conditions (a fair combination of mild ε' to permit adequate MW penetration and maximum ε'' for a high loss factor), which enables a field change near the natural frequency (Zhang et al., 2017). Volumetric heating is an important characteristic of MW technology, which allows for the rapid and uniform heating of the samples subjected to MAE.

3.2.2.2 Microwave Extraction

In biological matrices, MW extraction is characterized by the disruption of the cell structure aided by the volumetric heating that was previously described. Usually, dried biomass or fresh organic materials are used as raw materials for extraction. MW induces rapid heating and evaporation of traces of water within the cells. The accumulation of internal pressure leads to cell destruction and leaching out of the active constituents. The process improves the area of contact between the solid and liquid (solvent) phases, thus increasing solvent access to the desired components (Veggi et al., 2012). Furthermore, the rise in solvent temperature enhances the diffusion capacities of target components (Kaderides et al., 2019). The extraction process is also highly controlled by the characteristics of both the matrix and the solvent. Only materials or solvents with permanent dipoles get heated up by MW. In this context, the following solvents or their mixtures may be considered for MAE:

- polar solvents
- mixtures of non-polar–polar solvents and water, *in situ* water included
- apolar solvents or transparent MW solvents ($\varepsilon'' < 0.01$) such as hexane and chloroform

The first two are commonly applied for enhancing MW interaction and solute-solvent compatibility. The absorbed MW energy may superheat the solvent, favourably improving the solvent penetration into the solid matrix and the solubility of the target compounds (Macedo et al., 2021). The third is particularly useful for extracting thermolabile components from samples with high water content, in which the heat-sensitive solutes diffuse to the cold solvent, protecting them from degradation (Llompart et al., 2019). The synergistic effect of mass and heat transfer gradients working towards the outside of the cells (Figure. 3.1) accelerates the extraction process and improves the yield (Gomez et al., 2020). For plant tissues, higher temperatures attained from MW exposure can break down the ether linkages of cellulose and rapidly transform them into soluble fractions. The increase in temperature of the cell wall drives the drying-up and weakening of cellulose, thus contributing to cell rupture (Mandal et al., 2007). Through mass and heat transfer phenomena, the extraction process undergoes three main steps (Figure 3.1), namely: (i) the equilibrium phase, whereby a solute is desorbed and carried to the exterior of the particle at a constant rate controlled by the solubilization and partitioning phenomena; (ii) the intermediary transition phase, wherein the mass transfer resistance begins to manifest in the solid-liquid interface; and (iii) the slow diffusion phase, often viewed as the rate-controlling step of the process, in which the solute diffuses freely into the extraction solvent by natural or forced convection (Veggi et al., 2012). As illustrated in Figure 3.2, the extraction rate does not vary linearly with time due to the unsteady state solute concentration inside the solid across these phases. This is governed by a series of steps occurring during the leaching process, including (i) infiltration of the solvent into the solute-containing particle by diffusion; (ii) solubilization of the constituents limited by the nature of the solid; (iii) migration of the solutes to the solid matrix exterior by diffusion; (iv) convective transfer of the extracted solutes from the solid surface into the bulk solution; (v) transport of the extract relative to the solid; and (vi) separation and release of the extract and solid. Many other forces can participate in the extraction process, including van der Waals forces, driving forces, interstitial diffusion, and chemical bonding. The perceived manifestation of these phenomena may be correlated to the solubilization power, purity, and polarity of the solvent (Veggi et al., 2012).

Conventional extraction

Microwave-assisted extraction

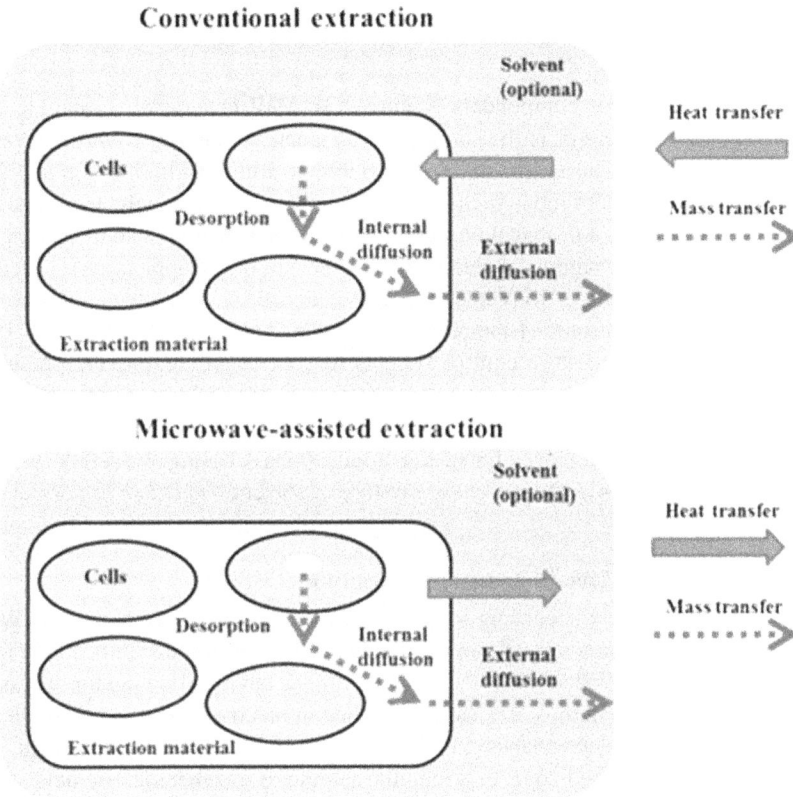

FIGURE 3.1 Basic heat and mass transfer mechanisms in microwave-assisted extraction and conventional solvent extraction. (From Gomez et al. (2020). With permission).

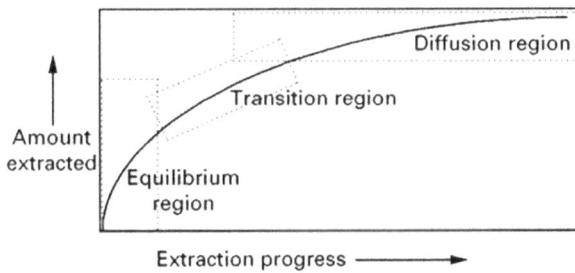

FIGURE 3.2 Extraction phases in microwave-assisted extraction. (From Raynie (2000). With permission).

The key factors that influence solute recovery from a matrix using MAE are the mass transfer rate of the component(s) from the matrix to the bulk solution and the nature of the solute–matrix interactions. Other important parameters include the solvent-to-feed ratio, processing temperature and time, and MW power. As such, a few process enhancements can be incorporated to assist in MAE. For one, stirring the sample mixture can accelerate the recovery by improving the desorption and dissolution of desired compounds associated with the matrix. This can induce convective transfer of the concentrated extract in the solvent phase and minimize mass transfer limitations due to insufficient solvent use (Veggi et al., 2012). Another addition is using external heating elements for the indirect heating of apolar solvents to extract non-polar compounds (Llompart et al., 2019).

3.2.3 INSTRUMENTATION

3.2.3.1 Main Components

A MW system has four major components (Figure 3.3): MW generator, waveguide, applicator, and circulator. The MW generator is the source of MW energy resulting from the acceleration of charges (Gomez et al., 2020; Thostenson and Chou, 1999). To attain the high power and frequencies needed for MW heating, vacuum tubes such as the magnetron are commonly used. The waveguide or transmission line serves as a propagation medium of the MW from the source to the applicator or resonance cavity where the sample is contained. The design of the applicator is crucial in MAE because it mediates the transfer of MW energy to the material. A circulator guides the MW path to be in the forward direction only (Llompart et al., 2019). As for accessories, MW systems are normally equipped with temperature control systems to control the heating (Kappe et al., 2008; Llompart et al., 2019). Temperature sensors have shifted from unreliable contact-type sensors to IR fibre-optical pyrometers (Kostas et al., 2017). Other components such as stirrers, vacuum vessels, and heating elements are also employed for special applications. Cooling and reflux systems may be needed for SFME set-ups. Lastly, the solvent or mixture of solvents is also an important component in MAE, except for SFME.

3.2.3.2 Classifications of MW Systems for Extraction

MW systems can be classified according to the design of the applicator or how MW energy is applied to the sample. The more versatile and commonly used multimode applicator (Figure 3.3) is characterized by the random dispersion of radiation within the cavity, allowing for the evenly irradiated cavity zones and sample(s); thus, it is usually applied in batch operations and in the processing of objects with large, complicated shapes. In a single-mode or focused system (Figure 3.3 (b)), the radiation is focused on a confined zone, exposing the sample to a higher electric field strength than in the first case. Concurrent extraction of up to 12, 24, or 40 samples is feasible in the former by incorporating a rotor with turntables, while this capability is limited in the latter (up to four samples

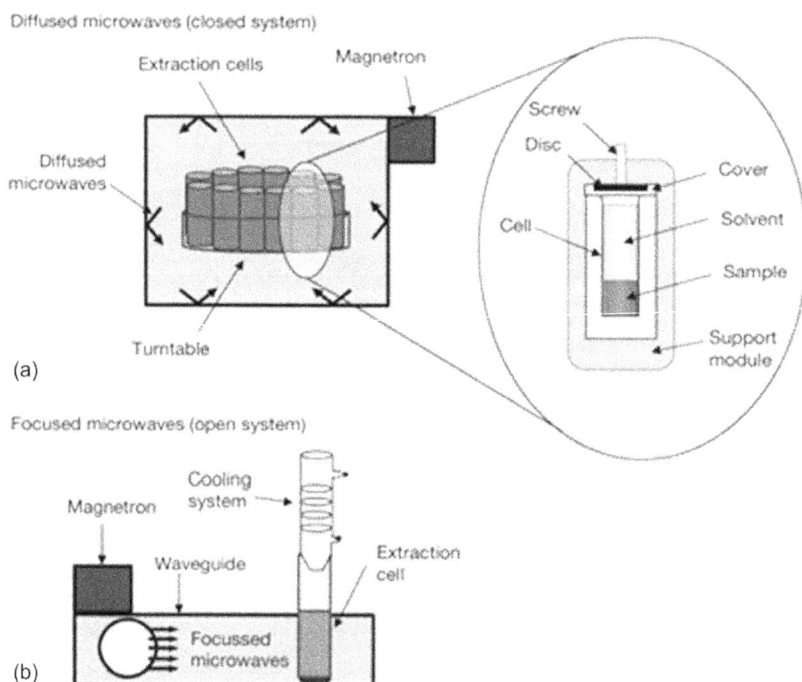

FIGURE 3.3 Multimode systems (a) and single-mode, focused systems (b). (From Llompart et al. (2019). With permission).

only). The MAE-focused system may be efficient in energy transfer; nevertheless, its throughput is limited (Gomez et al., 2020; Llompart et al., 2019).

Regarding vessel configuration, MW systems could be a closed-vessel or open-vessel type. In a closed-vessel type and pressurized MAE, the sample is sealed inside the vessel before exposure to MW irradiation at controlled high pressure and temperature. The open-vessel type is normally used in focused MAE systems wherein MW is applied at atmospheric pressure, although systems based on multimode configurations of this type are now available. The systems that combine high-pressure vessels and focused MW heating technology that can extract up to 24 samples in sequence have also surfaced for commercial applications.

3.3 ADVANTAGES OF MAE

3.3.1 Process Improvement

One of the most evident advantages of MAE is process acceleration, which is a combined result of rapid heating rate and the intense cellular rupture of the plant materials (Backes et al., 2018; Kaderides et al., 2019). For example, conventional SE usually takes 1–24 h while MAE only requires 10 min for optimum extraction of total phenolic content (TPC) of *Vernonia amygdalina* leaf (Alara et al., 2018), 1 min for polyphenols of *Myrtus communis* L. leaves (Dahmoune et al., 2015), 40 min for extraction of solasenol from tobacco leaves (Zhou and Liu, 2006), and 50–120 min for the extraction of orange peel oil (Razzaghi et al., 2019). In addition, MAE consumes less solvent than conventional methods. In MAE, instead of an external solvent, the sample moisture is of the essence in MW absorption, resulting in the expansion of plant cells and subsequent release of the target compounds from the matrix (Othman et al., 2021). However, a vacuum should be applied to lower the boiling point of water, further preserving the thermosensitive compounds. MAE is also superior to conventional methods in improving the yield and quality of the extracts. These features are attributed to cell destruction and instantaneous heat generation under MW, which accelerates the mass transfer and reduces thermal exposure. Evidence of cellular rupture has been observed in pomegranate peels (Kaderides et al., 2019), Chinese herbs (Wang et al., 2008), and tobacco leaves (Zhou and Liu, 2006) after MAE. Enhanced extraction yields were obtained when pectin from pomelo peels was extracted through MAE (Liew et al., 2016), phenolics from pomegranate peels (Kaderides et al., 2019), total phenolics from pineapple skin (Alias and Abbas, 2017), and antioxidants from *Veronia amygdalina* leaf (Alara et al., 2018). In the last two studies, the quality of the extracts was significantly improved, indicated by six times higher EC_{50} and 90 units fewer IC_{50} values, respectively (Alara et al., 2018; Alias and Abbas, 2017).

3.3.2 Environmental and Economic Impact

MAE is a sustainable technology from many angles. First, it reduces the amount of solvent used, especially the organic ones. Second, energy consumption is lower due to shortened processing time. The analysis of energy consumption revealed that significantly less energy is demanded by MAE compared to CE processes (del Pilar Sánchez-Camargo et al., 2021). Ciriminna et al. (2016) showed that to extract 1 g of essential oil from rosemary, 4.5 kWh is required by hydrodistillation (HD), while MAE needs only 0.25 kWh. The application of MW in MHG reduces energy consumption and minimizes CO_2 emissions and wastewater discharge.

3.4 LIMITATIONS OF MAE

3.4.1 Solvent Selection

The efficiency of MAE using polar solvents has been mentioned in several studies and this chapter's succeeding sections. On the contrary, non-polar solvents such as ether and hexane have low MW energy absorption and may eventually damage the magnetron beyond repair (Yu et al., 2007).

When MAE is used to extract non-polar compounds in biological materials, non-polar solvents are used as a transparent solvent to induce sudden temperature increase inside the cellular structure, which results in its rupture. *In situ* water contributes a great deal to this process (Wang et al., 2008). However, this remains unavailable most of the time as samples used in MAE are either pre-treated or dried. Therefore, external heating elements can also improve processes using transparent solvents (Llompart et al., 2019). Recent studies show that non-polar solvents can be solely used in MAE to extract substances with similar polarity. Yu et al. (2007) employed a MW absorbing solid medium called carbonyl iron powders, which enabled the extraction set-up to use ether, a non-polar solvent. This process has been found to be the most feasible rather than the use of polar or mixed solvents alone in extracting non-polar compounds. Other MAE equipment additions, such as a micro-solid phase extraction device that contains activated carbon (Wang et al., 2008) and absorption tubes containing graphite powder (Zhao et al., 2012), were used and proven effective in the extraction of organophosphorus pesticides from fresh fruit and vegetables using non-polar compounds.

3.4.2 FEASIBILITY IN THE INDUSTRIAL SCALE

The application of MW energy on a large scale involves higher technical complexity, and cost becomes a factor when the extraction capacity increases to 100 g. Aside from the initial investment costs, the maintenance cost is also another aspect to be looked at. Industries are also seeking safety and regulatory aspects to justify the applicability of MAE compared to conventional processes (Li et al., 2012; Ciriminna et al., 2016; Kala et al., 2016). Despite these drawbacks, substantial studies and processes have been proposed to advance the economic viability of MAE. Kala et al. (2016) has proposed a process (Figure 3.4) that requires minimal heating. In this instance, the moisture is increased before MAE, and the solvent is added under pressure to enhance the access of the solvent to the ruptured plant tissues. Other studies optimized process parameters to be used for large-scale applications.

3.5 IMPORTANT PROCESS PARAMETERS

MAE's efficiency depends on several factors, including MW power, solvent characteristics and its ratio with the sample, exposure time, presence of vacuum, and temperature, to name a few (Llompart et al., 2019). The interdependence may not be generalized across biological materials due to the diverse nature of the bioactive components. Thus, the selection of processing conditions must consider the desired responses from each raw material and MW system (Kaderides et al., 2019).

3.5.1 EFFECT OF MICROWAVE POWER

MW power affects the extraction rate and performance in both laboratory and industrial scales. This factor should be addressed through the extraction temperature and irradiation time to increase process efficiency (Ciriminna et al., 2016; Kala et al., 2016; Liew et al., 2016). Generally, an increase in MW power is translated to an increase in the extraction rate. Applied power increases the cell rupture and release of target compounds to the surrounding solvent (Kaderides et al., 2019). This also increases the temperature and decreases the solvent's viscosity (Chutia and Mahanta, 2021). However, the effect is not the same for different compounds as some researchers apply low MW power and extended extraction time to increase the purity of the extract (Liew et al., 2016; Kaderides et al., 2019). The excessive increase of MW power may lead to the destruction of the target compounds or the plant tissues due to a parallel increase in temperature (Kaderides et al., 2019). The increasing temperature at a high power level can incur solvent loss due to the boiling phenomenon (Haffizi et al., 2020; Dahmoune et al., 2015; Spigno and De Faveri, 2009). Alara et al. (2018) showed that recovery yields of TPC from *V. amygdalina* L. decreased when the power exceeded 500 W. Moreover, Haffizi et al. (2020) found that high power is not needed to obtain high yields in the extraction of essential and vegetable oils.

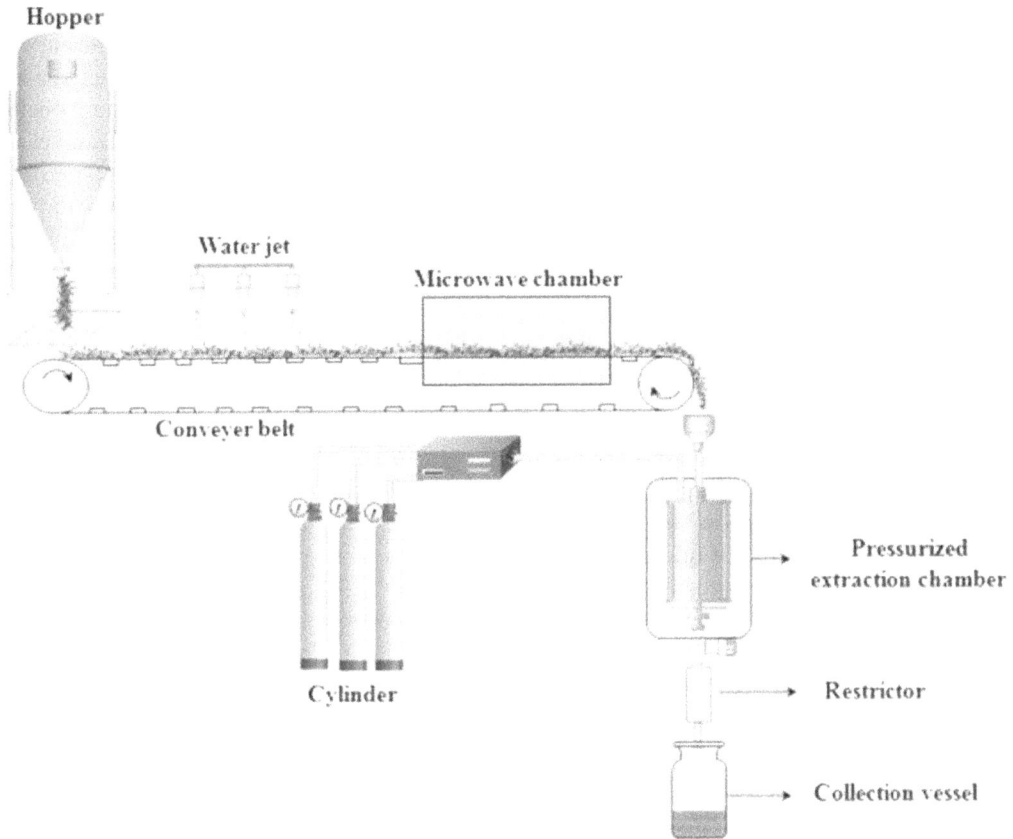

FIGURE 3.4 Industrial microwave-assisted extraction process proposed by Kala (2016). (From Kala et al. (2016). With Permission).

3.5.2 EXTRACTION SOLVENT

The type of extraction solvents can influence the diffusion of the compounds of interest, their penetration into the solid matrix, and the mass transfer of the process (Kaderides et al., 2019). In MAE, additional factors such as the dielectric constant, dielectric loss, and loss tangent of the solvents should be considered. As mentioned earlier, these parameters govern the absorption of MW energy, which further affects the heating rate in the materials (Routray and Orsat, 2012). Ethanol is a strong MW absorber at 2.45 GHz, with a loss tangent of 0.898, compared to methanol (0.522) and water (0.123). Therefore, ethanol has a higher capacity to convert MW energy into heat, increasing the rupturing of the plant materials and the infusion of target compounds (Kaderides et al., 2019). The efficiency of MW extraction can be increased by using binary or a mixture of solvents that help address solvent polarity. Ethanol is the most used solvent, while methanol has been prohibited in finished products, especially those used for food and pharmaceutical purposes (Spigno and De Faveri, 2009; Kala et al., 2016). Currently, eutectic or ionic solvents have been used to extract phenolic compounds. However, further investigation is required for practical applications (Kala et al., 2016).

3.5.3 SOLID/SOLVENT RATIO

In MAE, samples should be fully immersed in the solvent to increase the contact surface area, thus, augmenting the diffusivity and extraction yield (Haffizi et al., 2020). The optimum ratio of solid feed to solvent (FSR) balances the mass transfer of the target compound and the absorption of MW energy by the solvent (Othman et al., 2021). A further decrease of FSR poses inefficiency as it requires higher energy to heat up, cool down, and separate the solvent from the extracted compound. For a given extraction time and MW power, a low FSR may result in the MW energy being absorbed by the solvent itself, which hinders the internal heating of the solid matrix (Kazemi et al., 2019). This was evidenced in the extraction of carotenoids from passion fruit peel with an FSR of 1:10 (w/v) (Chutia and Mahanta, 2021), pectin from green pistachio hulls with an FSR of 1:15 (w/v) (Kazemi et al., 2019), and polysaccharides from melon peel with an FSR of 1:20.94 (w/v) (Golbargi et al., 2021). These ratios are consistent with the frequently employed range from 1:10 to 1:50 (w/v) across raw materials (Llompart et al., 2019). Optimizing FSR is also important to reduce the amount of harmful solvents used (such as in vegetable oils extraction) to increase process recovery and decrease production costs (Veggi et al., 2012).

3.5.4 EXTRACTION AND IRRADIATION TIME

Unlike CE, extraction and irradiation time in MAE is of the essence as it affects the quality of extracted compounds. In most studies, an increase in treatment time results in a corresponding increase in extraction yield as this allows the EM waves to significantly disrupt the plant cells and prolong the solid-to-solvent contact, thereby improving the diffusion of the desired compounds. This has been observed in the extraction of carotenoids from passion fruit peel (25 min) (Chutia and Mahanta, 2021), polysaccharides from melon peels (12.75 min) (Golbargi et al., 2021), pectin from pistachio hulls (165 s) (Kazemi et al., 2019), and TPC of *Vernonia amygdalina* leaf (8 min) (Alara et al., 2018). However, prolonged extraction and irradiation time significantly elevates the temperature of the solvent and the system, which degrades the extracted compounds. Extended treatment time can also result in the hydrolysis of target compounds, such as pectin extraction from pomelo peels (Liew et al., 2016).

3.5.5 SURFACE AREA

The surface area of the samples is critical in MAE as it is directly proportional to the extraction efficiency. Size reduction methods such as milling, homogenization, and grinding were usually attempted to increase the interaction between the samples and the solvent during extraction (Routray and Orsat, 2012).

3.5.6 TEMPERATURE

Temperature is a significant factor that is sometimes tied up with the applied power. Mild heating improves the diffusivity of the solvent to the plant matrix and softens the plant matrix and cell walls. This results in increased leaching of the target compounds to the solvent (Alara et al., 2018). An increase in temperature also decreases the solvent's viscosity, which improves extraction efficiency. However, beyond the optimum value, further temperature increases may lead to overheating and degradation of target compounds, thus decreasing the extraction yield (Routray and Orsat, 2012). If water is used as the only solvent, such as in the extraction of essential oils, the high boiling point of water may damage heat-sensitive compounds (Haffizi et al., 2020). The problem can be addressed with the help of a vacuum extraction system (Othman et al., 2021).

3.5.7 VACUUM EXTRACTION

The application of a vacuum can protect thermolabile and oxygen-sensitive compounds on top of the standalone MAE system. Under vacuum, the boiling temperature of solvents is reduced, and the extracted compounds are better preserved (Othman et al., 2021). It has been shown that retention of phenols is two times higher in the vacuumed MW extraction set-up than in the ambient one (Kala et al., 2016). The absence of oxygen or low oxygen conditions also contributes to reducing the degradation of oxygen-sensitive compounds. For example, the extraction yield of myricetin in Chinese herbs is 10% more at 40 MPa than under atmospheric conditions (Wang et al., 2008).

3.6 MAE OF BIOACTIVE COMPOUNDS FROM AGRO-INDUSTRIAL BYPRODUCTS

Agro-industrial byproducts were considered waste and have been under-utilized for a long time. However, these materials can be good sources of several bioactive compounds. To increase the sustainability of agricultural production, valuable constituents from agricultural byproducts can be recovered by extraction processes. So far, MAE has become one of the most explored extraction techniques for this particular purpose. This section highlights recent advances in applying MAE for recovering bioactive compounds from agricultural byproducts. A summary of past studies is provided in Table 3.1.

3.6.1 PHENOLIC COMPOUNDS

Phenolic compounds have gained much interest due to their health benefits. They can be taken in the form of nutraceutical or natural antioxidants in food products. Phenolic compounds are usually extracted by CE; nevertheless, these techniques suffer from several limitations, such as long extraction time and elevated temperature, which could adversely affect the bioactivity of the target compounds (Inoue et al., 2010). MAE can be a viable solution for improving the yield and quality of the phenolic extract. It was reported that the closed MW system using aqueous ethanol produced 1.7 times more TPC from pomegranate peel than those processed by ultrasound-assisted extraction (UAE). Moreover, the processing time was reduced by approximately 40% (Kaderides et al., 2019). MAE (416 W and 100°C) was also used for extracting phenolics from Vernonia *amygdalina* leaf with the optimum FSR of 1:8 (w/v) and extraction time of 8 min. It was noticed that TPC started to degrade at 10 min of heating at the optimum MW power. The process produced a 4.41% higher extraction yield than the Soxhlet extraction method (Alara et al., 2018).

In a different study, TPC decreased when the temperature was raised to 90°C and further decreased up to 120°C during the MAE of pineapple peels (Alias and Abbas, 2017). The optimum conditions (750 W, 60°C, and 50:50 (v/v) ethanol:water ratio) gave 7 times higher TPC and 4.7 times higher EC_{50} than Soxhlet extraction. Adding an enzyme to the MAE of olive pomace was proven to protect phenolic compounds (Macedo et al., 2021). The combination of enzyme and MAE resulted in the higher extractability of phenolics (341 mg GAE/g OP) at higher temperatures and a faster heating rate.

3.6.2 PLANT PIGMENTS

MAE has been applied to recover different natural pigments, such as anthocyanins, curcumin, carotenoids, safflower yellow, flavonoids, safflomin A, and lycopene (Martínez-Abad et al., 2020). These pigments have a wide range of applications, such as natural food colourants and antioxidants. Due to the heat sensitivity of the plant pigments, the extraction process should be carried out at low MW power and in an interval of less than 30 min. Varadharajan et al. (2017) applied a Box–Behnken design to optimize the MAE of anthocyanins from grape juice waste. The extraction yield of 1.325 mg/g was obtained at the power of 428.02 W, treatment time of 2.23 min, and FSR of 1:18.43 (w/v). The same study showed that power and exposure time were critical factors, and the process should not exceed 450 W and 2-min exposure to avoid degradation of the compounds. Jafari et al. (2019) have extracted anthocyanins from saffron tepals, and a yield of 1010 mg/g was achieved using a MW power of 360 W, extraction time of 9.3 min, and temperature

TABLE 3.1

List of Extracted Bioactive Compounds from Agro-industrial Byproducts through MAE

Extracted Compound	Raw Material	Solvent	Power (W)	Treatment Conditions	Product Applications	Reference
Phenolics	Pomegranate peels	50% ethanol; FSR 1:60 (w/v)	600 W	4 min	Functional foods, pharmaceutical, natural antioxidants for food products	Kaderides et al. (2019)
Phenolics	Olive pomace	Water, FSR 1:15 (w/v) with enzyme treatment (pectinase, cellulose and tannase)	600 W	60°C, 5 min	nutraceutical (antioxidant)	Macedo et al. (2021)
Phenolics	*Vernonia amygdalina* leaf	Distilled water, FSR 1:8 (w/v)	416 W	100°C, 8 min	Pharmaceuticals and natural food (antioxidant)	Alara et al. (2018)
Phenolics	Mango peels	60% ethanol, FSR 1:50 (w/v)	800 W	90 s	Pharmaceutical (antioxidant and anti-proliferative)	del Pilar Sánchez-Camargo et al. (2021)
Phenolics	Pineapple skins	50% ethanol, FSR 1:50 (w/v)	750 W	60°C	Natural antioxidants for food products	Alias and Abbas (2017)
Phenolics (Stilbenes)	Grape stems and canes	80% ethanol, FSR 1:100 (grapestems) and 1:125 (canes) (w/v)	750 W	125°C, 5 min	Natural food products, pharmaceuticals	Piñeiro et al. (2017)
Carotenoids	Passion fruit peel	86.9% olive oil, FSR 1:10 (w/v)	200 W	below 110°C, 25 min	Natural food products, pharmaceuticals	Chutia and Mahanta (2021)
Anthocyanin pigments	Fig peel	Ethanol (pH 3, citric acid), FSR 1:50 (w/v)	400 W	62.41°C, 5 min	Food (natural colourant), pharmaceuticals, cosmetics	Backes et al. (2018)
Soluble dietary fibre	Coffee silverskin (CSS)	Deionized water, FSR 1:20 (w/v)	1660 W (and 200 W ultrasound in UMAE)	80°C, 20 min	Health-promoting/functional foods, hydrocolloids (thickener, stabilizer, and gelling agents)	Wen et al. (2020)
Pectin	Banana peels	Distilled water (pH 3 using HCl), FSR 1:10 (w/v)	900 W	100 s, pulse ratio = 0.5 (IMAE)	Hydrocolloids (emulsifier, texturizer, thickener, stabilizer, gelling agent)	Swamy and Muthukumarappan (2017)

(Continued)

TABLE 3.1 (CONTINUED)
List of Extracted Bioactive Compounds from Agro-industrial Byproducts through MAE

Extracted Compound	Raw Material	Solvent	Power (W)	Treatment Conditions	Product Applications	Reference
Pectin	Green pistachio hull	Aqueous solution (1.5 pH), FSR 1:15 (w/v)	700 W	165 s	Hydrocolloids (thickener, stabilizer, gelling agent), pharmaceuticals (drug encapsulation)	Kazemi et al. (2019)
Pectin	Orange peel	Aqueous solution (pH 1.2), FSR 1:21.5 (w/v) with Tween-80 (8 g/l, w/v)	400 W	7 min	Food processing (thickener, stabilizer, gelling agent), pharmaceutical (prevent heart disease and gallstones)	Su et al. (2019)
Pectin	Pumpkin biomass Sugar beet pulp	35% hydrochloric acid (pH 2.5) Sulphuric acid (pH 1–2)	1200 W 150–250 W	2–10 min 2–4 min	Pharmaceuticals, cosmetics, food processing (gelling agent, stabilizer, fruit filling for bakery and confectionary products), prebiotics	Marić et al. (2018) Li et al. (2012)
Arabinan-rich pectic polysaccharides	Melon peels	Water, FSR 1:20.94 (w/v)	414.4 W	75°C, 12.75 min	Food and pharmaceutical supplement products	Golbargi et al. (2021)
Plant/algal polysaccharides	Loquat Snow chrysanthemum	FSR 1:40 (w/v) FSR 1:59 (w/v)	500 W	6.5 min	Nutraceuticals, pharmaceuticals, cosmetics, natural food, and beverage products	Mirzadeh et al (2021) Mirzadeh et al (2021)
Lipid	Microalgae	Water with 2.5% of imidazolium-based IL additive, FSR 1:250 (w/v)	700 W	60°C, 5 min	Biofuels	Krishnan et al. (2020)
Orange essential oil	Orange peel (73 ± 0.15% Moisture Content)	Distilled water, FSR 1:1 (w/w)	400 W	50 min	Natural food and beverage products, personal care and cosmetics, pharmaceutical and medical applications	Razzhagi et al. (2019)

(Continued)

TABLE 3.1 (CONTINUED)
List of Extracted Bioactive Compounds from Agro-industrial Byproducts through MAE

Extracted Compound	Raw Material	Solvent	Power (W)	Treatment Conditions	Product Applications	Reference
Rice bran oil	Rice bran	Hexane, FSR 1:1.6–1:2.3 (w/v), microwave heating then solvent washing	320 W	8–10 min	Edible oil	Pandey and Shrivastava (2018)
Chitosan	Purified shrimp waste (heads and scales) chitin	50% NaOH (w/v), FSR 1:10 (w/v)	1400 W	10 min	Food additives, functional food, dietary and nutraceutical products	Samar et al. (2013)
Chitosan	Demineralized and deproteinized shrimp shell waste	50% NaOH, 1:20 (w/v)	160–350 W	24 min	Fish food additive, biomedical applications (wound dressings, drug delivery systems)	El Knidri et al (2016)

of 48°C. The results also revealed that degradation and polymerization of anthocyanin occur at temperatures higher than 50°C. In industrial practices, the extraction of pigments is often implemented as the next procedure in the extraction of essential oils. For example, the extraction of carotenoids and other yellow pigments from lemon peel waste was done after microwave-assisted hydrodistillation (MAHD) of lemon essential oil (Martínez-Abad et al., 2020) or the extraction of flavonoids from orange pomace was accomplished after the extraction of essential oils (Petrotos et al., 2021). This extraction arrangement has improved the yield and quality of both the oils and the pigments compared to the extraction of only dried materials. However, not all research on pigment extraction favoured MAE compared to SE, UAE, and CE. This is especially evident when non-polar solvents are used to extract lipophilic substances. Chutia and Mahanta (2021) found that the MAE of carotenoids from passion fruit peel by olive oil and sunflower oil had a lower yield (86.9%) than that of UAE (91.4%). The difference is attributed to the dependence of MAE on the dielectric property of the materials. Oils have low MW absorption capacity due to their long-chain and non-polar nature, decreasing extraction efficiency. However, MAE is still a promising process for pigment extraction as it requires a significantly shorter time and lowers energy consumption (Elik et al., 2020).

3.6.3 POLYSACCHARIDES

Polysaccharides, the most abundant organic matter, play a huge role in maintaining the structure of agro-industrial commodities such as cellulose in plants, pectin in fruits, and chitin/chitosan in crustacean exoskeleton shells. These are mainly found in the parts of the commodities that are thrown away, such as the exoskeleton and plant peels, which are also extracted using various methods, including MAE. Swamy and Muthukumarappan (2017) showed that the yield of pectin extracted from banana peels was higher in MAE (2.58%) compared to that of a continuous process (2.18%). MAE also eliminates overheating through the balance of heat and mass transfer. It is also interesting that pectin extracted by MAE has a larger particle size, higher water holding capacity, higher degree of esterification, and higher galacturonic acid content. These characteristics are essential when pectin is applied to foodstuffs or medicines. Compared to other processes, MAE was also more efficient in yield and energy consumption (Misra and Yadav, 2020). MAE is particularly useful for extracting chitin or chitosan from shrimp waste, which normally requires extended time and harsh conditions. El Knidri et al. (2016) demonstrated that MAE produced chitosan within 24 min with increased Degree of Deacetylation (DD=82.73%), whereas CE took 6–7 h and only yielded chitosan with a DD of 81.5%. The process even obtained chitosan with a higher DD (95.19%) with MAE, using the optimum MW power of 1400 W and extraction time of 10 min.

3.6.4 ESSENTIAL OILS

Even with the use of non-polar compounds to extract essential oils, MAE is still more efficient than CE, demanding high amounts of solvents and a long extraction time. When four types of MAE systems were applied to extract essential oils from orange peels, MHG was found to be the most cost-efficient, followed by microwave steam distillation and microwave steam diffusion (MSDf) (Razzaghi et al., 2019).

Pandey and Shrivastava (2018) have created a two-step MAE for optimizing the oryzanol content in rice bran oil. In the first step, the rice bran was subjected to boiling hexane (67°C–69°C) with an 8–10 min extraction time and FSR of 1:1.6–1:2.3 (w/v). This increased localized pressure facilitates the access of the solvent when the solvent is introduced in the second stage. The process recovered more than 95% of oil compared to that of Soxhlet extraction. In addition, the phospholipid content, antioxidant activity, alpha-tocopherol content, and free fatty acid content of the extract were far more superior to CE.

3.7 CONCLUSION

The advances of MAE have helped address several challenges in recovering valuable constituents from agricultural wastes. Unlike conventional methods, MAE shortens the processing time, lowers energy consumption, and reduces the volume of organic solvent used for extraction. Additional benefits are the improvement of both yields and quality of extracted products. Further development should be directed towards upgrading MAE systems for different types of solvents and raw materials. In addition, as different families of bioactive compounds can be obtained from the same material, the versatility of MAE systems is important for their simultaneous extraction. The cost of equipment and relevant accessories remains an issue; hence, reducing the complexity of the commercial systems should be considered in the future.

REFERENCES

Alara, O., N. Abdurahman, S. A. Mudalip, and O. Olalere. 2018. Microwave-assisted extraction of *Vernonia amygdalina* leaf for optimal recovery of total phenolic content. *J. Appl. Res. Med. Arom. Plants* 10:16–24.

Alias, N. H., and Z. Abbas. 2017. Microwave-assisted extraction of phenolic compound from pineapple skins: The optimum operating condition and comparision with Soxhlet extraction. *Malays. J. Anal. Sci.* 21(3):690–699.

Backes, E., C. Pereira, L. Barros, et al. 2018. Recovery of bioactive anthocyanin pigments from *Ficus carica* L. peel by heat, microwave, and ultrasound based extraction techniques. *Int. Food Res. J.* 113:197–209.

Benmoussa, H., W. Elfalleh, S. He, M. Romdhane, A. Benhamou, and R. Chawech. 2018. Microwave hydro-diffusion and gravity for rapid extraction of essential oil from Tunisian cumin (*Cuminum cyminum* L.) seeds: Optimization by response surface methodology. *Ind. Crops Prod.* 124:633–642.

Chutia, H., and C. L. Mahanta. 2021. Green ultrasound and microwave extraction of carotenoids from passion fruit peel using vegetable oils as a solvent: Optimization, comparison, kinetics, and thermodynamic studies. *Innov. Food Sci. Emerg. Technol.* 67:102547.

Ciriminna, R., D. Carnaroglio, R. Delisi, S. Arvati, A. Tamburino, and M. Pagliaro. 2016. Industrial feasibility of natural products extraction with microwave technology. *ChemistrySelect* 1(3):549–555.

Dahmoune, F., B. Nayak, K. Moussi, H. Remini, and K. Madani. 2015. Optimization of microwave-assisted extraction of polyphenols from *Myrtus communis* L. leaves. *Food Chem.* 166:585–595.

del Pilar Sánchez-Camargo, A., D. Ballesteros-Vivas, L. M. Buelvas-Puello, et al. 2021. Microwave-assisted extraction of phenolic compounds with antioxidant and anti-proliferative activities from supercritical CO_2 pre-extracted mango peel as valorization strategy. *J. LWT* 137:110414.

El Knidri, H., R. El Khalfaouy, A. Laajeb, A. Addaou, and A. Lahsini. 2016. Eco-friendly extraction and characterization of chitin and chitosan from the shrimp shell waste via microwave irradiation. *Process Saf. Environ. Prot.* 104:395–405.

Elik, A., D. K. Yanık, and F. Göğüş. 2020. Microwave-assisted extraction of carotenoids from carrot juice processing waste using flaxseed oil as a solvent. *LWT* 123:109100.

Golbargi, F., S. M. T. Gharibzahedi, A. Zoghi, M. Mohammadi, and R. Hashemifesharaki. 2021. Microwave-assisted extraction of arabinan-rich pectic polysaccharides from melon peels: Optimization, purification, bioactivity, and techno-functionality. *Carbohydr. Polym.* 256:117522.

Gomez, L., B. Tiwari, and M. Garcia-Vaquero. 2020. Emerging extraction techniques: Microwave-assisted extraction. In: *Sustainable Seaweed Technologies.* Elsevier.

Haffizi, M., S. Sulaiman, D. N. Jimat, and A. Amid. 2020. A comparison of conditions for the extraction of vegetable and essential oils via microwave-assisted extraction. *Paper read at IOP Conference Series: Materials Science and Engineering.*

Inoue, T., S. Tsubaki, K. Ogawa, K. Onishi, and J.-i. Azuma. 2010. Isolation of hesperidin from peels of thinned Citrus unshiu fruits by microwave-assisted extraction. *Food Chem.* 123(2):542–547.

Jafari, S. M., K. Mahdavee Khazaei, and E. Assadpour. 2019. Production of a natural color through microwave-assisted extraction of saffron tepal's anthocyanins. *J. Nutr. Sci.* 7(4):1438–1445.

Kaderides, K., L. Papaoikonomou, M. Serafim, and A. M. Goula. 2019. Microwave-assisted extraction of phenolics from pomegranate peels: Optimization, kinetics, and comparison with ultrasounds extraction. *Chem. Eng. Process.* 137:1–11.

Kala, H. K., R. Mehta, K. K. Sen, R. Tandey, and V. Mandal. 2016. Critical analysis of research trends and issues in microwave assisted extraction of phenolics: Have we really done enough. *Trends Analyt. Chem.* 85:140–152.

Kappe, C. O., D. Dallinger, and S. S. Murphree. 2008. *Practical Microwave Synthesis for Organic Chemists: Strategies, Instruments, and Protocols*. John Wiley & Sons.

Kazemi, M., F. Khodaiyan, M. Labbafi, S. S. Hosseini, and M. Hojjati. 2019. Pistachio green hull pectin: Optimization of microwave-assisted extraction and evaluation of its physicochemical, structural and functional properties. *Food Chem.* 271:663–672.

Kostas, E. T., D. Beneroso, and J. P. Robinson. 2017. The application of microwave heating in bioenergy: A review on the microwave pre-treatment and upgrading technologies for biomass. *Renew. Sustain. Energ. Rev.* 77:12–27.

Krishnan, S., N. Abd Ghani, N. F. Aminuddin, et al. 2020. Microwave-assisted lipid extraction from Chlorella vulgaris in water with 0.5%–2.5% of imidazolium based ionic liquid as additive. *Renew. Energy* 149:244–252.

Li, H., Z. Zhao, C. Xiouras, G. D. Stefanidis, X. Li, and X. Gao. 2019. Fundamentals and applications of microwave heating to chemicals separation processes. *Renew. Sustain. Energ. Rev.* 114:109316.

Li, Y., M. Radoiu, A.-S. Fabiano-Tixier, and F. Chemat. 2012. From laboratory to industry: Scale-up, quality, and safety consideration for microwave-assisted extraction. In: *Microwave-Assisted Extraction for Bioactive Compounds*. Springer.

Liew, S. Q., G. C. Ngoh, R. Yusoff, and W. H. Teoh. 2016. Sequential ultrasound-microwave assisted acid extraction (UMAE) of pectin from pomelo peels. *Int. J. Biol. Macromol.* 93(A):426–435.

Llompart, M., C. Garcia-Jares, M. Celeiro, and T. Dagnac. 2019. Extraction microwave-assisted extraction. In: Encyclopedia of Analytical Science (3rd ed.) Academic Press.

Macedo, G. A., A. L. Santana, L. M. Crawford, S. C. Wang, F. F. Dias, and J. M. J. L. de Moura Bell. 2021. Integrated microwave-and enzyme-assisted extraction of phenolic compounds from olive pomace. *LWT* 138:110621.

Mandal, V., Y. Mohan, and S. J. P. r. Hemalatha. 2007. Microwave assisted extraction—An innovative and promising extraction tool for medicinal plant research. *Pharmacogn. Rev.* 1(1):7–18.

Marić, M., A. N. Grassino, Z. Zhu, F. J. Barba, M. Brnčić, and S. R. Brnčić. 2018. An overview of the traditional and innovative approaches for pectin extraction from plant food wastes and by-products: Ultrasound-, microwaves-, and enzyme-assisted extraction. *Trends Food Sci. Technol.* 76:28–37.

Martínez-Abad, A., M. Ramos, M. Hamzaoui, S. Kohnen, A. Jiménez, and M. C. Garrigós. 2020. Optimisation of sequential microwave-assisted extraction of essential oil and pigment from lemon peels waste. *J. Food Sci.* 9(10):1493.

Mirzadeh, M., A. K. Lelekami, and L. Khedmat. 2021. Plant/algal polysaccharides extracted by microwave: A review on hypoglycemic, hypolipidemic, prebiotic, and immune-stimulatory effect. *Carbohydr. Polym.* 266:118134.

Misra, N., and S. K. Yadav. 2020. Extraction of pectin from black carrot pomace using intermittent microwave, ultrasound and conventional heating: Kinetics, characterization and process economics. *Food Hydrocoll.* 102:105592.

Nguyen, L. T., I. Ahmad, and N. Y. Jayanath. 2020. Dielectric properties of selected seafood and their products. *Encycl. Mar. Biotechnol.* 2020:2867–2880.

Othman, S. N. S., A. N. Mustapa, and K. H. Ku Hamid. 2021. Extraction of polyphenols from Clinacanthus nutans Lindau (C. nutans) by vacuum solvent-free microwave extraction (V-SFME). *Chem. Eng. Commun.* 208(5):727–740.

Pandey, R., and S. L. Shrivastava. 2018. Comparative evaluation of rice bran oil obtained with two-step microwave assisted extraction and conventional solvent extraction. *J. Food Eng.* 218:106–114.

Petrotos, K., I. Giavasis, K. Gerasopoulos, C. Mitsagga, C. Papaioannou, and P. Gkoutsidis. 2021. Optimization of vacuum-microwave-assisted extraction of natural polyphenols and flavonoids from raw solid waste of the orange juice producing industry at industrial scale. *Molecules* 26(1):246.

Piñeiro, Z., Marrufo-Curtido, A., Vela, C., & Palma, M. 2017. Microwave-assisted extraction of stilbenes from woody vine material. *Food Bioprod. Process, 103*: 18–26.

Raynie, D. 2000. *Extraction, The Procter & Gamble Company*. Academic Press.

Razzaghi, S. E., A. Arabhosseini, M. Turk, et al. 2019. Operational efficiencies of six microwave based extraction methods for orange peel oil. *J. Food Eng.* 241:26–32.

Routray, W., and V. Orsat. 2012. Microwave-assisted extraction of flavonoids: A review: *Food Bioproc. Tech.* 5(2):409–424.

Samar, M. M., M. El-Kalyoubi, M. Khalaf, and M. Abd El-Razik. 2013. Physicochemical, functional, antioxidant and antibacterial properties of chitosan extracted from shrimp wastes by microwave technique. *Ann. Agric. Sci.* 58(1):33–41.

Spigno, G., and D. De Faveri. 2009. Microwave-assisted extraction of tea phenols: A phenomenological study. *J. Food Eng.* 93(2):210–217.

Stuerga, D. 2012. Microwave–materials interactions and dielectric properties: From molecules and macromolecules to solids and colloidal suspensions. In: *Microw. Org. Synth.*, 1–56. Wiley Online Library

Su, D.-L., P.-J. Li, S. Y. Quek, et al. 2019. Efficient extraction and characterization of pectin from orange peel by a combined surfactant and microwave assisted process. *Food Chem.* 286:1–7.

Swamy, G. J., and K. Muthukumarappan. 2017. Optimization of continuous and intermittent microwave extraction of pectin from banana peels. *Food Chem.* 220:108–114.

Thostenson, E., and T. Chou. 1999. Composites part A: Appl. *J. Sci. Manuf.* 30(9):1055–1071.

Varadharajan, V., S. Shanmugam, and A. Ramaswamy. 2017. Model generation and process optimization of microwave-assisted aqueous extraction of anthocyanins from grape juice waste. *J. Food Process Eng.* 40(3):e12486.

Veggi, P. C., J. Martinez, and M. A. A. Meireles. 2012. Fundamentals of microwave extraction. In: *Microwave-Assisted Extraction for Bioactive Compounds*. Springer.

Wang, J.-X., X.-H. Xiao, and G.-K. Li. 2008. Study of vacuum microwave-assisted extraction of polyphenolic compounds and pigment from Chinese herbs. *J. Chromatogr. A* 1198:45–53.

Wen, L., Zhang, Z., Zhao, M., Senthamaraikannan, R., Padamati, R. B., Sun, D. W., and Tiwari, B. K. 2020. Green extraction of soluble dietary fibre from coffee silverskin: impact of ultrasound/microwave-assisted extraction. *Int. J. Food Sci. Technol.* 55(5): 2242–2250.

Yu, Y., Z. M. Wang, Y. T. Wang, et al. 2007. Non-polar solvent microwave-assisted extraction of volatile constituents from dried Zingiber officinale Rosc. *Chin. J. Chem.* 25(3):346–350.

Zhang, Y., W. Guo, Z. Yue, et al. 2017. Rapid determination of 54 pharmaceutical and personal care products in fish samples using microwave-assisted extraction—Hollow fiber—Liquid/solid phase microextraction. *J. Chromatogr. B* 1051:41–53.

Zhao, X., X. Xu, R. Su, H. Zhang, and Z. Wang. 2012. An application of new microwave absorption tube in non-polar solvent microwave-assisted extraction of organophosphorus pesticides from fresh vegetable samples. *J. Chromatogr. A* 1229:6–12.

Zhou, H.-Y., and C.-Z. Liu. 2006. Microwave-assisted extraction of solanesol from tobacco leaves. *J. Chromatogr. A* 1129(1):135–139.

4 Ultrasound-Assisted Extraction of High Value Compounds from Agro-Industrial Byproducts

Anuradha Saini, Divyani Panwar, Parmjit S. Panesar, and Anjineyulu Kothakota

CONTENTS

4.1 INTRODUCTION

Numerous byproducts are generated globally from different agricultural industries because of inadequate facilities related to processing, storage, and transportation (Food and Agriculture Organization; FAO, 2011). In 2009, the FAO of the United Nations reported that approximately 32% (weight basis) of food had been wasted in the world, which accounts for about 24% of the total food produced (Lipinski et al., 2013; Panwar et al., 2020). Among all the food commodities, cereals contain the largest global share of food loss and waste in terms of caloric content (53%); however, fruits and vegetables are the largest sources of waste and loss based on weight (44%), as shown in Figures 4.1 and 4.2.

With changes in lifestyle and increased health concerns among people, functional compounds extracted from agro-industrial byproducts are replacing synthetic additives. The isolation of high-value components from agro-industrial byproducts (Table 4.1) and further use of these compounds is an attractive consideration for food safety and quality (Saini et al., 2019a). These components have different physiological benefits like anti-inflammatory, antiallergenic, antimicrobial,

DOI: 10.1201/9781003125679-4

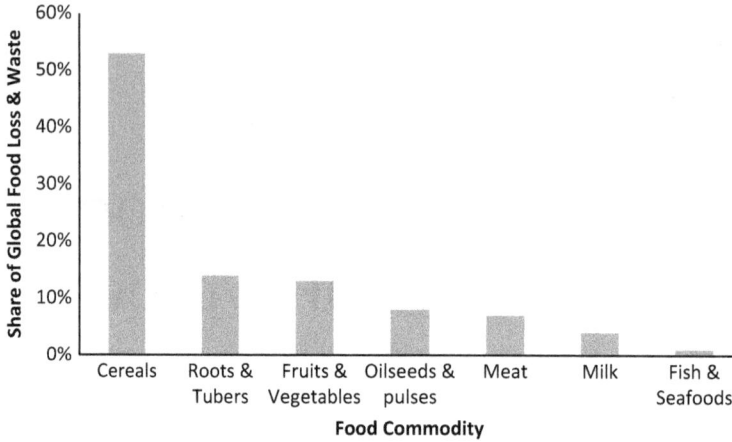

FIGURE 4.1   Global food loss and waste share of commodities on caloric basis (100% = 1.5 quadrillion calories). Sources: FAO, 2011; Lipinski et al., 2013.

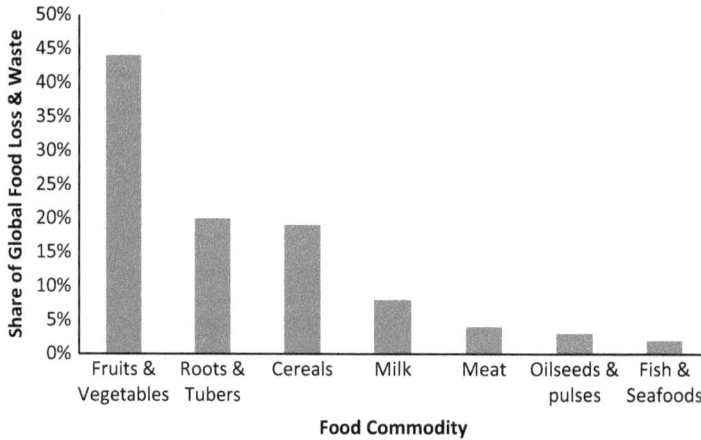

FIGURE 4.2   Global food loss and waste share of commodities on weight basis (100% = 1.3 billion tonnes). Sources: FAO, 2011; Lipinski et al., 2013.

antithrombotic, antioxidant, vasodilatory, and cardioprotective effects. In addition, these polyphenolic compounds attract consumers due to their therapeutic and nutritional value together, which is important for developing functional foods (Studdert et al., 2011).

Different methods have been used to extract high-value compounds from natural sources, such as the maceration technique, mechanical expelling, microwave-assisted extraction, supercritical fluid extraction, and solvent extraction. However, these methods have some limitations that include low yield of extraction in mechanical expelling, the need for a large quantity of solvent along with the long duration of solvent extraction, the necessity of an aqueous phase in microwave-assisted extraction, and the huge monetary investment required in the supercritical fluid extraction method (Samaram et al., 2015). Conversely, the ultrasound-assisted extraction (UAE) technique has overcome these limitations and become one of the most sustainable techniques for the extraction of byproducts from fruits, vegetables, cereals, pulses, meat, seafood, etc. This technique is effectively used for bioactive compounds extraction that relies on various factors such as time, temperature, power, solvent type, duty cycle, frequency, source matrix, liquid-solid ratio, etc. These factors should be controlled to obtain the maximum extraction yield. In addition, UAE is safe, economical,

TABLE 4.1

High-Value Compounds from Various Byproducts Obtained from Different Food Processing Industries

Bioactive Compounds	Byproducts	References
Byproducts from the fruit and vegetable industry		
Carotenoids (α-carotene, β-carotene, lycopene, lutein), Phenolics (epicatechin, cyanidin-3-rutinoside, gallic acid, caffeic acid, chlorogenic acid, *p*-coumaric acid), flavonoids (hesperidin, neohesperidin, naringin, rutin, quercetin, kaempferol)	Garlic seeds, carrot pomace, stones and peels (e.g., citrus fruits, guava, eggplant, tomato, plantain, grapes, sweetsop, starfruit, mango, watermelon, plum, pear, peach, hawthorn, pineapple, muskmelon, apple, avocado, banana, cherry, blueberry)	Ferarsa et al. (2018); Deng et al. (2012); Panwar et al. (2019); Wang et al. (2008); Saini et al. (2019b); Panwar et al. (2021a,b)
Byproducts from the cereals and pulses industry		
Total carotenoids, total anthocyanins, total flavonoids, total polyphenols, phenolic acids (ferulic acid, *p*-coumaric acid, caffeic acid), total flavonoid content, C-glycosylated flavonoids (vitexin, isovitexin), total phenols (free and bound flavonoids)	Bran, germ, shorts of wheat, germ meal, bran of corn, rice bran of different coloured rice varieties, rice husk, defatted rice bran, seed coat (peanut, pea, pigeon pea, and black cowpeas), husk (soybean, black gram and chickpea), germ, plumule, aleurone of black gram	Smuda et al. (2018); Gao et al. (2018); Ghasemzadeh et al. (2018); Attree et al. (2015); Balasundram et al. (2006); Niño-Medina et al. (2017); Girish et al. (2016); Gan et al. (2016); Saini et al. (2019c); Zhao et al. (2018)
Byproducts from the meat, poultry, and seafood industry		
Protein and peptides, chitin, polyunsaturated fatty acids, gelatine, collagen	Heads, gills, skin, trimmings, fins, frames, bones, viscera, blood, livers, sweetbreads, kidneys, oxtails, tripe, and roe of commercial fish species	Álvarez et al. (2018); Mora et al. (2014); Kim et al. (2013); Tu et al. (2015); Kjartansson et al. (2006)

Source: Saini and Panesar (2020).

easy to handle, and reproducible that can be used in ambient temperature and atmospheric pressure conditions (Chemat et al., 2011).

In the last decade, UAE has gained more popularity for various applications. In recent studies on extraction, preference has been given to non-conventional techniques (e.g., UAE) than conventional techniques of extraction. Similarly, organic solvents are being replaced in the extraction of bioactive compounds by smart or greener solvents (e.g., deep eutectic solvents) due to their higher extraction efficiency and their environmental friendliness (Saini and Panesar, 2020), for example, the UAE of pectin (96.25%) from pomelo peels using ChCl:malonic acid (Elgharbawy et al., 2019), polyphenols (34.54 mg GAE/g DW) from red grape pomace by sodium acetate: lactic acid (Patsea et al., 2017), and levulinic acid (40%) from rice husk through $C_4(Mim)_2][2HSO_4]$ (Ullah et al., 2019). Therefore, the UAE technique has great possibilities in extraction using smart solvents to deliver clean industrial processes with high-energy efficiency. This chapter provides information regarding the fundamentals of UAE, factors affecting the UAE process and its application in the recovery of bioactive compounds from different agro-industrial waste.

4.2 ULTRASOUND-BASED EXTRACTION

UAE is the technique in which an acoustic wave in the frequency range of 20 or 25 kHz is used to extract high-value compounds from different sources such as fruits, vegetables, cereals, pulses, milk, meat, and their byproducts (Figure 4.3). The acoustic cavitation was the main driving force for extraction through which a series of compressions and rarefactions takes place in extracting a solvent, causing bubble formation (Shirsath et al., 2012), as shown in Figure 4.4. The ultrasound effect is related to the magnitude of the frequency. For instance, physical effects dominate when the frequency is lower than 20–100 kHz, whereas chemical effects dominate in the frequency range of 200–500 kHz (Balasundram, 2006). Generally, the cavitation phenomenon led to high shear forces, which increased the mass transfer of bioactive compounds across the cellular membranes of the plant matrix into the solution (Rabelo et al., 2016; Sharmila et al., 2016). However, various studies revealed that the working principle of the ultrasonication technique acted on the combined mechanisms of sonoporation, capillarity, erosion, detexturation, and fragmentation (Chemat et al., 2016).

The major advantages of using UAE as an extraction technique are as follows: (i) variety of solvents used, such as aqueous, non-aqueous, deep eutectic solvents, and ionic liquids; (ii) the process is feasible at ambient temperature and pressure; (iii) it reduces unit operations; (iv) reduced time, energy, and power use with a higher yield; and (v) quick return of investment.

4.3   FACTORS AFFECTING UAE

There are numerous factors such as power, frequency, solvent type, and temperature that directly affect the yield of extraction and efficiency of the UAE process (Tiwari, 2015). Furthermore, the UAE extraction technique was used to achieve the highest extraction yield with minimum consumption of power and monetary resources (Esclapez et al., 2011). Therefore, to develop a green extraction process with reliable results, it is essential to consider the effect of these factors and optimize them to get the desired results. The effects of these factors are discussed in the following subsections.

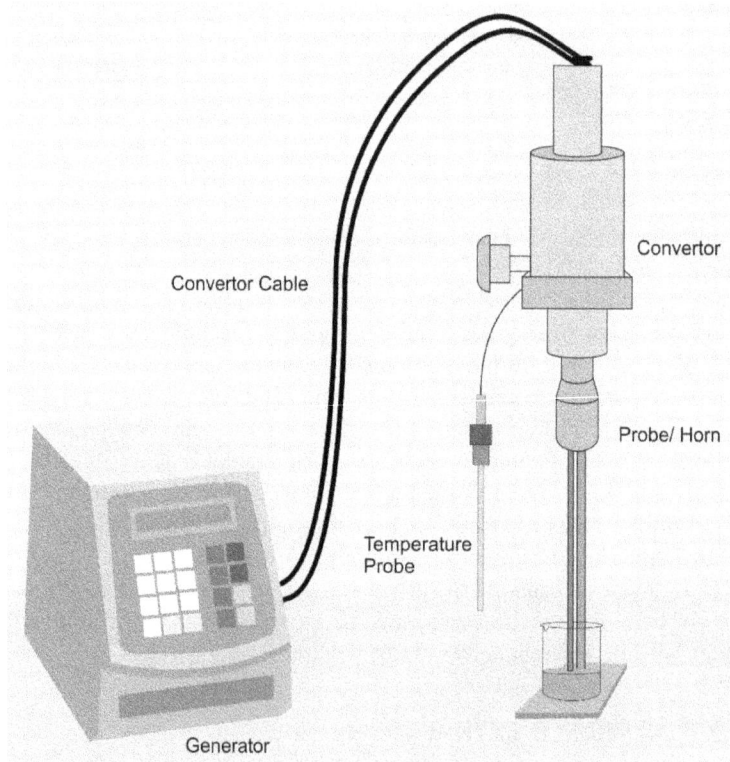

FIGURE 4.3   Schematic representation of the probe sonicator.

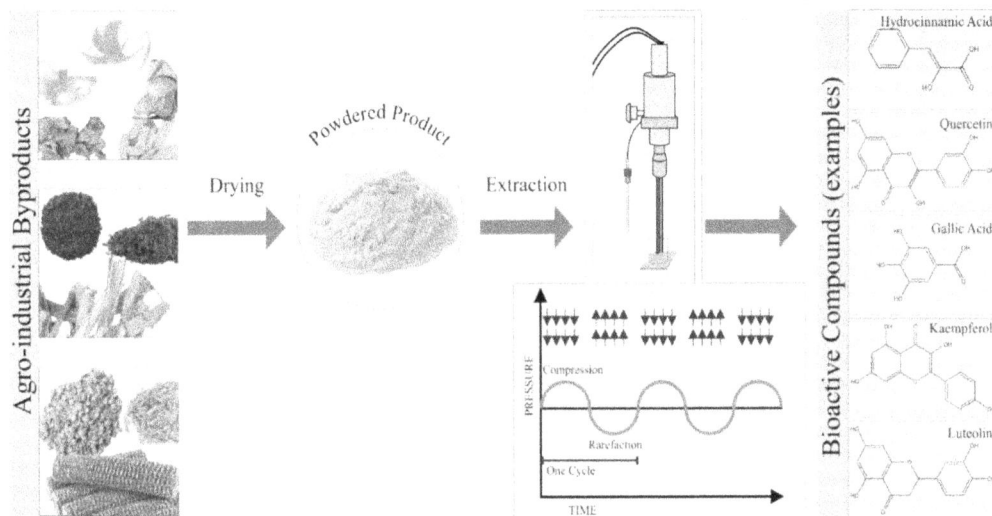

FIGURE 4.4 General procedure for the ultrasound-assisted extraction of bioactive compounds from agro-industrial waste.

4.3.1 ULTRASONIC POWER

Ultrasonic power (expressed in W) is one key parameter that affects the UAE process. The power dissipated during the UAE process is also expressed in terms of amplitude percentage in the range of 0–100% (Palma and Barroso, 2002). Numerous studies have been reported that high ultrasonic power induced a severe shear force on the raw material, which caused major alterations in the medium; therefore, its optimization was required to achieve the desired extraction efficiency (Chemat et al., 2017). It has been shown that with increased ultrasonic power, the cavitation effect and yield of the UAE process increased to a certain point. This was probably due to the increased cavitation effect and collapse of bubbles with increased power. Moreover, it has been noticed that the size of the bubble is proportional to the ultrasonic power (Suslick et al., 1987; Suslick and Price, 1999). Therefore, increased bubble size also intensified its impact on bubble collapse, which provoked fragmentation and the formation of pores in the tissue and further enhanced diffusivity to give a high extraction yield (Maran and Priya, 2014). Besides this, UAE (with probe) phenomenon was based on the generation of vibrations that increased solvent penetration, the contact area between the solvent and raw material, as well as disruption of tissues for a higher extraction yield (Lou et al., 2010).

However, after increasing power to a certain level, the yield of the UAE process started to decline. This can be explained by the increased number of bubbles formed after increasing the power to a high level. An increase in the formation of bubbles caused collision among them, which then led to deformation and collapse in a non-spherical pattern, thereby decreasing the level of the cavitation effect (Dezhkunov et al., 2005). Moreover, high power enhanced agitation and led to the erosion of the ultrasound probe, which reduced the final yield as erosion left unwanted residue in the product (Tiwari, 2015). In addition, the layer of cavitation bubbles surrounding the tip of the ultrasound probe hampered the energy transmission into the extraction medium, which reduced the extraction yield (Contamine et al., 1995). A study on the interactive effect of ultrasonic power along with time and temperature on waste spent coffee grounds reported that with increased power from 100 W to 200 W, an increase in the total phenolic yield was observed; however, after increasing the power above 250 W, the yield was found to have decreased (Al-Dhabi et al., 2017). Similarly, in UAE, a high yield of pectin from tomato peel was obtained at 600 W power, and a further reduction in pectin was observed at 750 W (Sengar et al., 2020). Moreover, a linear relationship between ultrasound power and extraction yield of polyphenols was found from the pomace of grapes (González-Centeno et al., 2014), curcumin from *Curcuma amada* (Shirsath et al., 2017), protein from chicken bone (Dong et al., 2019), and high-value compounds from pomegranate peel (Kumar and Rao, 2020).

4.3.2 ULTRASONIC FREQUENCY

Ultrasonic frequency (expressed in Hz) is another important parameter in the UAE process that must be considered for optimization to achieve the maximum extraction yield. Frequency strongly affected the extraction process because of its direct influence on the number, size, and distribution of cavitation bubbles as well as the resistance to mass transfer (Wu and Nyborg, 2008; Brotchie et al., 2009). It was reported that at lower frequencies, larger bubbles were formed, causing violent bubble implosions, which led to the generation of high mechanical and shear forces, and therefore a high cavitation effect and extraction yield were observed (Esclapez et al., 2011). However, with increased ultrasound frequency, a decline in the cavitation effect was observed. This was explained by the fact that the duration of the refraction phase was inversely proportional to the frequency (Mason and Lorimer, 2002). Therefore, with high frequency, the length of the refraction phase was short, thereby leading to a short compression-refraction cycle (duration required for the formation and growth of the cavitation bubble), which explained the small size of the cavitation bubbles and consequently fewer implosion effects (Kanthale et al., 2008; Rutkowska et al., 2017). The extraction of high-value compounds from pomegranate peel using two frequencies (37 and 80 kHz) was performed, and the maximum extraction yield was observed at 37 kHz (Machado et al., 2019). Similarly, varying ultrasound frequencies of 100, 80, and 45 kHz were analyzed in the UAE of flavonoids from the leaves and bark of *Abiesnephrolepis* and showed that the maximum yield was obtained at 45 kHz (Wei et al., 2020). Besides, various studies have also indicated the effect of frequency (e.g., 20 kHz) on UAE for antioxidant recovery from pomegranate peels (Pan et al., 2012), pectin from sisal waste (Yang et al., 2018), alkaloids from potato peel (Hossain et al., 2014), β-carotene from carrot waste (Mirheli and Dinani, 2018), and phenolics from mango peel (Guandalini et al., 2019), etc.

4.3.3 TEMPERATURE

Temperature is known to be a key variable during the UAE process as it strongly affects the raw material characteristics along with the solvent's properties. It is strongly advised to choose an extraction temperature based on the raw material and type of compound to be extracted. Therefore, control of temperature is required to inhibit the deterioration of temperature-sensitive compounds (Hemwimol et al., 2006). Moreover, an increase in the yield was observed using UAE with an increase in temperature, but a further rise in temperature reduced the extraction yield. The increased yield with increased temperature is probably due to the increased desorption and solute-solvent solubility along with decreased viscosity of the solvent, which enhanced solvent diffusivity in the tissue matrix (Celli et al., 2015; Hossain et al., 2012). The further increase in temperature reduced the extraction yield due to increased vapour pressure and reduced surface tension in the cavitation bubbles, which caused a reduction in the cavitation effects (Tiwari et al., 2015). A similar trend was stated in a study where pectin extraction from custard apple peel was done using UAE in the temperature range of 50–80°C and reported that the yield increased with temperature from 50°C to 65°C and further yield was decreased beyond 65°C (Shivamathi et al., 2019). However, some studies on the significant increase in extraction yield with increased temperature using UAE have been reported. For example, a major surge in the amount of phenolics and β-glucan extracted from defatted oat bran with an increase in ultrasonic temperature from 20°C to 90°C has been reported (Chen et al., 2018). Similarly, a positive linear effect of extraction temperature on the amount of polyphenols recovered from peach palm waste has been observed (Da Silva et al., 2020).

4.3.4 Ultrasonic Treatment Time

Time is a very important factor that should be optimized to obtain the desired extraction yield. With an increase in ultrasonic time, the extraction yield was improved due to the increased cavitation effect, which caused swelling, fragmentation, and pore formation in the plant cell walls that, in turn, led to more exposure of the solute to the solvent that helped its release into the extraction medium (Mason, 1996). However, longer durations of extraction created undesirable structural changes in the desired compounds and thereby resulted in a decrease in extraction yield (Ma et al., 2008). The effect of extraction time (1.59–18.41 min) on the extraction of lycopene from tomato waste using UAE indicated that the extraction yield of lycopene was observed to increase up to 10 min and reduced after an increase in time (Rahimi and Mikani, 2019). Similarly, the amount of pectin recovered from sunflower byproducts was observed to increase up to an extraction time of 30 min, and then it decreased significantly due to overexposure (Ezzati et al., 2020). Moreover, a positive linear relationship between ultrasonic treatment time and extraction yield of antioxidants from rice bran (Tabaraki and Nateghi, 2011), β-carotene from carrot waste (Mirheli and Dinani, 2018), phenolic compounds from olive waste cake (Khalili and Dinani, 2018), and pectin from eggplant peel has been observed (Kazemi et al., 2019).

4.3.5 Type of Solvent

A variety of solvents like acetone, ethanol, methanol, and acidified water are available to extract compounds using UAE. However, the solvent must be selected according to the target compound's solubility. Besides, the cost, toxicity, surface tension, availability, viscosity, and solvent vapour pressure must also be considered. These physical properties have been observed to affect cavitation (Li et al., 2004; Wen et al., 2018). Therefore, selecting a suitable solvent to extract the compounds from raw materials is the key step in the UAE process. It has been reported that solvents like ethanol and methanol were widely used for the extraction process as they increased the rate of solubility and diffusion of compounds by lowering the dielectric constant of the solvent. However, researchers have reported that the use of these solvents at high concentrations caused the cell wall proteins to denature as well as the denaturation and collapse of the matrix cells, which caused difficulties in the extraction process (Garcia-Castello et al., 2015). Therefore, recent studies are being focused on the utilization of water-ethanol mixtures for the extraction processes (Kumar et al., 2021). For example, in a study, the maximum yield of bioactive compounds extracted from *Citrus sinensis* peel by UAE was observed using a mixture of 50% ethanol in water (Montero-Calderon et al., 2019). The results were similar in the polyphenol extraction and anthocyanins extracted from olive leaves and black chokeberry waste, respectively, where a solvent mixture of 50% ethanol in water gave the highest yields during UAE (Şahin and Şamli, 2013; Alessandro et al., 2014). Recently, successful use of vegetable oils (soy and sunflower oil) in the form of solvents has also been reported in UAE of carotenoids from pomegranate oils (Goula et al., 2017).

4.3.6 Solvent to Solid Ratio

The solvent to solid ratio is another critical factor in the UAE process. It has been reported that the yield was increased with an increased solvent to solid ratio during UAE, but after a certain point, the yield started to decline. When the solvent–solid ratio was low, the solution was observed to be highly viscous, which reduced the cavitation effects. Furthermore, with an increase in the solvent to solid ratio, the concentration and viscosity of the solution decreased, which caused an increase in the diffusivity of the solute into the solvent, which resulted in an increased cavitation effect and extraction yield (Stanisavljević et al., 2007). Besides, a high ratio of solvent to solid also improved the phenomenon of mass transfer due to the high concentration difference between the solid and solvent (Oreopoulou et al., 2019). However, after a certain solvent–solid ratio was reached, the extraction yield started to decline because of the improved cavitation effect, which led to the degradation of the desired compounds and also generated undesirable elements (Samaram et al., 2015; Maran et al., 2017). In pomegranate peels, the pectin extraction yield was improved with a change in the solvent–solid ratio from 10:1 to 15:1; however, a further increase in the solvent–solid ratio resulted in a significant decrease in the pectin yield (Moorthy et al., 2015). Pectin extraction from walnut processing waste by UAE had a maximum pectin yield at a solvent–solid ratio of 15 (w/v), and a further increase in the solvent–solid ratio decreased the yield (Asgari et al., 2020).

4.4 UAE FROM VARIOUS AGRO-INDUSTRIAL BYPRODUCTS

UAE is an important extraction technique that aims to reduce its energy consumption and extraction time to lower the final cost. Various studies have reported the recovery of high-value compounds from agro-industrial byproducts using UAE, comparative studies of UAE with other extraction techniques, and the optimization of process parameters to achieve a higher yield.

4.4.1 UAE FROM BYPRODUCTS OF THE FRUIT AND VEGETABLE INDUSTRY

Different byproducts are produced through processing fruits and vegetables, which include peels, seeds, pomace, rind, etc., and they are important sources of bioactive compounds (Table 4.2). In the last few decades, many studies revealed that the concentration of bioactive compounds in byproducts (e.g., peels) was higher than the edible portions of vegetables and fruits (Galanakis, 2012; Banerjee et al., 2017). For instance, polyphenol concentration was found to be higher (15%) in citrus peels (e.g., grapes, lemons, oranges), stone (e.g., mango), and seeds of jackfruit and avocado rather than in the edible flesh (Soong and Barlow, 2004; Gorinstein et al., 2001).

Extraction of functional compounds from grapefruit (*Citrus paradise* L.) and malta (*Citrus sinensis*) peel was found to be higher using UAE than conventional extraction methods (Nishad et al., 2019a, b). Further, the extraction yield of flavonoids and polyphenols from citrus species (*Citrus limetta*) cv. "mousambi" and (*Citrus reticulata*) cv. "kinnow" were also gathered at higher levels with UAE than maceration (Saini et al., 2019b).

In the case of phenolic compounds, different solvents (ethanol, methanol, water, acetone) were used at various concentrations (70%, 80%, and 100%). Among all solvents, the highest yield of total phenolic content (TPC) was found to be 15.25 g GAE (Gallic acid equivalents)/100 g, using acetone (80%) at a power of 56.71 W and temperature 48°C for 40 min (Nipornram et al., 2018). However, with the same solvent, the TPC from mandarin (*Citrus reticulate* Blanco cv) peel was 3.08 g GAE/100 g at 50.93 W power (Singanusong et al., 2015).

Mango seed kernel is an important by-product generated from mango processing industries and can be valorized by extracting the bioactive compounds. The yield of total polyphenols in mango kernel was 6.67% at 45°C temperature for 27 min in the aqueous two-phase UAE. The concentration of polyphenols was found to be higher by combining the aqueous two-phase separation with UAE than conventional UAE (Gao et al., 2012).

Commercially, citrus peels and apple pomace have been used to produce pectin, which is a versatile food ingredient used in food products including jams, jelly, yoghurt drinks, and ice cream. Pectin is the structure-building polysaccharide used for gelation, stabilization, or as a thickening agent. However, cacao pod husk (Muñoz-Almagro et al., 2019), mango peels (Mugwagwa and Chimphango, 2019), sugar beet pulp, watermelon rind, red and black carrot pomace, etc. (Misra and Yadav, 2020; Dranca and Oroian, 2018) are newer sources that have also been recently explored to extract high-value compounds using UAE. During harvesting, around 60–75% of broccoli produce is wasted worldwide, and it might be used as a good substitute for commercial pectin. In a recent study, the pectin recovered from a broccoli stalk where the pectic fraction had a yield of 18%, which comprised 75% galacturonic acid with 56% of the degree of methyl-esterification, and 1.1% acetyl content (Petkowicz and Williams, 2020).

TABLE 4.2

Various High-Value Compounds Extracted from Fruit and Vegetable Byproducts Using the Ultrasound-Assisted Method

Source	Solvent/Enzyme	Operating Conditions	Examples Output	References
Grapefruit (*Citrus paradise* L.) peel	Ethanol (70%)	Amplitude: 71.11%, Time: 32 min, Ratio: 1:39.63	TF: 2.7653 mg QE/g TPC: 21.1671 mg GAE/g Quercetin: 120.97 ppm Naringin: 42,038.59 ppm Phloridzin dihydrate: 86.82 ppm	Nishad et al. (2019a)
Blueberry wine waste (*Vaccinium ashei*)	Ethanol	Time: 23.67 min, Liquid/solid: 21.70 ml/g, Temp.: 61.3°C	TPC: 16.41 mg GAE/g DW	He et al. (2016)
Lemon (*C. limon*) seeds	Ethanol (95%)	Temp.: 56°C, Time: 56 min, Power: 300 W, Liquid/solid: 30:1	Limonoids: 6.02% Nomolin: 277.20 mg/g Limonin: 715.40 mg/g Obakunone: 47.40 mg/g	Yu et al. (2017)
Sugar beet molasses	Ethanol (57–63%)	Temp.: 41°C–48°C, Time: 66–73 min,	Antioxidant activity: 16.66 mg/g TPC: 0.1736 mg GAE/ml	Chen et al. (2015)
Lemon (*Citrus limon* L.) peel	Methanol (80%)	Time: 60 min, Ratio: 1:6 g/ml; Temp.: 48°C	TF: 4.52 mg catequin equivalent/100 g TPC: 67.17 mg GAE/100 g	Jagannath and Biradar (2019)
Grape skin (*Vitis vinifera* L.)	Water	Power: 150 W/L, Frequency: 40 kHz, Time: 25 min	TPC: 32.31 mg GAE/g FW	González-Centeno et al. (2014)
Carrot pomace	Ethanol	Time: 17 min, Ethanol concentration: 48%, Temp.: 34°C	Total phenols: 316.92 µg GAE/g	Jabbar et al. (2015)
Grapefruit peel	Acidified water	Ultrasound power: 12.56 W/cm²; Temp.: 67.7°C, Ratio: 3:150 g/ml, Time: 28 min	Pectin: 27.34%	Wang et al. (2015)
Orange peel	Ethanol (80%)	Power: 150 W, Time: 30 min, Temp.: 40°C, Ratio: 0.25:1 g/ml	Naringin: 70.3 mg/100 g	Khan et al. (2010)
Blackberry leaves	Methanol (61–64%)	Temp.: 66°C–68°C, Time: 105–117 min	TPC: 80.19 mg GAE/g Dry weight (DW)	Aybastıer et al. (2013) *(Continued)*

TABLE 4.2 (CONTINUED)
Various High-Value Compounds Extracted from Fruit and Vegetable Byproducts Using the Ultrasound-Assisted Method

		Examples		
Source	Solvent/Enzyme	Operating Conditions	Output	References
Malta (*Citrus sinensis*) peel	Ethanol (70%)	Amplitude: 70.89%, Solvent–solid: 40 ml/g, Time: 35 min	TPC: 1,590 mg GAE/100 g TFC: 104.99 mg QE/100 g Phloridzin dihydrate: 935.476 ppm Catechin: 238.18 ppm Naringin: 10, 796.53 ppm Caffeic acid: 47.083 ppm	Nishad et al. (2019b)
Mandarin (*Citrus unshiu*) peel	ChCl-Lea-MU (80%)	Temp.: 50°C, Time: 25 min, Power: 200 W	Glycosides of flavonoids (GoFs): 4707 mg/100 g TF: 6582 mg/100 g	Xu et al. (2019)
Lemon (*Citrus limon* L.), Grapefruit (*Citrus paradisi* L.) and Orange (*Citrus sinensis* L.) seeds meal	Water	Frequency: 20 kHz, Time: 20 min, Temp.: ≤40°C, Power: 280 W	Polymethoxylated flavonoids (PMFs): 1875 mg/100 g Soluble dietary fibre: 4.95–7.95%; Insoluble dietary fibre: 75.95–82.24%	Karaman et al. (2017)
Orange peel	Limonene	Temp.: 20°C, Ultrasound power: 208 W/cm², Time: 5 min	Total carotenoids: 1.13 mg β-carotene/100 ml	Boukroufa et al. (2017)
Orange peel	Acetone (75.79%)	Amplitude: 65.94%, Time: 8.33 min	TPC: 1357 mg GAE/100 g	Dahmoune et al. (2014)
Kinnow (*Citrus reticulata*) peel	Acetone (100 %)	Ratio: 1:3 g/ml at ambient conditions	TF: 440 mg CE/100 g TPC: 2800 mg GAE/100 g	Saini et al. (2019c)
Citrus (*Citrus unshiu* Marc) peel	Methanol (80 %)	Sample weight: 2 g (i) Temp.: 40°C for 1 h (ii) Temp.: 15°C for 1 h	Phenolic compounds (μg/g): (i) Caffeic (50.9), *p*-Coumaric (97.9), Ferulic (1187.6), Sinapic (218.2), Protocatechuic (15), *p*-Hydroxy-benzoic (31.2), Vanillic (40.1) (ii) Caffeic (97.5), *p*-Coumaric (177.3), Ferulic (2226.8), Sinapic (219.8), Protocatechuic (19.4), *p*-Hydroxy-benzoic (45.6), Vanillic (41.5)	Ma et al. (2009)

(Continued)

TABLE 4.2 (CONTINUED)
Various High-Value Compounds Extracted from Fruit and Vegetable Byproducts Using the Ultrasound-Assisted Method

Source	Solvent/Enzyme	Operating Conditions	Examples Output	References
Sour Orange (*Citrus aurantium* L.) peel	Acidified water	Power: 150 W, Time: 10 min, Temp.: <30°C, pH 1.5, Ratio: 1:20 g/ml	TPC: 39.95 mg GAE/g of pectin Galacturonic acid: 65.3% Pectin yield: 28.07% Sugars: 0.4% (74% galactose)	Hosseini et al. (2019)
Grapefruit (G), orange (O), mandarin (M), and lemon (L) peel	Ethanol (100%)	Amplitude: 30%, Time: 5 min Ratio: 0.25:15 g/ml	Limonene: 13.6 mg/g (G); 12.8 mg/g (O); 30.1 mg/g (M); 1.4 mg/g (L)	Omar et al. (2013)
Orange (*Citrus sinensis*) peel	Ethanol (50%)	Power: 400 W, Temp.: 40°C, Time: 30 min, Ratio: 1:10 g/ml	Total carotenoids: 0.63 mg β-carotene/100 g	Montero-Calderon et al. (2019)
Orange peel	Water	Ultrasound power: 0.956 W/cm², Temp.: 59.83°C, Time: 30 min, particle size: 2 cm², Ratio: 1:10 g/ml	TPC: 50.02 mg GAE/100 g	Boukroufa et al. (2015)
Sweet lime (*Citrus limetta*) peel	Acetone (100 %)	Ratio: 1:3 g/ml at ambient conditions	TF: 21.99 mg CE/g TPC: 2.07 mg GAE/g	Saini et al. (2019c)
Mandarin peel	Methanol (80%)	Frequency: 35 kHz, Temp.: 45°C, Time: 60 min, Ratio: 1:20 g/ml	TPC: 32.48 mg GAE/g	Safdar et al. (2017)
Grapefruit peel	Acidified water	Temp.: 70°C, Time: 25 min	Pectin: 17.92%	Bagherian et al. (2011)
Grapefruit peel	Water	Frequency: 40 kHz, Ratio: 1:40 g/ml	TF: 93–190 mg Rutin equivalent (RE)/100 g Hesperidin: 62–109 mg/100 g Naringin: 13–49 mg/100 g	Van Hung et al. (2020)
Lemon peel	Ethanol (63.93%)	Amplitude: 77.79%, Time: 15.05 min	TPC: 15.22 mg GAE/g	Dahmoune et al. (2013)
Kinnow mandarin (*Citrus reticulata*) peel	Methanol (100%)	Amplitude: 32.8%, Temp.: 43.14°C, Time: 33.71 min, Ratio: 1:6.16 g/ml	Maximum lutein yield: 2.97 mg/100 g	Saini et al. (2021)
Mandarin peel	Acetone (80%)	Power: 56.71 W, Temp.: 48°C, Time: 40 min	Hesperidin: 64.4484 mg/g TPC: 152.5664 mg GAE/g	Nipornram et al. (2018)

(Continued)

TABLE 4.2 (CONTINUED)
Various High-Value Compounds Extracted from Fruit and Vegetable Byproducts Using the Ultrasound-Assisted Method

Source	Solvent/Enzyme	Operating Conditions	Output	References
			Examples	
Grapefruit peel	Ethanol (99.7%)	Time: 2 h, Ratio: 1:10	Hesperidin: 0.0714 mg/g Naringin: 14 mg/g	Wu et al. (2007)
Mandarin peel	Acetone (80%)	Power: 50.93 W	TPC: 30.8361 mg GAE/g Hesperidin: 13.74 mg/g TF: 25.3983 mg Quercetin equivalent (QE)/g	Singanusong et al. (2015)
Sweet orange peel	Ethanol (70%)	Amplitude: 70.89%, Ratio: 1:40 g/ml, Time: 35 min	Naringin: 10,796.53 ppm Phloridzin Dihydrate: 935.476 ppm Caffeic acid: 47.083 ppm	Nishad et al. (2019b)
Lime peel	Acetone (80%)	Power: 50.93 W	Naringin: 0.5339 mg/g	Singanusong et al. (2015)
Orange peel	ILs[BMIM][CL]	Amplitude: 80%, Ratio: 1:3 g/ml Power: 200 W, Time: 5 min	Total carotenoids: 0.032 β-carotene/g	Murador et al. (2019)

With the recent developments in extraction, neoteric solvents (e.g., ionic solvents, deep eutectic solvents) are replacing organic solvents as they have a negative effect on the environment and high intake of total process energy. For instance, total carotenoids were extracted from orange peel (*Citrus sinensis*) using Chlorinated 1-Butyl-3-methylimidazolium (BMIM-Cl) and acetone. It was reported that ionic liquid along with UAE (3.21 mg β-carotene/100 g) showed a higher yield of carotenoids than the conventional method (0.78 mg β-carotene/100 g) using acetone (Murador et al., 2019). Similarly, pectin from pomelo peel was extracted using natural deep eutectic solvents (NaDESs), and the highest yield of pectin (96.25%) was found using ChCl: malonic acid (Elgharbawy et al., 2019). Therefore, these solvents are new alternatives for obtaining high recovery of polyphenols from citrus peels.

4.4.2 UAE from Cereals and Pulses Byproducts

During the milling of cereal and pulses, the bran, germ, shorts, husk, germ meal, etc. are generated as byproducts, which are a potential source of high-value compounds. These byproducts contain different bioactive compounds (phytochemicals) such as terpenes, phenolic compounds, stilbenoids, lignans, dietary fibres (β-glucan), phytosterols, saponins, vitamins, and enzyme inhibitors (Table 4.3). These compounds have bioactivity which affect the host through the interaction of various compounds in living tissue to perform a specific function

The UAE of total phenolics (366.1–1924.9 mg/100g) and flavonoids (139.3–681.6 mg/100g) from cereal (wheat, rice, and corn) milling byproducts such as bran, germ, and shorts have been reported (Smuda et al., 2018). Moreover, a high yield of phenolics (14.90 ± 0.70 mg GAE/g) and flavonoids (3.08 ± 0.17 mg CE (Catechin equivalent)/g) has been achieved in rice husk using UAE (Gao et al., 2018). Bioactive compounds like lignin (4.75 mg/100 g), betaine (868 mg/100 g), phytosterols (158 mg/100 g), and choline (172 mg/100 g) were also present in wheat bran (Fardet, 2010). Besides this, UAE was also applied to extract valuable compounds from coloured rice bran and different yields of total flavonoids (72–96%), total phenolic (72–92%), and vitamins like γ-tocopherol, α-tocopherol, and γ-tocotrienol was observed in the range of 10.76–28.43, 13.12–67.30, and 16.91–53.09 mg/g DM (Dry matter), respectively (Huang and Lai, 2016).

On the other hand, byproducts (e.g., seed coat, hull) from pulses also contain a significant amount of bioactive compounds that provide beneficial effects to human beings. During the milling of pulses, the seed coat is the main by-product and rich source of bioactive compounds. Polyphenols were extracted from the seed coats of kidney beans, including petunidin 3-glucoside (0–0.17 mg/g db), cyaniding 3,5-diglucoside (0–0.04 mg/g db), delphinidin 3-glucoside (0–2.61 mg/g db), pelargonidin 3-glucoside (0–0.59 mg/g db), and cyaniding 3-glucoside ranging from 0 to 0.12 mg/g db (Choung et al., 2003). In defatted lentil samples, TPC in esterified, free, and insoluble forms was found in the ranges of 2.32–21.54, 1.37–5.53, and 2.55–17.51 mg GAE/g, respectively (Alshikh et al., 2015).

Cereal and pulse byproducts are being used for extracting valuable compounds for their efficient utilization. UAE can minimize waste and maximize the utilization of natural resources through extraction, which are generated during the processing of food commodities. For example, the UAE technique has been used to extract polyphenols and monomeric anthocyanins from purple and black rice bran. It was revealed that UAE showed high concentrations of phenolic and anthocyanins from purple and black rice bran in comparison with conventional methods (Das et al., 2017). Similarly, UAE results in 18–29% higher free phenolic content compared with the orbital shaker. In addition, the UAE technique is faster compared to the orbital shaker in terms of exposure time and is applicable to matrices with different compositions (Souza et al., 2018). Therefore, the UAE method indicates the potential for the extraction of antioxidant compounds from rice bran at low temperatures within a short extraction time. In addition, rice bran can be used in food additives and cosmeceutical products (Gam et al., 2018).

TABLE 4.3

Various High-Value Compounds Extracted from Cereal and Pulse Byproducts Using the Ultrasound-Assisted Method

		Examples		
Source	Solvent/Enzyme	Operating Conditions	Output	References
Purple rice bran (*Oryza sativa* L.)	Ethanol (31.17 %)	Temp: 31.7°C, pH: 2.39, Time: 15.77 min	TPC: 1978.67 mg GAE/100 g Anthocyanin: 34.86 mg cynanidin-3-glucoside/l	Das et al., 2017
Wheat bran	Ethanol 64% (v/v)	Temperature: 60°C, Time: 25 min	TPC: 3.12 mg GAE/g	Wang et al., 2008
Black rice bran	Ethanol (23.78 %)	Temp: 35.97°C, pH: 2.52, Time: 22.89 min	TPC: 2232.87 mg GAE/100 g Anthocyanin: 31.95 mg cynanidin-3-glucoside/l	Das et al., 2017
Rice bran	Ethanol (65–67%)	Temp: 51–54°C, Time: 40–45 min	TPC: 2.37–6.35 mg GAE/g of dry sample	Tabaraki and Nateghi, 2011
Rice bran	Ethanol (74 %)	Temp: 94.9°C, Time: 41.6 min,	TFC: 1.63 mg QE/g DM TPC: 2.78 mg GAE/g DM	Gam et al., 2018
Red sword bean seed coat (*Canavalia gladiata* (Jacq.) DC)	Hydroethanol (60.2%)	Temp: 50°C, Time: 18.4 min, Ultrasound power: 400 W, Liquid/solid ratio: 29.3 ml/g	Trolox equivalent antioxidant capacity (TEAC): 755.98 ± 10.23 μmol Trolox/g Dry weight (DW) TFC: 4.46 ± 0.15 mg CE/g DW TPC: 59.62 ± 2.77 mg GAE/g DW	Zhou et al., 2019
Mung bean (*Vignaradiata*) seed coat	Ethanol (37.6%)	Temp: 70°C, Time: 46.1 min, Solvent/material ratio: 35.1:1 ml/g, Ultrasound power: 500 W	TEAC: 178.28 ± 7.39 μmol Trolox/g DW TPC: 33.91 ± 1.06 mg GAE/g DW TPC: 15.06 ± 1.11 mg CE/g DW	Zhou et al., 2017
Red corn cob	Ethanol (55%)	Time: 120 min, Solid/liquid ratio: 30	TPC: 527.33 ± 103.79 GAE mg/100 g	Hernández et al., 2018

Wheat bran consists of high levels of bioactive compounds that have significant roles in the prevention of diabetes, cardiovascular disease, and cancer. The UAE technique was applied to extract phenolic acids like *p*-coumaric, *t*-ferulic, caffeic, and vanillic, and it was revealed that the temperature, time duration, and type of solvent affected the concentration of phenolic acids in the extracts. Besides, ethanol was the most promising solvent over methanol and acetone (Cengiz et al., 2020). Furthermore, wheat bran that contains arabinoxylan, a major dietary fibre component, has been extracted using ultrasound-assisted enzymatic extraction. UAE enhanced the efficacy of enzymatic treatment to produce a higher extraction yield (Wang et al., 2014).

The UAE technique was used for the extraction of natural antioxidants from mung bean seed coats and produced an antioxidant capacity higher than conventional methods. The antioxidant compounds, which included vitexin, *p*-coumaric acid, isovitexin, gallic acid, catechin, and rutin, were identified in the seed coat extract (Zhou et al., 2017). TPC and Total flavonoids content (TFC) have been found in higher concentrations in the hull compared to the whole mung bean and cotyledon. Moreover, UAE extracts a higher content of total polyphenols along with total flavonoids than conventional solvent extraction using acetone (Singh et al., 2017). The recovery of polyphenols from the red sword bean (*Canavalia gladiata* (Jacq.) DC.) seed coat has indicated that the recovered extract's antioxidant activity using UAE was greater than maceration and Soxhlet extraction. Compounds like digallic acid, trigalloyl hexoside, gallic acid, methyl gallate, and digalloyl hexoside were also found in the red sword bean seed coat extract (Zhou et al., 2019).

4.4.3 UAE FROM MEAT, POULTRY, AND SEAFOOD BYPRODUCTS

Animal-derived food processing produces a significant amount of byproducts that include blood, viscera, bones, heads, frames, fins, gills, trimmings, skin, livers, brains, hearts, kidneys, oxtails, tripe (stomach of cattle), and roes of commercial fish types. These byproducts consist of peptides, protein, polyunsaturated fatty acids, chitin, collagen, gelatine, etc. Extensive research has been carried out on byproducts from plant-derived foods than animal-derived foods for the extraction of bioactive compounds using UAE. Alkali, as well as the ultrasound-assisted alkali method, was used to extract liver protein from chicken, and the results showed that the yield of protein obtained by ultrasound-assisted alkali treatment increased by 55.4% over alkali treatment. Besides this, improved functional attributes such as hydrophobicity and water/oil holding capacity, etc. of chicken liver protein isolate were achieved by ultrasound-assisted alkali extraction, which indicated the potential of UAE in the food industry (Zou et al., 2017). Moreover, the haemoglobin from chicken blood obtained by processing chicken using ultrasonic treatment can be a good substitute for synthetic polymeric flocculants (Garcia et al., 2015). The yield of collagen from chicken lung was found to be 31.25% at 150 W. In addition, the extracted collagen had higher thermal stability and better solubility and fibril forming capacity with ultrasound (Zou et al., 2020). Moreover, collagen from cattle tendons was extracted using pepsin and the ultrasonic-pepsin tandem method. The results indicated that the ultrasonic-pepsin tandem method could increase the extraction efficiency of collagen without compromising collagen quality (Ran and Wang, 2013).

The comparative study of high-value compounds extraction using UAE from shrimp byproducts with the use of organic solvents and green, deep eutectic solvents revealed a higher yield of astaxanthin (146 μg/g) compared to organic solvents, which had a yield of 102 μg/g (Zhang et al., 2014). Minerals like iron were also extracted from cow, sheep, and chicken liver samples using ultrasound-assisted deep eutectic solvent technique and the recovery of minerals (e.g., iron) was higher than 95% (Yilmaz and Soylak, 2015).

4.4.4 UAE FROM DAIRY BYPRODUCTS

The processing of milk to produce several products such as ghee, cheese, yoghurt, ice cream, butter, condensed milk, etc. produces a significant volume of byproducts. These byproducts include whey, ghee residue, skim milk, and buttermilk, which have numerous uses in the food industry. Dairy byproducts have a high nutritive value and are used to produce beverages, sports drinks, flavoured milk, sweets, chocolates, and cookies. However, these are a rich source of high-value compounds. For example, sweet and acid whey contains various bioactive components such as protein, lactose, minerals, etc. Whey protein consists of 20% of the total proteins present in milk, whereas the remaining protein in milk constitutes 80% casein. In addition, a sufficient amount of lysine- and sulphur-containing amino acids were present in whey protein (Siso, 1996). Whey can be used as animal feed, but due to the richness of calcium and lactose in it, it is difficult for some animals to metabolize (Di Giacomo et al., 1997). Therefore, these valuable compounds can be recovered from the byproducts for use in other applications.

A study on lactose recovery from paneer whey was conducted using different pre-treatments like thermal, sonication, and thermosonication. It was revealed that maximum recovery of lactose (94.59% ± 1.26%) was obtained using combined pre-treatment of ultrasound and heating (Khaire and Gogate, 2018). In another study, ultrasound-assisted ultrafiltration of whey for the recovery of lactose was performed, and the maximum yield of permeates (96%) and lactose (94.3%) was reported using dual-frequency ultrasound (Khaire et al., 2019). The recovery of proteins from whey was achieved using ultrasound in spray drying and ultrafiltration. The use of ultrasound in spray drying and ultrafiltration provided an improvement in membrane processing (e.g., lower fouling) and spray drying (e.g., smaller particle size) along with higher recovery and better quality of products (Prabhuzantye et al., 2019). Moreover, the effect of various ultrasonic reactors was studied for lactose extraction with crystallization. The ultrasonic bath at 25 kHz exhibited a greater lactose yield (90.3% ± 0.3%) under 20 min of sonication than the ultrasonic horn (20 kHz), which had a lower yield (88.8% ± 0.6%) with 24 h standing time (Khaire and Gogate, 2020). However, the lactose was extracted from whey with an ultrasound-assisted technique using antisolvent crystallization. The maximum recovery of lactose (~98%) was found with the ultrasonic horn with ~97% purity than lactose (~94% with 92% purity) obtained using an ultrasonic bath (Gajendragadkar and Gogate, 2017). Therefore, the UAE technique increased the recovery yield of lactose from the whey.

4.5 SUMMARY AND FUTURE PROSPECTS

The UAE technique has considerable potential to isolate functional compounds from agro-industrial waste, and this technique provides higher extraction efficiency and other benefits than conventional methods. However, the efficiency of the ultrasound technique is remarkably dependent on the extraction method, equipment design, and different factors such as power, frequency, temperature, solvent, etc. Most of the studies in UAE were carried out on bath and probe sonicators (laboratory level) that may not be suitable for the commercial level. Therefore, there is a need to develop a suitable system in which high-quality bioactive compounds can be obtained at the commercial level.

Further, extraction by using a combination of ultrasound and smart solvents (deep eutectic solvents) has shown a higher yield of extraction compared to organic solvents alone at the lab scale. However, there is a requirement for further research in UAE on bioactive compounds using smart solvents for industrial applicability.

REFERENCES

Al-Dhabi, N.A., Ponmurugan, K., and Jeganathan, P.M. 2017. Development and validation of ultrasound-assisted solid-liquid extraction of phenolic compounds from waste spent coffee grounds. *Ultrasonics Sonochemistry* 34: 206–213.

Alshikh, N., de Camargo, A.C., and Shahidi, F. 2015. Phenolics of selected lentil cultivars: Antioxidant activities and inhibition of low-density lipoprotein and DNA damage. *Journal of Functional Foods* 18: 1022–1038.

Álvarez, C., Lélu, P., Lynch, S.A., and Tiwari, B.K. 2018. Optimised protein recovery from mackerel whole fish by using sequential acid/alkaline isoelectric solubilization precipitation (ISP) extraction assisted by ultrasound. *LWT-Food Science and Technology* 88: 210–216.

Asgari, K., Labbafi, M., Khodaiyan, F., Kazemi, M., and Hosseini, S.S. 2020. High-methylated pectin from walnut processing wastes as a potential resource: Ultrasound assisted extraction and physicochemical, structural and functional analysis. *International Journal of Biological Macromolecules* 52: 1274–1282.

Attree, R., Du, B., and Xu, B. 2015. Distribution of phenolic compounds in seed coat and cotyledon, and their contribution to antioxidant capacities of red and black seed coat peanuts (*Arachis hypogaea* L.). *Industrial Crops and Products* 67: 448–456.

Aybastıer, Ö., Is, E., and Saliha, S. 2013. Optimization of ultrasonic-assisted extraction of antioxidant compounds from blackberry leaves using response surface methodology. *Industrial Crops and Products* 44: 558–565.

Bagherian, H., Ashtiani, F.Z., Fouladitajar, A., and Mohtashamy, M. 2011. Comparisons between conventional, microwave- and ultrasound-assisted methods for extraction of pectin from grapefruit. *Chemical Engineering and Processing: Process Intensification* 50: 1237–1243.

Balasundram, N., Sundram, K., and Samman, S. 2006. Phenolic compounds in plants and agri-industrial by-products: Antioxidant activity, occurrence, and potential uses. *Food Chemistry* 99(1): 191–203.

Banerjee, J., Singh, R., Vijayaraghavan, R., MacFarlane, D., Patti, A.F., and Arora, A. 2017. Bioactives from fruit processing wastes: Green approaches to valuable chemicals. *Food Chemistry* 225: 10–22.

Boukroufa, M., Boutekedjiret, C., and Chemat, F. 2017. Development of a green procedure of citrus fruits waste processing to recover carotenoids. *Resource-Efficient Technologies* 3(3): 252–262.

Boukroufa, M., Boutekedjiret, C., Petigny, L., Rakotomanomana, N., and Chemat, F. 2015. Bio-refinery of orange peels waste: A new concept based on integrated green and solvent free extraction processes using ultrasound and microwave techniques to obtain essential oil, polyphenols and pectin. *Ultrasonics Sonochemistry* 24: 72–79.

Brotchie, A., Grieser, F., and Ashokkumar, M. 2009. Effect of power and frequency on bubble-size distributions in acoustic cavitation. *Physical Review Letters* 102(8): 084302.

Celli, G.B., Ghanem, A., and Brooks, M.S.L. 2015. Optimization of ultrasound-assisted extraction of anthocyanins from haskap berries (*Lonicera caerulea* L.) using response surface methodology. *Ultrasonics Sonochemistry* 27: 449–455.

Cengiz, M.F., Babacan, Ü.,Akıncı, E., Keşci, S., and Kaba, A. 2020. Extraction of phenolic acids from ancient wheat bran samples by ultrasound application. *Journal of Chemical Technology and Biotechnology*. https://doi.org/10.1002/jctb.6519.

Chemat, F., Rombaut, N., Sicaire, A., Meullemiestre, A., and Abert-vian, M. 2017. Ultrasound assisted extraction of food and natural products. Mechanisms, techniques, combinations, protocols and applications. A review. *Ultrasonics Sonochemistry* 34: 540–560.

Chemat, F., Zill-e-Huma, and Khan, M.K. 2011. Applications of ultrasound in food technology: Processing, preservation and extraction. *Ultrasonics Sonochemistry* 18(4): 813–835.

Chen, C., Wang, L., Wang, R., Luo, X., Li, Y., Li, J. et al. 2018. Ultrasound-assisted extraction from defatted oat (*Avena sativa* L.) bran to simultaneously enhance phenolic compounds and β-glucan contents: Compositional and kinetic studies. *Journal of Food Engineering* 222: 1–10.

Chen, M., Zhao, Y., and Yu, S. 2015. Optimisation of ultrasonic-assisted extraction of phenolic compounds, antioxidants, and anthocyanins from sugar beet molasses. *Food Chemistry* 172: 543–550.

Choung, G., Choi, B.R., An, Y.N., Chu, Y.H., and Cho, Y.S. 2003. Anthocyanin in profile of Korean cultivated kidney bean (*Phaseolus vulgaris* L.). *Journal of Agricultural and Food Chemistry* 51(24): 7040–7043.

Contamine, R.F., Wilhelm, A.M., Berlan, J., and Delmas, H. 1995. Power measurement in sonochemistry. *Ultrasonics Sonochemistry* 2(1): S43–S47.

D'Alessandro, L.G., Dimitrov, K., Vauchel, P., and Nikov, I. 2014. Kinetics of ultrasound assisted extraction of anthocyanins from *Aronia melanocarpa* (black chokeberry) wastes. *Chemical Engineering Research and Design* 92(10): 1818–1826.

da Silva Donadone, D.B., Giombelli, C., Silva, D.L.G., Stevanato, N., da Silva, C., and Bolanho Barros, B.C. 2020. Ultrasound-assisted extraction of phenolic compounds and soluble sugars from the stem portion of peach palm. *Journal of Food Processing and Preservation* 44(9): e14636.

Dahmoune, F., Boulekbache, L., Moussi, K., Aoun, O., Spigno, G., and Madani, K. 2013. Valorization of *Citrus limon* residues for the recovery of antioxidants: Evaluation and optimization of microwave and ultrasound application to solvent extraction. *Industrial Crops and Products* 50: 77–87.

Dahmoune, F., Moussi, K., Remini, H., Belbahi, A., Aoun, O., Spigno, G., and Madani, K. 2014. Optimization of ultrasound-assisted extraction of phenolic compounds from *Citrus sinensis* l. peels using response surface methodology. *Chemical Engineering* 37: 889–894.

Das, A.B., Goud, V.V., and Das, C. 2017. Extraction of phenolic compounds and anthocyanin from black and purple rice bran (*Oryza sativa* L.) using ultrasound: A comparative analysis and phytochemical profiling. *Industrial Crops and Products* 95: 332–341.

Deng, G.F., Shen, C., Xu, X.R., Kuang, R.D., Guo, Y.J., Zeng, L.S. et al. 2012. Potential of fruits waste as natural resources of bioactive compounds. *International Journal of Molecular Sciences* 13(7): 8308–8323.

Dezhkunov, N.V., Francescutto, A., Ciuti, P., and Sturman, F. 2005. Temperature dependence of cavitation activity at different ultrasound intensity. In: *Session of the Russian Acoustical Society XVI*, pp. 74–77. RSA. https://arts.units.it/handle/11368/2559311

Di Giacomo, G., Del Re, G., and Spera, D. 1997. Milk whey treatment with recovery of valuable products. *Desalination* 108(1–3): 273–276.

Dong, Z.Y., Li, M.Y., Tian, G., Zhang, T.H., Ren, H., and Quek, S.Y. 2019. Effects of ultrasonic pretreatment on the structure and functionality of chicken bone protein prepared by enzymatic method. *Food Chemistry* 299: 125103.

Dranca, F. and Oroian, M. 2018. Extraction, purification and characterization of pectin from alternative sources with potential technological applications. *Food Research International* 113: 327–350.

Elgharbawy, A.A.M., Hayyan, A., Hayyan, M., Mirghani, M.E.S., Salleh, H.M., Rashid, S.N. et al. 2019. Natural deep eutectic solvent-assisted pectin extraction from pomelo peel using sonoreactor: Experimental optimization approach. *Processes* 7(7): 1–19.

Esclapez, M.D., García-Pérez, J.V., Mulet, A., and Cárcel, J.A. 2011. Ultrasound-assisted extraction of natural products. *Food Engineering Reviews* 3(2): 108.

Ezzati, S., Ayaseh, A., Ghanbarzadeh, B., and Heshmati, M.K. 2020. Pectin from sunflower by-product: Optimization of ultrasound-assisted extraction, characterization, and functional analysis. *International Journal of Biological Macromolecules* 165(A): 776–786.

FAO. 2011. *Global Food Losses and Food Waste – Extent, Causes and Prevention*. FAO.

Fardet, A. 2010. New hypotheses for the health-protective mechanisms of whole-grain cereals: What is beyond fibre? *Nutrition Research Reviews* 23(1): 65–134.

Ferarsa, S., Zhang, W., Moulai-Mostefa, N., Ding, L., Jaffrin, M.Y., and Grimi, N. 2018. Recovery of anthocyanins and other phenolic compounds from purple eggplant peels and pulps using ultrasonic-assisted extraction. *Food and Bioproducts Processing* 109: 19–28.

Gajendragadkar, C.N. and Gogate, P.R. 2017. Ultrasound assisted intensified recovery of lactose from whey based on anti-solvent crystallization. *Ultrasonics Sonochemistry* 38: 754–765.

Galanakis, C.M. 2012. Recovery of high added-value components from food wastes: Conventional, emerging technologies and commercialized applications. *Trends in Food Science and Technology* 26(2): 68–87.

Gam, D.H., Jo, J.M., Jung, H.J., and Kim, J.W. 2018. Optimization of extraction conditions of antioxidant activity and bioactive compounds from rice bran by response surface methodology. *Applied Chemistry for Engineering* 29: 726–733.

Gan, R.Y., Deng, Z.Q., Yan, A.X., Shah, N.P., Lui, W.Y., Chan, C.L., and Corke, H. 2016. Pigmented edible bean coats as natural sources of polyphenols with antioxidant and antibacterial effects. *LWT-Food Science and Technology* 73: 168–177.

Gao, Y., Guo, X., Liu, Y., Fang, Z., Zhang, M., Zhang, R. et al. 2018. A full utilization of rice husk to evaluate phytochemical bioactivities and prepare cellulose nanocrystals. *Scientific Reports* 8(1): 10482.

Gao, Y., Wang, Z., Luo, L., Dai, J., and Li, D. 2012. Optimization of ultrasound-assisted aqueous two-phase extraction of polyphenols from mango seed kernel. *Transactions of the Chinese Society of Agricultural Engineering* 28: 255–261.

Garcia, R.A., Clevenstine, S.M., and Piazza, G.J. 2015. Ultrasonic processing for recovery of chicken erythrocyte hemoglobin. *Food and Bioproducts Processing* 94: 1–9.

Garcia-Castello, E.M., Rodriguez-Lopez, A.D., Mayor, L., Ballesteros, R., Conidi, C., and Cassano, A. 2015. Optimization of conventional and ultrasound assisted extraction of flavonoids from grapefruit (*Citrus paradisi* L.) solid wastes. *LWT - Food Science and Technology* 64(2): 1114–1122.

Ghasemzadeh, A., Karbalaii, M.T., Jaafar, H., and Rahmat, A. 2018. Phytochemical constituents, antioxidant activity, and anti-proliferative properties of black, red, and brown rice bran. *Chemistry Central Journal* 12(1): 17.

Girish, T.K., Kumar, K.A., and Rao, U.J.S.P. 2016. C-glycosylated flavonoids from black gram husk: Protection against DNA and erythrocytes from oxidative damage and their cytotoxic effect on HeLa cells. *Toxicology Reports* 3: 652–663.

González-centeno, M.R., Knoerzer, K., Sabarez, H., Simal, S., Rosselló, C., and Femenia, A. 2014. Effect of acoustic frequency and power density on the aqueous ultrasonic-assisted extraction of grape pomace (*Vitis vinifera* L.)—A response surface approach. *Ultrasonics Sonochemistry* 21(6): 2176–2184.

Gorinstein, S., Martín-Belloso, O., Park, Y.S., Haruenkit, R., Lojek, A., Čí ž, M. et al. 2001. Comparison of some biochemical characteristics of different citrus fruits. *Food Chemistry* 74(3): 309–315.

Goula, A.M., Ververi, M., Adamopoulou, A., and Kaderides, K. 2017. Green ultrasound-assisted extraction of carotenoids from pomegranate wastes using vegetable oils. *Ultrasonics Sonochemistry* 34: 821–830.

Guandalini, B.B.V., Rodrigues, N.P., and Marczak, L.D.F. 2019. Sequential extraction of phenolics and pectin from mango peel assisted by ultrasound. *Food Research International* 119: 455–461.

He, B., Zhang, L., Yue, X., Liang, J., Jiang, J., Gao, X., and Yue, P. 2016. Optimization of ultrasound-Assisted Extraction of phenolic compounds and anthocyanins from blueberry (*Vaccinium ashei*) wine pomace. *Food Chemistry* 204: 70–76.

Hemwimol, S., Pavasant, P., and Shotipruk, A. 2006. Ultrasound-assisted extraction of anthraquinones from roots of *Morinda citrifolia*. *Ultrasonics Sonochemistry* 13(6): 543–548.

Hernández, M., Ventura, J., Castro, C., Boone, V., Rojas, R., Ascacio-Valdés, J., and Martínez-Ávila, G. 2018. UPLC-ESI-QTOF-MS 2-based identification and antioxidant activity assessment of phenolic compounds from red corn cob (*Zea mays* L.). *Molecules* 23(6): 1425.

Hossain, M.B., Brunton, N.P., Patras, A., Tiwari, B., O'donnell, C.P., Martin-Diana, A.B., and Barry-Ryan, C. 2012. Optimization of ultrasound assisted extraction of antioxidant compounds from marjoram (*Origanum majorana* L.) using response surface methodology. *Ultrasonics Sonochemistry* 19(3): 582–590.

Hossain, M.B., Tiwari, B.K., Gangopadhyay, N., O'Donnell, C.P., Brunton, N.P., and Rai, D.K. 2014. Ultrasonic extraction of steroidal alkaloids from potato peel waste. *Ultrasonics Sonochemistry* 21(4): 1470–1476.

Hosseini, S.S., Khodaiyan, F., Kazemi, M., and Najari, Z. 2019. Optimization and characterization of pectin extracted from sour orange peel by ultrasound assisted method. *International Journal of Biological Macromolecules* 125: 621–629.

Huang, Y.P. and Lai, H.M. 2016. Bioactive compounds and antioxidative activity of colored rice bran. *Journal of Food and Drug Analysis* 24(3): 564–574.

Jabbar, S., Abid, M., Wu, T., Hashim, M.M., Saeeduddin, M., Hu, B., Lei, S., and Zeng, X. 2015. Ultrasound-assisted extraction of bioactive compounds and antioxidants from carrot pomace: A response surface approach. *Journal of Food Processing and Preservation* 39(6): 1878–1888.

Jagannath, A. and Biradar, R. 2019. Comparative evaluation of Soxhlet and ultrasonics on the structural morphology and extraction of bioactive compounds of lemon (*Citrus limon* l.) peel. *Journal of Food Chemistry and Nanotechnology* 5(3): 56–64.

Karaman, E., Yılmaz, E., & Tuncel, N. B. (2017). Physicochemical, microstructural and functional characterization of dietary fibers extracted from lemon, orange and grapefruit seeds press meals. *Bioactive Carbohydrates and Dietary Fibre*, 11, 9–17.

Kanthale, P., Ashokkumar, M., and Grieser, F. 2008. Sonoluminescence, sonochemistry (H_2O_2 yield) and bubble dynamics: Frequency and power effects. *Ultrasonics Sonochemistry* 15(2): 143–150.

Kazemi, M., Khodaiyan, F., and Hosseini, S.S. 2019. Eggplant peel as a high potential source of high methylated pectin: Ultrasonic extraction optimization and characterization. *LWT-Food Science and Technology* 105: 182–189.

Khaire, R.A. and Gogate, P.R. 2018. Intensified recovery of lactose from whey using thermal, ultrasonic and thermosonication pretreatments. *Journal of Food Engineering* 237: 240–248.

Khaire, R.A. and Gogate, P.R. 2020. Understanding the role of different operating modes and ultrasonic reactor configurations for improved sonocrystallization of lactose. *Chemical Engineering and Processing-Process Intensification*. https://doi.org/10.1016/j.cep.2020.108212.

Khaire, R.A., Sunny, A.A., and Gogate, P.R. 2019. Ultrasound assisted ultrafiltration of whey using dual frequency ultrasound for intensified recovery of lactose. *Chemical Engineering and Processing-Process Intensification* 142. 107581.

Khalili, F. and Dinani, S.T. 2018. Extraction of phenolic compounds from olive-waste cake using ultrasonic process. *Journal of Food Measurement and Characterization* 12(2): 974–981.

Khan, M.K., Abert-Vian, M., Fabiano-Tixier, A.S., Dangles, O., and Chemat, F. 2010. Ultrasound-assisted extraction of polyphenols (flavanone glycosides) from orange (*Citrus sinensis* L.) peel. *Food Chemistry* 119(2): 851–858.

Kim, H.K., Kim, Y.H., Park, H.J., and Lee, N.H. 2013. Application of ultrasonic treatment to extraction of collagen from the skins of sea bass *Lateolabrax japonicas*. *Fisheries Science* 79(5): 849–856.

Kjartansson, G.T., Zivanovic, S., Kristbergsson, K., and Weiss, J. 2006. Sonication-assisted extraction of chitin from shells of fresh water prawns (*Macrobrachium rosenbergii*). *Journal of Agricultural and Food Chemistry* 54(9): 3317–3323.

Kumar, A. and Rao, P.S. 2020. Optimization of pulsed-mode ultrasound assisted extraction of bioactive compounds from pomegranate peel using response surface methodology. *Journal of Food Measurement and Characterization* 14(6): 3493–3507.

Kumar, K., Srivastav, S., and Sharanagat, V.S. 2021. Ultrasound assisted extraction (UAE) of bioactive compounds from fruit and vegetable processing by-products: A review. *Ultrasonics Sonochemistry* 70: 105325.

Li, H., Pordesimo, L., and Weiss, J. 2004. High intensity ultrasound-assisted extraction of oil from soybeans. *Food Research International* 37(7): 731–738.

Lipinski, B., Hanson, C., Lomax, J., Kitinoja, L., Waite, R., and Searchinger, T. 2013. Reducing food loss and waste. *World Resources Institute Working Paper*, pp. 1–40.

Lou, Z., Wang, H., Zhang, M., and Wang, Z. 2010. Improved extraction of oil from chickpea under ultrasound in a dynamic system. *Journal of Food Engineering* 98(1): 13–18.

Ma, Y.Q., Chen, J.C., Liu, D.H., and Ye, X.Q. 2009. Simultaneous extraction of phenolic compounds of citrus peel extracts: Effect of ultrasound. *Ultrasonics Sonochemistry* 16(1): 57–62.

Ma, Y.Q., Ye, X.Q., Fang, Z.X., Chen, J.C., Xu, G.H., and Liu, D.H. 2008. Phenolic compounds and antioxidant activity of extracts from ultrasonic treatment of Satsuma mandarin (*Citrus unshiu* Marc.) peels. *Journal of Agricultural and Food Chemistry* 56(14): 5682–5690.

Machado, A.P.D.F., Sumere, B.R., Mekaru, C., Martinez, J., Bezerra, R.M.N., and Rostagno, M.A. 2019. Extraction of polyphenols and antioxidants from pomegranate peel using ultrasound: Influence of temperature, frequency and operation mode. *International Journal of Food Science and Technology* 54(9): 2792–2801.

Maran, J.P. and Priya, B. 2014. Ultrasound-assisted extraction of polysaccharide from *Nephelium lappaceum* L. fruit peel. *International Journal of Biological Macromolecules* 70: 530–536.

Maran, J. P., Manikandan, S., Nivetha, C. V., & Dinesh, R. (2017). Ultrasound assisted extraction of bioactive compounds from Nephelium lappaceum L. fruit peel using central composite face centered response surface design. *Arabian Journal of Chemistry*, 10, S1145–S1157.

Mason, T.J. 1996. Sonochemistry. *Ultrasonics Sonochemistry* 3: 8253.

Mason, T.J. and Lorimer, J.P. 2002. *Applied Sonochemistry: The Uses of Power Ultrasound in Chemistry and Processing*, (Vol. 10). Wiley-Vch, Weinheim

Mirheli, M. and Dinani, S.T. 2018. Extraction of β-carotene pigment from carrot processing waste using ultrasonic-shaking incubation method. *Journal of Food Measurement and Characterization* 12(3): 1818–1828.

Misra, N.N. and Yadav, S.K. 2020. Extraction of pectin from black carrot pomace using intermittent microwave, ultrasound and conventional heating: Kinetics, characterization and process economics. *Food Hydrocolloids* 102: 105592.

Montero-Calderon, A., Cortes, C., Zulueta, A., Frigola, A., and Esteve, M.J. 2019. Green solvents and ultrasound-Assisted Extraction of bioactive orange (*Citrus sinensis*) peel compounds. *Scientific Reports* 9(1): 1–8.

Moorthy, I.G., Maran, J.P., Muneeswari, S., Naganyashree, S., and Shivamathi, C.S. 2015. Response surface optimization of ultrasound assisted extraction of pectin from pomegranate peel. *International Journal of Biological Macromolecules* 72: 1323–1328.

Mora, L., Reig, M., and Toldrá, F. 2014. Bioactive peptides generated from meat industry by-products. *Food Research International* 65: 344–349.

Mugwagwa, L.R. and Chimphango, A.F.A. 2019. Box-Behnken design based multi-objective optimisation of sequential extraction of pectin and anthocyanins from mango peels. *Carbohydrate Polymers* 219: 29–38.

Muñoz-Almagro, N., Valadez-Carmona, L., Mendiola, J.A., Ibáñez, E., and Villamiel, M. 2019. Structural characterisation of pectin obtained from cacao pod husk. Comparison of conventional and subcritical water extraction. *Carbohydrate Polymers* 217: 69–78.

Murador, D.C., Braga, A.R.C., Martins, P.L., Mercadante, A.Z., and de Rosso, V.V. 2019. Ionic liquid associated with ultrasonic-assisted extraction: A new approach to obtain carotenoids from orange peel. *Food Research International* 126: 108653.

Niño-Medina, G., Muy-Rangel, D., and Urías-Orona, V. 2017. Chickpea (*Cicer arietinum*) and soybean (*Glycine max*) hulls: By-products with potential use as a source of high value-added food products. *Waste and Biomass Valorization* 8(4): 1199–1203.

Nipornram, S., Tochampa, W., Rattanatraiwong, P., and Singanusong, R. 2018. Optimization of low power ultrasound-assisted extraction of phenolic compounds from mandarin (*Citrus reticulata* Blanco cv. Sainampueng) peel. *Food Chemistry* 241: 338–345.

Nishad, J., Saha, S., Dubey, A.K., Varghese, E., and Kaur, C. 2019a. Optimization and comparison of non-conventional extraction technologies for *Citrus paradisi* L. peels: A valorization approach. *Journal of Food Science and Technology* 56(3): 1221–1233.

Nishad, J., Saha, S., and Kaur, C. 2019b. Enzyme- and ultrasound-assisted extractions of polyphenols from *Citrus sinensis* (cv. Malta) peel: A comparative study. *Journal of Food Processing and Preservation* 43(8): e14046.

Omar, J., Alonso, I., Garaikoetxea, A., and Etxebarria, N. 2013. Optimization of focused ultrasound extraction (FUSE) and supercritical fluid extraction (SFE) of citrus peel volatile oils and antioxidants. *Food Analytical Methods* 6(4): 1244–1252.

Oreopoulou, A., Tsimogiannis, D., and Oreopoulou, V. 2019. Extraction of polyphenols from aromatic and medicinal plants: An overview of the methods and the effect of extraction parameters. In: *Polyphenols in Plants*, ed. R.R. Watson, pp. 243–259. Academic Press.

Palma, M., and Barroso, C.G. 2002. Ultrasound-assisted extraction and determination of tartaric and malic acids from grapes and winemaking by-products. *Analytica Chimica Acta* 458(1): 119–130.

Pan, Z., Qu, W., Ma, H., Atungulu, G.G., and McHugh, T.H. 2012. Continuous and pulsed ultrasound-assisted extractions of antioxidants from pomegranate peel. *Ultrasonics Sonochemistry* 19(2): 365–372.

Panwar, D., Panesar, P.S., and Chopra, H.K. 2021a. Recent trends on the valorization strategies for the management of citrus by-products. *Food Reviews International* 37(1): 91–120.

Panwar, D., Saini, A., Panesar, P.S., and Chopra, H.K. 2021b. Unraveling the scientific perspectives of citrus by-products utilization: Progress towards circular economy. *Trends in Food Science and Technology* 111: 549–561.

Panwar, D., Panesar, P.S., Singla, G., Krishania, M., and Thakur, A. 2020. Recovery of nutrients and transformations of municipal/domestic food waste. In: *Waste Valorisation: Waste Streams in a Circular Economy*, eds. C.S.K. Lin, G. Kaur, C. Li, X. Yang and C.V. Stevens, pp. 109–159. John Wiley & Sons.

Patsea, M., Stefou, I., Grigorakis, S., and Makris, D.P. 2017. Screening of natural sodium acetate-based low-transition temperature mixtures (LTTMs) for enhanced extraction of antioxidants and pigments from red vinification solid wastes. *Environmental Processes* 4(1): 123–135.

Petkowicz, C.L. and Williams, P.A. 2020. Pectins from food waste: Characterization and functional properties of a pectin extracted from broccoli stalk. *Food Hydrocolloids* 107. 105930.

Prabhuzantye, T., Khaire, R.A., and Gogate, P.R. 2019. Enhancing the recovery of whey proteins based on application of ultrasound in ultrafiltration and spray drying. *Ultrasonics Sonochemistry* 55: 125–134.

Rabelo, R.S., Machado, M.T.C., Martínez, J., and Hubinger, M.D. 2016. Ultrasound assisted extraction and Nano filtration of phenolic compounds from artichoke solid wastes. *Journal of Food Engineering* 178: 170–180.

Rahimi, S. and Mikani, M. 2019. Lycopene green ultrasound-assisted extraction using edible oil accompany with response surface methodology (RSM) optimization performance: Application in tomato processing wastes. *Microchemical Journal* 146: 1033–1042.

Ran, X.-G. and Wang, L.-Y. 2013. Use of ultrasonic and pepsin treatment in tandem for collagen extraction from meat industry by-products. *Journal of the Science of Food and Agriculture* 94(3): 585–590.

Rutkowska, M., Namieśnik, J., and Konieczka, P. 2017. Ultrasound-assisted extraction. In: *The Application of Green Solvents in Separation Processes*, eds. F.P. Pereira and M. Tobiszewski, pp. 301–324. Elsevier.

Safdar, M.N., Kausar, T., Jabbar, S., Mumtaz, A., Ahad, K., and Saddozai, A.A. 2017. Extraction and quantification of polyphenols from kinnow (*Citrus reticulate* L.) peel using ultrasound and maceration techniques. *Journal of Food and Drug Analysis* 25(3): 488–500.

Şahin, S. and Şamlı, R. 2013. Optimization of olive leaf extract obtained by ultrasound-assisted extraction with response surface methodology. *Ultrasonics Sonochemistry* 20(1): 595–602.

Saini, A. and Panesar, P.S. 2020. Beneficiation of food processing by-products through extraction of bioactive compounds using neoteric solvents. *LWT-Food Science and Technology* 134. 110263.

Saini, A., Panesar, P.S., and Bera, M.B. 2019a. Valorization of fruits and vegetables waste through green extraction of bioactive compounds and their nanoemulsions-based delivery system. *Bioresources and Bioprocessing* 6(1): 26.

Saini, A., Panesar, P.S., and Bera, M.B. 2019b. Comparative study on the extraction and quantification of polyphenols from citrus peels using maceration and ultrasonic technique. *Current Research in Nutrition and Food Science* 7(3): 678–685.

Saini, A., Panesar, P.S., and Bera, M.B. 2021. Valuation of *Citrus reticulata* (kinnow) peel for the extraction of lutein using ultrasonication technique. *Biomass Conversion and Biorefinery* 11(5): 2157–2165.

Saini, A., Panwar, D., Panesar, P.S., and Bera, M.B.2019c. Bioactive compounds from cereal and pulse processing by products and their potential health benefits. *Austin Journal of Nutrition & Metabolism* 6: 1068.

Samaram, S., Mirhosseini, H., Tan, C.P., Ghazali, H.M., Bordbar, S., and Serjouie, A. 2015. Optimisation of ultrasound-assisted extraction of oil from papaya seed by response surface methodology: Oil recovery, radical scavenging antioxidant activity, and oxidation stability. *Food Chemistry* 172: 7–17.

Sengar, A.S., Rawson, A., Muthiah, M., and Kalakandan, S.K. 2020. Comparison of different ultrasound assisted extraction techniques for pectin from tomato processing waste. *Ultrasonics Sonochemistry* 61: 104812.

Sharmila, G., Nikitha, V.S., Ilaiyarasi, S., Dhivya, K., Rajasekar, V., Kumar, N.M., Muthukumaran, K., and Muthukumaran, C. 2016. Ultrasound assisted extraction of total phenolics from *Cassia auriculata* leaves and evaluation of its antioxidant activities. *Industrial Crops and Products* 84: 13–21.

Shirsath, S., Sonawane, S., and Gogate, P. 2012. Intensification of extraction of natural products using ultrasonic irradiations—A review of current status. *Chemical Engineering and Processing: Process Intensification* 53: 10–23.

Shirsath, S.R., Sable, S.S., Gaikwad, S.G., Sonawane, S.H., Saini, D.R., and Gogate, P.R. 2017. Intensification of extraction of curcumin from *Curcuma amada* using ultrasound assisted approach: Effect of different operating parameters. *Ultrasonics Sonochemistry* 38: 37–445.

Shivamathi, C.S., Moorthy, I.G., Kumar, R.V., Soosai, M.R., Maran, J.P., Kumar, R.S., and Varalakshmi, P. 2019. Optimization of ultrasound assisted extraction of pectin from custard apple peel: Potential and new source. *Carbohydrate Polymers* 225: 115240.

Singanusong, R., Nipornram, S., Tochampa, W., and Rattanatraiwong, P. 2015. Low power ultrasound-assisted extraction of phenolic compounds from mandarin (*Citrus reticulata* Blanco cv. Sainampueng) and lime (*Citrus aurantifolia*) peels and the antioxidant. *Food Analytical Methods* 8(5): 1112–1123.

Singh, B., Singh, N., Thakur, S., and Kaur, A. 2017. Ultrasound assisted extraction of polyphenols and their distribution in whole mung bean, hull and cotyledon. *Journal of Food Science and Technology* 54(4): 921–932.

Siso, M.G. 1996. The biotechnological utilization of cheese whey: A review. *Bioresource Technology* 57(1): 1–11.

Smuda, S.S., Mohsen, S.M., Olsen, K., and Aly, M.H. 2018. Bioactive compounds and antioxidant activities of some cereal milling by-products. *Journal of Food Science and Technology* 55(3): 1134–1142.

Soong, Y.Y. and Barlow, P.J. 2004. Antioxidant activity and phenolic content of selected fruit seeds. *Food Chemistry* 88(3): 411–417.

Souza, M.M., Silva, B.D., Costa, C.S., and Badiale-Furlong, E. 2018. Free phenolic compounds extraction from Brazilian halophytes, soybean and rice bran by ultrasound-assisted and orbital shaker methods. *Anais da Academia Brasileira de Ciências* 90(4): 3363–3372.

Stanisavljević, I.T., Lazić, M.L., and Veljković, V.B. 2007. Ultrasonic extraction of oil from tobacco (*Nicotiana tabacum* L.) seeds. *Ultrasonics Sonochemistry* 14(5): 646–652.

Studdert, V., Gay, C., and Blood, D. 2011. *Saunders Comprehensive Veterinary Dictionary*. Elsevier Health Sciences.

Suslick, K.S. and Price, G.J. 1999. Applications of ultrasound to materials chemistry. *Annual Review of Materials Science* 29(1): 295–326.

Suslick, K.S., Casadonte, D.J., Green, M.L.H., and Thompson, M.E. 1987. Effects of high intensity ultrasound on inorganic solids. *Ultrasonics* 25(1): 56–59.

Tabaraki, R. and Nateghi, A. 2011. Optimization of ultrasonic-assisted extraction of natural antioxidants from rice bran using response surface methodology. *Ultrasonics Sonochemistry* 18(6): 1279–1286.

Tiwari, B.K. 2015. Ultrasound: A clean, green extraction technology. *TrAC Trends in Analytical Chemistry* 71: 100–109.

Tu, Z.C., Huang, T., Wang, H., Sha, X.M., Shi, Y., Huang, X.Q., Man, Z.Z., and Li, D.J. 2015. Physico-chemical properties of gelatin from bighead carp (*Hypophthalmichthys nobilis*) scales by ultrasound-assisted extraction. *Journal of Food Science and Technology* 52(4): 2166–2174.

Ullah, Z., Man, Z., Khan, A.S., Muhammad, N., Mahmood, H., Ghanem, O.B. et al. 2019. Extraction of valuable chemicals from sustainable rice husk waste using ultrasonic assisted ionic liquids technology. *Journal of Cleaner Production* 220: 620–629.

Van Hung, P., Yen Nhi, N.H., Ting, L.Y., and Lan Phi, N.T. 2020. Chemical composition and biological activities of extracts from pomelo peel by-products under enzyme and ultrasound-assisted extractions. *Journal of Chemistry*. https://doi.org/10.1155/2020/1043251.

Wang, J., Sun, B., Cao, Y., Tian, Y., and Li, X. 2008. Optimisation of ultrasound-assisted extraction of phenolic compounds from wheat bran. *Food Chemistry* 106(2): 804–810.

Wang, J., Sun, B., Liu, Y., and Zhang, H. 2014. Optimisation of ultrasound-assisted enzymatic extraction of arabinoxylan from wheat bran. *Food Chemistry* 150: 482–488.

Wang, W., Ma, X., Xu, Y., Cao, Y., Jiang, Z., Ding, T., Ye, X., and Liu, D. 2015. Ultrasound-assisted heating extraction of pectin from grapefruit peel: Optimization and comparison with the conventional method. *Food Chemistry* 178: 106–114.

Wang, Y.C., Chuang, Y.C., and Hsu, H.W. 2008. The flavonoid, carotenoid and pectin content in peels of citrus cultivated in Taiwan. *Food Chemistry* 106(1): 277–284.

Wei, M., Zhao, R., Peng, X., Feng, C., Gu, H., and Yang, L. 2020. Ultrasound-assisted extraction of taxifolin, diosmin, and quercetin from *abiesnephrolepis* (Trautv.) maxim: Kinetic and thermodynamic characteristics. *Molecules* 25(6): 1401.

Wen, C., Zhang, J., Zhang, H., Dzah, C.S., Zandile, M., Duan, Y. et al. 2018. Advances in ultrasound assisted extraction of bioactive compounds from cash crops–A review. *Ultrasonics Sonochemistry* 48: 538–549.

Wu, J. and Nyborg, W.L. 2008. Ultrasound, cavitation bubbles and their interaction with cells. *Advanced Drug Delivery Reviews* 60(10): 1103–1116.

Wu, T., Guan, Y., and Ye, J. 2007. Determination of flavonoids and ascorbic acid in grapefruit peel and juice by capillary electrophoresis with electrochemical detection. *Food Chemistry* 100(4): 1573–1579.

Xu, M., Ran, L., Chen, N., Fan, X., Ren, D., and Yi, L. 2019. Polarity-dependent extraction of flavonoids from citrus peel waste using a tailor-made deep eutectic solvent. *Food Chemistry* 297: 124970.

Yang, Y., Wang, Z., Hu, D., Xiao, K., and Wu, J.Y. 2018. Efficient extraction of pectin from sisal waste by combined enzymatic and ultrasonic process. *Food Hydrocolloids* 79: 189–196.

Yilmaz, E. and Soylak, M. 2015. Ultrasound assisted-deep eutectic solvent extraction of iron from sheep, bovine and chicken liver samples. *Talanta* 136: 170–173.

Yu, H., Wang, C., Deng, S., and Bi, Y. 2017. Optimization of ultrasonic-assisted extraction and UPLC-TOF/MS analysis of limonoids from lemon seed. *LWT-Food Science and Technology* 84: 135–142.

Zhang, H., Tang, B., and Row, K.H. 2014. A green deep eutectic solvent-based ultrasound-assisted method to extract astaxanthin from shrimp by products. *Analytical Letters* 47(5): 742–749.

Zhao, G., Zhang, R., Dong, L., Huang, F., Liu, L., Deng, Y. et al. 2018. A comparison of the chemical composition, in vitro bioaccessibility and antioxidant activity of phenolic compounds from rice bran and its dietary fibres. *Molecules* 23(1): E202.

Zhou, Y., Xu, X.Y., Gan, R.Y., Zheng, J., Li, Y., Zhang, J.J. et al. 2019. Optimization of ultrasound-assisted extraction of antioxidant polyphenols from the seed coats of red sword bean (*Canavalia gladiate* (Jacq.) DC.). *Antioxidants* 8(7): 200.

Zhou, Y., Zheng, J., Gan, R.Y., Zhou, T., Xu, D.P., and Li, H.B. 2017. Optimization of ultrasound-assisted extraction of antioxidants from the mung bean coat. *Molecules* 22(4): 638.

Zou, Y., Li, P.P., Zhang, K., Wang, L., Zhang, M.H., Sun, Z.L. et al. 2017. Effects of ultrasound-assisted alkaline extraction on the physiochemical and functional characteristics of chicken liver protein isolate. *Poultry Science* 96(8): 2975–2985.

Zou, Y., Yang, H., Zhang, X., Xu, P., Jiang, D., Zhang, M. et al. 2020. Effect of ultrasound power on extraction kinetic model, and physicochemical and structural characteristics of collagen from chicken lung. *Food Production, Processing and Nutrition* 2(1): 3.

5 Adding Values to Agro-Industrial Byproducts for the Bioeconomy in Vietnam

Son Chu-Ky, Nguyen-Thanh Vu, Quyet-Tien Phi,
Tuan Pham Anh, Kim-Anh To, Le-Ha Quan,
Tien-Thanh Nguyen, Hong-Nga Luong, Thu-Trang Vu,
Tien-Cuong Nguyen, Tuan-Anh Pham, Thanh-Ha Le,
Ngoc Tung Quach, and Chinh-Nghia Nguyen

CONTENTS

5.1 INTRODUCTION

A bioeconomy based on the effective use of biomass was affirmed by the World Economic Forum in 2010 as a mandatory strategic development orientation to ensure energy security and sustainable development while protecting the environment. A biorefinery is understood as the cascading use of biomass and an important component of the bioeconomy. The main goal of a biorefinery is to exploit biomass resource efficiency through the full use of resources. It also includes process redesign to minimize bioresource use and valorize the waste into useful products.

DOI: 10.1201/9781003125679-5

A biorefinery is an opportunity, but it is also a responsibility for countries with abundant biomass resources, particularly developing countries. In Asia, India and China, two countries with large biomass resources, have made spectacular transitions and ranked on the list of countries investing in biorefinery. Rich in biomass resources from agriculture, food production, and forestry, Vietnam has great opportunities for biomass technology development, simultaneously using resources for energy production and sustainable development. Efficient use of biomass will help Vietnam become less dependent on conventional energy sources, reduce carbon emissions and environmental pollution, and increase income for farmers participating in the biomass value chain (Um and Hanley, 2008). Despite the abundance of biomass in the country, industrial crop byproducts are the most important sources of biomass, thanks to the crop residue that remains in the field after harvest and the more important wastes released from subsequent industrial processing.

This chapter will discuss the uses of byproducts from industrial processing lines in Vietnam.

5.2 SHRIMP-BASED BYPRODUCTS

5.2.1 CURRENT STATUS AND CHALLENGES OF THE SHRIMP INDUSTRY

Global food processing produces more than 20 million tonnes of shrimp waste annually, mainly consisting of the head and shell. Shrimp waste is generally one of the biggest sources of chitin, with a yield of 15–40% of the dry shrimp shell weight. Current technologies for treating shrimp waste to recover chitin mainly depend on chemicals and available existing equipment, which leads to pollution in major shrimp processing countries such as India, Thailand, Bangladesh, and Vietnam.

Vietnam is a major world shrimp exporter. The total shrimp production in Vietnam in 2017 was 700 million tonnes and reached 800 million tonnes in 2019 (Nguyen, Nguyen, and Jolly, 2019), achieving a 7% annual growth. Vietnam ranked the second leading shrimp exporter globally after China (Anderson, Valderrama, and Jory, 2019). Shrimp intensification farming applied with agriculture innovation is strongly encouraged for the UN sustainable development goals. It is the economic driver towards rural poverty mitigation and intensively contributes to exports. The shrimp farmed in the country takes place on a total surface of 700,000 hectares, mainly located in the Mekong Delta, namely in the Ca Mau, Bac Lieu, Soc Trang, Ben Tre, and Kien Giang provinces, which account for approximately 89% of shrimp produced in Vietnam (Nguyen, Nguyen, and Jolly, 2019). Shrimp processing releases a huge amount of solid residue, accounting for 48–56% of whole shrimp cultured (Sachindra, Bhaskar, and Mahendrakar, 2006). This residue is an attractive source for recovering valuable components, mainly chitin, protein, and carotenoids.

Generally, shrimp solid waste is mainly composed of the cephalothorax, carapace, and tail, representing 48–56% of the whole shrimp. This by-product contains significant valuable components, including protein (35–50 wt%), chitin (15–25 wt%), minerals (10–15 wt%) on a dry basis, and carotenoids (Sachindra, Bhaskar, and Mahendrakar, 2006). Hence, the residue from shrimp production is not a by-product but a valuable feedstock. Many studies have been carried out in order to optimize the recovery of these components.

To remove calcium carbonate and protein in the shrimp shell, technologies often use high concentrations of hydrochloric acid (HCl) and sodium hydroxide (NaOH), in which the recovery of 1 tonne of chitin produces more than 400 m^3 of wastewater with a high ammonia content and salinity leading to difficulties in wastewater management. In recent innovative technologies, biotechnology-based processes, such as fermentation, and enzyme-catalyzed in combination with fermentation is preferred as it reduces chemical pollution and is expected to lead to zero-waste treatment. By using sustainable processes, calcium, astaxanthin, lipoprotein, polypeptides, and amino acids can be recovered along with chitin and protein.

The scheme for recovering the value from shrimp waste is illustrated in Figure 5.1.

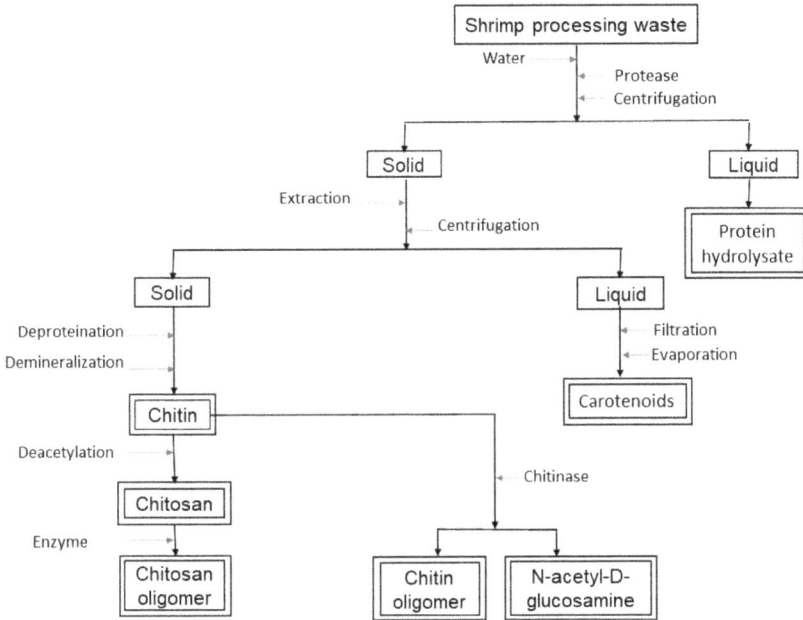

FIGURE 5.1 Biorefinery scheme for shrimp waste to recover chitin, protein, and carotenoids (adapted from Synowiecki and Al-Khateeb, 2000).

5.2.2 Process Sections of the Shrimp Biorefinery and Its Topology

5.2.3 Development of a Biorefinery Approach for Shrimp Processing in Vietnam

In the seafood processing industry, the proportion of frozen crustacean products for export accounts for 70–80% of the processing capacity. Annually, processing plants discharge a large amount of shrimp waste residue of over 300,000 tonnes. According to the General Department of Fisheries, in 2017, the total output of shrimp in Vietnam reached over 600,000 tonnes. Consumption of large amounts of shrimp raw materials from seafood processing factories has resulted in large quantities of byproducts, including heads, shells, and tails. In addition, there are broken shrimp bodies, shrimp that have been peeled wrong, or shrimp with colour changes. Depending on the processing method and the species, the amount of by-product from *Penaeus monodon* is 40% and from *Macrobrachium rosenbergii* is 60% of the raw material purchased. For peeled shrimp and pulled back shrimp, the tail and shell of the shrimp account for about 25% of the shrimp's weight. For white leg shrimp, the shrimp head makes up 28% of the body and the shell weight is 9%, so the total amount of shrimp by-product is 37% of the entire shrimp. Many studies show that the proportion of shrimp waste ranges from 30% to 70% per body (Figure 5.2).

Previously, shrimp waste was mainly used as raw materials for processing animal feed, poultry, fertilizer, etc. Later, with more research, high-value products from shrimp waste, such as chitin and chitosan, have been made available through chemical methods and effective and safe methods like electrochemical and biological methods. In addition, during the recovery of chitin and chitosan, other valuable constituents, including carotenoids, proteins, and minerals (Ca, P, K, Mg, Mn, and Fe) in shrimp waste (especially shrimp heads) have been studied.

In Vietnam, the production of chitin and chitosan from shrimp byproducts has been growing relatively recently; however, the technology is still limited. Process studies as well as methodological improvements have been made on a large scale at trial but have not yet been widely transferred to an industrial scale. Current domestic production bases are mostly inorganic (demineralization with HCl solution, deacetylation and deproteinization with NaOH solution, and decolourization with

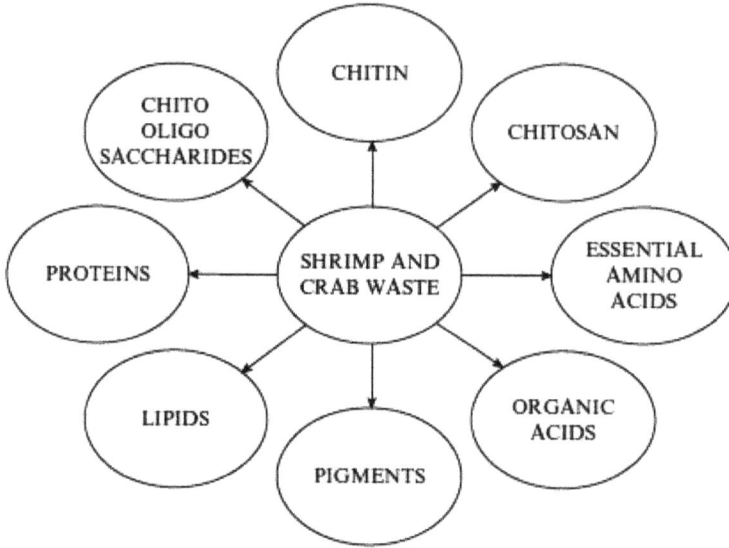

FIGURE 5.2 Value-added products that can be recovered from shrimp and crab waste.

NaCl or strong oxidizable compounds). Although chitin production at the industrial scale is mainly based on traditional chemical methods, at Vietnam Food Company, there are a number of processes to improve biochemical applications.

5.3 SOY MILK BY-PRODUCT

Soy milk residue (okara) is a by-product of the soy milk and tofu processing industry. Crude soybean meal is an insoluble white or yellow substance (depending on the type of soybean) obtained after filtration during soy milk production. Soybeans (*Glycine max*) have the highest protein content of all legumes (40–42%). They also contain 35–40% glucose and 18–20% fat. Each kilogram of soybeans made into soy milk or tofu will produce about 1.1 kg of bean residue, of which moisture is 76–80% (Liu 1997), moisture 70–90% (Urbanski and Lang, 2016). The relatively high water content creates favourable conditions for micro-organisms to cause rapid spoilage. Therefore, the use of bean residue in food processing is still limited; most of it is used as fertilizer and animal feed. However, the dry matter composition of soy milk residue contains about 27% protein, 10% fat, 42% insoluble fibre, and 12% soluble fibre (O'Toole, 1999). These ingredients can be added when processing food products but may result in unappealing taste due to the denatured protein content. There have been studies to improve the functional properties of soybean meal protein by fermentation and the use of hydrolytic enzymes (Urbanski and Lang, 2016; Kasai et al., 2004).

5.3.1 CHEMICAL COMPOSITION OF SOY MILK BYPRODUCTS

According to the publication by Li Bo et al. (2012), the chemical composition of okara depends on the soaking stage and milk extracted from soybeans. It also depends on soybean varieties and production methods but includes 46.3% fibre, 17.8% protein, 5.9% lipid, 3.9% ash, 2.6% reducing sugar, 0.22% flavone, and 67% moisture. Several other soy components are also likely to be present in soybean meal, including isoflavones, lignans, phytosterols, counestans, saponins, and phytates.

The composition of soy milk residue calculated on a dry basis is presented in Table 5.1 (Rashad et al., 2010) and of wet soy milk residue (Table 5.2) (Li et al. 2013).

The amino acid composition of wet soy milk residue protein is almost lower than that of amino acids in soybean seeds, except for some amino acids such as cysteine, threonine, and glycine (Puechkamut and Panyathitipong, 2012; Sarwar et al., 1983).

TABLE 5.1

Composition of Soy Milk Residue by Dry Matter

Nutrition composition	Fibre	Protein	Lipid	Carbohydrate	Ash
Percentage (% mass)	12.2	25.52	12	32.6	3.96

TABLE 5.2

Composition of Wet Soy Milk Residue

Moisture %	Protein %	Lipid %	Fibre %	Carbohydrate %	Ash %	References
81.0	4.8	3.6	3.3	6.4	0.8	Ohno, Ano, and Shoda, 1996
84.5	4.7	1.5	1.5	7.0	0.4	O'Toole, 1999
85.0	3.6	1.4	9.2	–	–	Zhu et al., 2008
84.5	4.2	1.5	3.52	5.78	0.55	Fafaungwithayakul, Hongsprabhas, and Hongsprabhas, 2011

5.3.2 UTILIZATION OF SOY MILK BYPRODUCTS

For every 1,000 litres of soy milk produced, an average of 250 kg of soybean residue is released into the environment. An estimated 14 million tonnes of soybean meal are generated worldwide every year. In Japan alone, nearly one million tonnes of okara is emitted annually from the tofu processing industry. In Vietnam, the amount of soybean residue generated by the soy milk industry is also increasing. With the total capacity of Vinasoy's factories reaching 390 million litres per year, it will generate more than 40,000 tonnes of soybean meal per year. In addition, soybean residues are also discharged from many other soy milk production companies and small tofu production facilities across the country. With an annual consumption of 2 million tonnes of soybeans (according to the Vietnam import and export report 2020;), it is estimated that 2.2 million tonnes of soybean residues are generated.

Okara is a nutritious product with a high protein, fatty oil, fibre, and mineral composition, along with monosaccharides and oligosaccharides. Due to its high water and protein content, okara is easily decomposed and rots naturally when not stored in proper conditions. Okara is perishable and has a shelf life of only one day. Unless the residue is frozen or dried, the bean residue's high moisture content and rich nutrients will produce an unpleasant rancid odour that reduces its use-value over time (Waliszewski, Pardio, and Carreon, 2002). Many Asian countries have found a use for okara as a source of nutrition in many food items such as soups, salads, baked goods, and fermented food products such as tempeh (Liu, 1997; O'Toole, 1999; Riaz, 2006). Many studies use okara in biological processes for food, medicine, and other products such as

- creating an aldose reductase inhibitor using *Aspergillus* sp. (Fujita, Funako, and Hayashi, 2004);
- cultivation of lingzhi, *Ganoderma lucidum* (Hsieh and Yang, 2004);
- producing β-fructofuranosidase (Hayashi et al., 1992);
- making food by using *Bacillus subtilis* to enhance the value of bean residue and its antioxidant activity (Zhu et al., 2008);
- a substrate for citric acid fermentation by *Aspergillus niger* sp. and *Aspergillus terreus* sp. (Khare, Jha, and Gandhi, 1995);
- fermentation to produce antioxidant activity with *Bacillus natto* (Hattori et al., 1995);
- using yeast to ferment soybean residue to improve the antioxidant capacity and some nutrients of raw materials (Rashad et al., 2011);

- biosynthesis of antibiotics in the form of lipopeptide and iturin A used in agriculture using recombinant *Bacillus subtills* sp. (Ohno, Ano, and Shoda, 1996);
- a medium for cellulose biosynthesis by solid fermentation of *Bacillus subtilis* (Shu-bin et al., 2012); and
- a protease to hydrolyze okara to produce dipeptides and tripeptides with antioxidant activity (Yokomizo, Takenaka, and Takenake, 2002).

It was reported in the literature (Urbanski and Lang 2016) that the addition of 10% and 20% dry okara to flour reduced the volume of the loaf by 18.7% and 32.2%, respectively, and increased the hardness from 0.24 ± 0.04 Pa (control sample, no okara) to 0.44 ± 0.07 Pa and 0.9 ± 0.05 Pa, respectively. Similarly, Bowles and Demiate (2006) investigated the addition of okara to French bread at 0%, 5%, 10%, and 15% and reported sensory analysis using the taste preference method. Bread with a lower okara content is preferred over bread containing 15% okara (Bowles and Demiate, 2006). Ana Aguado studied the use of soybean meal as a substitute for wheat gluten (Aguado, 2010).

While Asian countries have found many ways to use soybean meal with added value, most Western countries and the United States consider it agricultural processing waste of very little value (Riaz, 2006). The most common application is animal feed. To utilize the nutrients in okara, while reducing waste and being competitive, the soy milk and tofu industries must develop new products or ingredients. Therefore, enzymatic hydrolysis of okara's protein can be an alternative to obtain value-added products. Protein hydrolyzate is used to improve or modify the chemical, functional, and organoleptic properties of a protein without affecting its nutritional value (Gonzàlez-Tello et al., 1994). Hydrolyzed proteins, especially dipeptides and tripeptides, have greater nutritional value and are absorbed more efficiently by the gastrointestinal tract than pure proteins and free amino acids (Bhaskar et al., 2007). In addition, proteolysis can release bioactive peptides, defined as short chains of amino acids (between 2 and 20 units) that give physiological benefits to the organism. The biological activity of peptides can be described by their antibacterial, antitumor, immunomodulatory, antithrombotic, antioxidant, or antihypertensive properties (Clare and Swaisgood, 2000).

In a previous study, the antioxidant activities of soybean and okara hydrolysates obtained by fermentation with *Aspergillus oryzae* or by using proteolytic enzymes (Neutrase and Flavourzyme) were analyzed (Ngoc and Ha, 2017). 2,2-Diphenyl-1-picrylhydrazyl DPPH radical scavenging assay was used to determine the antioxidant activities of hydrolysates. The concentration of peptides required for DPPH free radical scavenging for IC50 value (50%) was used to evaluate the antioxidant activity of peptides obtained from hydrolysates. The results showed that when fermented with *A. oryzae*, the okara hydrolysate had higher antioxidant activity than the soybean hydrolysate, with IC50 values of 0.447 mg/ml and 3.95 mg/ml, respectively. The hydrolyzed okara obtained from hydrolysis using Neutrase had higher antioxidant activity than the one obtained from hydrolysis using Flavourzyme, with IC50 values of 0.200 and 0.407 mg/ml, respectively. Different peptide fractions obtained from the hydrolysates using a cut-off membrane (10, 3, and 1 kDa) possessed different antioxidant activities. The <1 kDa peptide fraction exhibited the highest antioxidant activity with an IC50 value of 0.158 mg/ml.

5.4 CEREAL, NUT, LEGUME AND STARCHY ROOT BYPRODUCTS

5.4.1 CEREAL BYPRODUCTS

Rice (*Oryza sativa*) is one of the main cereal crops and the most common staple foods for most of the world's population, especially in Asian countries (Bird et al., 2000). Approximately 600 million tonnes are harvested worldwide annually (Chen et al., 2012). Frequently, rice is eaten in cooked form by humans to obtain various nutrients as well as to supplement their caloric intake (Kim et al., 2011). The milling of paddy rice has a 70% yield of rice (endosperm) as its major product, although there are some unconsumed portions of the rice produced, such as the rice husk (20%), rice bran

(8%), and rice germ. Vietnam is the world's fifth-largest rice-producing country contributing to 7% of global export by value and 6% of its agricultural GDP.

By-products from the rice milling process have high amounts of nutrients when compared to white rice itself. Rice straw, rice hull, broken rice, rice germ, rice bran, rice bran oil, and wax are the by-products of the rice industry (Hoi et al., 2003). These by-products usually have basic applications in their original form but now can be used as raw materials for different value-added research or in food applications with functional properties. Most of the rice byproducts, including rice husk and rice bran, are used as animal feed. Brewer's rice, which is a mixture of broken rice, rice bran, and rice germ, is also used almost exclusively in the production of beer. In recent years, rice byproducts have received increased attention as functional foods due to their phenolic based compounds and high amounts of vitamins, minerals, and fibre, which can help lower cholesterol and have anti-atherogenic activity. The properties of the rice byproducts that might contribute to this effect include tocotrienol, γ-oryzanol, and fibre. The effects of rice byproducts on the reduction of diabetic risk have been shown by many studies. Black rice bran extracts reduce the progression of dietary cholesterol-induced atherosclerotic plaque development and cholesterol plasma levels in rabbits while decreasing serum triglycerides and total cholesterol levels in mice. Some components of rice bran oil that are mainly responsible for the cholesterol-lowering effects in rice bran oil's unsaponifiable fraction (4.2%) include tocols (tocopherols and tocotrienols), β-sitosterol, γ-oryzanol, and unsaturated fatty acids (Esa, Ling, and Peng, 2013).

5.4.2 Starchy Root Byproducts

5.4.2.1 Byproducts from Cassava Starch Processing

Global production of cassava (*Manihot esculenta* Crantz) has nearly doubled in the past 30 years, reaching about 290 million tonnes/year in 2014–2018. More than 50% of cassava is produced in Africa, while 30% is produced in Asia and 14% in Latin America. Cassava is a globally important industrial crop for the production of starch, modified starch, alternative sweeteners, and derivatives used in food, pharmaceuticals, paper, textile, and other industries, while it caters for foods in Africa. International trade in cassava is growing based on industrial demand for cassava starch, its derivatives, and bioethanol. There is a need to open a new market for cassava, develop substitutes for petroleum materials, and improve the overall value of the cassava processing industry in Africa.

Cassava in Vietnam has made spectacular development from a side crop to a key industrial commodity, Vietnam is the second-largest cassava starch exporter with total exports of US$2 million in 2017 Being the third-largest crop per area after rice and corn, the cultivation of cassava in Vietnam has remained unchanged at 550,000 hectares, with a current yield of 20–40 tonnes/hectare; 10 million tonnes of cassava was harvested in 2018. This year (2021), Vietnam exported 1.75 million tonnes of cassava starch and 0.66 million tonnes of cassava chips, equivalent to 72.5% and 7.51% of total national cassava production (Carriere and Inglett, 2003). During the harvest and processing into starch, a huge amount of solid residues is discharged on the field and in the cassava starch factories: cassava leaves (6–30% of the cassava root), cassava stalks and cassava peel (42% of the cassava root), and cassava bagasse (1.4 tonnes) (Huong et al., 2018). In addition, there is a large volume of effluent that contains mainly starch but is suitable for biogas production to create energy for the factory (Schirmer, 2014) (Figure 5.3).

5.4.2.2 Byproducts from Sweet Potato

Sweet potato plays an important role in the development of agricultural economy. Sweet potato is one of the most widely grown food crops in the world. It is cultivated in about 111 countries. In Vietnam, sweet potato production in 2019 reached 1.4 million tonnes. The processing of the root tuber into starch, alcoholic beverages, ethanol, flour, purees, and other commodities generates byproducts that currently have very little use and normally add to the waste going into landfills and bodies of water. Other important byproducts of sweet potato production are vines and leaves from

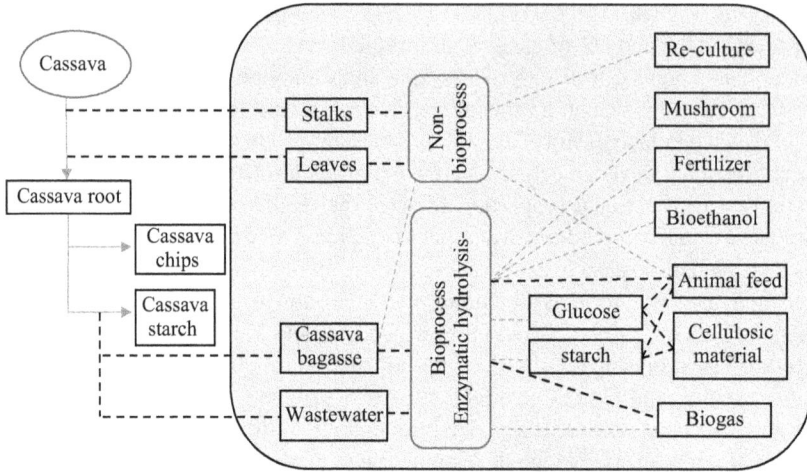

FIGURE 5.3   Schematic diagram of cassava waste stream.

the agricultural phase. Vines and leaves are a good source of protein, vitamins, minerals, pigments, and polyphenols and have the potential to be used for human consumption after cooking. Also, the leaves and vines contain important levels of polyphenols and carotenoids that could be recovered and used as antioxidants and colourants, respectively. However, except for small-scale farms that use this by-product to feed farm animals, vines and leaves have very little application. Sweet potato peels produced from industrial processes are responsible for an enormous amount of waste. It has been estimated that approximately 30% of the raw material coming into sweet potato canning operations is discarded and not used for human consumption. If they are used at all, the peels are typically used for either fertilizer or animal food. The total protein content of sweet potato is about 5% of the dry weight, and the predominant soluble protein is sporamin. Sweet potato protein exhibits good solubility and emulsifying properties and thus is a candidate as a functional ingredient in processed food applications. One use for some of the wastewater from sweet potato alcoholic fermentation is to manufacture fibrous paper pots that can be used to grow seedlings. An important volume of research has been conducted in the last few decades about the utilization of sweet potato byproducts, which indicates that sweet potato byproducts have the potential for the recovery of functional proteins, phenolic compounds, soluble fibre, anthocyanins, and carotenoids. Due to its carbohydrate-rich nature, sweet potato byproducts could be used as a substrate for ethanol and lactic acid production or biogas by anaerobic digestion.

Potato tubers contain 20% starch and other substances such as sugars, nitrogenous compounds, mineral salts, organic acids, etc.). Traces of these substances are retained in the pulp during starch processing. The remaining portion, together with water, is transferred to the wastewater. Sweet potato wastewater causes problems with environmental pollution. There are two methods to utilize wastewater from sweet potatoes.

5.4.3   LEGUME BYPRODUCTS

Legumes include various kinds of beans, such as soybean, mung bean, black bean, kidney bean, etc. Byproducts that are either a direct or indirect result of legume processing or utilization have economic as well as nutritional importance. The main byproducts that come directly from legume processing are husks, obtained by milling and decortication, and starch, obtained by protein extraction from air classification or slurry centrifugation. Those byproducts are obtained indirectly from parts of the legume plant. Vines, stems, straw, leaves, and pods are part of typical leguminous crop plants. They are eaten either by man or animals. Several legumes' seed coats or husks contain a relatively low amount of protein (7–11%) and a high amount of fibre (20–32%). The nutritional importance of husks as a ruminant feed source is apparent. The husk and pods of pigeon pea (*C. cajan*) are fed to cattle. In addition, the milling of pigeon pea for the preparation of dhal provides as much as 15% of the total yield for cattle feed. The husk is a major portion of this, with dhal powder and small broken peas, contributing the remainder. Also, byproducts from milling chickpea (*C. arietinum*) are fed to livestock. The various parts of leguminous plants can provide an additional source of food and feed. Accordingly, dried vines from chickpea plants are used as cattle feed and green parts of pigeon pea and black gram (*P. rrrungo*) plants provide excellent fodder, while broad beans and cowpeas are also grown for hay and silage. The roots of cowpeas are eaten in the same way as sweet potatoes in East Africa. Stalks from pigeon pea plants are used for various purposes, including roofing and basket-making. In addition, old woody parts of this plant are used for fuel, and charcoal is made from the thick stem.

Soybeans are normally processed to obtain protein isolates or other end-products such as soymilk and tofu. During this process, a fibre-rich residue called okara is removed. Large quantities of byproducts generated during plant food processing create an economic and environmental problem due to their high volume and elimination costs. Today, they are considered a promising source of functional compounds. After isolation of the main constituent, there are abundant remains that are an inexpensive material that has been undervalued until now, as it was only used as fuel or fertilizer. During the extraction of oil from soybeans, a substantial number of byproducts are produced. Specifically, these byproducts include soybean gums and soapstocks. These byproducts become potentially valuable to the manufacturer when they are efficiently recovered and processed.

5.4.4 Nut Byproducts

The cashew nut is among the most important agro-industrial crops in India, Brazil, Vietnam, and African countries. The global production of cashew nuts in 2005 was 3.21 million tonnes, which increased to 4.44 million tonnes in 2013. The cashew nut contains kernels (35–45%) and shells (55–65%) of the whole nut. The honeycomb structure of the cashew nut shell comprises a dark reddish-brown liquid (15–30%) called cashew nut shell liquid (CNSL). Cashew nut as a whole contains beneficial fatty acids, biologically active amino acids, selenium, phytosterols, tocopherols, high starchy content, alkyl-phenols, and a polysaccharide profile that is industrially and nutritionally important. Byproducts from cashew nut processing were reported to be used in various applications such as bioethanol production, biodiesel production, surface coating agent, dye, pesticide, larvicide, anti-termite products, adhesive resins, and rubber. While there are a large number of studies on the utilization of various agro-industrial residues through biotechnological interventions, there is a scarcity of coalesced data relevant to cashew nut waste generation and its diverse applications. The cashew is found in two parts: the nut and the peduncle. The nut is of greater economic interest. The cashew nut is a dry fruit with high economic value and is enjoyed as a culinary item. The peduncle, i.e., the cashew apple, is a pseudo-fruit with a short shelf life, so it is generally consumed in local areas as fresh fruit. The cashew apple is widely accepted by people due to its mild astringent flavour. Approximately 85% of the annual production of cashew apples is wasted, which is around 1 million tonnes. Cashew apples were used to prepare an array of products, including cashew apple juice, jam, jellies, ice creams, and other products prepared in laboratories like burgers, pastries, cakes, cereal bars, etc. The cashew nut is covered with a thin antioxidant-rich layer of reddish-brown colour, known as the testa. The testa constitutes about 1–3% of the total cashew nut weight It is an excellent source of hydrolysable tannins such as catechin, epicatechin, epigallocatechin, and polymeric proanthocyanidins. It is comprised of phenolic acids like syringic acid, gallic acid, and p-coumaric acids as the major constituents. The concentration of catechin and epicatechin were found as 5.70 and 4.46 g/kg of dry matter, respectively (Viswanath et al., 2016). Approximately 25% tannin is present in the outer skin of the cashew nut, possessing similar properties as wattle bark used in the leather industry (Viswanath et al., 2016). CNSL is a viscous greenish-yellow to reddish-brown colour liquid found in the soft honeycomb of the cashew nut shell, a pericarp fluid of the cashew nut. It is a renewable product produced as a by-product of the cashew nut processing industry. CNSL constitutes about one-third of the total nut weight or 30–35% of the nut shell weight, depending upon the extraction method. Other value-added uses of cashew nut byproducts include biofuel production and stabilization, coating and laminates, agriculture support (pesticides, larvicides, anti-termite properties), and production of biocomposites (adhesives, resins, and rubber).

5.5   DRIED DISTILLERS GRAINS FROM THE ETHANOL INDUSTRY

The ethanol industry is one of the fastest-growing industries in the world. Ethanol is used in many fields such as biofuel, medicine, food, and many other industries. There are several potential byproducts in ethanol distillation from starchy materials such as CO_2, distiller grain, and thin stillage. Distiller grain is a by-product after the distillation. The wet distillers grains (WDG) is the wet solid (with a moisture content of about 70–89%) obtained after centrifuging the whole stillage. The WDG can be dried to a moisture level below 10%; then, it is called dried distillers grain (DDG). The concentrated thin stillage (supernatant from the centrifugation of the whole stillage) contains soluble ingredients and can be mixed with WDG before drying to obtain dried distillers grains with solubles (DDGS). It is estimated that from 100 kg of corn material, 40.2 l of ethanol, 32.3 kg of DDGS, and 32.3 kg of CO_2 are produced (Schingoethe, 2006).

Total alcohol production worldwide has increased in recent years. In 2019, production was 109.8 billion litres, in which the United States and Brazil were the two largest alcohol-producing countries (59.7 billion litres and 32.4 billion litres, respectively) (RFA, 2019). Overall, about 87 million tonnes of DDGS were produced, including 50 million tonnes from the US. In 2019, the US export volume of DDGS reached 10 million tonnes, mainly for animal feed, of which 35% was imported by southeast Asian countries (about 3.5 million tonnes). The main importing countries of US DDGS were Mexico (about 1.8 million tonnes), South Korea and Vietnam (about 1.2 million tonnes each), Indonesia (0.91 million tonnes), Thailand (0.87 million tonnes), and Turkey, Japan, Canada, Philippines, and New Zealand (in total about 2 million tonnes of DDGS)

The proximate composition of DDGS is influenced by the type of starchy materials and production technology. Most of the starch is converted by yeast into ethanol and CO_2. After the distillation process, the percentage of nutrients present in the DDGS, such as proteins, fats, crude fibre, ash, etc., mostly increases about two to three times compared to the original materials. Protein in corn or wheat DDGS accounts for about 30–40% (Liu 2011), including protein yeasts (up to 20% of total DDGS). The residual starch accounts for about 5.1–6.2% of the DDGS dry matter. Residual starch in the DDGS consists of amylopectin encapsulated in the protein and cellulose network of the cell wall, which are resistant to hydrolysis by amylases (Sajilata, Singhal, and Kulkarni, 2006). However, most of the residual starch of DDGS is short-chain amylose helical complexes with lipids that occur immediately after starch gelatinization (Srichuwong and Jane, 2011). Fibre is also the main component in DDGS and accounts for about 40–60% of the dry matter of DDGS (Liu 2011). The fibre of DDGS comes from materials and β-glucan of yeast.

Thanks to the residual nutrients, DDGS is mainly used for animal feed for livestock, poultry, and aquaculture, with the mixing ratio in the formula varying depending on the animal.

In order to valorize the value of DDGS, especially its protein and fibre sources, many studies have successfully applied this by-product to a number of food products with the maximum recommended dose of DDGS to ensure acceptable sensory perception for consumers; such products include sugar cookies (15%), blended foods corn-soy-milk (2.5%); spaghetti (max 10%), pita bread (max 15%), extruded snacks (max 20%), high energy biscuit (max 50%), and Chinese steam bread (max 15%) (Mohammadi Shad, Venkitasamy, and Wen, 2021).

DDGS is also a rich source of carbon and nitrogen, which can be pretreated by soaking the DDGS in aqueous ammonia and subsequent enzymatic/dilute acid hydrolysis, ammonia fibre expansion, dilute acid or steam explosion to increase the digestibility and availability for micro-organisms.

Many recent studies have used DDGS as a substrate source for microbial fermentation to produce alcohol, butanol, organic acids, acetone, methanol, hydrogen, and many other value-added products (Iram, Cekmecelioglu, and Demirci, 2020).

The source of fibre in DDGS was also converted into xylo-oligosaccharides and used as a prebiotic (Mohammadi Shad, Venkitasamy, and Wen, 2021).

The enrichment of protein in DDGS has also been considered. This enrichment is based on the properties of the protein fraction of DDGS or raw materials. DDGS is initially defatted with hexane. Protein in defatted DDGS can be extracted by various solvents, including NaOH, ethanol, and reducing agents (Cookman and Glatz, 2009). Soluble proteins were precipitated by pH adjustment. For DDGS from corn or wheat, a solvent combining 0.5 M NaOH, 50% ethanol, and $NaHSO_3$ (reducing agent) result in the best protein extraction efficiency, which can increase the percentage of protein in the original DDGS from 30–40% to 60–70% (Chatzifragkou et al., 2015; Cookman and Glatz, 2009; Bandara, Chen, and Wu, 2011). In addition, protease enzyme can be used to obtain products containing fewer non-protein components (Bandara, Chen, and Wu, 2011; Cookman and Glatz, 2009).

In Vietnam, the ethanol production capacity in 2015 was 20 million litres/year and 100 million litres/year for beverage and biofuel, respectively; rice and cassava are common materials used for potable and biofuel ethanol, respectively. Distiller grain, mainly DDG, is also generated with relatively large output.

The proximate composition of rice-based DDG and cassava-based DDG from a few ethanol factories in Vietnam were analyzed (Taranu et al., 2019). Interestingly, rice-based DDG contains a high percentage of protein, up to 70.4% (in dry matter). Meanwhile, in cassava-based DDG, protein accounts for 12% and fibre 33% of dry matter of DDG (Taranu et al., 2019).

In fact, these DDG byproducts are mostly used for animal feed purposes despite the high protein content. This is a valuable source of vegetative protein, which can be fully utilized to improve the protein content of food products.

In our recent study, protein (accounting for 78% dry matter) in rice-based DDG from an ethanol factory was directly extracted. The solvent combination of 0.5 M NaOH and 0.05 M $NaHSO_3$ has been shown to be the best for protein extraction. However, this did not enrich the protein in the initial DDG (protein content was lowered to 65% due to the extraction of the non-protein component). In contrast, the use of NaOH 10 mM has increased protein content up to 87% of dry matter by washing out the non-protein components such as starch from rice-based DDG. Protein-enriched DDG was used up to 15% as an ingredient for cookies without negatively affecting the taste or colour of this product (Nguyen, Nguyen, and Chu-Ky, 2021).

In another study, using the same rice-based DDG, the extraction and hydrolysis of protein from this source were investigated to convert this protein source into digestible forms such as peptides and amino acids for food application. The results showed that 0.5 M NaOH solution combined with 0.05 M $NaHSO_3$ yielded an extraction of 78% after 2 h at 70°C. The hydrolysis of extracted protein was performed at 60°C, pH 7.5 for 4 h with enzyme concentration of 0.8% dry matter of DDG of Alcalase 2.4 l (Novozymes) and resulted mostly in low molecular products (<10 kDa) with a recovery yield of 89.79%. In this digestible form, the potential application of rice-based DDG protein in food production is therefore advanced (Do et al., 2021).

The protein in cassava-based DDG was improved through solid-state fermentation (SSF) using *Trichoderma harzianum* and *Yarrowia lipolytica* W29, and was conducted in 8 and 5 days of fermentation, respectively. Under optimal conditions, the crude protein of cassava DDG fermented by mould and yeast increased from 11.84% Dry matter (DM) for the unfermented sample to 15.29% and 14.06% DM, respectively. The in vitro protein digestibility, which allows us to estimate the protein availability for intestinal absorption after digestion, was improved significantly from 82.5% to 89.2% and 86.9% for *T. harzianum* and *Y. lipolytica* fermentation. Besides increasing the nutritional value, the SSF showed a clear effect in reducing the cyanide content of raw cassava DDG from 62.3 to 24.3 to 53.6 mg/kg DM. The results indicated that the protein enrichment of this bioethanol by-product using mould and yeast fermentation could be very promising as a cheap and abundant source of essential amino acids for animal feed ingredients in Vietnam (Vuong et al., 2021).

5.6   SINGLE-CELL PROTEIN FROM BIOETHANOL VINASSE

Since the beginning of the 21st century, due to issues related to global climate changes, resource depletion, national energy security, and geopolitical instability, biofuel production has increased. According to the US Department of Energy, in 2019, the world production of bioethanol reached 109.9 billion litres (https://afdc.energy.gov/data/10331). The US and Brazil are the two dominant players, with global production shares of 54.4% and 29.5%, respectively. The vast majority of bioethanol in the US is produced from corn, while in Brazil, sugarcane is the main feedstock. The remaining 16.1% bioethanol share is derived from sugar beet, cassava, sweet sorghum, etc. Vinasse, the remaining liquid after distillation of a fermented mixture to obtain ethanol, is the main by-product of the ethanol industry. For each litre of bioethanol, 12–14 litres of vinasse are produced (Leme and Seabra, 2017). Thus, the annual global vinasse by-product is estimated at 1.3–1.5 billion metric tonnes. In our calculation, one cassava ethanol factory with an average capacity of 100 million litres per year would produce 5000 m^3 of vinasse every day, which contains 400 tonnes of carbon, 2 tonnes of nitrogen, 1 tonne of phosphorous, and 5.5 tonnes of potassium. If not handled properly, the amount of vinasse produced would cause a big environmental issue.

Currently, corn vinasse is evaporated to produce DDGS and serves as an inexpensive protein additive in livestock diets. DDGS production accounts for 21.8% of the revenue of the bioethanol industry in the US. On the other hand, the nutritional value of vinasse from sugarcane ethanol is not as significant to justify the cost of evaporation (Rabelo, da Costa, and Vaz Rossel, 2015). Similarly, distillers dried grains obtained from cassava has a protein content of merely 12% (Taranu et al., 2019). These types of vinasse are often used for irrigation, supplying water and fertilizers to crops, or biogas production (Walter et al., 2011).

Considering vinasse's richness in organic matter, attempts have been made to use it for single-cell protein (SCP) production. The process would not only provide a cheap source of protein suitable to supplement animal diets but also reduce the organic load for subsequent wastewater treatment. A range of yeast and fungi strains have been tested for the biodegradation of sugarcane distillery wastewater and biomass accumulation (Table 5.3). Using a medium containing 75% sugarcane vinasse supplemented with an exogenous source of nitrogen and phosphorous, the fungus *Rhizopus oligosporus* reduced 80% of Chemical oxygen demand (COD) and produced biomass containing 50% protein (Nitayavardhana et al., 2013). The fungus *Rh. oligosporus* is widely used as a starter culture for tempeh production and is thus, considered safe. Oyster mushrooms *Pleurotus sajor caju*, *P. ostreatus*, *P. albidus*, and *P. flabellatus* were cultivated in sugarcane vinasse and used as a complementary diet for *Danio rerio* fish. The highest biomass obtained was 16.27 g/L-1 after 15 days of cultivation. When offered to the fish at a 2% (medium/body weight) rate for 28 days, fish survival was 100%, and no significant biomass loss was observed. As calculated by Nitayavardhana and Khanal (2010), a typical 25-million-gallon-per-year (94.6 million litres) ethanol plant could produce approximately 30,596 dry tonnes of fungal biomass and generate additional revenue of US$ 9.5 million annually. Although attempts at producing bacterial and algae SCP have been made (Pires et al., 2016; dos Santos et al., 2016), fungi and yeasts are considered the better candidates. Algae fermentation requires several weeks to complete; meanwhile, the pH of the vinasse needs to be adjusted to neutral for bacterial fermentation. Yeasts and fungi are also advantageous due to their relatively lower intracellular nucleic acid content compared to bacteria and thus a better fit as a nutrient additive for livestock. In terms of product recovery, filamentous fungi can be economically obtained by filtration; meanwhile, bacterial and yeast biomass requires more expensive centrifugation hardware. For the production of fungal biomass, airlift bioreactors are preferable. Airlift bioreactors have a simple design and provide good mixing and oxygen transfer but with low energy

TABLE 5.3
Fungal Biomass Accumulation and Pollutant Removal of Sugarcane Vinasse by Yeasts and Fungi (adapted from Chuppa-Tostain et al., 2020)

Strain	COD reduction, %	Mineral reduction, %	Final pH	Biomass, g/l
Arthroderma otae MUCL 41713	59.22	36.99	6.91	18.11
Aspergillus alutaceus MUCL 39539	69.23	73.49	7.91	21.70
Aspergillus flavus MUCL 19006	70.86	20.88	8.72	19.77
Aspergillus niger MUCL 19001	70.11	77.57	8.31	21.25
Aspergillus oryzae MUCL 19009	65.98	66.62	8.86	24.35
Aspergillus parasiticus MUCL 14491	74.60	53.66	8.46	24.48
Aspergillus terreus MUCL 38960	76.53	66,00	9,00	29.19
Candida dubliniensis MUCL 41201	45.98	32.18	8.45	10.29
Candida glabrata MUCL 29833	57.34	26.62	8.12	12.43
Candida tropicalis MUCL 29893	50.41	20.75	8.42	15.21
Clavispora lusitaniea MUCL 29855	54.72	28.35	7.39	28.56
Colletotrichum graminicola MUCL 44764	57.75	39.23	7.94	11.79
Cryptococcus albidus MUCL 30400	44.89	30.29	8.13	12.47
Fennellia flavipes MUCL 38811	58.65	61.46	8.74	17.98
Flavodon flavus MUCL 38427	28.99	37.48	6.17	14.98
Fusarium proliferatum MUCL 43482	34.44	53.83	6.37	6.38
Fusarium sporotrichioides MUCL 6133	55.13	43.81	8.25	8.60
Galactomyces geotrichum MUCL 43077	56.95	46.39	8.26	6.79
Gibberella fujikuroi MUCL 42883	37.89	36.58	6.76	4.12
Gibberella zeae MUCL 42841	55.80	33.52	8.04	20.90
Issatchenkia orientalis MUCL 29849	48.13	49.56	8.08	25.41
Komagataella pastoris MUCL 31260	69.70	56.84	7.99	9.10
Penicillium rugulosum MUCL 41583	62.48	56.28	8.72	2.36
Penicillium verrucosum MUCL 28674	62.09	49.05	9.03	8.00
Phanerochaete chrysosporium MUCL 38489	23.51	70.49	7.01	17.00
Pichia angusta MUCL 27761	49.52	20.78	6.53	4.04
Pichia guilliermondii MUCL 29837	54.78	28.21	7.54	12.23
Pichia jadinii MUCL 30058	40.91	46.33	8.21	14.67
Pseudozyma antarctica MUCL 47637	51.33	22.11	8.90	0.75
Rhizopus microsporus MUCL 31005	67.32	52.04	8.85	14.66
Saccharomyces cerevisiae MUCL 39449	55.84	41.17	8.88	11.64
Thanatephorus cucumeris MUCL 43254	65.28	44.63	6.66	16.32
Trametes hirsuta MUCL 40169	74.01	62.86	7.80	29.40
Trametes versicolor MUCL 44890	73.64	39.05	7.79	25.96

consumption. Shear forces during airlift fermentation are minimal and facilitate fungal pellet formation. Although the production of SCP from vinasse is economically attractive, currently, there is no report on the commercial realization of the process.

REFERENCES

Aguado, Ana. 2010. "Development of Okara Powder as a Gluten Free Alternative to All Purpose Flour for Value Added Use in Baked Goods." https://drum.lib.umd.edu/handle/1903/11281

Anderson, James L, Diego Valderrama, and Darryl E Jory. 2019. "GOAL 2019: Global Shrimp Production Review." https://www.globalseafood.org/adv°Cate/goal-2019-global-shrimp-production-review/.

Bandara, Nandika, Lingyun Chen, and Jianping Wu. 2011. "Protein Extraction from Triticale Distillers Grains." *Cereal Chemistry Journal* 88(6): 553–59. https://doi.org/10.1094/CCHEM-03-11-0026.

Bhaskar, Nimisha, Vinod K Modi, K Govindaraju, Cheruppanpullil Radha, and R G Lalitha. 2007. "Utilization of Meat Industry by Products: Protein Hydrolysate from Sheep Visceral Mass." *Bioresource Technology* 98(2): 388–94. https://doi.org/10.1016/j.biortech.2005.12.017.

Bird, Anthony R, Takashi Hayakawa, Yustinus Marsono, James M Gooden, Ian R Record, Raymond L Correll, and David L Topping. 2000. "Coarse Brown Rice Increases Fecal and Large Bowel Short-Chain Fatty Acids and Starch but Lowers Calcium in the Large Bowel of Pigs." *The Journal of Nutrition* 130(7): 1780–87. https://doi.org/10.1093/jn/130.7.1780.

Bowles, Simone, and Ivo Motin Demiate. 2006. "Physicochemical Characterization of the Soymilk by Product - Okara." *Food Science and Technology* 26(3): 652–59. https://doi.org/10.1590/S0101-2061200 6000300026.

Carriere, C J, and George E Inglett. 2003. "Constitutive Analysis of the Nonlinear Viscoelastic Properties of Cellulosic Fiber Gels Produced from Corn or Oat Hulls." *Food Hydrocolloids* 17(5): 605. https://doi.org /10.13141/jve.vol6.no1.pp4-12.

Chatzifragkou, Afroditi, Ondrej Kosik, Parvathy Chandran Prabhakumari, Alison Lovegrove, Richard A Frazier, Peter R Shewry, and Dimitrios Charalampopoulos. 2015. "Biorefinery Strategies for Upgrading Distillers' Dried Grains with Solubles (DDGS)." *Process Biochemistry* 50(12): 2194–207. https://doi.org /10.1016/j.pr°Cbio.2015.09.005.

Chen, Ming-Hsuan, Suk Hyun Choi, Nobuyuke Kozukue, Hyun-Jeong Kim, and Mendel Friedman. 2012. "Growth-Inhibitory Effects of Pigmented Rice Bran Extracts and Three Red Bran Fractions against Human Cancer Cells: Relationships with Composition and Antioxidative Activities." *Journal of Agricultural and Food Chemistry* 60(36): 9151–61. https://doi.org/10.1021/jf3025453.

Chuppa-Tostain, Graziella, Melissa Tan, Laetitia Adelard, Alain Shum-Cheong-Sing, Jean-Marie François, Yanis Caro, and Thomas Petit. 2020. "Evaluation of Filamentous Fungi and Yeasts for the Biodegradation of Sugarcane Distillery Wastewater." *Microorganisms*. https://doi.org/10.3390/microorganisms8101588.

Clare, D A, and H E Swaisgood. 2000. "Bioactive Milk Peptides: A Prospectusl." *Journal of Dairy Science* 83(6): 1187–95. https://doi.org/10.3168/jds.S0022-0302(00)74983-6.

Cookman, Drew J, and Charles E Glatz. 2009. "Extraction of Protein from Distiller's Grain." *Bioresource Technology* 100(6): 2012–17. https://doi.org/10.1016/j.biortech.2008.09.059.

Do, Thi-Thanh-Huong, Gia-Long Nguyen, Son Chu-ky, and Tien-Thanh Nguyen. 2021. "Extraction and Hydrolysis of Protein from Industrial Rice-Based Dried Distiller's Grain toward Food Application." *Journal of Vietnam Agricultural Science and Technology* 01: 122.

Esa, Norhaizan Mohd, Tan Bee Ling, and Loh Su Peng. 2013. "By-Products of Rice Processing: An Overview of Health Benefits and Applications." *Journal of Rice Research* 1: 107. https://doi.org10.4172/jrr.1000107.

Fafaungwithayakul, Natdanai, Pranithi Hongsprabhas, and Parichat Hongsprabhas. 2011. "Effect of Soy Soluble Polysaccharide on the Stability of Soy-Stabilised Emulsions During in Vitro Protein Digestion." *Food Biophysics* 6(3): 407–15. https://doi.org/10.1007/s11483-011-9216-1.

Fujita, Tomoyuki, Tomoyoshi Funako, and Hideo Hayashi. 2004. "8-Hydroxydaidzein, an Aldose Reductase Inhibitor from Okara Fermented with *Aspergillus* sp. HK-388." *Bioscience, Biotechnology, and Biochemistry* 68(7): 1588–90. https://doi.org/10.1271/bbb.68.1588.

Gonzàlez-Tello, P, F Camacho, E Jurado, M P Páez, and E M Guadix. 1994. "Enzymatic Hydrolysis of Whey Proteins: I. Kinetic Models." *Biotechnology and Bioengineering* 44(4): 523–28. https://doi.org/10.1002 /bit.260440415.

Hattori, T, H Ohishi, T Yokota, H Ohoami, and K Watanabe. 1995. "Antioxidative Effect of Crude Antioxidant Preparation from Soybean Fermented by Bacillus natto." *LWT - Food Science and Technology* 28(1): 135–38. https://doi.org/10.1016/S0023-6438(95)80025-5.

Hayashi, Sachio, Katsuhiro Matsuzaki, Toshikazu Kawahara, Yoshiyuki Takasaki, and Kiyohisa Imada. 1992. "Utilisation of Soybean Residue for the Production of β-Fructofuranosidase." *Bioresource Technology* 41(3): 231–3. doi.org/10.1016/0960-8524(92)90007-k.

Hoi, Bui Duc, Le Hong Khanh, Mai Van Le, Le Thi Cuc, Hoang Thi Ngoc Chau, Le Ngoc Tu, and Hong-Nga Luong. 2003. *Cereal, Starchy Roots and Legume Processing Technology*. Hanoi University of Science and Technology Publishing House.

Hsieh, Chienyan, and Fan-Chiang Yang. 2004. "Reusing Soy Residue for the Solid-State Fermentation of Ganoderma lucidum." *Bioresource Technology* 91(1): 105–9. https://doi.org/10.1016/S0960 -8524(03)00157-3.

Huong, Duong Thu, Pham Kim Dang, Tran Hiep, Ngo Thi Huyen Trang, Nguyen Thi Nguyet, and Vu Van Hanh. 2018. "Protein Enrichment of Cassava Residue by Simultaneous Saccharification and Fermentation." *Vietnam Journal of Agriculture Sciences* 16(3): 207–14.

Iram, Attia, Deniz Cekmecelioglu, and Ali Demirci. 2020. "Distillers' Dried Grains with Solubles (DDGS) and Its Potential as Fermentation Feedstock." *Applied Microbiology and Biotechnology* 104(14): 6115–28. https://doi.org/10.1007/s00253-020-10682-0.

Kasai, Naoya, Aoi Murata, Hiroshi Inui, Tatsuji Sakamoto, and Rokibul Isra Kahn. 2004. "Enzymatic High Digestion of Soybean Milk Residue (Okara)." *Journal of Agricultural and Food Chemistry* 52(18): 5709–16. https://doi.org/10.1021/jf035067v.

Khare, S K, Krishna Jha, and A P Gandhi. 1995. "Citric Acid Production from Okara (Soy-Residue) by Solid-State Fermentation." *Bioresource Technology* 54(3): 323–25. https://doi.org/10.1016/0960-8524(95)00155-7.

Kim, Sung Phil, Jun Young Yang, Mi Young Kang, Jun Cheol Park, Seok Hyun Nam, and Mendel Friedman. 2011. "Composition of Liquid Rice Hull Smoke and Anti-Inflammatory Effects in Mice." *Journal of Agricultural and Food Chemistry* 59(9): 4570–81. https://doi.org/10.1021/jf2003392.

Leme, Rodrigo Marcelo, and Joaquim E A Seabra. 2017. "Technical-Economic Assessment of Different Biogas Upgrading Routes from Vinasse Anaerobic Digestion in the Brazilian Bioethanol Industry." *Energy* 119: 754–66. https://doi.org/10.1016/j.energy.2016.11.029.

Li, Bo, Meiying Qiao, and Fei Lu. 2012. "Composition, Nutrition, and Utilization of Okara (Soybean Residue)." *Food Reviews International* 28(3): 231–52. https://doi.org/10.1080/87559129.2011.595023.

Li, Shuhong, Dan Zhu, Kejuan Li, Yingnan Yang, Zhongfang Lei, and Zhenya Zhang. 2013. "Soybean Curd Residue: Composition, Utilization, and Related Limiting Factors." *ISRN Industrial Engineering* 2013: 423590. https://doi.org/10.1155/2013/423590.

Liu, KeShun. 1997. *Soybeans: Chemistry, Technology and Utilization*. Springer US.

Liu, KeShun. 2011. "Chemical Composition of Distillers Grains, a Review." *Journal of Agricultural and Food Chemistry* 59(5): 1508–26.

Mohammadi Shad, Zeinab, Chandrasekar Venkitasamy, and Zhiyou Wen. 2021. "Corn Distillers Dried Grains with Solubles: Production, Properties, and Potential Uses." *Cereal Chemistry* 98(5): 999–1019. https://doi.org/10.1002/cche.10445.

Ngoc, Ngo Minh, and Quan Le Ha. 2017. "Antioxidant Activities of Hydrolysates Originated from Soybean and Soy Milk Residue." *Vietnam Journal of Science and Technology* 55(5A): 134–42.

Nguyen, Tien-Thanh, Thi-Thao Nguyen, and Son Chu-Ky. 2021. "Enrichment of Protein Content in Rice-Based Dried Distillers' Grain from Food-Ethanol Factory by Chemical Method." *JST: Engineering and Technology for Sustainable Development* 31(4).

Nguyen, Tram A, Kim A Nguyen, and Curtis Jolly. 2019. "Is Super-Intensification the Solution to Shrimp Production and Export Sustainability?" *Sustainability*. https://doi.org/10.3390/su11195277.

Nitayavardhana, Saoharit, Kerati Issarapayup, Prasert Pavasant, and Samir Kumar Khanal. 2013. "Production of Protein-Rich Fungal Biomass in an Airlift Bioreactor Using Vinasse as Substrate." *Bioresource Technology* 133: 301–6. https://doi.org/10.1016/j.biortech.2013.01.073.

Nitayavardhana, Saoharit, Prachand Shrestha, Mary L Rasmussen, Buddhi P Lamsal, J (Hans) van Leeuwen, and Samir Kumar Khanal. 2010. "Ultrasound Improved Ethanol Fermentation from Cassava Chips in Cassava-Based Ethanol Plants." *Bioresource Technology* 101(8): 2741–47.

O'Toole, Desmond K. 1999. "Characteristics and Use of Okara, the Soybean Residue from Soy Milk Production A Review." *Journal of Agricultural and Food Chemistry* 47(2): 363–71. https://doi.org/10.1021/jf980754l.

Ohno, Akihiro, Takashi Ano, and Makoto Shoda. 1996. "Use of Soybean Curd Residue, Okara, for the Solid State Substrate in the Production of a Lipopeptide Antibiotic, Iturin A, by Bacillus subtilis NB22." *Process Biochemistry* 31(8): 801–6. https://doi.org/10.1016/S0032-9592(96)00034-9.

Pires, Josiane F, Gustavo M R Ferreira, Kelly C Reis, Rosane F Schwan, and Cristina F Silva. 2016. "Mixed Yeasts Inocula for Simultaneous Production of SCP and Treatment of Vinasse to Reduce Soil and Fresh Water Pollution." *Journal of Environmental Management* 182: 455–63. https://doi.org/10.1016/j.jenvman.2016.08.006.

Puechkamut, Yuporn, and Woralak Panyathitipong. 2012. "Characteristics of Proteins from Fresh and Dried Residues of Soy Milk Production." *Agriculture and Natural Resources* 46(5): 804–811.

Rabelo, Sarita Cândida, Aline Carvalho da Costa, and Carlos Eduardo Vaz Rossel. 2015. "Industrial Waste Recovery." In: *Sugarcane: Agricultural Production, Bioenergy and Ethanol*, edited by Fernando Santos, Aluízio Borém, and Celso B T - Sugarcane Caldas, 365–81. San Diego: Academic Press. https://doi.org/10.1016/B978-0-12-802239-9.00017-7.

Rashad, M M, E A Mahmoud, M H Abdou, and U M Nooman. 2011. "Improvement of Nutritional Quality and Antioxidant Activities of Yeast Fermented Soybean Curd Residue." *African Journal of Biotechnology* 10(30): 5504–13.

Rashad, Mona, Mahmoud Sitohy, Samy Fanous Sharobeem, Abeer Mahmoud, Mohamed Nooman, and Amr Alkashef. 2010. "Production, Purification and Characterization of Extracellular Protease from Candida guilliermondii Using Okara as a Substrate." *Advances in Food Sciences* 32(June): 100–9.

RFA. 2019. *Annual World Fuel Ethanol Production*. https://ethanolrfa.org/statistics/annual-ethanol -production/.

Riaz, Mian N. 2006. *Soy Applications in Food*. Boca Raton, FL: CRC Press.

Sachindra, N M, N Bhaskar, and N S Mahendrakar. 2006. "Recovery of Carotenoids from Shrimp Waste in Organic Solvents." *Waste Management* 26(10): 1092–98. https://doi.org/10.1016/j.wasman.2005.07.002.

Sajilata, M G, Rekha S Singhal, and Pushpa R Kulkarni. 2006. "Resistant Starch–A Review." *Comprehensive Reviews in Food Science and Food Safety* 5(1): 1–17. https://doi.org/10.1111/j.1541-4337.2006.tb00076.x.

Santos, Raquel Rezende dos, Ofélia de Queiroz Fernandes Araújo, José Luiz de Medeiros, and Ricardo Moreira Chaloub. 2016. "Cultivation of *Spirulina maxima* in Medium Supplemented with Sugarcane Vinasse." *Bioresource Technology* 204: 38–48. https://doi.org/10.1016/j.biortech.2015.12.077.

Sarwar, G D, A Christensen, A J Finlayson, Mendel Friedman, L R Hackler, S L Mackenzie, P L Pellett, and R Tkachuk. 1983. "Inter- and Intra-Laboratory Variation in Amino Acid Analysis of Food Proteins." *Journal of Food Science* 48(2): 526–31. https://doi.org/10.1111/j.1365-2621.1983.tb10781.x.

Schingoethe, David J. 2006. "Utilization of DDGS by Cattle." In: 27th Western Nutrition Conference, Winnipeg, MB, 61–74.

Schirmer, Matthias. 2014. "Biomass and Waste as Renewable and Sustainable Energy Source in Vietnam." *Journal of Vietnamese Environment* 6(1): 4–12.

Shu-bin, Li, Zhou Ren-chao, Li Xia, Chen Chu-yi, and Yang Ai-lin. 2012. "Solid-State Fermentation with Okara for Production of Cellobiase-Rich Cellulases Preparation by a Selected Bacillus subtilis Pa5." *African Journal of Biotechnology* 11(11): 2720–30.

Srichuwong, Sathaporn, and Jay-lin Jane. 2011. "Methods for Characterization of Residual Starch in Distiller's Dried Grains with Solubles (DDGS)." *Cereal Chemistry Journal* 88(3): 278–82. https://doi.org/10.1094 /CCHEM-11-10-0155.

Synowiecki, Józef, and Nadia Ali Abdul Quawi Al-Khateeb. 2000. "The Recovery of Protein Hydrolysate during Enzymatic Isolation of Chitin from Shrimp Crangon Crangon Processing Discards." *Food Chemistry* 68(2): 147–52. https. doi.org/10.1016/S0308-8146(99)00165-X.

Taranu, Ionelia, Tien-Thanh Nguyen, Pham Dang, Mihail Gras, Gina Pistol, Daniela Marin, Mircea-Catalin Rotar, et al. 2019. "Rice and Cassava Distillers Dried Grains in Vietnam: Nutritional Values and Effects of Their Dietary Inclusion on Blood Chemical Parameters and Immune Responses of Growing Pigs." *Waste and Biomass Valorization* 10(11): 3373–82.

Um, Byung-Hwan, and Thomas R Hanley. 2008. "A Comparison of Simple Rheological Parameters and Simulation Data for Zymomonas mobilis Fermentation Broths with High Substrate Loading in a 3-L Bioreactor." *Applied Biochemistry and Biotechnology* 145(1): 29–38. https://doi.org/10.1007/s12010-007 -8105-z.

Urbanski, Gregory E, and Kevin W Lang. 2016. *Enzymatic Hydrolysates of Okara*. US9295274B1, Issued 2016.

Vuong, Mai-Dinh, Nguyen-Tien Thanh, Chu-Ky Son, and Waché Yves. 2021. "Protein Enrichment of Cassava-Based Dried Distiller's Grain by Solid State Fermentation Using Trichoderma harzianum and Yarrowia lipolytica for Feed Ingredients." *Waste and Biomass Valorization* 12(7): 3875–88. https://doi.org/10 .1007/s12649-020-01262-4.

Waliszewski, K N, V Pardio, and E Carreon. 2002. "Physicochemical and Sensory Properties of Corn Tortillas Made from Nixtamalized Corn Flour Fortified with Spent Soymilk Residue (Okara)." *Journal of Food Science* 67(8): 3194–97. https://doi.org/10.1111/j.1365-2621.2002.tb08881.x.

Walter, Arnaldo, Paulo Dolzan, Oscar Quilodrán, Janaína G de Oliveira, Cinthia da Silva, Fabrício Piacente, and Anna Segerstedt. 2011. "Sustainability Assessment of Bio-Ethanol Production in Brazil Considering Land Use Change, GHG Emissions and Socio-Economic Aspects." *Energy Policy* 39(10): 5703–16. https://doi.org/10.1016/j.enpol.2010.07.043.

Yokomizo, Atsushi, Yoko Takenaka, and Tetsuo Takenake. 2002. "Antioxidative Activity of Peptides Prepared from Okara Protein." *Food Science and Technology Research* 8(4): 357–59. https://doi.org/10.3136/fstr .8.357.

Zhu, Y P, J F Fan, Y Q Cheng, and L T Li. 2008. "Improvement of the Antioxidant Activity of Chinese Traditional Fermented Okara (Meitauza) Using Bacillus subtilis B2." *Food Control* 19(7): 654–61. https://doi.org/10.1016/j.foodcont.2007.07.009.

6 Production of Enzymes from Agro-Industrial Byproducts

Rupinder Kaur, Parmjit S. Panesar, and Gisha Singla

CONTENTS

6.1 INTRODUCTION

Since antiquity, the activities of human beings have led to the generation of varied quantities of wastes that are under-utilized. In addition, the rapid growth of the food and the agriculture sectors has also added to the annual surplus waste generation globally, which is called "agro-industrial wastes". These are defined as the wastes caused due to the diverse agricultural practices or processing of agricultural or animal products (Mirabella et al., 2014).

Generally, these wastes are nutritious in nature and that are largely made up of carbohydrates, protein, polyphenols, and other components (Marinari et al., 2000; Yadav et al., 2009). Most of these residues are either untreated or disposed of in bins due to the lack of proper infrastructure. Moreover, other disposal techniques to date include burning, dumping, or land-fill. Being a rich source of nutrients, disposal of these residues poses serious environmental issues. Besides, various stringent governmental and environmental rules and legislation have aggravated the problem of waste disposal (Babbar and Oberoi, 2014; Bos and Hamelinck, 2014). Hence, there is a need to develop an economical, robust, and eco-friendly technique for the successful utilization of these residues, which has become a major worldwide concern.

As per the new directives, the term "biorefinery" has come into existence. The waste generated from one industry can be used as an input for another, leading to the valorization of wastes

DOI: 10.1201/9781003125679-6

89

into value-added products (Ravindran and Jaiswal, 2016). Being rich in nutrients, these are not considered wastes, rather cost-effective substrates for the generation of novel products. Above all, the availability of complex components that represents carbon and nitrogen sources, facilitates the growth of microorganisms. Therefore, exploitation of the different agro-industrial residues via microbial fermentation to develop highly valued components, especially food-grade enzymes, is a sustainable option and a promising area.

In an industrial era, enzymes have indefinite applications in various sectors. These are the biological catalysts that catalyze biochemical reactions lower the activation energy of any particular reaction. In contrast to any chemical processes, enzymes are advantageous because of their biodegradability, eco-friendliness, and cost-effectiveness (Machielsen et al., 2007; Panesar et al., 2010b). Despite its applications in various sectors, since ancient times, these had versatile potential to prepare diverse food commodities like cheese, kefir, alcoholic beverages, fermented products, and many more (Panesar et al., 2016). Natural sources of enzymes include plants, animals, and microbes; however, microbes are widely investigated for the production of enzymes in contrast to the other two because of their ability to use cheap lignocellulosic residues as substrates. Besides this, other techno-economic benefits of microbial sources, such as faster growth, higher yield, no effect of seasonal variation, and ease of genetic manipulation, have also facilitated their use in enzyme production (Aunstrup, 1974).

Several species of fungi, bacteria, and yeast have been explored to produce enzymes using inorganic substrates, such as glycerol, starch, glucose, and other salts, but this only added to the overall cost of production. Consequently, to circumvent this issue, a variety of cheap lignocellulosic components have been tested as potential raw materials for microbial growth by researchers worldwide in the past decade that would aid in improving the economics of the process and provide a solution for the effective utilization of waste besides meeting the global demand of enzymes (Salihu et al., 2011).

Hence, this chapter presents a comprehensive discussion of the various agro-industrial residues generated, their characteristics and composition, biotechnological strategies for the fabrication of different food-grade enzymes, technical challenges during production, and their commercialization potential.

6.2   LIGNOCELLULOSE RESIDUES

Industrial and agricultural activities generate ample quantities of waste per year, which is anticipated to increase in the coming decades due to the population's growth and subsequent rise in food production (Food and Agriculture Organization (FAO), 2018; Nayak and Bhushan, 2019). On average, 1.3 billion tonnes, which accounts for one-third of the produce, are wasted worldwide annually. Amongst this produce, fruits and vegetables, tubers and roots represent 40–50% of global food wastage (Searle, 2013). Asia ranks first in global waste production, of which China is the lead producer (82.8 million tonnes), trailed by Indonesia, Japan, Philippines, and Vietnam (Lin et al., 2013). Further, the annual residue production in Asian countries is likely to enhance upto 416 million tonnes from 278 million tonnes between 2005 and 2025 (Melikoglu et al., 2013). After Asia, the United States is the second-largest food-waste producer, followedby Europe, Africa, and Oceania, as depicted in Figure 6.1 (Cherubin et al., 2018).

6.2.1   Types and Composition of Lignocellulosic Residues

The lignocellulosic residues are classed into two categories: industrial residues and agriculture residues. Agricultural wastes are distributed into process and field residues (Figure 6.2). Field residues represent those generated during the crop harvesting process, and this may be comprised of seeds, leaves, stalks, etc. On the other hand, process residues include peels, pulp, husks, and more produced during crop processing. Generally, the differentiation of agriculture residues is carried out based on availability and characteristics (Zafar, 2014). "Industrial wastes" refers to those wastes

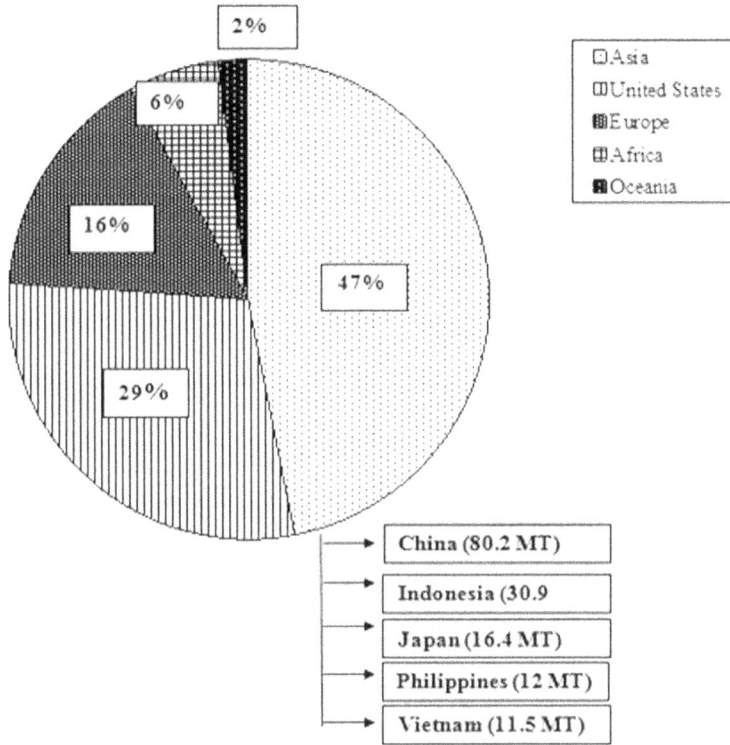

Legend:
- Asia
- United States
- Europe
- Africa
- Oceania

2%
6%
16%
47%
29%

China (80.2 MT)

Indonesia (30.9

Japan (16.4 MT)

Philippines (12 MT)

Vietnam (11.5 MT)

FIGURE 6.1   Generation of agro-industrial waste worldwide (MT: million tonnes).

produced from the food-processing industry and are made up of bran, pomace, bagasse, and effluents released from the industry every year (Anwar et al., 2014; da Silva, 2016).

Agro-industrial residues are rich in lignocellulosic materials, such as cellulose, hemicellulose,
and lignin (Figure 6.3). Amongst them, cellulose is the main component present in these residues.
In contrast to agriculture residues, industrial wastes are also composed of various parts cellulose,
lignin, hemicellulose, ash, moisture, nitrogen, as well as carbon, which includes a huge amount
of Chemical Oxygen Demand (COD), Biological Oxygen Demand (BOD), and suspended solids responsible for environmental pollution if untreated (Rudra et al., 2015). For example, corn
cob, pomace, wheat straw, and wheat bran are comprised of approximately 20–30% hemicellulose, 40–50% cellulose, 10–25% lignin, and others (Voragen et al., 2009; Ahmed et al., 2011).
Similarly, oil cakes generated from the oil-processing industry include a high amount of fat, oil,
and dissolved and suspended solids, which are also a major source of pollution (Ramachandran
et al., 2007).

Cellulose, the copious homopolymer, comprises hundreds and thousands of glucose monomers
bonded with β 1-4 linkages in a linear fashion and has an Molecular Weight Cut Off (MWCO) of
approximately 100 kDa (Himmel et al., 2007). This polymer is responsible for the rigidness in the
cell wall of plants owing to the presence of three hydroxyl groups, which form the various intramolecular or intermolecular hydrogen bonds (Zhou et al., 2016).

The second-most abundant component, hemicellulose, is a heteropolymer with an MWCO
of <30 kDa (Calvo Flores and Dobado, 2010). The structure of hemicellulose varies with the sugar
unit (pentoses, hexoses, hexuronic acid, etc.) and length of the chain and branching units. Thus, these
hemicelluloses can be recognized as xylans, galactans, an mannans, and xyloglucans (Ebringerová
et al., 2005). Lignin is a hydrophobic branched polymer comprised of aliphatic, phenolic hydroxyls,
carboxyl, carbonyl and methoxyl groups linked by ether bonds that provides lignin with a very

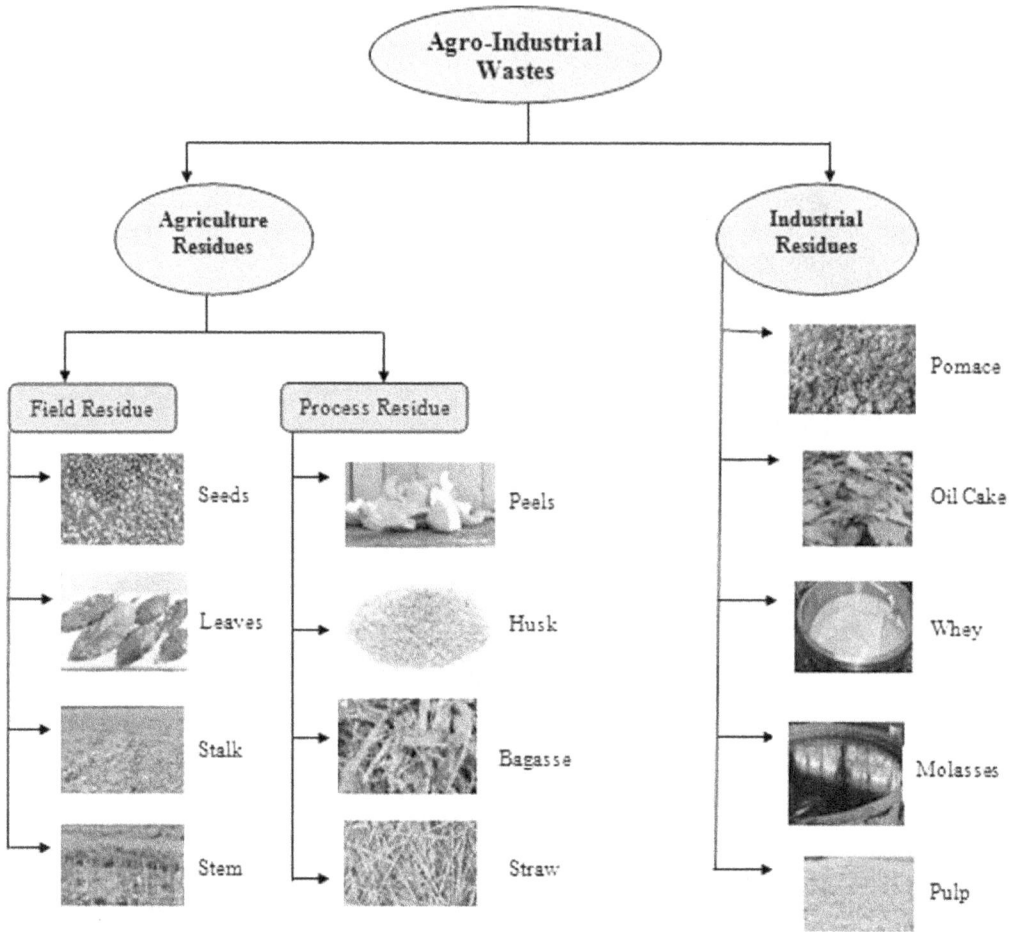

FIGURE 6.2 Types of agro-industrial wastes.

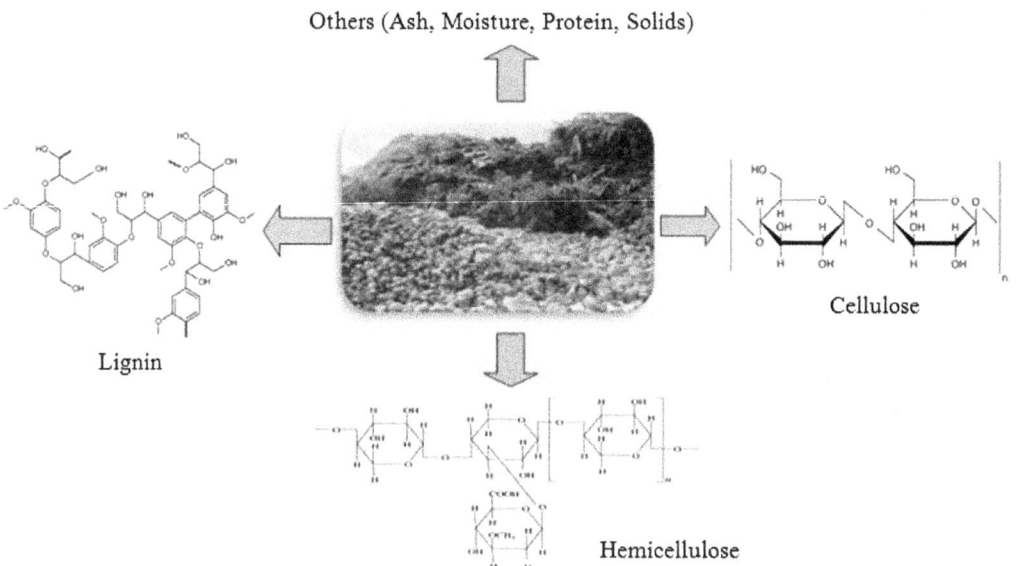

FIGURE 6.3 Components of agro-industrial residues.

complex structure (Davin and Lewis, 2005; Geneau-Sbartaï et al., 2008). The composition of lignin varies with the plant and the environment (Rico-García et al., 2020).

The composition and concentration of these lignocellulosic components in agro-industrial residues differ with the tissue, species, and plant maturity (Zhou et al., 2016). The concentration of these components in various agro-industrial residues is represented in Table 6.1.

6.3   BIOPROCESS TECHNOLOGY FOR ENZYME PRODUCTION

Studies have been conducted on the valorization strategies of various agro-industrial residues into high-value components in the past decade. Research is also being carried out on viable bioprocess technology to utilize these lignocellulosic materials as cheap substrates for the production of enzymes (Ravindran and Jaiswal, 2016). Enzymes act as biological catalysts that can be used in various industrial sectors. Yet, having a substantial demand, these are one of the most expensive catalysts as they enhance the overall cost of the production process by almost 50%. Thus, to cope with the high economy of the process, biotechnological strategies have been developed to exploit the diverse agro-industrial residues generated globally to produce commercial enzymes via fermentation technique (Ravindran et al., 2018).

Industrially, production is executed either by submerged fermentation (SmF) or by solid state fermentation (SSF). The latter is preferred over the former concerning the effective valorization of agro-industrial residues, thereby improving the economics of the production (Pandey et al., 2000a). Different biotechnological tools and techniques employed during enzyme production from diverse lignocellulosic biomasses are discussed in the following sections. The general plan for commercial enzyme production is represented in Figure 6.4. Moreover, a summary of the various enzymes produced from various wastes using different microbial sources is listed in Table 6.2.

6.3.1   AMYLASE

Amylase belongs to the hydrolase family of enzymes that cleaves the glycosidic linkage in the starch. Amylase is classed into two major groups, namely, α-amylase and glucoamylase (Anto et al., 2006). α-Amylase (EC 3.2.1.1) is responsible for breaking glucose units, especially the glycosidic bonds present in the amylase chain, leading to the generation of maltose, maltotriose, and glucose (Azad et al., 2009). Among all the enzymes, this enzyme has great industrial significance in the area of food, paper, textiles, detergents, and clinical as well as analytical chemistry, accounting for almost 30% of the global enzyme market share (Kandra, 2003; de Souza and de Oliveira Magalhães, 2010).

Although different microbial species have been reported for fermentation, fungal species like *Aspergillus, Penicillium,* and *Rhizopus* are considered the main producers of this enzyme (Pandey et al., 2000b). In the context of agro-industrial wastes, several residues (wheat bran, mustard oil cake, banana waste, apple pomace, rice husk, cassava waste, sugar beet pulp, coconut oil cake, tea waste, rapeseed cake, rice bran, peanut meal, wheat straw, etc.) have been explored as cheap substrates for enzyme production (Gupta et al., 2003).

Dual substrates, such as cassava waste and groundnut shell, have been tested for their potential during amylase production from *Bacillus* sp. under SSF conditions, which indicated the maximum yield (866 U/ml) at pH 7.0, 60 h, using 2 g/l groundnut shell and cassava waste, concentrations of both substrates is same, that is 2 g/L. Of groundnut shell and 2 g/L cassava waste was used during enzyme production. (Selvam et al., 2016). Cassava is rich in starch, whereas ground nut has a high oil content. Thus, the combination of these wastes as substrates had a synergistic effect that led to higher enzyme yields. Moreover, pre-treatment of these substrates with NaOH played an important role by increasing the nutrient accessibility to the bacteria, hence enhancing the enzyme yield.

A comparative study was undertaken of the various residues such as corncob, orange peel, sugar cane, rice straw, olive oil cake, and banana peel to explore their potency as substrates for amylase production from *A. awamori* (Karam et al., 2017). The maximum yield was obtained

TABLE 6.1

Composition of Agro-Industrial Residues

Types of Residues	Cellulose (%, w/w)	Hemicellulose (%, w/w)	Lignin (%, w/w)	Ash	Moisture	Solids	References
Rice straw	39.2	23.5	36.1	12.4	6.58	98.62	El-Tayeb et al. (2012)
Wheat straw	32.9	24	8.9	6.7	7	95.6	Martin et al. (2012)
Oat straw	39.4	27–38	16–19	6–8	–	–	Martin et al. (2012)
Barley straw	33.8	21.9	13.8	11	–	–	Nigam et al. (2009)
Sugarcane straw	36–41	21–31	16–26	–	–	–	Zhao et al. (2009); Moutta et al. (2012)
Banana straw	53	29	15	–	–	–	Silveira et al. (2008)
Rye straw	38	31	19	2–5	–	–	Hon (2000); Nigam et al. (2009)
Canola straw	22	17	18	–	–	–	Adapa et al. (2011)
Wheat bran	10–15	30	5.6	3.9	8.20	–	Beaugrand et al. (2004); Onilude et al. (2012)
Rice bran	9–12.8	8.7–11.4	25.63	3.4	–	–	Pandey et al. (1994); Ramezanzadeh et al. (2000)
Barley bran	23	32.7	21.4	–	–	–	Cruz et al. (2000)
Corn bran	34	39	49	–	–	–	Graminha et al. (2008)
Rice husk	35	25	20	–	–	–	Abbas and Ansumali (2010)
Coconut husk	24–43	3–12	25–45	–	–	–	Carrijo et al. (2002)
Coffee husk	43	7	9	–	–	–	Gouvea et al. (2009)
Corn cob	45	35	15	1.36	–	–	Prasad et al. (2007)
Sugarcane bagasse	30.2	56.7	13.4	1.9	4.8	91.66	El-Tayeb et al. (2012); Motte et al. (2013)
Walnut shell	36	28	43	–	–	–	Tufail et al. (2018)
Almond shell	38	29	30	–	–	–	Li et al. (2018)
Pistachio shell	43	25	16	–	–	–	Li et al. (2018)
Brewers spent grain	16.8	28.4	27.8	4.6	–	–	Mussatto et al. (2008)
Orange peel	9.21	10.5	0.84	3.5	–	11.86	Rivas et al. (2008)
Potato peel	35	5	4	–	–	–	Choonut et al. (2014)
Pineapple peel	22	75	3	–	91	93.6	Paepatung et al. (2009); Dos Santos et al. (2013)
Banana peel	12	10	3	–	–	–	Nigam et al. (2009)

(Continued)

TABLE 6.1 (CONTINUED)
Composition of Agro-Industrial Residues

Types of Residues	Composition						References
	Cellulose (%,w/w)	Hemicellulose (%,w/w)	Lignin (%,w/w)	Ash	Moisture	Solids	
Apple pomace	43	23	20	1.1	–	–	Dhillon et al. (2012); Nawirska and Kwaśniewska (2005)
Tomato pomace	9	5	5	–	–	–	Ververis et al. (2007)
Olive oil cake	31	21	26	4.2	–	–	Maymone et al. (1961); Akgül et al. (2013)
Sunflower oil cake	25	12	8	6.6	–	–	Rodríguez et al. (2008)
Oil palm cake	64	15	5	–	–	–	Rodríguez et al. (2008)
Corn stalk	61	19–24	7–9	10.8	6.40	97.78	Graminha et al. (2008);El-Tayeb et al. (2012)
Sorghum stalk	17	25	11	–	–	–	Graminha et al. (2008)
Soya stalk	35	25	20	10.39	11.84	–	Motte et al. (2013); Szymańska-Chargot et al. (2017)
Sunflower stalk	42	30	13	11.17	–	–	Motte et al. (2013); Szymańska-Chargot et al. (2017)
Whey	–	–	–	0.5–0.7	93–95	8	Panesar et al. (2007)

The composition of the various lignocellulosic materials is expressed as %, dry weight

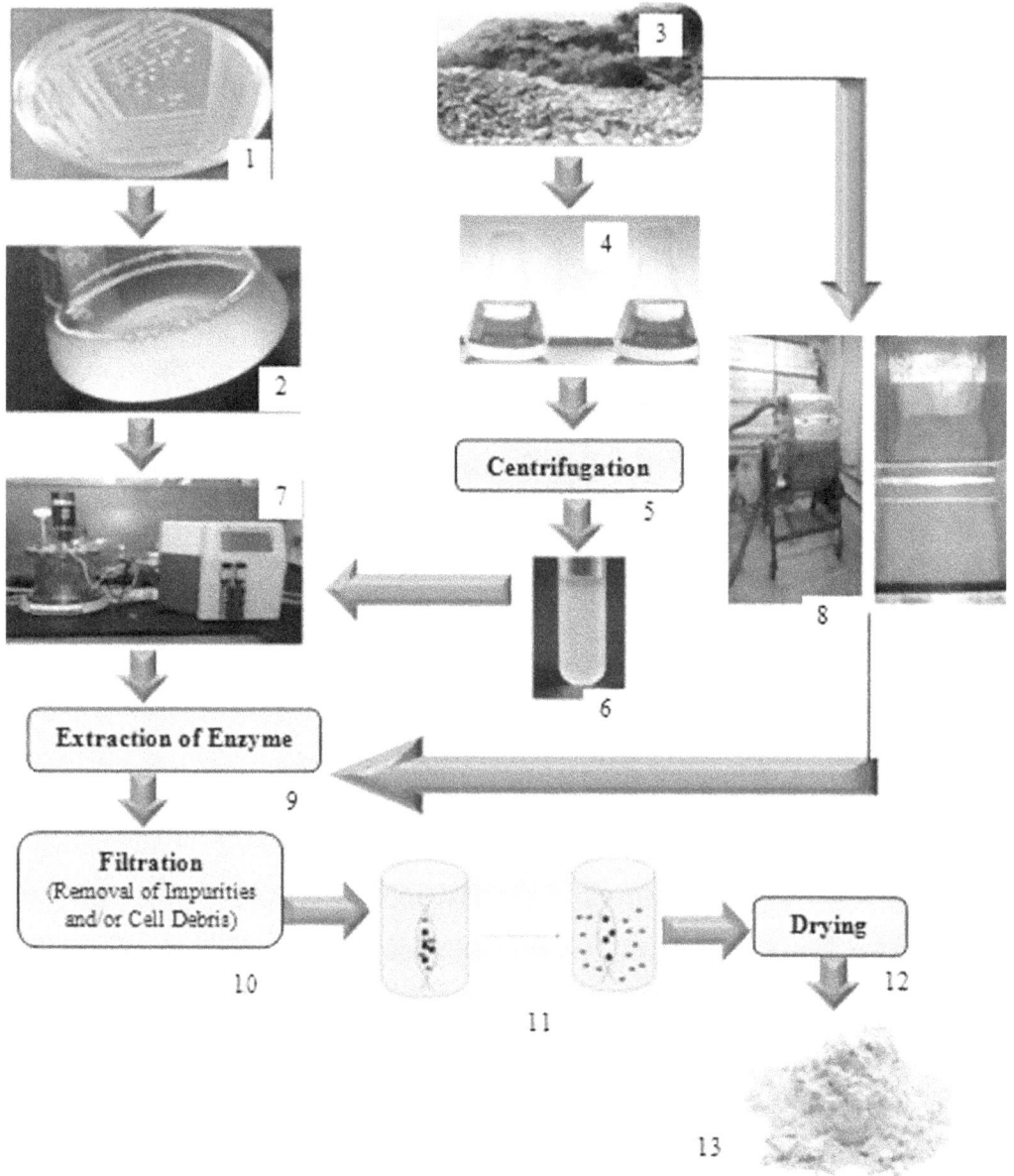

FIGURE 6.4 Schematic representation of the industrial production of enzymes. 1. Isolation of pure culture, 2. Inoculum preparation, 3. Agricultural and industrial wastes (lignocellulosic biomass), 4. Pre-treatment of lignocellulosic biomass (hydrolysis), 5. Centrifugation, 6. Separation of supernatant, 7. Enzyme production through SmF in bioreactors, 8. Enzyme production through SSF in trays, 9. Extraction of enzyme (cell disruption or other method), 10. Filtration (for removal of cell debris or impurities or media), 11. Purification of enzyme, 12. Drying of enzyme solution, 13. Enzyme powder.

from olive oil cake, followed by sugar cane, banana peel, corncob, rice straw, and orange peel. Similarly, using a co-culture technique, five residues were tested by both SmF and SSF (Ahmed et al., 2017). With the SmF technique, a combination of *B. licheniformis* and *B. subtilis* enhanced the amylase activity by 176.3% and 329.2%, respectively, in contrast with the mono-cultures using watermelon rind as the substrate. In addition, two-step statistical approaches also improved the yield by 10.3-fold. Fermentation carried out with the co-culture method is

TABLE 6.2

Production of the Enzymes from Various Lignocellulosic Residues

Enzymes	Lignocellulosic Biomass	Microbial Source	References
Amylase	Mango waste	*Bacillus* sp.	Saleh et al. (2020)
	Cassava bagasse	*Bacillus* sp.	Gois et al. (2020)
	Molasses	*A. niger* EFRL-FC-024	Jokhio et al. (2020)
	Wheat bran	*A. oryzae* F-923	Fadel et al. (2020)
	Potato peel	*Bacillus subtilis*	Obi et al. (2019)
	Bread waste	*Rhizopus oryzae*	Benabda et al. (2019)
	Wheat bran	*Bacillus* sp. KR1	Pranay et al. (2019)
Pectinase	Yellow mombin pulp	*A. niger* IOC 4003	da Câmara Rochaet al. (2020)
	Date fruit waste	*B. licheniformis* KIBGE-IB3	Aslam et al. (2020)
	Tobacco stalk	*B. tequilensis* CAS-MEI-2-33	Zhang et al. (2019)
	Mosambi peels	*Schizophyllum commune*	Mehmood et al. (2019)
	Orange peel+wheat bran+corn steep liquor	*A. niger* ATCC 9642	Borszcz et al. (2017)
	Orange peel+coconut fibre	*B. subtilis* SAV-21	Kaur and Gupta (2017)
	Orange peel	*Rhizopus* sp. C4	Handa et al. (2016)
Protease	Groundnut oil cake B22	*B. subtilis* B22	Elumalai et al. (2020)
	Shrimp waste	*Haloferax lucentensis* GUBF-2 MG076078	Goankar and Furtado (2020)
	Wheat bran	*A. oryzae* DRDFS13	Mamo et al. (2020)
	Pineapple crown	*Rhizopus microsporus* var. *chinensis*	Lizardi-Jimenez et al. (2019)
	Bread waste	*Rhizopus oryzae*	Benabda et al. (2019)
	Cottonseed cake	*B. subtilis*	Singh and Bajaj (2016)
	Wheat bran+rice bran	*Brevibacterium luteolum* MTCC 5982	Rao et al. (2017)
Lipase	Wheat bran	*A. terreus* NRRL-255	De Azevedo et al. (2020)
	Molasses+ Corn Steep Liquor (CSL)+Olive mill waste	*Diutina rugosa*	Maria de Fátima et al. (2020)
	De-oiled castor seed cake	*Acinetobacter* sp. UBT1	Patel et al. (2020)
	Mustard oil cake	*B. licheniformis* MTCC 3244, *B. coagulans* MTCC 10305	Prabha et al. (2019)
	Olive mill wastewater	*Magnusiomyces capitatus*	Salgado et al. (2020)
	Soybean meal	*A. niger*	Prabaningtyas et al. (2018)
	Rice bran	*A. niger*	Costa et al. (2017)
Lactase	Cheese whey	*Enterobacter ludwigii* MCC 3423	Alikunju et al. (2018)
	Deproteinized cheese whey	*A. oryzae* CCT 0977	Viana et al. (2018)
	Whey powder	*K. lactis*	You et al. (2017)
	Rice straw+peanut pod	*Bacillus* sp.	Kazemi et al. (2017)
	Wheat straw+peanut pod	*A. niger* ATCC 9142	Kazemi et al. (2016)
	Gingelly oil cake+Ground nut oil cake	*A. foetidus* MTCC 6322	Rao and Rishikesh (2019)
	Wheat bran+CSL	*A. tubingensis*	Raol et al. (2015)
	Potato starch waste	*Lactococcus lactis*	Bhanwar and Ganguli (2014)

advantageous over the others with respect to improved substrate utilization, high productivity and prevention from contamination (Saini et al., 2016). Compared with the other wastes, watermelon rind is cheap, non-toxic, and is comprised of various nutrients (25.9% carbohydrates, 4% moisture, and 8% ash) that make it a suitable candidate for microbial fermentation (Sundarram and Murthy, 2014).

Co-production of two enzymes (protease and amylase) using potato peel powder from thermophilic bacteria *Anoxybacillus rupiensis* was reported, revealing maximum protease and amylase activities as 26.2 and 64.9 U/ml, respectively (Tuysuz et al., 2020). Incubation times higher than 48 h led to a decrease in both the enzyme's activities. This is due to depletion in the nutritional components in the substrate and metal ions supporting enzyme production. Besides, the higher temperature at a prolonged time might have resulted in the denaturation of the enzyme. Further, de-escalation in the activity is the result of proteolytic activity. The study was carried out in both sterile and non-sterile conditions, but higher yields were obtained under sterile conditions. Although in non-sterile conditions, no contamination was observed, still less microbial growth and enzyme activities were obtained. This may be because potato peels contain some inhibitory components that disintegrate during sterilization. Another explanation may be the separation of starch and protein from cellulose and lignin during sterilization that facilitated their consumption by microbes (Qureshi et al., 2016). Potato peels are known to be rich in protein and starch. The high starch content in these peels supports amylase production and the production of bacterial protease because bacterial protease is produced in starch-rich materials (Khosravi-Darani et al., 2008; Bretón-Toral et al., 2016). Simultaneous preparation of two or more enzymes from a single substrate is a novel technique and is accepted by researchers worldwide with respect to the reduction in energy and time consumption as well as reduced labour requirements (Mukhtar and Ikram Ul-Haq, 2012) hence this technique has become the recent thrust area of research (Chugh et al., 2016; García-Pérez et al., 2018).

6.3.2 PECTINASE

Pectinase (EC 3.1.1.11) catalyzes the disruption of pectin molecules present in plants (Mojsov, 2016) through depolymerization (hydrolase and lyase) and de-esterification (esterase) reactions (Pedrolli et al., 2009). Hydrolase hydrolyzes the glycosidic linkages of pectin, and lyase catalyzes the disintegration of the glycosidic bonds present in the pectic acid, resulting in the production of unsaturated galacturonic acid and methyl galacturonate via trans-elimination reactions. Likewise, pectin esterase results in the de-esterification of methyl ester bonds in pectin (Garg et al., 2016). On a global scale, the market sale of pectinase accounts for approximately 25% (Sethi et al., 2016a). This enzyme has a wide array of potency in various sectors like clarifying fruit juice, fermenting tea and coffee, the pulp and paper industry, processing fruits and vegetables, oil extraction, retting vegetable fibres, and more (Schwan and Wheals, 2004).

The main industrial source of pectinase is micro-organisms; nevertheless, about 50% of the pectinase is produced from fungi as well as yeast and 35% from bacteria. The remaining 15% of the enzyme is obtained from the plants (Anisa and Girish, 2014). The most common species of fungi used for pectinase production are *Aspergillus*, *Rhizopus*, and *Penicillium* (Ramos-Ibarra et al., 2017).

Among all the substrates investigated, rice bran resulted in maximum pectinase activity (39.1 U/ml/min) from *A. niger* IBT-7 (Abdullah et al., 2018). Owing to the rice's rich source of minerals like phosphorus, potassium, and calcium, rice bran is an excellent substrate for enzyme production. Besides, it also provides anchorage to the fungus that helps in the substrate's assimilation and utilization, leading to higher enzyme production (Chutmanop et al., 2008). Pectinase production using different cheap residues, such as coffee pulp, orange peel, sugarcane bagasse, lemon peel, and banana peel as potential carbon sources and wheat bran as nitrogen sources, were investigated using *A. tamarii* and the SSF technique (Shanmugavel et al., 2018). The results indicated that a combination of wheat bran and sugarcane bagasse (3:1) resulted in higher enzyme yields (101.05 U/ml).

Similarly, pectinase production has been carried out with *Bacillus* sp. using various other agro-industrial wastes like apple pomace, wheat straw, orange peel, wheat bran, sugarcane bagasse, and many other substrates (Embaby et al., 2014; Bibi et al., 2016; Jahan et al., 2017; Kaur and Gupta, 2017; Kuvvet et al., 2019). Pineapple stem extract has also been explored for enzyme production with *B. subtilis* BKDS1 andSmF conditions (Kavuthodi and Sebastian, 2018). During fermentation, the metal ion Ca^{2+} and $CaCO_3$ have a profound impact on the production as $CaCl_2$ stimulates pectinase activity (Oumer and Abate, 2017). Moreover, $CaCO_3$ acts as an enzyme regulator since the enzyme could be denatured in acidic conditions (Zou et al., 2014). Further, when compared to a shaker, the enzyme production rate was faster in the fermenter at 24 h; the maximum activity of 1437.723 U/ml was achieved in a fermenter, whereas activity of 1510.391 U/ml was obtained in a shaker at 48 h. The overall study proves pineapple stem to be an efficient substrate

A novel cost-effective strategy known as the two-layered strategy was developed by Qadir et al. (2020) in which a co-culture of yeast (*Saccharomyces cerevisiae* MK-157 and *Geotrichum candidum* AA15) after immobilization on corn cob was employed for enzyme production using orange peels as the raw material; further optimization of the significant factors was conducted with the Box–Behnken and Plackett–Burman (PB) design. Although orange peels have been widely used for enzyme production from free cells until now, it had not been explored as a substrate for immobilized co-culture cells on corncobs. Pre-treatment of the corn cob is required prior to immobilization due to its smooth structure; cells cannot adhere to its surface (Genisheva et al., 2011). Thus, an alkaline pre-treatment is a prerequisite since its opens up the pores, making it feasible for cell attachment leading to higher immobilization efficiency (Lee et al., 2012). Co-culturing of yeast cells in immobilized form is an effective and economical process as it allows complex interactions between the two strains that allow them to ferment a heterogenous substrate.

6.3.3 Protease

Peptidyl-peptide hydrolase, or as it is commonly known, protease (EC 3.4.21-24), is one of the most pivotal and abundant classes of enzyme that cleave the peptide bond through the hydrolysis reaction, thus allowing the fusion of the amino acid sequence present in the polypeptide chain resulting in the formation of protein (Sawant and Nagendran, 2014). It is an enzyme of commercial importance that contributes to approximately 60% of the global market share (Sharma et al., 2017) and has potential in different areas such as baking, food, pharmacy, detergents, photography, cosmetics, and other sectors (Abdel-Naby et al., 2017).

Depending on the pH, Protease enzyme are segregated into acidic (pH 2–5), neutral (pH 7), and alkaline (pH 8–11) groups (Mukhtar and Haq, 2008). Various species of bacteria, yeast, and molds, especially of the genus *Bacillus* (Qureshi et al., 2011), *Rhizopus* (Benabda et al., 2019), and *Penicillium* (Benmrad et al., 2018), have been explored for protease production using both SmF and SSF techniques.

For example, using 3 g/l coffee pulp waste in combination with 2 g/l corn cob as the substrate, a protease yield of 920 U/ml was achieved after 60 h of fermentation at pH 8 and 37°C temperature with *Bacillus* sp. BT MASC 3 after optimizing with the statistical technique (Kandasamy et al., 2016). Both substrates (coffee pulp waste and corn cob) are cheap as well as readily available, and utilization of these substrates indicates feasibility for the economical production of protease. Further, optimizations of the process parameters by statistical models have been widely used in several biotechnological processes owing to statistical optimization techniques suitability and applicability. Similarly, among the various substrates tested for their efficiency in enzyme production, chickling vetch peels supported enzyme production and resulted in a maximum yield of about 499.99 ± 11 U/ml and 5266.8 ± 202.5 U/gds in both liquid static surface culture and SSF technique, respectively, using *A. terreus* NCFT 4269.10(Sethi et al., 2016b). According to the kinetic studies, supplementation of riboflavin as the fermentation medium led to the maximum growth of *A. terreus*. Although both the fermentation techniques exhibited efficiency during bioprocess, SSF proved to be superior since protein degradation and catabolite repression by proteases was absent in SSF compared to SmF.

A study conducted by Hammami et al. (2018) revealed that in a comparative study of the various substrates, the maximal activity of alkaline protease was acquired from wheat bran with *B. mojavensis* under SmF conditions at 200 rpm. It was found that the enzyme activity and the biomass accelerated with simultaneous enhancement in substrate concentration. Further, in the absence of a yeast extract as a nitrogen source, an enhanced activity of approximately 1000 U/ml was obtained from wheat bran. As per the chemical composition, wheat bran is comprised of a high concentration of carbohydrates (60.48%), proteins (21.13%), ash (11.66%), and lipids (5.45%). Hence, wheat bran is a complete medium and constitutes the desired components essential for growth. In addition, in contrast within organic sugars like glucose, maltose, and sucrose, wheat bran is considered a suitable substrate that inhibits catabolite repression and elevates enzyme production (Sinsuwan et al., 2015). In the same way, under SSF fermentation, *Neocosmospora* sp. N1 grown on wheat bran supported the fabrication of alkaline protease (Matkawala et al., 2019). Interestingly, supplementation of wheat bran with custard apple seed powder (4:1) led to a 1.88-fold increase in the activity. This may be due to the presence of the porous structure in both these substrates that facilitates the adsorption of fungus onto the substrates for better nutrient assimilation and growth. Besides, the protein content of wheat bran as well as custard apple seed powder is 21.13% and 57.61%, respectively. Moreover, the C:N ratio of the wheat bran and custard apple seed powder substrate mixture is 1.58, which is less than wheat bran; thus, these substrates favoured enzyme activity.

In another study, the optimization of enzyme production by Evolutionary Operation (EVOP) factorial design technique resulted in an elevated enzyme activity using a mixture of wheat bran and soybean meal (4:1) from *Rhizopus oryzae* (SN5)/NCIM 1447 under SSF (Negi et al., 2020). Further optimization with an artificial neural network enhanced the enzyme yield by using a combination of wheat bran and soybean meal in a 70:30 ratio at 30°C. The texture, porosity, as well as nutritional composition of the wheat bran supported microbial growth. However, incorporating soybean meal into the medium enhanced the yield by 17% since it is rich in protein. However, with the increase in the concentration of soybean meal, a decline in the activity was seen that may be attributed to the inhibitors present in soybean meal.

6.3.4 LIPASE

Lipase or triacylglycerol hydrolase (EC 3.1.1.3) hydrolyzes the ester linkages of triglycerides to free fatty acids, glycerol, diacylglycerol, and monoacylglycerol (Treichel et al., 2010). Apart from this, lipase is also responsible for catalyzing various trans-esterification, esterification, and inter-esterification reactions both in non-aqueous and micro-aqueous media (Knappmann et al., 2017). Owing to this versatility, lipase has numerous applications in the food industry, detergents, biodiesel production, cosmetics, biosensors, pharmaceuticals, etc. (Hemlata et al., 2016; Memarpoor-Yazadi et al., 2017; Narwal and Pundir, 2017; Selvakumar and Sivashanmugam, 2017). Amongst all the microbial sources, fungal lipases are widely used (Musa and Adebayo-Tayo, 2012).

Several residues have been investigated during lipase production under both fermentation conditions: SSF and SmF. However, wheat bran has proven to bean efficient raw material in submerged fermentation from *A. flavus* at pH 7.15 by using sequential experimental design as an optimization technique (Colla et al., 2016). Besides wheat bran, soyabean oil was used as an inducer that enhanced enzyme production. In contrast with olive oil, soybean oil was used as an inducer as it was cheaper and easily available. Moreover, higher enzyme levels from wheat bran with submerged fermentation were possible because after boiling, it provided a homogenous culture medium.

Jatropha press cake, which comprises almost 40% protein, was investigated as a substrate for enzyme production using *Rhizomucor miehei* CBS 260.62 and *A. niger* 6516, which revealed its efficiency as both fungi exhibited good growth and high levels of lipase activity (Ilmi et al., 2017). Pre-treatment of press cake with alkaline 2.5% NaOH for 45 min aided in the enhanced productivity of the enzyme, which may be due to the increase in the amount of reducing sugar and protein that might be available for microbial attack, facilitating their growth and lipase activity. The study also indicated that the addition of maltodextrin and dextrose in the fermentation medium supported microbial growth but inhibited lipase production due to the catabolite repression of sugars. Catabolite repression is one of the well-known phenomena found in the *Aspergillus* species. Basically, the CREA is a protein which is known as Carbon Catabolite Responsive Element present in Aspergillus species as well as other microorganisms protein plays a vital role in initiating this repression and the gene encoding this protein *cre*A is widely present in the various species of *Aspergillus* (Ruijter and Visser, 1997). In the case of *R. miehei*, catabolite repression has not been described; however, inhibition of lipase in the presence of glucose has been reported (Davranov and Khalameizer, 1997). The inhibition of Lipase takes place at the transcriptional level, and a similar mechanism might have taken place in both fungal species in the presence of glucose and maltodextrin.

Combined statistical approaches, such as one-factor-at-a-time design, PB design and face-centre-centre-composite design, have been successfully employed to optimize the production of halophilic lipase from *Fusarium solani* NFCCL 4084 with palm oil mill effluent (POME) as production medium (Geoffry and Achur, 2018). Higher levels of lipase activity were recorded that may be credited to the presence of saturated fatty acids in POME (Das et al., 2017). Nevertheless, after an optimum value, the yield was reduced either because of the minerals required in minor quantities or due to the presence of palm oil that might have interfered with the oxygen diffusion and aeration (Habib et al., 1997). Above all, supplementation of Tween-80 also enhanced the production; hence it has the potential to be used both as an inducer and surfactant during enzyme production because of the presence of oleic acid (Salihu et al., 2016).

Above all the substrates used, mango waste (peel, tegument, and kernel) was used either separately or in combination for SmF with *Yarrowia lipolytica* (Pereira et al., 2019). Amongst the peel, tegument, and kernel, the tegument gave the highest enzyme activity, and the incorporation of yeast extract in the medium elevated the activity (3412 U/ml) at pH 5.0, 27.9°C, and 187 rpm. With in the same parameters, except for agitation at 600 rpm, the maximum yield of 2500 U/ml was obtained in a four-litre bioreactor. The fibrous tegument constitutes lignin, hemicellulose, and cellulose favouring cell growth, metabolism, and further enzyme activity (Henrique et al., 2013).

A comparative study was undertaken of the different bioreactor designs (plugged flow bed with forced-water saturated aeration, pilot-scale rotating drum, and static flatbed) to achieve the utmost yield in SSF from *A. brasiliensis* by means of malt bagasse as a substrate (Eichler et al., 2020). Using a static bed bioreactor, maximal lipase production was obtained under the optimum conditions of pH 7.7, supplementation of 11.3% soyabean oil and 32.7°C. Within similar optimal parameters, a plugged flow bed with forced-water saturated aeration bioreactor demonstrated superior performance and achieved a higher yield. The static flatbed and plugged flow bed with forced-water saturated aeration bioreactor have similar conformation, and during the validation experiments, similar behaviour was observed. Moreover, solid-state beds with forced aeration system bioreactors are advantageous over others in terms of better heat transfer capacity and the ability to remove metabolic heat as well as carbon dioxide from the substrate (Mitchell et al., 2000). The lowest yield was obtained by using a rotating drum bioreactor; for homogenization of the substrates, mechanical mixing was carried out for 30 min after every 24 h, which might have disrupted the fungus (Mitchell et al., 2006).

6.3.5 β-GALACTOSIDASE

Lactase, the common name for β-galactosidase (EC 3.2.1.23), is classed in the glycoside hydrolase family of enzymes. This enzyme catalyzes the hydrolysis of lactose, resulting in glucose and galactose formation. Besides this, another important reaction carried out by this enzyme is the transgalactosylation reaction, wherein the biotransformation of lactose is carried out in prebiotics, such as galacto-oligosaccharides and lactulose (Nagy et al., 2001; Panesar et al., 2006). Owing to these catalytic properties, lactase is of commercial importance and has varied applications in food, dairy, fermentation, and other industrial areas (Pawlak-Szukalska et al., 2014).

Lactase is ubiquitous in nature and is naturally present in plants, animal organs, as well as microbes, the latter being the preferred source of the enzyme (Haider and Husain, 2007). The most common microbial sources employed for enzyme production are *Kluyveromyces fragilis*, *K. lactis*,

A. niger, A. oryzae, and *Bacillus* sp. However, the commercial preparations of lactase are carried out by *Aspergillus* and *Kluyveromyces* species. The source of the lactase has a profound impact on its properties (such as molecular weight, amino acid chain length, the position of the active site, temperature, and optimum pH levels and stability), structure, and specificity (Mlichova and Rosenberg, 2006; Panesar et al., 2010a). Though this enzyme has potential industrial application, the high cost hinders its application. Therefore, diverse approaches have been explored to curtail the production cost by using inexpensive substrates.

Several agro-industrial wastes were compared as potential sources during lactase production using halotolerant fungi *A. tubingensis* GR1 under solid-state conditions (Raol et al., 2015). Using a statistical technique (central composite design), a yield of 15,936 U/gds was recorded using wheat bran supplemented with corn steep liquor. Wheat bran acts as an inducer for the expression of galactosidase in several microbial species. Further, the presence of high hemicellulose content, sugar, ammonium ions, and vitamins in wheat bran as well as in corn steep liquor led to the elevated levels of enzyme yield (Awan et al., 2011).

Screening of the two agro-industrial residues, rice straw and wheat straw, during enzyme production with *Bacillus* sp. under SSF indicated rice straw as an efficient substrate for enzyme production (2689 U/mg) after 72 h (Kazemi et al., 2017). However, a combination of wheat straw (30%) and peanut pod (70%) positively impacted lactase production, and there was an increase in the enzyme yield by 1.369-fold. In a similar study, *B. licheniformis* ATCC 12759 was cultivated on rice bran, indicating improved lactase activity (Akcan, 2018). Further, Erlenmeyer flasks of various capacities (100, 250, 500, and 1000 ml) were used to investigate the scale-up process of enzyme production. Although the results were satisfactory, there was a decline in the yield with a subsequent increase in substrate concentration because of reduced aeration. This indicates that rice bran comprises various nutrients that act as a good carbon and nitrogen source. An ideal substrate is a prerequisite during fermentation as it would provide all the vital nutrients for microbial growth and gain higher yields (Singh and Soni, 2001).

A. niger strain was grown in three separate media: soybean residue, soya okara, and soymilk, to determine the efficiency of these residues in lactase production (Martarello et al., 2019). The results indicated that 2% (w/v) soybean residue gave the highest yield of 24.64 IU/ml. The reduction in the enzyme activity in the soya okara medium may be due to the presence of phytic acids that act as chelating agents that bind the minerals present in the residue, making it impossible for assimilation by fungus (O'Toole, 1999). Further optimization studies by statistical techniques revealed that apart from all the parameters, pH and temperature had a profound effect on the activity. Khayati et al. (2014) studied β-galactosidase production by *B. licheniformis* from various cheap substrates under solid-state using a statistical experimental design. The statistical technique known as Taguchi Orthogonal Array Design was used for the optimization of process parameters during beta galactosidase production and the study indicated that among the various factors, the concentration of pea pod, incubation time, inducer (lactose), and C:N ratio had a profound influence on the enzyme activity.

6.3.5.1 Case Study: Utilization of Agro-Industrial Wastes for the Production of Lactase from *Rhizomucor pusillus*

A study was carried out by Kaur et al. (2018) to investigate the efficiency of various lignocellulose biomass (apple pomace, rice husk, wheat bran, sugarcane bagasse, kinnow peel, and rice bran) used as carbon sources during lactase production with *Rhizomucor pusillus* under SSF conditions:5 g of the respective substrates were taken, supplemented with 3% (w/w) $(NH_4)_2SO_4$. Water and deproteinized whey were added to the fermentation medium as two different moistening agents. Extraction was carried out by mixing the medium in the buffer solution (sodium citrate) and shaking the mixture for about 1 h. The mixture was further filtered, centrifuged, and the aliquot was taken to check the enzyme's activity. Enzymatic assay was determined per the procedures defined by Réczey et al. (1992).

Among the different substrates, wheat bran gave a maximum yield (4.55 IU/gds) when deproteinized whey was used as a moistening agent in contrast with the distilled water, which led to lower

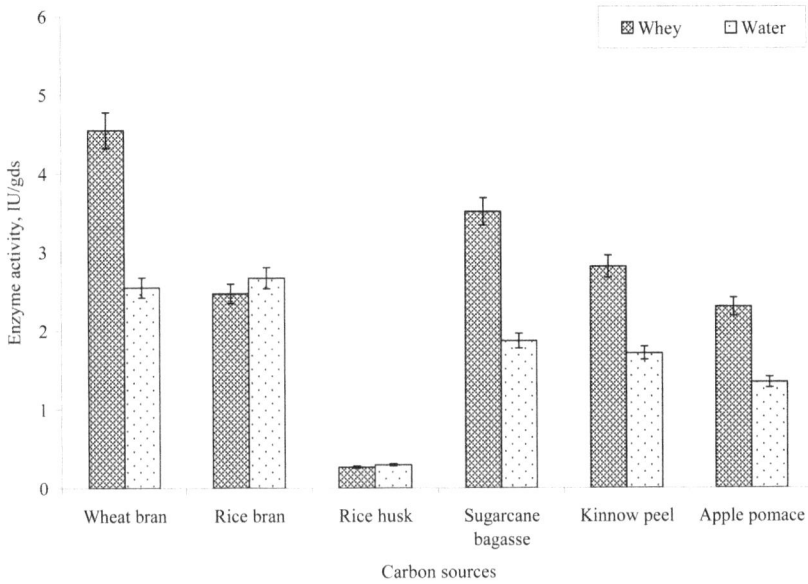

FIGURE 6.5   Effect of the various agro-industrial residues on β-galactosidase production with *Rhizomucor pusillus.*

enzyme activity (2.55 IU/gds). However, the highest activity of 2.67 IU/gds was obtained using water as a moistening agent with rice bran as a substrate (Figure 6.5).

From the preliminary optimization studies, a lactase yield of 101.89 IU/gds was obtained under optimal conditions from 15 g wheat bran containing 50% moisture by using deproteinized whey (50%) and supplemented with corn steep liquor (7%,v/w) and magnesium sulphate (0.5%, w/w). The other optimal process parameters were5.5 pH, 1×10^6 spores/ml inoculum size, 50°C temperature, and 7 days of fermentation time.

Although studies have been reported using both fermentation processes (SmF and SSF), solid-state is the favoured technique since it facilitates the use of inexpensive substrates during any bio-process. Besides, it is environmentally friendly, and it is known as the natural habitat of fungi, which is favoured for the growth of microbes and valorization of cheap residues into high-value components making the whole process cost-effective (Leite et al., 2019).

6.4   CHALLENGES IN THE ENZYME PRODUCTION FROM LIGNOCELLULOSE BIOMASS

Although numerous biotechnological strategies have been developed, studies are being undertaken to use cheap agricultural and industrial wastes as carbon and nitrogen sources for the cost-effective production of enzymes by SmF and SSF. Still, some technical barriers in both these fermentation techniques hinder the industrial production of enzymes for both SmF and SSF.

6.4.1   SmF

In contrast to SSF, SmF poses fewer technical challenges. Nonetheless, a major issue that arises during the utilization of the agro-industrial residues is their heterogenous nature, which leads to reduced enzyme productivity, and can only be overcome by initial pre-treatment of the substrates. Besides this, the other major issue is the problem faced during the sterilization process, again because of the wastes' heterogeneous nature (Hansen et al., 2015).

6.4.2 SSF

SSF is one of the preferred methods for the fabrication of enzymes using agro-industrial wastes since it leads to higher yields and helps maintain the economy of the process. Yet, there are some challenges in heat dispersal, biomass determination, and scale-up that limit its industrial application.

6.4.2.1 Heat Dispersal

During the course of the reaction, owing to the compact nature of the residues, low thermal conductivities, high heat generation, and poor heat removal are observed, which influences not only the metabolic activities of the micro-organisms, such as growth, spore formation, etc. but also affect product formation, because high heat may lead to denaturation of the enzyme (Pandey, 2003). Consequently, to evade this problem, various alterations have been carried out in the design of the bioreactor, such as using column reactors or packed-bed reactors for fermentation (Dilipkumar et al., 2013). These reactors allow enhanced mass and heat transfer, continuous gas monitoring, and high substrate loading capacity (Thomas et al., 2013). Several alterations have also been carried out in these reactors (multiplex gas sampler, counter-current bioreactor, and aerated moving bed bioreactor) that facilitate not only continuous production but also optimization studies in the SSF process (Pliego-Sandoval et al., 2012; Piedrahita-Aguirre et al., 2014).

6.4.2.2 Estimation of Biomass

Biomass determination is another crucial factor during the fermentation process to understand the physiological characteristics of the microbial cell and the kinetic study of any bioprocess. However, the separation of biomass and its estimation impedes the SSF process. Traditional estimation methods involve the determination of biological activities, measurement of biological components, light scattering, DNA assay, light reflectance, and rate of carbon dioxide production that are quite tedious and time-consuming (Kennedy et al., 1992; Murthy et al., 1993). In recent times, major breakthroughs involve developing novel techniques for biomass determination: coupling of the computational method and dynamic imaging. They are preferred over the other methods in terms of cost and accuracy of the results (Yingyi et al., 2012).

6.4.2.3 Scaling Up the Process

SSF fermentations are generally carried out in trays or flasks that involve the utilization of a few grams of the substrate. Thus, scaling-up SSF process is difficult due to the inadequate space and poor control of aeration as well as temperature. Accordingly, to industrialize the SSF process, several changes have been made to the bioreactor design (fixed-bed bioreactors, rotary drums, aerated trays), mixing strategies (intermittent mixing), mode of fermentation (batch, fed-batch, continuous) to achieve maximum titres of enzyme from the agro-industrial residues (Figueroa-Montero et al., 2011; Thomas et al., 2013; Flodman and Noureddini, 2013; Da Silveira et al., 2014).

For instance, higher pectinase activity (1840 U/kg/h) was obtained using sugarcane bagasse (3 kg) and wheat bran (27 kg) in a 40-cm-height bed after 10 h in a packed-bed reactor (Pitol et al., 2016). Similarly, using a column tray reactor, higher yields of pectinase were reported operating at an air flow rate of 194 ml/min (Ruiz et al., 2012). Recently, fixed-bed bioreactors with forced aerators were used to produce various enzymes and have been proven to be a promising alternative to commercialize the SSF process (Castro et al., 2015).

6.5 GLOBAL STATUS

Enzymes are the biological catalysts involved in various chemical reactions required in life. Their ability to perform specific chemical biotransformation reactions has promoted their use in large-scale industrial processes. Their industrial applications account for almost 80% of the global market, and the demand for industrial enzymes is on an incessant rise (Van Oort, 2010; Kumar and Singh, 2013).

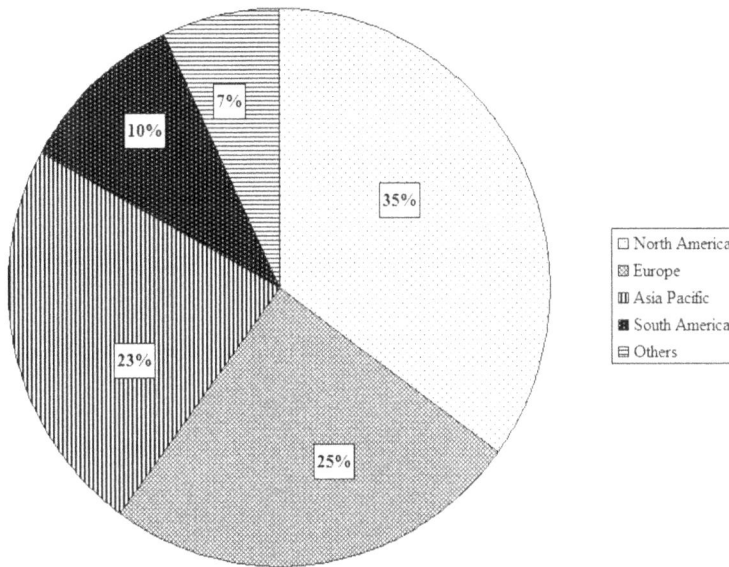

FIGURE 6.6 Worldwide enzyme market by region (Source: https://www.marketsandmarkets.com/Market-Reports/enzyme-market-46202020.html).

Globally, the market size in 2019 was valued to be US$9.9 billion, which is predicted to increase at a Compound Annual Growth Rate of 7.1% from 2020 to 2027 (https://www.grandviewresearch.com/industry-analysis/enzymes-industry). The growth of the market size is driven by the high demand for the enzymes used in the pharmaceutical sector due to the increase in the pervasiveness of chronic disease and the rise in the need for renewable energy sources, such as biofuels. Apart from this, the expansion of the food and beverage sector and the requirement for improved food flavour, texture, and quality are major factors for enhancement in the market size of enzymes (https://www.gminsights.com/industry-analysis/enzymes-market).

In 2018, North America had the biggest share in the global enzyme market, followed by Europe, and this is foreseen to rise in the forecast period (Figure 6.6). Various technological advancements in this region that allowed the use of enzymes in various sectors have boosted the market size. In addition, the pharmaceutical sector is also witnessing a rise in enzyme use, and European region has also observed a surge in the feed industry that led to the highest market share in the world (http://www.marketsandmarkets.com/Market-Reports/enzyme-market-46202020.html). On the other hand, the Asia-Pacific share in the global enzyme market is also estimated to rise in the upcoming years due to the rapid occurrence of chronic diseases.

Some of the global players in the enzyme sector include Novozymes, BASF SE. DuPont, Amano Enzyme Inc., AB Enzyme GbmH, Koninklijke DSM N.V., Advanced Enzymes Technologies Ltd., Nexgen Biotechnologies, Inc., Chr. Hansen Holding A/S, and Hayashibara Company. Examples of some of the important enzymes available in the market are depicted in Table 6.3.

6.6 CONCLUSION AND FUTURE PERSPECTIVE

Globally, due to the expansion in the agricultural and food-processing sector, enormous quantities of waste, known as agro-industrial residues, pose serious environmental threats if left untreated. Hence, effective utilization of these residues into high-valued components is the need of the hour. In order to combat this issue, various academic and industrial sectors have focused their attention on the utilization of these wastes for the development of cost-effective strategies, and this area of research has been in the hotspot recently because of the presence of

TABLE 6.3

Examples of Commercially Available Enzymes

S. No	Enzymes	Companies
1.	Amylase	Novozymes, Denmark
		DuPont, USA
		Royal DSM N.V., Netherlands
		AB Enzymes, Germany
		Maps Enzymes Limited, Gujrat, India
2.	Protease	Novozymes, Denmark
		Royal DSM N.V., Netherlands
		AB Enzymes, Germany
		Amano Enzymes, Japan
		DuPont, USA
		Maps Enzymes Limited, Gujrat, India
3.	Lipase	DuPont, USA
		Novozymes, Denmark
		Amano Enzymes, Japan
		Maps Enzymes Limited, Gujrat, India
4.	Pectinase	Novozymes, Denmark
		DuPont, USA
		AB Enzymes, Germany
		Royal DSM N.V., Netherlands
		Shandong Sukahan Bio-Technology Co.,Ltd., China
5.	Lactase	Novozymes, Denmark
		DuPont, USA
		Amano Enzymes, Japan
		DSM Food Specialities, Netherlands
		Enzyme Development Corporation (EDC), New York, USA

abundant and easily available waste. Various biotechnological tools and techniques have been employed to successfully utilize these residues in products of commercial worth, especially enzymes.

Since ancient times, enzymes have been an integral part of human life, and enzyme technology is growing rapidly worldwide. Further, the market size of enzymes is quite large, and the global market size is forecasted to grow at a very rapid rate in the upcoming years as these enzymes have potential in various food, pharmaceutical, and other industrial sectors. Thus, the production of the enzymes from various inexpensive agro-industrial residues has gained tremendous impetus over the past few years. Different fermentation techniques such as SmF and SSF have been investigated in bioprocessing to attain higher yields. However, in contrast with the latter, the former technique is widely preferred with respect to several techno-economic benefits. Despite this, the industrial production of enzyme by SSF is hindered due to various challenges like heat dispersal, bioreactor configuration, etc. Thus, there is a need to develop robust bioreactor designs, and computerization can overcome the issue of scaling-up and lead to the development of an economical process. Further, it is foreseen that a combination of the utilization of agro-industrial residues and several allied technologies will solve the problem of environmental pollution and contribute to sustainable agricultural techniques. In addition, advancements in biotechnological techniques, such as genetic engineering, *in silico* characterization, micro-array technique, enzyme engineering, and several other techniques could be practised to tailor the properties of micro-organisms or enzymes as per demand.

REFERENCES

Abbas, A., and Ansumali, S. 2010. Global potential of rice husk as a renewable feedstock for ethanol biofuel production. *BioEnergy Research* 3(4): 328–334.

Abdel-Naby, M.A., Ahmed, S.A., Wehaidy, H.R. et al. 2017. Catalytic, kinetic and thermodynamic properties of stabilized *Bacillus stearothermophilus* alkaline protease. *International Journal of Biological Macromolecules* 96: 265–271.

Abdullah, R., Farooq, I., Kaleem, A. et al. 2018. Pectinase production from *Aspergillus niger* IBT-7 using solid state fermentation. *Bangladesh Journal of Botany* 47(3): 473–478.

Adapa, P.K., Tabil, L.G., Schoenau, G.J. et al. 2011. Quantitative analysis of lignocellulosic companents of non-treated and stream exploded barley, canola, oat and wheat straw using fourier transform infrared spectroscopy. *Journal of Agricultural Science and Technology B* 1: 177–188.

Ahmed, I., Zia, M.A., Iftikhar, T. et al. 2011. Characterization and detergent compatibility of purified protease produced from *Aspergillus niger* by utilizing agro wastes. *BioResources* 6(4): 4505–4522.

Ahmed, S.A., Mostafa, F.A., Helmy, W.A. et al. 2017. Improvement of bacterial α-amylase production and application using two steps statistical factorial design. *Biocatalysis and Agricultural Biotechnology* 10: 224–233.

Akcan, N. 2018. Cultural conditions optimization for production of β-galactosidase from *Bacillus licheniformis* ATCC 12759 under solid-state fermentation. *Turkish Journal of Biochemistry* 43(3): 240–247.

Akgül, M., Korkut, S., Çamlıbel, O. et al. 2013. Some chemical properties of luffa and its suitability for medium density fiberboard (MDF) production. *BioResources* 8(2): 1709–1717.

Alikunju, A.P., Joy, S., Rahiman, M. et al. 2018. A statistical approach to optimize cold active β-galactosidase production by an arctic sediment pscychrotrophic bacteria, *Enterobacter ludwigii* (MCC 3423) in cheese whey. *Catalysis Letters* 148(2): 712–724.

Anisa, S.K., and Girish, K. 2014. Pectinolytic activity of *Rhizopus* sp. and *Trichoderma viride*. *International Journal of Research in Pure and Applied Microbiology* 4(2): 28–31.

Anto, H., Trivedi, U.B., and Patel, K.C. 2006. Glucoamylase production by solid-state fermentation using rice flake manufacturing waste products as substrate. *Bioresource Technology* 97(10): 1161–1166.

Anwar, Z., Gulfraz, M., and Irshad, M. 2014. Agro-industrial lignocellulosic biomass a key to unlock the future bio-energy: A brief review. *Journal of Radiation Research and Applied Sciences* 7(2): 163–173.

Aslam, F., Ansari, A., Aman, A. et al. 2020. Production of commercially important enzymes from *Bacillus licheniformis* KIBGE-IB3 using date fruit wastes as substrate. *Journal of Genetic Engineering and Biotechnology* 18(1): 1–7.

Aunstrup, K. 1974. Industrial production of proteolytic enzymes. *Industrial Aspects in Biochemistry* 30(1): 23–46.

Awan, M.S., Tabbasam, N., Ayub, N. et al. 2011. Gamma radiation induced mutagenesis in *Aspergillus niger* to enhance its microbial fermentation activity for industrial enzyme production. *Molecular Biology Reports* 38(2): 1367–1374.

Azad, M.A.K., Bae, J.H., Kim, J.S. et al. 2009. Isolation and characterization of a novel thermostable α-amylase from Korean pine seeds. *New Biotechnology* 26(3–4): 143–149.

Babbar, N., and Oberoi, H.S. 2014. Enzymes in value-addition of agricultural and agro-industrial residues. In: *Enzymes in Value-Addition of Wastes*, eds. S.K. Brar, and M. Verma, pp. 29–50. Nova Publishers.

Beaugrand, J., Reis, D., Guillon, F. et al. 2004. Xylanase-mediated hydrolysis of wheat bran: Evidence for subcellular heterogeneity of cell walls. *International Journal of Plant Sciences* 165(4): 553–563.

Benabda, O., M'hir, S., Kasmi, M. et al. 2019. Optimization of protease and amylase production by *Rhizopus oryzae* cultivated on bread waste using solid-state fermentation. *Journal of Chemistry* 2019: 1–9

Benmrad, M.O., Emna, M., Mouna, B.E. et al. 2018. Production, purification, and biochemical characterization of serine alkaline protease from *Penicillium chrysogenium* strain x5 used as excellent bio-additive for textile processing. *International Journal of Biological Macromolecules* 119: 1002–1016.

Bhanwar, S., and Ganguli, A. 2014. α-Amylase and β-galactosidase production on potato starch waste by *Lactococcus lactis* subsp *lactis* isolated from pickled yam. *Journal of Scientific and Industrial Research* 73(5): 324–330.

Bibi, N., Ali, S., and Tabassum, R. 2016. Statistical optimization of pectinase biosynthesis from orange peel by *Bacillus licheniformis* using submerged fermentation. *Waste and Biomass Valorization* 7(3): 467–481.

Borszcz, V., Boscato, T.R.P., Antunes, A. et al. 2017. Recovery of pectinase obtained by solid-state cultivation of agro-industrial residues. *Industrial Biotechnology* 13(3): 141–148.

Bos, A., and Hamelinck, C. 2014. *Greenhouse Gas Impact of Marginal Fossil Fuel Use*. Project Number: BIENL14773 2014.

Bretón-Toral, A., Trejo-Estrada, S.R., and McDonald, A.G. 2016. Lactic acid production from potato peel waste, spent coffee grounds and almond shells with undefined mixed cultures isolated from coffee mucilage from Coatepec Mexico. *Fermentation Technology* 6: 1–6.

Calvo-Flores, F.G., and Dobado, J.A. 2010. Lignin as renewable raw material. *ChemSusChem* 3(11): 1227–1235.

Carrijo, O.A., Liz, R.S.D., and Makishima, N. 2002. Fiber of green coconut shell as an agricultural substrate. *Horticultura Brasileira* 20(4): 533–535.

Castro, A.M., Castilho, L.R., and Freire, D.M. 2015. Performance of a fixed-bed solid-state fermentation bioreactor with forced aeration for the production of hydrolases by *Aspergillus awamori*. *Biochemical Engineering Journal* 93: 303–308.

Cherubin, M.R., Oliveira, D.M.D.S., Feigl, B.J. et al. 2018. Crop residue harvest for bioenergy production and its implications on soil functioning and plant growth: A review. *Scientia Agricola* 75(3): 255–272.

Choonut, A., Saejong, M., and Sangkharak, K. 2014. The production of ethanol and hydrogen from pineapple peel by *Saccharomyces cerevisiae* and *Enterobacter aerogenes*. *Energy Procedia* 52: 242–249.

Chugh, P., Soni, R., and Soni, S.K. 2016. Deoiled rice bran: A substrate for co-production of a consortium of hydrolytic enzymes by *Aspergillus niger* P-19. *Waste and Biomass Valorization* 7(3): 513–525.

Chutmanop, J., Chuichulcherm, S., Chisti, Y. et al. 2008. Protease production by *Aspergillus oryzae* in solid-state fermentation using agroindustrial substrates. *Journal of Chemical Technology and Biotechnology: International Research in Process, Environmental and Clean Technology* 83(7): 1012–1018.

Colla, L.M., Primaz, A.L., Benedetti, S. et al. 2016. Surface response methodology for the optimization of lipase production under submerged fermentation by filamentous fungi. *Brazilian Journal of Microbiology* 47(2): 461–467.

Costa, T.M., Hermann, K.L., Garcia-Roman, M. et al. 2017. Lipase production by *Aspergillus niger* grown in different agro-industrial wastes by solid-state fermentation. *Brazilian Journal of Chemical Engineering* 34(2): 419–427.

Cruz, J.M., Domínguez, J.M., Domínguez, H. et al. 2000. Preparation of fermentation media from agricultural wastes and their bioconversion into xylitol. *Food Biotechnology* 14(1–2): 79–97.

da Câmara Rocha, J., da Silva Araújo, J., de Paiva, W.K.V. et al. 2020. Yellow mombin pulp residue valorization for pectinases production by *Aspergillus niger* IOC 4003 and its application in juice clarification. *Biocatalysis and Agricultural Biotechnology* 30: 101876.

da Silva, L.L. 2016. Adding value to agro-Industrial wastes. *Industrial Chemistry* 2(2): e103.

Da Silveira, C.L., Mazutti, M.A., and Salau, N.P. 2014. Modeling the microbial growth and temperature profile in a fixed-bed bioreactor. *Bioprocess and Biosystems Engineering* 37(10): 1945–1954.

Das, A., Bhattacharya, S., Shivakumar, S. et al. 2017. Coconut oil induced production of a surfactant-compatible lipase from *Aspergillus tamarii* under submerged fermentation. *Journal of Basic Microbiology* 57(2): 114–120.

Davin, L.B., and Lewis, N.G. 2005. Lignin primary structures and diligent sites. *Current Opinion in Biotechnology* 16(4): 407–415.

Davranov, K., and Khalameizer, V.B. 1997. Current state of the study of microbial lipases. *Chemistry of Natural Compounds* 33(2): 113–126.

de Azevedo, W.M., de Oliveira, L.F.R., Alcântara, M.A. et al. 2020. Turning cacay butter and wheat bran into substrate for lipase production by *Aspergillus terreus* NRRL-255. *Preparative Biochemistry and Biotechnology* 50(7): 1–.

de Souza, P.M., and de Oliveira Magalhães. 2010. Application of microbial α-amylase in industry- A review. *Brazilian Journal of Microbiology* 41(4): 850–861.

Dhillon, G.S., Brar, S.K., Kaur, S. et al. 2012. Lactoserum as a moistening medium and crude inducer for fungal cellulase and hemicellulase induction through solid-state fermentation of apple pomace. *Biomass and Bioenergy* 41: 165–174.

Dilipkumar, M., Rajamohan, N., and Rajasimman, M. 2013. Inulinase production in a packed bed reactor by solid state fermentation. *Carbohydrate Polymers* 96(1): 196–199.

dos Santos, R.M., Neto, W.P.F., Silvério, H.A. et al. 2013. Cellulose nanocrystals from pineapple leaf, a new approach for the reuse of this agro-waste. *Industrial Crops and Products* 50: 707–714.

Ebringerová, A., Hromádková, Z., and Heinze, T. 2005. Hemicellulose. *Advances in Polymer Science* 186: 1–67.

Eichler, P., Bastiani, D.C., Santos, F.A. et al. 2020. Lipase production by *Aspergillus brasiliensis* in solid-state cultivation of malt bagasse in different bioreactors configurations. *Anais da Academia Brasileira de Ciências* 92(suppl 2):1–15.

El-Tayeb, T.S., Abdelhafez, A.A., Ali, S.H. et al. 2012. Effect of acid hydrolysis and fungal biotreatment on agro-industrial wastes for obtainment of free sugars for bioethanol production. *Brazilian Journal of Microbiology* 43(4): 1523–1535.

Elumalai, P., Lim, J.M., Park, Y.J. et al. 2020. Agricultural waste materials enhance protease production by *Bacillus subtilis* B22 in submerged fermentation under blue light-emitting diodes. *Bioprocess and Biosystems Engineering*, 43: 1–10.

Embaby, A.M., Masoud, A.A., Marey, H.S. et al. 2014. Raw agro-industrial orange peel waste as a low cost effective inducer for alkaline polygalacturonase production from *Bacillus licheniformis* SHG10. *SpringerPlus* 3(1): 327.

Fadel, M., AbdEl-Halim, S., Sharada, H. et al. 2020. Production of glucoamylase, α-amylase and cellulase by *Aspergillus oryzae* F-923 cultivated on wheat bran under solid state fermentation. *Journal of Advances in Biology and Biotechnology* 23(4): 8–22.

Food and Agriculture Organization (FAO). 2018. *The Future of Food and Agriculture-Alternative Pathways to 2050*. FAO.

Figueroa-Montero, A., Esparza-Isunza, T., Saucedo-Castañeda, G. et al. 2011. Improvement of heat removal in solid-state fermentation tray bioreactors by forced air convection. *Journal of Chemical Technology and Biotechnology* 86(10): 1321–1331.

Flodman, H.R., and Noureddini, H. 2013. Effects of intermittent mechanical mixing on solid-state fermentation of wet corn distillers grain with *Trichoderma reesei*. *Biochemical Engineering Journal* 81: 24–28.

Gaonkar, S.K., and Furtado, I.J. 2020. Valorization of low-cost agro-wastes residues for the maximum production of protease and lipase haloextremozymes by *Haloferax lucentensis* GUBF-2 MG076078. *Process Biochemistry* 101: 72–88.

García-Pérez, T., López, J.C., Passos, F. et al. 2018. Simultaneous methane abatement and PHB production by *Methylocystis hirsuta* in a novel gas-recycling bubble column bioreactor. *Chemical Engineering Journal* 334: 691–697.

Garg, G., Singh, A., Kaur, A. et al. 2016. Microbial pectinases: An ecofriendly tool of nature for industries. *3 Biotech* 6(1): 1–13.

Geneau-Sbartaï, C., Leyris, J., Silvestre, F. et al. 2008. Sunflower cake as a natural composite: Composition and plastic properties. *Journal of Agricultural and Food Chemistry* 56(23): 11198–11208.

Genisheva, Z., Mussatto, S.I., Oliveira, J.M. et al. 2011. Evaluating the potential of wine-making residues and corn cobs as support materials for cell immobilization for ethanol production. *Industrial Crops and Products* 34(1): 979–985.

Geoffry, K., and Achur, R.N. 2018. Optimization of novel halophilic lipase production by *Fusarium solani* strain NFCCL 4084 using palm oil mill effluent. *Journal of Genetic Engineering and Biotechnology* 16(2): 327–334.

Gois, I.M., Santos, A.M., and Silva, C.F. 2020. Amylase from *Bacillus* sp. produced by solid state fermentation using cassava bagasse as starch source. *Brazilian Archives of Biology and Technology* 63: 1–9.

Gouvea, B.M., Torres, C., Franca, A.S. et al. 2009. Feasibility of ethanol production from coffee husks. *Biotechnology Letters* 31(9): 1315–1319.

Graminha, E.B.N., Gonçalves, A.Z.L., Pirota, R.D.P.B. et al. 2008. Enzyme production by solid-state fermentation: Application to animal nutrition. *Animal Feed Science and Technology* 144(1–2): 1–22.

Gupta, R., Gigras, P., Mohapatra, H. et al. 2003. Microbial α-amylases: A biotechnological perspective. *Process Biochemistry* 38(11): 1599–1616.

Habib, M.A.B., Yusoff, F.M., Phang, S.M. et al. 1997. Nutritional values of chironomid larvae grown in palm oil mill effluent and algal culture. *Aquaculture* 158(1–2): 95–105.

Haider, T., and Husain, Q. 2007. Preparation of lactose free milk by using ammonium sulphate fractionated proteins from almonds. *Journal of the Science of Food and Agriculture* 87(7): 1278–1283.

Hammami, A., Bayoudh, A., Abdelhedi, O. et al. 2018. Low-cost culture medium for the production of proteases by *Bacillus mojavensis* SA and their potential use for the preparation of antioxidant protein hydrolysate from meat sausage by-products. *Annals of Microbiology* 68(8): 473–484.

Handa, S., Sharma, N., and Pathania, S. 2016. Multiple parameter optimization for maximization of pectinase production by *Rhizopus* sp. C4 under solid state fermentation. *Fermentation* 2(2): 10.

Hansen, G.H., Lübeck, M., Frisvad, J.C. et al. 2015. Production of cellulolytic enzymes from ascomycetes: Comparison of solid state and submerged fermentation. *Process Biochemistry* 50(9): 1327–1341.

Hemlata, B., Uzma, Z., and Tukaram, K. 2016. Substrate kinetics of thiol activated hyperthermostable alkaline lipase of *Bacillus sonorensis* 4R and its application in bio-detergent formulation. *Biocatalysis and Agricultural Biotechnology* 8: 104–111.

Henrique, M.A., Silvério, H.A., Neto, W.P.F. et al. 2013. Valorization of an agro-industrial waste, mango seed, by the extraction and characterization of its cellulose nanocrystals. *Journal of Environmental Management* 121: 202–209.

Himmel, M.E., Ding, S.Y., Johnson, D.K. et al. 2007. Biomass recalcitrance: Engineering plants and enzymes for biofuels production. *Science* 315(5813): 804–807.

Hon, D.N.S. 2000. Pragmatic approaches to utilization of natural polymers: Challenges and opportunities. In *Natural Polymers and Agrofibers Based Composites* (Eds. E. Frollini, A. Leao and A. H. C. Mattoso) CIP-Brasil pp. 1–14.

https://www.gminsights.com/industry-analysis/enzymes-market.

https://www.grandviewresearch.com/industry-analysis/enzymes-industry.

https://www.marketsandmarkets.com/Market-Reports/enzyme-market-46202020.html.

Ilmi, M., Hidayat, C., Hastuti, P. et al. 2017. Utilisation of Jatropha press cake as substrate in biomass and lipase production from *Aspergillus niger* 65I6 and *Rhizomucor miehei* CBS 360.62. *Biocatalysis and Agricultural Biotechnology* 9: 103–107.

Jahan, N., Shahid, F., Aman, A. et al. 2017. Utilization of agro waste pectin for the production of industrially important polygalacturonase. *Heliyon* 3(6): e00330.

Jokhio, M.A., Naqvi, H.A., Yasmin, A. et al. 2020. Starch hydrolyzing enzyme production from *Aspergillus niger* EFRL-FC-024 using molasses as carbon source. *Pure and Applied Biology* 10(1): 262–271.

Kandasamy, S., Muthusamy, G., Balakrishnan, S. et al. 2016. Optimization of protease production from surface-modified coffee pulp waste and corncobs using *Bacillus* sp. by SSF. *3 Biotech* 6(2): 167.

Kandra, L. 2003. α-Amylases of medical and industrial importance. *Journal of Molecular Structure: THEOCHEM* 666: 487–498.

Karam, E.A., Wahab, W.A.A., Saleh, S.A. et al. 2017. Production, immobilization and thermodynamic studies of free and immobilized *Aspergillus awamori* amylase. *International Journal of Biological Macromolecules* 102: 694–703.

Kaur, R., Panesar, P.S., and Singh, R.S. 2018. Utilization of agro-industrial residues for the production of β-galactosidase using fungal isolate under solid state fermentation conditions. *Acta Alimentaria* 47(2): 162–170.

Kaur, S.J., and Gupta, V.K. 2017. Production of pectinolytic enzymes pectinase and pectin lyase by *Bacillus subtilis* SAV-21 in solid state fermentation. *Annals of Microbiology* 67(4): 333–342.

Kavuthodi, B., and Sebastian, D. 2018. Biotechnological valorization of pineapple stem for pectinase production by *Bacillus subtilis* BKDS1: Media formulation and statistical optimization for submerged fermentation. *Biocatalysis and Agricultural Biotechnology* 16: 715–722.

Kazemi, S., Khayati, G., and Faezi-Ghasemi, M. 2016. β-galactosidase production by *Aspergillus niger* ATCC 9142 using inexpensive substrates in solid-state fermentation: Optimization by orthogonal arrays design. *Iranian Biomedical Journal* 20(5): 287.

Kazemi, S., Khayati, G., and Ghasemi, M.F. 2017. Optimization of β-galactosidase production from a native *Bacillus* spp. isolated from northern soils in Iran. *Indian Journal of Biotechnology* 16(4): 557–562.

Kennedy, M.J., Thakur, M.S., Wang, D.I.C. et al. 1992. Estimating cell concentration in the presence of suspended solids: A light scatter technique. *Biotechnology and Bioengineering* 40(8): 875–888.

Khayati, G., Anvari, M., and Kazemi, S. 2014. Peanut pod-an inexpensive substrate for β-galactosidase production by *Bacillus* sp. in solid-state fermentation: Process evaluation and optimization by Taguchi design of experimental (DOE) methodology. *Minerva Biotecnologica* 26(4): 301–307.

Khosravi-Darani, K., Falahatpishe, H.R., and Jalali, M. 2008. Alkaline protease production on date waste by an alkalophilic *Bacillus* sp. 2-5 isolated from soil. *African Journal of Biotechnology* 7(10): 1536–1542.

Knappmann, I., Lehmkuhl, K., Köhler, J. et al. 2017. Lipase-catalyzed kinetic resolution as key step in the synthesis of enantiomerically pure σ ligands with 2-benzopyran structure. *Bioorganic and Medicinal Chemistry* 25(13): 3384–3395.

Kumar, A., and Singh, S. 2013. Directed evolution: Tailoring biocatalysts for industrial applications. *Critical Reviews in Biotechnology* 33(4): 365–378.

Kuvvet, C., Uzuner, S., and Cekmecelioglu, D. 2019. Improvement of pectinase production by co-culture of *Bacillus* spp. using apple pomace as a carbon source. *Waste and Biomass Valorization* 10(5): 1241–1249.

Lee, S.E., Lee, C.G., Lee, H.Y. et al. 2012. Preparation of corncob grits as a carrier for immobilizing yeast cells for ethanol production. *Journal of Microbiology and Biotechnology* 22(12): 1673–1680.

Leite, P., Silva, C., Salgado, J.M. et al. 2019. Simultaneous production of lignocellulolytic enzymes and extraction of antioxidant compounds by solid-state fermentation of agro-industrial wastes. *Industrial Crops and Products* 137: 315–322.

Li, X., Liu, Y., Hao, J. et al. 2018. Study of almond shell characteristics. *Materials* 11(9): 1782.

Lin, C.S.K., Pfaltzgraff, L.A., Herrero-Davila, L. et al. 2013. Food waste as a valuable resource for the production of chemicals, materials and fuels. Current situation and global perspective. *Energy and Environmental Science* 6(2): 426–464.

Lizardi-Jiménez, M.A., Ricardo-Díaz, J., Quiñones-Muñoz, T.A. et al. 2019. Fungal strain selection for protease production by solid-state fermentation using agro-industrial waste as substrates. *Chemical Papers* 73(10): 2603–2610.

Machielsen, R., Dijkhuizen, S., and van der Oost, J. 2007. Improving enzyme performance in food applications. In: *Novel Enzyme Technology for Food Applications*, ed. R., Rastall, pp. 16–42. CRC Press.

Mamo, J., Kangwa, M., Fernandez-Lahore, H.M. et al. 2020. Optimization of media composition and growth conditions for production of milk-clotting protease (MCP) from *Aspergillus oryzae* DRDFS13 under solid-state fermentation. *Brazilian Journal of Microbiology*, 51(2): 571–584. 1

Maria de Fátima de. Freitas, M., Cavalcante, L.S., Gudiña, E.J. et al. 2020. Sustainable lipase production by *Diutina rugosa* NRRL Y-95 through a combined use of agro-industrial residues as feedstock. *Applied Biochemistry and Biotechnology*, 193: 589–605.

Marinari, S., Masciandaro, G., Ceccanti, B. et al. 2000. Influence of organic and mineral fertilisers on soil biological and physical properties. *Bioresource Technology* 72(1): 9–17.

Martarello, R.D.A., Cunha, L., Cardoso, S.L. et al. 2019. Optimization and partial purification of beta-galactosidase production by *Aspergillus niger* isolated from Brazilian soils using soybean residue. *AMB Express* 9(1): 81.

Martin, J.G.P., Porto, E., Corrêa, C.B. et al. 2012. Antimicrobial potential and chemical composition of agroindustrial wastes. *Journal of Natural Products* 5: 27–36.

Matkawala, F., Nighojkar, S., Kumar, A. et al. 2019. Enhanced production of alkaline protease by *Neocosmospora* sp. N1 using custard apple seed powder as inducer and its application for stain removal and dehairing. *Biocatalysis and Agricultural Biotechnology* 21: 101310.

Maymone, B., Battaglini, A., and Tiberio, M. 1961. Ricerche sul valore nutritivo della sansa d'olive. *Alimentazione Animale* 5(4): 219–250.

Mehmood, T., Saman, T., Irfan, M. et al. 2019. Pectinase production from *Schizophyllum commune* through central composite design using citrus waste and its immobilization for industrial exploitation. *Waste and Biomass Valorization* 10(9): 2527–2536.

Melikoglu, M., Lin, C.S.K., and Webb, C. 2013. Analysing global food waste problem: Pinpointing the facts and estimating the energy content. *Central European Journal of Engineering* 3(2): 157–164.

Memarpoor-Yazdi, M., Karbalaei-Heidari, H.R., and Khajeh, K. 2017. Production of the renewable extremophile lipase: Valuable biocatalyst with potential usage in food industry. *Food and Bioproducts Processing* 102: 153–166.

Mirabella, N., Castellani, V., and Sala, S. 2014. Current options for the valorization of food manufacturing waste: A review. *Journal of Cleaner Production* 65: 28–41.

Mitchell, D.A., Krieger, N., and Berovic, M.M. 2006. *Solid-State Fermentation Bioreactors*. Springer.

Mitchell, D.A., Krieger, N., Stuart, D.M. et al. 2000. New developments in solid-state fermentation: II. Rational approaches to the design, operation and scale-up of bioreactors. *Process Biochemistry* 35(10): 1211–1225.

Mlichova, Z., and Rosenberg, M. 2006. Current trends of β-galactosidase application in food technology. *Journal of Food and Nutrition Research* 45: 47–54.

Mojsov, K.D. 2016. *Aspergillus* enzymes for food industries. In: *New and Future Developments in Microbial Biotechnology and Bioengineering: Aspergillus System Properties and Applications*, ed. V.K.Gupta, pp. 215–220. Elsevier.

Motte, J.C., Trably, E., Escudié, R. et al. 2013. Total solids content: A key parameter of metabolic pathways in dry anaerobic digestion. *Biotechnology for Biofuels* 6(1): 164.

Moutta, R.O., Chandel, A.K., Rodrigues, R.C.L.B. et al. 2012. Statistical optimization of sugarcane leaves hydrolysis into simple sugars by dilute sulfuric acid catalyzed process. *Sugar Technology* 14(1): 53–60.

Mukhtar, H., and Haq, I. 2008. Production of alkaline protease by *Bacillus subtilis* and its application as a depilating agent in leather processing. *Pakistan Journal of Botany* 40(4): 1673–1679.

Mukhtar, H., and Ikram-Ul-Haq. 2012. Concomitant production of two proteases and alpha-amylase by a novel strain of *Bacillus subtilis* in a microprocessor controlled bioreactor. *Brazilian Journal of Microbiology* 43(3): 1072–1079.

Murthy, M.R., Thakur, M.S., and Karanth, N.G. 1993. Monitoring of biomass in solid state fermentation using light reflectance. *Biosensors and Bioelectronics* 8(1): 59–63.

Musa, H., and Tayo, B.C.A. 2012. Screening of microorganisms isolated from different environmental samples for extracellular lipase production. *AU Journal of Technology* 15(3): 179–186.

Mussatto, S.I., Fernandes, M., Milagres, A.M. et al. 2008. Effect of hemicellulose and lignin on enzymatic hydrolysis of cellulose from brewer's spent grain. *Enzyme and Microbial Technology* 43(2): 124–129.

Nagy, Z., Keresztessy, Z., Szentirmai, A. et al. 2001. Carbon source regulation of β -galactosidase biosynthesis in *Penicillium chrysogenum*. *Journal of Basic Microbiology: an International Journal on Biochemistry, Physiology, Genetics, Morphology, and Ecology of Microorganisms* 41(6): 351–362.

Narwal, V., and Pundir, C.S. 2017. An improved amperometric triglyceride biosensor based on co-immobilization of nanoparticles of lipase, glycerol kinase and glycerol 3-phosphate oxidase onto pencil graphite electrode. *Enzyme and Microbial Technology* 100: 11–16.

Nawirska, A., and Kwaśniewska, M. 2005. Dietary fibre fractions from fruit and vegetable processing waste. *Food Chemistry* 91(2): 221–225.

Nayak, A., and Bhushan, B. 2019. An overview of the recent trends on the waste valorization techniques for food wastes. *Journal of Environmental Management* 233: 352–370.

Negi, S., Jain, S., and Raj, A. 2020. Combined ANN/EVOP factorial design approach for media screening for cost-effective production of alkaline proteases from *Rhizopus oryzae* (SN5)/NCIM-1447 under SSF. *AMB Express* 10(1): 1–9.

Nguyen, T.A.D., Kim, K.R., Han, S.J., et al. 2010. Pretreatment of rice straw with ammonia and ionic liquid for lignocelluloses conversion to fermentable sugars. *Bioresource Technology* 101(19): 7432–7438.

Nigam, P.S., Gupta, N., and Anthwal, A. 2009. Pre-treatment of agro-industrial residues. In: *Biotechnology for Agro-Industrial Residues Utilization*, eds. P.S.Nigam, and A. Pandey, pp. 13–33. Springer.

Obi, C.N., Okezie, O., and Ezugwu, A.N. 2019. Amylase production by solid state fermentation of agro-industrial wastes using *Bacillus* species. *European Journal of Nutrition and Food Safety*, 9(4): 408–414.

Onilude, A.A., Fadaunsi, I.F., and Garuba, E.O. 2012. Inulinase production by *Saccharomyces* sp. in solid state fermentation using wheat bran as substrate. *Annals of Microbiology* 62(2): 843–848.

Oumer, O.J., and Abate, D. 2017. Characterization of pectinase from *Bacillus subtilis* strain Btk 27 and its potential application in removal of mucilage from coffee beans. *Enzyme Research* 2017: 1–7.

Paepatung, N., Nopharatana, A., and Songkasiri, W. 2009. Bio-methane potential of biological solid materials and agricultural wastes. *Asian Journal on Energy and Environment* 10(1): 19–27.

Pandey, A. 2003. Solid-state fermentation. *Biochemical Engineering Journal* 13(2–3): 81–84.

Pandey, A., Nigam, P., Soccol, C.R. et al. 2000b. Advances in microbial amylases. *Biotechnology and Applied Biochemistry* 31(2): 135–152.

Pandey, A., Selvakumar, P., and Ashakumary, L. 1994. Glucoamylase production by *Aspergillus niger* on rice bran is improved by adding nitrogen sources. *World Journal of Microbiology and Biotechnology* 10(3): 348–349.

Pandey, A., Soccol, C.R., and Mitchell, D. 2000a. New developments in solid state fermentation: I- bioprocesses and products. *Process Biochemistry* 35(10): 1153–1169.

Panesar, P.S., Kaur, R., Singla, G. et al. 2016. Bio-processing of agro-industrial wastes for production of food-grade enzymes: Progresses and prospects. *Applied Food Biotechnology* 3(4): 208–227.

Panesar, P.S., Kennedy, J.F., Gandhi, D.N. et al. 2007. Bioutilisation of whey for lactic acid production. *Food Chemistry* 105(1): 1–14.

Panesar, P.S., Kumari, S., and Panesar, R. 2010a. Potential applications of immobilized β-galactosidase in food processing industries. *Enzyme Research* 2010:1–16.

Panesar, P.S., Marwaha, S.S., and Chopra, H.K. 2010b. *Enzymes in Food Processing: Fundamentals and Potential Applications*. IK Publishers.

Panesar, P.S., Panesar, R., Singh, R.S. et al. 2006. Microbial production, immobilization and applications of β -D-galactosidase. *Journal of Chemical Technology and Biotechnology* 81(4): 530–543.

Patel, R., Prajapati, V., Trivedi, U. et al. 2020. Optimization of organic solvent-tolerant lipase production by *Acinetobacter* sp. UBT1 using deoiled castor seed cake. *3 Biotech* 10(12): 1–13.

Pawlak-Szukalska, A., Wanarska, M., Popinigis, A.T. et al. 2014. A novel cold-active β-d-galactosidase with transglycosylation activity from the Antarctic *Arthrobacter* sp. 32cB–Gene cloning, purification and characterization. *Process Biochemistry* 49(12): 2122–2133.

Pedrolli, D.B., Monteiro, A.C., Gomes, E. et al. 2009. Pectin and pectinases: Production, characterization and industrial application of microbial pectinolytic enzymes. *Open Biotechnology Journal*, 3: 9–18.

Pereira, A.D.S., Fontes-Sant'Ana, G.C., and Amaral, P.F. 2019. Mango agro-industrial wastes for lipase production from *Yarrowia lipolytica* and the potential of the fermented solid as a biocatalyst. *Food and Bioproducts Processing* 115: 68–77.

Piedrahita-Aguirre, C.A., Bastos, R.G., Carvalho, A.L. et al. 2014. The influence of process parameters in production of lipopeptide iturin A using aerated packed bed bioreactors in solid-state fermentation. *Bioprocess and Biosystems Engineering* 37(8): 1569–1576.

Pitol, L.O., Biz, A., Mallmann, E. et al. 2016. Production of pectinases by solid-state fermentation in a pilot-scale packed-bed bioreactor. *Chemical Engineering Journal* 283: 1009–1018.

Pliego-Sandoval, J., Amaya-Delgado, L., Mateos-Díaz, J.C. et al. 2012. Multiplex gas sampler for monitoring respirometry in column-type bioreactors used in solid-state fermentation. *Biotechnology and Biotechnological Equipment* 26(3): 3031–3038.

Prabaningtyas, R.K., Putri, D.N., Utami, T.S. et al. 2018. Production of immobilized extracellular lipase from *Aspergillus niger* by solid state fermentation method using palm kernel cake, soybean meal, and coir pith as the substrate. *Energy Procedia* 153: 242–247.

Prabha, S., Verma, G., Pandey, S. et al. 2019. Utilization of agro-industrial by-products for production of lipase using mix culture batch process. *Bioscience Biotechnology Research Communications* 12(3): 748–756.

Pranay, K., Padmadeo, S.R., and Prasad, B. 2019. Production of amylase from *Bacillus subtilis* sp. strain KR1 under solid state fermentation on different agrowastes. *Biocatalysis and Agricultural Biotechnology* 21: 101300.

Prasad, S., Singh, A., and Joshi, H.C. 2007. Ethanol as an alternative fuel from agricultural, industrial and urban residues. *Resources, Conservation and Recycling* 50(1): 1–39.

Qadir, F., Ejaz, U., and Sohail, M. 2020. Co-culturing corncob-immobilized yeasts on orange peels for the production of pectinase. *Biotechnology Letters* 42(9): 1743–1753.

Qureshi, A.S., Bhutto, M.A., Khushk, I. et al. 2011. Optimization of cultural conditions for protease production by *Bacillus subtilis* EFRL 01. *African Journal of Biotechnology* 10(26): 5173–5181.

Qureshi, A.S., Khushk, I., Ali, C.H. et al. 2016. Coproduction of protease and amylase by thermophilic *Bacillus* sp. BBXS-2 using open solid-state fermentation of lignocellulosic biomass. *Biocatalysis and Agricultural Biotechnology* 8: 146–151.

Ramachandran, S., Singh, S.K., Larroche, C. et al. 2007. Oil cakes and their biotechnological applications–A review. *Bioresource Technology* 98(10): 2000–2009.

Ramezanzadeh, F.M., Rao, R.M., Prinyawiwatkul, W. et al. 2000. Effects of microwave heat, packaging, and storage temperature on fatty acid and proximate compositions in rice bran. *Journal of Agricultural and Food Chemistry* 48(2): 464–467.

Ramos-Ibarra, J.R., Miramontes, C., Arias, A. et al. 2017. Production of hydrolytic enzymes by solid-state fermentation with new fungal strains using orange by-products. *Revista Mexicana de Ingeniería Química* 16(1): 19–31.

Rao, M.R.K., and Rishikesh, T.V. 2019. Production and purification of β-galactosidase from *Aspergillus foetidus* MTCC 6322 using solid-state fermentation. *Drug Invention Today* 11(5):1218–1219.

Rao, R.R., Vimudha, M., Kamini, N.R. et al. 2017. Alkaline protease production from *Brevibacterium luteolum* (MTCC 5982) under solid-state fermentation and its application for sulfide-free unhairing of cowhides. *Applied Biochemistry and Biotechnology* 182(2): 511–528.

Raol, G.G., Raol, B.V., Prajapati, V.S. et al. 2015. Utilization of agro-industrial waste for β-galactosidase production under solid state fermentation using halotolerant *Aspergillus tubingensis* GR1 isolate. *3 Biotech* 5(4): 411–421.

Ravindran, R., Hassan, S.S., Williams, G.A. et al. 2018. A review on bioconversion of agro-industrial wastes to industrially important enzymes. *Bioengineering* 5(4): 93.

Ravindran, R., and Jaiswal, A.K. 2016. Exploitation of food industry waste for high-value products. *Trends in Biotechnology* 34(1): 58–69.

Réczey, K., Stålbrand, H., Hahn-Härerdal, B. et al. 1992. Mycelia-associated β-galactosidase activity in microbial pellets of *Aspergillus* and *Penicillium* strains. *Applied Microbiology and Biotechnology* 38(3): 393–397.

Rico-García, D., Ruiz-Rubio, L., Pérez-Alvarez, L. et al. 2020. Lignin-based hydrogels: Synthesis and applications. *Polymers* 12(1): 81.

Rivas, B., Torrado, A., Torre, P. et al. 2008. Submerged citric acid fermentation on orange peel autohydrolysate. *Journal of Agricultural and Food Chemistry* 56(7): 2380–2387.

Rodríguez, G., Lama, A., Rodríguez, R. et al. 2008. Olive stone an attractive source of bioactive and valuable compounds. *Bioresource Technology* 99(13): 5261–5269.

Rudra, S.G., Nishad, J., Jakhar, N. et al. 2015. Food industry waste: Mine of nutraceuticals. *International Journal of Environmental Science and Technology* 4(1): 205–229.

Ruijter, G.J., and Visser, J. 1997. Carbon repression in *Aspergilli*. *FEMS Microbiology Letters* 151(2): 103–114.

Ruiz, H.A., Rodríguez-Jasso, R.M., Rodríguez, R. et al. 2012. Pectinase production from lemon peel pomace as support and carbon source in solid-state fermentation column-tray bioreactor. *Biochemical Engineering Journal* 65: 90–95.

Saini, J.K., Singhania, R.R., Satlewal, A. et al. 2016. Improvement of wheat straw hydrolysis by cellulolytic blends of two *Penicillium* spp. *Renewable Energy* 98: 43–50.

Saleh, F., Hussain, A., Younis, T. et al. 2020. Comparative growth potential of thermophilic amylolytic *Bacillus* sp. on unconventional media food wastes and its industrial application. *Saudi Journal of Biological Sciences* 27(12): 3499–3504.

Salgado, V., Fonseca, C., da Silva, T.L. et al. 2020. Isolation and identification of *Magnusiomyces capitatus* as a lipase-producing yeast from olive mill wastewater. *Waste and Biomass Valorization* 11(7): 3207–3221.

Salihu, A., Alama, M.Z., Ismail Abdul Karim, M.I. et al. 2011. Optimization of lipase production by *Candida cylindracea* in palm oil mill effluent based medium using statistical experimental design. *Journal of Molecular Catalysis B: Enzymatic* 69(1–2): 66–73.

Salihu, A., Bala, M., and Alam, M.Z. 2016. Lipase production by *Aspergillus niger* using sheanut cake: An optimization study. *Journal of Taibah University for Science* 10(6): 850–859.

Sawant, R., and Nagendran, S. 2014. Protease: An enzyme with multiple industrial applications. *World Journal of Pharmacy and Pharmaceutical Sciences* 3(6): 568–579.

Schwan, R.F., and Wheals, A.E. 2004. The microbiology of cocoa fermentation and its role in chocolate quality. *Critical Reviews in Food Science and Nutrition* 44(4): 205–221.

Searle, S.M.C. 2013. *Availability of Cellulosic Residues and Wastes in the EU*. ICCT.

Selvakumar, P., and Sivashanmugam, P. 2017. Optimization of lipase production from organic solid waste by anaerobic digestion and its application in biodiesel production. *Fuel Processing Technology* 165: 1–8.

Selvam, K., Selvankumar, T., Rajiniganth, R. et al. 2016. Enhanced production of amylase from *Bacillus* sp. using groundnut shell and cassava waste as a substrate under process optimization: Waste to wealth approach. *Biocatalysis and Agricultural Biotechnology* 7: 250–256.

Sethi, B., Satpathy, A., Tripathy, S. et al. 2016a. Production of ethanol and clarification of apple juice by pectinase enzyme produced from *Aspergillus terreus* NCFT 4269.10. *International Journal of Biological Research* 4(1): 67–73.

Sethi, B.K., Jana, A., Nanda, P.K. et al. 2016b. Thermostable acidic protease production in *Aspergillus terreus* NCFT 4269.10 using chickling vetch peels. *Journal of Taibah University for Science* 10(4): 571–583.

Shanmugavel, M., Vasantharaj, S., Yazhmozhi, A. et al. 2018. A study on pectinases from *Aspergillus tamarii*: Toward greener approach for cotton bioscouring and phytopigments processing. *Biocatalysis and Agricultural Biotechnology* 15: 295–303.

Sharma, K.M., Kumar, R., Panwar, S. et al. 2017. Microbial alkaline proteases: Optimization of production parameters and their properties. *Journal of Genetic Engineering and Biotechnology* 15(1): 115–126.

Silveira, M.L.L., Furlan, S.A., and Ninow, J.L. 2008. Development of an alternative technology for the oyster mushroom production using liquid inoculum. *Food Science and Technology* 28(4): 858–862.

Singh, H., and Soni, S.K. 2001. Production of starch-gel digesting amyloglucosidase by *Aspergillus oryzae* HS-3 in solid state fermentation. *Process Biochemistry* 37(5): 453–459.

Singh, S., and Bajaj, B.K. 2016. Bioprocess optimization for production of thermoalkali-stable protease from *Bacillus subtilis* K-1 under solid-state fermentation. *Preparative Biochemistry and Biotechnology* 46(7): 717–724.

Sinsuwan, S., Jangchud, A., Rodtong, S. et al. 2015. Statistical optimization of the production of NaCl-tolerant proteases by a moderate halophile, *Virgibacillus* sp. SK37. *Food Technology and Biotechnology* 53(2): 136–145.

Sundarram, A., and Murthy, T.P.K. 2014. α-amylase production and applications: A review. *Journal of Applied and Environmental Microbiology* 2(4): 166–175.

Szymańska-Chargot, M., Chylińska, M., Gdula, K. et al. 2017. Isolation and characterization of cellulose from different fruit and vegetable pomaces. *Polymers* 9(10): 495.

Thomas, L., Larroche, C., and Pandey, A. 2013. Current developments in solid-state fermentation. *Biochemical Engineering Journal* 81: 146–161.

Toole, D.K. 1999. Characteristics and use of okara, the soybean residue from soy milk production a review. *Journal of Agricultural and Food Chemistry* 47(2): 363–371.

Treichel, H., de Oliveira, D., Mazutti, M.A. et al. 2010. A review on microbial lipases production. *Food and Bioprocess Technology* 3(2): 182–196.

Tufail, T., Saeed, F., Imran, M. et al. 2018. Biochemical characterization of wheat straw cell wall with special reference to bioactive profile. *International Journal of Food Properties* 21(1): 1303–1310.

Tuysuz, E., Gonul-Baltaci, N., Omeroglu, M.A. et al. 2020. Co-production of amylase and protease by locally isolated thermophilic bacterium *Anoxybacillus rupiensis* T2 in sterile and non-sterile media using waste potato peels as substrate. *Waste and Biomass Valorization*, *11*: 6793–6802.

Van Oort, M. 2010. Enzymes in food technology – Introduction. In: *Enzymes in Food Technology*, eds. R.J.Whitehurst, and M.Van Oort, pp. 1–17. Wiley-Blackwell.

Ververis, C., Georghiou, K., Danielidis, D. et al. 2007. Cellulose, hemicelluloses, lignin and ash content of some organic materials and their suitability for use as paper pulp supplements. *Bioresource Technology* 98(2): 296–301.

Viana, C.D.S., Pedrinho, D.R., Ito Morioka, L.R. et al. 2018. Determination of Cell permeabilization and beta-galactosidase extraction from *Aspergillus oryzae* CCT 0977 grown in cheese whey. *International Journal of Chemical Engineering* 2018: 1–6.

Voragen, A.G., Coenen, G.J., Verhoef, R.P. et al. 2009. Pectin, a versatile polysaccharide present in plant cell walls. *Structural Chemistry* 20(2): 263.

Yadav, S., Yadav, P.K., Yadav, D. et al. 2009. Purification and characterization of pectin lyase produced by *Aspergillus terricola* and its application in retting of natural fibers. *Applied Biochemistry and Biotechnology* 159(1): 270–283.

Yingyi, D., Wang, L., and Chen, H. 2012. Digital image analysis and fractal-based kinetic modelling for fungal biomass determination in solid-state fermentation. *Biochemical Engineering Journal* 67: 60–67.

You, S., Chang, H., Yin, Q. et al. 2017. Utilization of whey powder as substrate for low-cost preparation of β-galactosidase as main product, and ethanol as by-product, by a litre-scale integrated process. *Bioresource Technology* 245(A): 1271–1276.

Zafar, S. 2014. Waste management, waste-to-energy. https://www.bioenergyconsult.com/tag/waste-to-energy.

Zhang, G., Li, S., Xu, Y. et al. 2019. Production of alkaline pectinase: A case study investigating the use of tobacco stalk with the newly isolated strain *Bacillus tequilensis* CAS-MEI-2-33. *BMC Biotechnology* 19(1): 45.

Zhao, X., Peng, F., Cheng, K. et al. 2009. Enhancement of the enzymatic digestibility of sugarcane bagasse by alkali–peracetic acid pretreatment. *Enzyme and Microbial Technology* 44(1): 17–23.

Zhou, X., Broadbelt, L.J., and Vinu, R. 2016. Mechanistic understanding of thermochemical conversion of polymers and lignocellulosic biomass. *Advances in Chemical Engineering* 49: 95–198.

Zou, M., Guo, F., Li, X. et al. 2014. Enhancing production of alkaline polygalacturonate lyase from *Bacillus subtilis* by fed-batch fermentation. *PLOS ONE* 9(3): e90392.

7 Bioactive Proteins and Peptides from Agro-Industrial Waste
Extraction and Their Industrial Applications

Nuntarat Boonlao, Thatchajaree Mala,
Sushil Koirala, and Anil Kumar Anal

CONTENTS

7.1 INTRODUCTION

Due to increased consumer demands, the rapid growth of the food-processing industry has generated high amounts of agro-industrial wastes. The Food and Agriculture Organization of the United Nations estimates that 1.6 billion tonnes of food are wasted per year for the human nutrition globally. Of this, fruits, vegetables, and tubers represent 40–50% of the total quantity of food wasted (Freitas et al., 2021). Recent studies and research provide evidence of bioactive value-added compounds in the food waste, further heightening the value of these wastes, which were otherwise considered to have no value (Prajapati, Koirala, and Anal, 2021). Agro-industrial wastes such as seeds (Han et al., 2018), peels (Mala, Sadiq, and Anal, 2021), skin (Bayram et al., 2021), feathers (Prajapati, Koirala, and Anal, 2021), eggshell (Jain and Anal, 2018), pulp (Osorio et al., 2021), culled banana and banana pseudostem (Shrestha et al., 2021; Shrestha et al., 2018), and press cakes (Sadh et al., 2017) have been studied for the valorization of food waste with promising results. These wastes are sources of

DOI: 10.1201/9781003125679-7

Bioactive peptides

FIGURE 7.1   BPs with perceived health benefits.

a broad spectrum of metabolites with health benefits, such as peptides, phenolic compounds, antioxidants, antihypertensives, and antimicrobials (Figure 7.1) (Lemes et al. 2016). *Waste to wealth* instantly became a mantra for research by utilizing this waste as a raw material to obtain high value-added products with the potential to contribute to global food sustainability. In such a framework, one key factor in ensuring environmental safety, achieving zero-hunger (Sustainable Development Goal (SDG) 2), and sustainable consumption and production (SDG 12) is the valorization of these food wastes.

Agro-industrial waste generated from various food-processing sectors is generally disposed to landfills or incinerated, which causes numerous environmental hazards. Microbial proliferation starts once the food is wasted or dumped, which is necessary for its degradation. However, greenhouse gas emissions, toxic compounds, and the possibility of the growth of pathogens present huge problems when all agro-industries are considered under the same umbrella (Ng et al., 2020). The situation in developing countries is even worse. Pertinent issues like lack of viable alternatives, policy paralysis, lack of data, and lack of organized waste removal system aggravate the condition, which further increases greenhouse gas emissions, energy loss, health crisis, and affects the quality of life and wellbeing of society (Ravindaran et al., 2018)

Food waste generated from agro-industries contains bioactive compounds that need to be extracted for value. However, the process of extraction or recovery of such compounds is a challenging task, with only a few technologies studied so far for their optimization, recovery, purification, and potential scale up. Green technology based innovative processes such as ultrasound assisted extraction (UAE), microwave assisted extraction (MAE), supercritical fluid extraction (S_pFE), and enzyme assisted extraction (EAE) can reduce these limitations to a great extent for a final product with high quality and purity (Soquetta, Terra, and Bastos, 2018). The advantages of these technologies are associated with reduced energy consumption, chemical-free process, and short operation time (Sik et al., 2020). Hence, this chapter discusses the potential of bioactive proteins and peptides as potential future foods and describes the latest trends in innovative technologies for the sustainable extraction of these compounds from agro-industrial waste with commercial success stories.

7.2   BIOACTIVE PROTEINS AND PEPTIDES FROM AGRO-INDUSTRIAL WASTE

Most residues from agro-industrial waste are a valuable source of bioactive compounds useful for food, pharmaceuticals, and cosmetics. Agro-industrial wastes, for instance, seeds, peels, pulps, pits, press cake, bones, heads, skin, and blood, consist of a relatively high protein content derived from various plant and animal origins (Görgüç, Gençdağ, and Yılmaz, 2020; Brandelli, Sala, and Kalil, 2015; Harnedy and FitzGerald, 2012). These wastes are proteinaceous, which means they are composed of proteins, peptides, and amino acids. Protein-rich materials are a source for the generation of bioactive peptides (Harnedy and FitzGerald, 2012). Agro-industrial wastes used to generate bioactive peptides are presented in Table 7.1.

Bioactive peptides are encrypted inside bioactive proteins and are in a specific sequence of protein fragments containing 2–20 amino acid subunits with a molecular weight lower than 3 kDa (Chalamaiah et al., 2019). Bioactive peptides provide potential physiological functions in the body, for instance, immunomodulatory, antibacterial, and antihypertensive activity. Some bioactive peptides may also display various techno-functionalities (Harnedy and FitzGerald, 2012). Bioactive peptides can be obtained by several techniques, including physical, chemical, and biological, such as UAE or MAE, pulsed electric field (PEF), acid or alkaline hydrolysis, enzymatic hydrolysis, fermentation, gastrointestinal digestion, etc. (Ulug, Jahandideh, and Wu, 2021; Chalamaiah, Yu, and Wu, 2018). Bioactive peptides show differences in bio-functionalities depending upon their amino acid composition, sequence, and chemical structure of the peptides from different protein precursors (Lemes et al., 2016).

7.2.1 BIOACTIVE PROTEINS/PEPTIDES FROM PLANT WASTE

Bioactive peptides from plant origins are becoming popular due to consumer awareness of health, cost, food safety concerns, animal welfare, and environmental sustainability. Soybean meal (SBM), a by-product of soybean oil extraction, is considered a cheap protein source that is widely used in the food and feed industry. SBM exhibits a high amount of protein, balanced amino acid composition, and high lysine content compared with other vegetable protein sources (Chi and Cho 2016). There was an improvement in the bioactivity of bioactive peptides obtained from *Bacillus*-fermented SBM, particularly their antioxidant and antihypertensive activity (Dai et al., 2017). Olive pomace or olive oil cake is mainly produced in olive oil production. These residues are solid waste that includes olive pulp and stone. Olive stone is also known as a cheap source of protein (22%). The olive seed protein hydrolysate obtained by enzymatic hydrolysis exhibits antioxidant and antihypertensive properties; their efficacy depends on the types of enzymes used in hydrolysis (Esteve, Marina, and García, 2015). Cereal is another source of bioactive peptides. Cereal production generates a huge amount of co-product. Rice bran is a major waste (25%) from white rice production, and it composes 11–17% protein. The rice bran hydrolysate by ultrasonicate-aided enzymatic hydrolysis exhibits an enhancement in bioactivity in terms of antioxidant and lipase inhibitory activity in which the lipase inhibitory effect helps reduce the gastrointestinal absorption of fat (Fathi et al., 2021). 2,5-Diketopiperazines (DKPs) are the active peptide sequences released from cereal grain. Cyclic dipeptides exist in several foods formed during the roasting of cocoa, malt, beer, and coffee and are responsible for a specific bitter flavour and antioxidant activities (Kumar et al., 2012). A huge number of cereal proteins contain byproducts of corn starch processing, especially zein protein. This protein is rich in proline residues and is the main component of corn endosperm proteins. Zein peptides/hydrolysates possess an anti-inflammatory activity by preventing NF-κB (nuclear factor kappa light chain enhancer of activated B cells) activation (Liang et al., 2020).

7.2.2 BIOACTIVE PROTEINS/PEPTIDES FROM ANIMAL WASTE

About 10–20 million tonnes a year of waste generated from the processing industry are obtained from fish and meat processing (including poultry, pig, and cattle) (Jayathilakan et al., 2012). Numerous byproducts are generated in poultry and meat processing, such as heads, skin, blood, bones, feather, and viscera. Animal wastes mostly compose insoluble and hard-to-degrade structural proteins such as keratin, elastin, and collagen, the main components in bone, organs, and hard

TABLE 7.1

Sources and Functional Properties of Bioactive Proteins/Peptides Obtained from Several Agro-Industrial Wastes/Byproducts

Source	Production methods	Amino acid sequence	Health benefits and functional properties	References
Plant-based agro-industrial waste				
Soybean meal	Thermal hydrolysis	FPFPRPPHQK HPEREPQQPGE EQDERQFPFP	Antimicrobial activity	Freitas et al., 2019
Olive seed	Enzyme hydrolysis by Alcalase and thermolysin	VLDTGLAGA LVVDGEGY	Antioxidant and Antihypertensive activity	Esteve, Marina, and García, 2015
Rice bran	Enzyme hydrolysis by trypsin	YSK	Antioxidant and Antihypertensive activity	Wang et al., 2017
	Enzyme hydrolysis by pepsin	LRRHASEGGHGPHW EKLLGKQDKGVIIRA SSFSKGVQRAAF	Antimicrobial, lipopolysaccharide-neutralizing, and angiogenic activities	Taniguchi et al., 2017
	Enzyme hydrolysis by papain and trypsin	VAGAEDAAK AAVQGQVEK GGHELSK CQHHHDQWK	Antioxidant activity	Wattanasiritham et al., 2016
Palm kernel cake	Enzyme hydrolysis by papain	AWFS WAF LPWRPATNVF	Antioxidant activity	Zarei et al., 2014
Animal-based agro-industrial waste				
Eggshell membrane	Enzyme hydrolysis by alkaline protease, pepsin, chymotrypsin, neutral protease and papain	ESYHLPR NVIDPPIYAR MFAEWQPR LLFAMTKPK MLKMLPFK	Antioxidant activity	Zhao et al., 2019
Bovine bone	Enzyme hydrolysis by collagenase	GPIGPVGAR GPIGSRGPS GPQGIAGQR GPVGAAGPS GQAGVMG	Antioxidant activity	Song et al., 2021

(Continued)

TABLE 7.1 (CONTINUED)
Sources and Functional Properties of Bioactive Proteins/Peptides Obtained from Several Agro-Industrial Wastes/Byproducts

Source	Production methods	Amino acid sequence	Health benefits and functional properties	References
Spent hen carcass	Enzyme hydrolysis by thermoase	VRP LKY VRY KYKA LKYKA	Antihypertensive activity	Fan and Wu, 2021
Chicken foot	Physical extraction	LPGAKGEPGDFTYDSILKGE LLSSIEAEVNPPSDL GYEVGFDAEYYR	Antihypertensive and antidiabetic activity	Malison, Arpanutud, and Keeratipibul, 2021
Porcine plasma	Enzyme hydrolysis by protease	LEPVIGT	Antioxidant activity	Zhan et al., 2021
Broiler skin	Enzyme hydrolysis by elastase	GAHTGPRKPKPR GPGFDVR ADASVLPK	Antioxidant activity	Nadalian et al., 2019
Marine-based agro-industrial waste				
Shrimp (*Oratosquilla woodmasoni*) waste	Enzyme hydrolysis by thermolysin and pepsin	NGVAA	Antioxidant and antihypertensive activity	Joshi et al., 2020
Shrimp shell	Enzyme hydrolysis	FNPPGVMEHFETATR FDPPGVMEHFETATR AGFAGDDAPR	Antihypertensive and antioxidant activity	Ambigaipalan and Shahidi, 2017
Squid skin	Enzyme hydrolysis by Esperase, pepsin, and pancreatin	GRGSVPAHypGP	Antihypertensive activity	Alemán, Gómez-Guillén, and Montero, 2013
Alaska pollock frame	Enzyme hydrolysis	NGMTY NGLAP WT	Immunomodulating activity	Hou et al., 2012
Salmon fin	Enzyme hydrolysis	PAY FLNEFLHV	Anti-inflammatory activity Antioxidant activity	Ahn, Cho, and Je, 2015 Ahn, Kim, and Je, 2014
Salmon processing byproducts	Enzyme hydrolysis	VWDPPKFD FEDYVPLSCF FNVPLYE	Antihypertensive activity	Ahn et al., 2012

(Continued)

TABLE 7.1 (CONTINUED)
Sources and Functional Properties of Bioactive Proteins/Peptides Obtained from Several Agro-Industrial Wastes/Byproducts

Source	Production methods	Amino acid sequence	Health benefits and functional properties	References
Salmon skin	Enzyme hydrolysis by trypsin	GLP NLP GL LP	Antihypertensive activity	Lee, Jeon, and Byun, 2014
Pacific cod skin	Enzyme hydrolysis by pepsin, trypsin, and α-chymotrypsin	LLMLDNDLPP	Antihypertensive and antioxidant activity	Himaya et al., 2012
Silver carp muscle	Enzyme hydrolysis by papain	LVPVAVF	Antioxidant activity	Wang et al., 2021

Note: A: alanine; R: arginine; N: asparagine; D: aspartic acid; C: cysteine; Q: glutamine; E: glutamic acid; G: glycine; H: histidine; I: isoleucine; L: leucine; K: lysine; M: methionine; F: phenylalanine; P: proline; S: serine; T: threonine; W: tryptophane; Y: tyrosine; V: valine; Hyp: 4-hydroxyproline.

tissues. These wastes are known as a potent protein source that can be hydrolyzed into bioactive ingredients (Brandelli, Sala, and Kalil, 2015). Chicken feather is an abundant source of protein containing more than 82% curd protein, of which 91% is predominantly keratin (Lasekan, Abu Bakar, and Hashim, 2013). Keratin can be hydrolyzed into keratin hydrolysate by the activity of keratinolytic micro-organisms. *Bacillus* sp. RCM-SSR-102, a bacteria producing keratinase, hydrolyzes chicken feathers into feather keratin hydrolysate with antioxidant and antityrosinase activity, which is useful for the pharmaceutical and cosmetic industry (Kshetri et al., 2020). Prajapati et al. (2021) hydrolyzed chicken feathers by fermentation with *Bacillus amyloliquefaciens* KB1 and obtained feather keratin hydrolysate with functionality improvement such as antioxidant activity, *in vitro* digestibility, and water- and oil-holding capacity. An egg is a food product with high nutrition; thus, there is a huge demand for egg consumption worldwide. Byproducts from egg processing include the eggshell and eggshell membrane, which contribute to 11% of the total egg weight. However, the egg shell membrane is considered a valuable source of protein (Tsai et al. 2006). The fermentation of eggshell membrane using lactic acid bacteria could generate hydrolysates with excellent techno-functional properties, for instance, solubility, emulsifying properties, and foam properties.

Furthermore, these hydrolysates also demonstrated remarkable bioactive properties, including antioxidant, antibacterial, and antihypertensive activities (Jain and Anal, 2017). Poultry skin, a by-product of poultry meat, is considered an underutilized by-product due to its high fat content; thus, it is usually removed. However, chicken skin contains 10–22% of proteins, known as a good source of elastin, which is often utilized to produce functional foods, cosmetics, and regenerative medicine (Yusop et al., 2016). Elastin, a protein in connective tissue, exists together with collagen present in the skin, dermis, tendon, ligaments, vascular wall, blood vessels, aorta, and lungs, contributing to organ elasticity (Daamen et al., 2001). Elastin hydrolysates from poultry skin consist of especially hydrophobic amino acids (valine and proline), which possess antioxidant and angiotensin I-converting enzyme (ACE)-inhibitory activities. Thus, elastin from poultry skin is a natural antioxidant and antihypertensive ingredient useful in foods, beverages, and the cosmetics industry (Nadalian et al., 2019; Yusop et al., 2016). Porcine plasma, a by-product from slaughterhouses, comprises functional proteins that can be hydrolyzed into antioxidant peptides (Zhan et al., 2021).

7.2.3 BIOACTIVE PROTEINS AND PEPTIDES FROM MARINE WASTES

Fish byproducts are generated in a huge amount from the fishing industry, which is considered waste or underutilized material. The generated fish byproducts include heads, fins, bones, skin, trimming, viscera, and muscle. These wastes consist of proteins and lipids, which can be utilized to produce added-value products. Fishery byproducts are protein-rich materials containing 8–35% crude protein, usually utilized in low market products, for example, in animal feed, fish feed, and fertilizer (Tacias-Pascacio et al., 2021; Sila and Bougatef, 2016). Different types of protein exist in fish byproducts. Three main groups of proteins are found in skeletal muscles, including myofibrillar, stromal, and sarcoplasmic proteins (Xiong, 2018), among which proteins such as myosin, actin, actinin, α-actinin, calpains, cathepsins, albumin, creatine kinase, aldolase, and collagen are predominant among others. The diversity of proteins in fish residues contributes to the production of bioactive peptides with specific amino acid profiles (Tacias-Pascacio et al., 2021). Silver carp (*Hypophthalmichthys molitrix*) is a freshwater fish with a world production of around 4,700,000 tonnes in 2017. Silver carp muscle, a by-product of the surimi-making process, has been hydrolyzed into bioactive hydrolysates with different value-added bioactive peptides (Chalamaiah et al., 2012). The hydrolysates from silver carp muscle via Alcalase and papain hydrolysis have a molecular weight of <1 kDa with high antioxidant activity due to the presence of hydrophobic amino acids and a large proportion of peptides with low molecular weight (Wang et al., 2021). Salmon processing generates a large amount of byproducts, such as heads, intestines, frames, and pectoral fins. In particular, the pectoral fin by-product consists of high proteins, which is known as a potent source of bioactive peptides. The peptic hydrolysate from salmon pectoral fin by-product with antioxidant and anti-inflammatory activities was successfully generated by pepsin hydrolysis (Je et al., 2013). This hydrolysate exhibited a hepatoprotective effect against ethanol-induced oxidative stress in Sprague–Dawley rats. Indeed, the hydrolysate of the salmon pectoral fin by Alcalase hydrolysis also demonstrated the ACE-inhibitory effect, which is useful in the food and pharmaceutical industry for its benefits against hypertension and related diseases (Ahn et al., 2012). Crustacean aquaculture is known as the biggest seafood production globally, which is also a protein-rich source. In 2020, the global production of shrimp was 5.03 million tonnes and is predicted to grow to 7.28 million tonnes by 2025 (IMARC, 2020). Shrimp byproducts (approximately 50–60%) such as the head, viscera, and shell are generated as a solid waste during shrimp processing (Senphan and Benjakul, 2012). Shrimp waste consists of various bioactive compounds, for instance, proteins/peptides, lipids, minerals, vitamins, chitin/chitosan, enzymes, and pigments (Nirmal et al., 2020). Shrimp waste contains large quantities of proteins, which consist of 45% of the total proteins in shrimp (Kannan et al., 2011). More than 70% of proteins were recovered as a protein hydrolysate when shrimp waste was hydrolyzed by proteolytic enzymes (Gildberg et al., 2011). The shrimp waste hydrolysates contain several essential and non-essential amino acids with diverse functional and biological properties (Nirmal et al., 2020).

7.3   BIOLOGICAL ACTIVITIES AND HEALTH BENEFIT EFFECTS OF BIOACTIVE PEPTIDES

Bioactive peptides (BPs) exhibit varying biological activities depending on the sequence, composition, chain length, charge characteristic, molecular weight, and hydrophobic/ hydrophilic characteristic of amino acids forming peptides (Liu et al., 2016; Agyei et al., 2016; Li and Yu, 2015). Due to their hydrophobic and cationic characteristics, peptides such as bactericidal/permeability-increasing protein, protegrins, and cationic peptide fractions from flaxseed with the smallest amino acid sequence exhibited mainly antimicrobial effects (Gholizadeh and Kohnehrouz, 2013; Udenigwe and Aluko, 2012). Proline and valine, hydrophobic amino acids, exhibit antihypertensive properties. The fractions with cysteine- and methionine-containing peptides show antioxidant activity (Görgüç, Gençdağ, and Yılmaz, 2020), while histidine and glycine exhibit antioxidant and immunomodulatory activities, respectively (Agyei et al., 2016). The amino acid sequences that correspond to the biological activities of bioactive proteins/peptides obtained from various agro-industrial wastes/ byproducts are presented in Table 7.1. Bioactive proteins/peptides possess several biological activities: antioxidant, antihypertensive, cholesterol-lowering, antimicrobial, anticancer, immunomodulatory, and opioid.

7.3.1   Antioxidant Activity

The existence of reactive oxygen species (ROS), including superoxide anion (O_2-), hydrogen peroxide (H_2O_2), hydroxyl radical ($\cdot OH$), and other free radicals, occur due to an excessive oxidative reaction in the human body. Excessive ROS has harmful effects on human cells, leading to certain diseases, such as cardiovascular diseases, diabetes, hypertension, neurodegenerative diseases, inflammation diseases, and cancer (Fan et al., 2012). Indeed, unregulated oxidative reactions can also occur in the food system, impacting the product's quality attributes such as food flavour, colour, and texture (Ngoh and Gan, 2016). Consequently, antioxidants have been widely utilized to remove the harmful ROS and free radicals in the food system and pharmaceuticals. Bioactive peptides from various protein-enriched sources that possess antioxidant activity are gaining interest as a natural antioxidant substitute for synthetic antioxidants (Wong et al., 2019). The antioxidant activity of BPs strongly relies on their structure, composition, and hydrophobicity. Proline, valine, histidine, glycine, phenylalanine, cysteine, tyrosine, leucine, isoleucine, methionine, threonine, lysine, and glutamic acid are examples of amino acids that contribute to antioxidant properties (Górska-Warsewicz et al., 2018). BPs with antioxidant properties generally contain 4–16 amino acids with a molecular weight ranging from 0.4 to 2 kDa. The sequence or length of amino acids obtained from enzymatic hydrolysis and through the gastrointestinal tract influences the route of action, which may enhance the antioxidant activity in vivo (Toldrá et al., 2018).

The antioxidant activity of peptides can slow or prevent oxidative processes by donating electrons and stabilizing free radicals (Lemes et al., 2016). Antioxidant molecules can inhibit the activity of free radicals by two mechanisms known as hydrogen atom transfer and single electron transfer. Antioxidant peptides containing tyrosine play a role through hydrogen atom transfer, whereas those containing histidine, cysteine, and tryptophan act mainly through single electron transfer (Esfandi, Walters, and Tsopmo, 2019). Peptides containing aromatic amino acids, for example, phenylalanine and tyrosine, act as the potent radical-scavenging BPs because of their ability to easily donate a proton to electron-deficient radicals (Ajibola et al., 2011). Indeed, histidine-containing peptides act as a bioactive antioxidant peptide by donating hydrogen, lipid peroxyl radical trapping, or the metal ion-chelating ability of the imidazole group (Khan et al., 2014; Walters, Esfandi, and Tsopmo, 2018). On the other hand, the antioxidant activity of cysteine exhibits independent activity because the sulfhydryl group (-SH) in cysteine can interact directly with radicals (Sarmadi and Ismail, 2010). Peptides with antioxidant activity can protect the human cell from oxidative-stress-associated diseases (Zhu et al., 2008). Furthermore, they also play a crucial role in the food industry by slowing or preventing the oxidation of macromolecules like proteins and lipids that can lead to quality deterioration (Nwachukwu and Aluko, 2019).

7.3.2 Antihypertensive Activity

Hypertension can lead to cardiovascular and cerebrovascular diseases, such as stroke, atherosclerosis, and myocardial infarction (Li et al., 2016). ACE is responsible for increasing blood pressure by catalyzing the conversion of angiotensin I to angiotensin II. As a result, certain compounds with ACE-inhibitory effects are considered a potent antihypertension therapeutic to lower blood pressure, such as lisinopril, captopril, alacepril, and enalapril (Huang et al., 2011). Nonetheless, long-term use of these synthetic drugs has undesirable side effects for some patients (Balti et al., 2015). For this reason, BPs from biological sources that possess ACE inhibition have been gaining interest. The ACE-inhibition characteristic of peptides is related to their chain length, composition, and sequence. In general, peptides with ACE-inhibitory activity are short-chain peptides that consist of 2–12 amino acids in which large peptides cannot bind to the active sites of ACE. On the other hand, long-chain peptides could exhibit prominent ACE activity, which is usually dependent on amino acid type than the length of peptides. Negatively charged acidic amino acids (asparagine and glutamic acid) containing peptides can chelate zinc atoms necessary for ACE activity (Kaur et al., 2021). The ACE-inhibitory property of peptides also relies on a specific amino acid at the C-terminus and N-terminus. The existence of aromatic amino acids at the C-end and hydrophobic amino acids at the N-end enhances the ACE-inhibitory effect of BPs (Acquah, Stefano, and Udenigwe, 2018). Guang and Phillips (2009) reported the influence of peptides containing phenylalanine, proline, leucine, isoleucine, tyrosine, tryptophan, valine, arginine, and lysine on ACE binding. Piovesana et al. (2018) revealed that the presence of positively charged amino acids (particularly an alkyl group at the C-terminus) and hydrophobic amino acids in peptides could also perform the ACE-inhibitory activity.

7.3.3 Antimicrobial Activity

Antimicrobial peptides are usually small (containing 20–46 amino acids), basic (mostly containing lysine and arginine), and amphipathic (Toldrá et al., 2018). They are mostly rich in hydrophobic amino acids, including valine, tryptophan, phenylalanine, leucine, and isoleucine. These peptides consist of cationic amino acids, which possess a net positive charge between +2 and +9 (Haney and Hancock, 2013). Antimicrobial peptides can fight several pathogenic micro-organisms, including

Antibacterial peptides

Antibacterial peptide: peptide that exhibit antibacterial properties

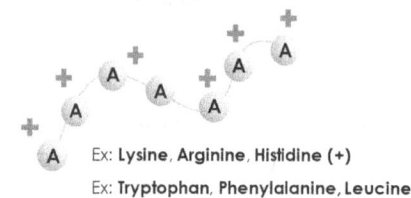

1. Direct membrane damage
2. Cellular interruption
3. Immunomodulatory

Lipopolysaccharide: **gram negative**
lipoteichoic acids : **gram positive**

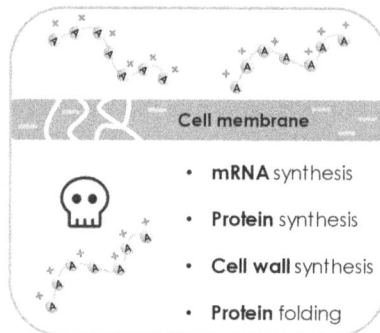

A Ex: Lysine, Arginine, Histidine (+)

Ex: **Tryptophan, Phenylalanine, Leucine**

- **Cationic residues** → Reduce hydrophobicity
- **Hydrophobicity** → hemolytic of red blood cells
- **Ratio (1:1, 1:2) between Cationic & Uncharged**

Cell membrane

- **mRNA** synthesis
- **Protein** synthesis
- **Cell wall** synthesis
- **Protein** folding

FIGURE 7.2   Mechanism of action of antimicrobial peptide against foodborne pathogens.

bacteria, fungi, and viruses (Treffers et al., 2005). Antimicrobial peptides act against pathogenic micro-organisms *via* the electrostatic interaction of peptides and their cell membrane (Figure 7.2). Antimicrobial peptides with a positive charge initially act on anionic lipids located on the membrane surface. The peptide is then infused with the lipid bilayer of the cell membrane, resulting in the replacement of lipids (Fjell et al., 2012). The ability of antimicrobial peptides to disrupt microbial cells can be achieved by forming ion channels or transmembrane pores. This leads to an imbalance of cellular contents, thereby modulating the process of replication, transcription, and translation of the DNA sequence via binding to specific intracellular targets, hindering the multiplication and growth of microbial cells (Zhao et al., 2012).

7.3.4 IMMUNOMODULATORY AND ANTI-INFLAMMATORY ACTIVITY

Immune function is considered the main regulation to control and prevent chronic diseases. Cancer is one of the immune deficiency diseases which currently become a serious issue globally. On the other hand, there are scarce effective strategies available to modulate the immune response. Most immunomodulatory pharmaceuticals are not appropriate for chronic and preventive purposes (Duarte et al., 2006). Peptides with immunomodulatory activity are responsible for promoting the human immune system by regulating cytokine, enhancing immune cell functions and natural killer cell activity, developing an immune system regulated by ROS, and antibody synthesis (Hou et al., 2012; Udenigwe and Aluko, 2012). At this moment, the exact mechanism of the peptides with immunomodulatory activity is not fully understood. However, the immunomodulatory activity of peptides may act via the activation of macrophages, stimulation of phagocytosis, enhancement of leukocytes, stimulation of immune modulators such as cytokines, nitric oxide, and immunoglobulins, and stimulation of NK cells (Görgüç, Gençdağ, and Yılmaz, 2020).

7.4 NOVEL EXTRACTION TECHNIQUES APPLIED TO AGRO-INDUSTRIAL WASTES

Current advancements in the extraction of bioactive compounds from agro-industrial wastes are directly connected with environmentally friendly techniques. Due to the need for large quantities of organic solvents, significant energy consumption, and lengthy processing times, traditional techniques are generally unsuitable. Currently, extraction techniques that are more environmentally friendly are being substituted for extraction techniques that are less environmentally beneficial. Green extraction techniques are increasingly adaptable, more selective in their product recovery, and can generate a greater sustainable industry (Belwal et al., 2020). Numerous research utilizing green extraction techniques have been developed in recent years. Among these developing techniques are UAE, MAE, pulsed electric field extraction (PEF), and enzyme assisted extraction; subcritical fluid extraction (S_bFE) and supercritical fluid extraction (S_pFE) should be noted (Figure 7.3). There have been a number of recent advancements in environmentally friendly extraction strategies for agricultural and industrial wastes, which are summarized in Table 7.2.

7.4.1 ULTRASOUND-ASSISTED EXTRACTION

UAE is one of the non-conventional extraction techniques under preliminary testing as a step towards industrial application (Freitas et al., 2021). Ultrasound waves rupture the cell wall, allowing the release of targeted bioactive compounds to increase the rate of diffusion and mass transfer, as well as to provide for more solvent penetration into the matrix permeated by ultrasound radiation (Chakraborty, Uppaluri, and Das, 2020). The combination of specific impacts with cavitation, mechanical agitation, and a lowered extraction temperature and duration leads to the UAE as a technique that uses a minimal amount of solvent and energy while extracting thermosensitive compounds (Mena-García et al., 2019). Figure 7.4 illustrates a schematic diagram of the ultrasonic extraction process.

Plant nutraceutical extraction using UAE is considered a time-saving technique that improves both production and product quality by reducing the amount of time needed (Gallo, Ferrara, and Naviglio, 2018). In addition to being a cost-effective and beneficial technology, this technique uses a basic laboratory that does not necessarily require the application of costly equipment (Khawli et al., 2019). The extraction efficiency of the UAE is greatly affected by the variables of ultrasonic wave frequency, temperature, sample type, sonication period, and solvent properties (Alirezalu et al., 2020; Mala, Sadiq, and Anal, 2021).

7.4.2 MICROWAVE-ASSISTED EXTRACTION

Compared to UAE, MAE is a relatively recent process. Additionally, a large number of research laboratories have investigated these applications, illustrating the great extraction potential of this non-conventional technique (Freitas et al., 2021). A kind of electromagnetic radiation in the microwave spectrum contains frequencies varying from 300 MHz to 300 GHz (Chen et al., 2017). When compared to conventional techniques, MAE provides numerous advantages, including extraction selectivity, lower installation costs, simplicity of maintenance, quicker extraction period, less preparatory time, and lower electrical energy demand (Vinatoru et al., 2017).

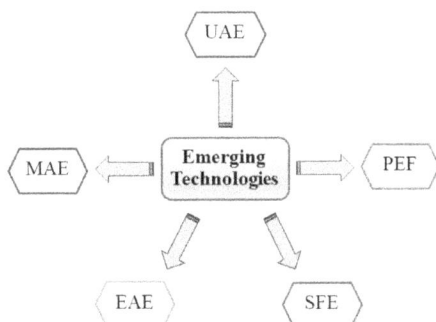

FIGURE 7.3 Emerging technologies for bioactive compound extraction.

TABLE 7.2

Current Implementations of Ecologically Friendly Extraction Techniques for the Extraction of Bioactive Proteins and Peptides from Agro-Industrial Wastes

Agro-industrial waste	Extraction techniques	Extracted compounds	Major findings	References
Cocoa bean shell waste	MAE	Proteins	Optimum MAE parameters were found at 5 min, pH 12, 97°C, with a ratio of solid and liquid at 0.04 g/ml.	Mellinas, Jiménez, and Garrigós, 2020
Squid muscle	Subcritical water extraction	Amino acids peptides	At 250°C, the maximum yields of free and structural amino acids reached 421.53 and 380.58 mg/100 g.	Asaduzzaman and Chun, 2015
Poultry wastes	Subcritical water extraction	Amino acids	The optimal variables for the highest yield of the amino acid were 533 K, 28 min, and 0.02% H_2SO_4.	Zhu et al., 2010
Rapeseed cake	EAE	Protein	The optimum conditions were 1% (w/w) of the enzyme, water-to-cake ratio (w/w) at 6:1 and 80 min.	Niu et al., 2012
Pineapple byproducts	MAE UAE	Bioactive peptides	The optimized extraction conditions were 99.96%, 26.83 min, and 20.96 ml/g for UAE and 100 W, 8.99 min, and 1:8 g/ml for MAE.	Mala, Sadiq, and Anal, 2021
Yeast waste from beer production	PEF	Protein	With the selected variables of electric fields at 10 kV/cm, pulse number at 8 and liquid-to-solid ratio at 40:1, the highest extraction yield of protein reached 2.788%.	Liu, 2012
Rapeseed stems and leaves	PEF	Protein	Extraction of polyphenols with a current density of 5 kV/cm can preserve the protein content.	Yu et al., 2015
Chicken meat waste	PEFs with mechanical processing	Protein	A two-step extraction process achieves the recovery of 12 ± 2% of chicken biomass waste. The protein content in the extracted waste contained 78 ± 8 mg/ml.	Ghosh et al., 2019
Microalgae biomass waste	PEF	Protein	Protein extraction yield (w/w% of total protein) increased from 0 to 41.6% while using PEF at 350°C and 25 MPa for a duration of 15 min.	Guo et al., 2019
Sea bream and sea bass fish byproducts	PEF	Peptides/hydrolysates	Gills produced a maximum concentration of antioxidants and amino acids among the byproducts.	Franco et al., 2020

(Continued)

TABLE 7.2 (CONTINUED)

Current Implementations of Ecologically Friendly Extraction Techniques for the Extraction of Bioactive Proteins and Peptides from Agro-Industrial Wastes

Agro-industrial waste	Extraction techniques	Extracted compounds	Major findings	References
Tuna skin and collagen	Subcritical water extraction	Peptides	As evidenced by the molecular weight pattern of the hydrolysate, peptides and free amino acids in the hydrolysate were related to functional properties.	Ahmed and Chun, 2018
Scallop viscera by-product	Subcritical water extraction	Proteins and amino acids	Protein patterns were found at 14, 29, 36, 45, and 66 kDa, and the highest amino acid was glycine.	Tavakoli and Yoshida, 2006
Sunflower by-product	MAE	Protein and amino acid	Chlorogenic acid (CGA)-rich extracts and a residual solid with a high content of protein (26%) and essential amino acids were detected after treatment.	Náthia-Neves and Alonso, 2021
Sesame bran	EAE and UAE	Protein	The optimum condition was 836 W, 43°C, 98 min, pH at 9.8, and enzyme concentration at 1.248 AU/100 g.	Görgüç, Bircan, and Yılmaz, 2019
Sugar cane (*Saccharum* L.) molasses	S$_p$FE	Amino acids	The optimal condition was 316bar, 50°C and 76 min.	Varaee et al., 2019
Sugar beet (*Beta vulgaris* L.) leaves	EAE	Protein	The optimized condition was 54.25°C, 81.35 min, and the solvent-to-solid ratio at 27.65 ml/g. The protein yield was increased from 43.27% to 79.01% with the assistance of enzyme extraction.	Akyüz and Ersus, 2021

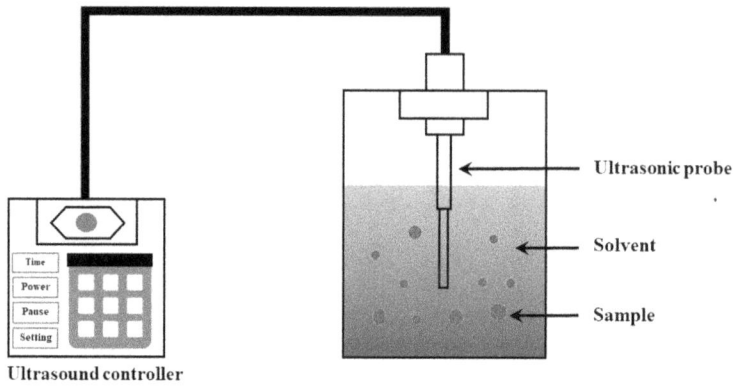

FIGURE 7.4 Schematic diagram of ultrasound equipment.

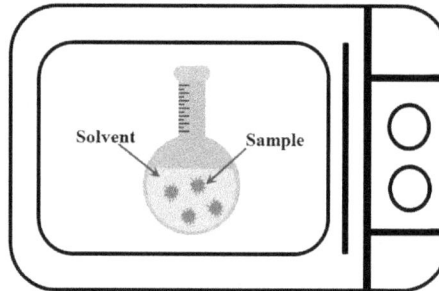

FIGURE 7.5 A schematic representation of the MAE system.

MAE is a green extraction technology that uses solvents to extract bioactive plant compounds from primary sources (Zhang et al., 2019). The microwave influence occurs as a result of electromagnetic energy being absorbed and converted to heat energy. The increase in temperature inside the matrix stimulates internal moisture evaporation, resulting in increased pressure impacting the cell walls. When the matrix is physically changed, the pores enlarge, and the process of using solvents becomes easier, thereby boosting the effectiveness of mass transfer and extraction (Bruno et al., 2019; Wen et al., 2019). The MAE technique is presented schematically in Figure 7.5.

7.4.3 PULSED ELECTRIC FIELD EXTRACTION

This innovative non-thermal processing technology is made possible with PEF treatment (Koubaa et al., 2015). For the past decade, PEF technology has been used to improve the effectiveness of extraction, drying, pressing, and diffusion (Vorobiev et al., 2004). An electric field affects the structure of the cell membrane, which is the primary mechanism for the PEF technique (Bekhit et al., 2016). The effectiveness of this technique is attributable to a number of factors, including the process temperature, the field strength, the pulse intensity, the individual energy input, and the sample properties (Gómez et al., 2019). This technique works at a field strength of roughly 500 V/cm and 1000 V/cm for approximately 10^{-4} to 10^{-2} seconds, and it produces a small level of temperature while destroying plant cell membranes (Fincan and Dejmek, 2002).

Electrolysis treats a material by subjecting it to intermittent pulsed electrical voltage, in this case by putting it between two electrodes with a high voltage and a grounded electrode. PEF treatment is typically performed at room temperature (Barbosa-Canovas et al., 2000; Vorobiev and Lebovka, 2010). When cells are subjected to a certain electric field, their membranes become permeabilized.

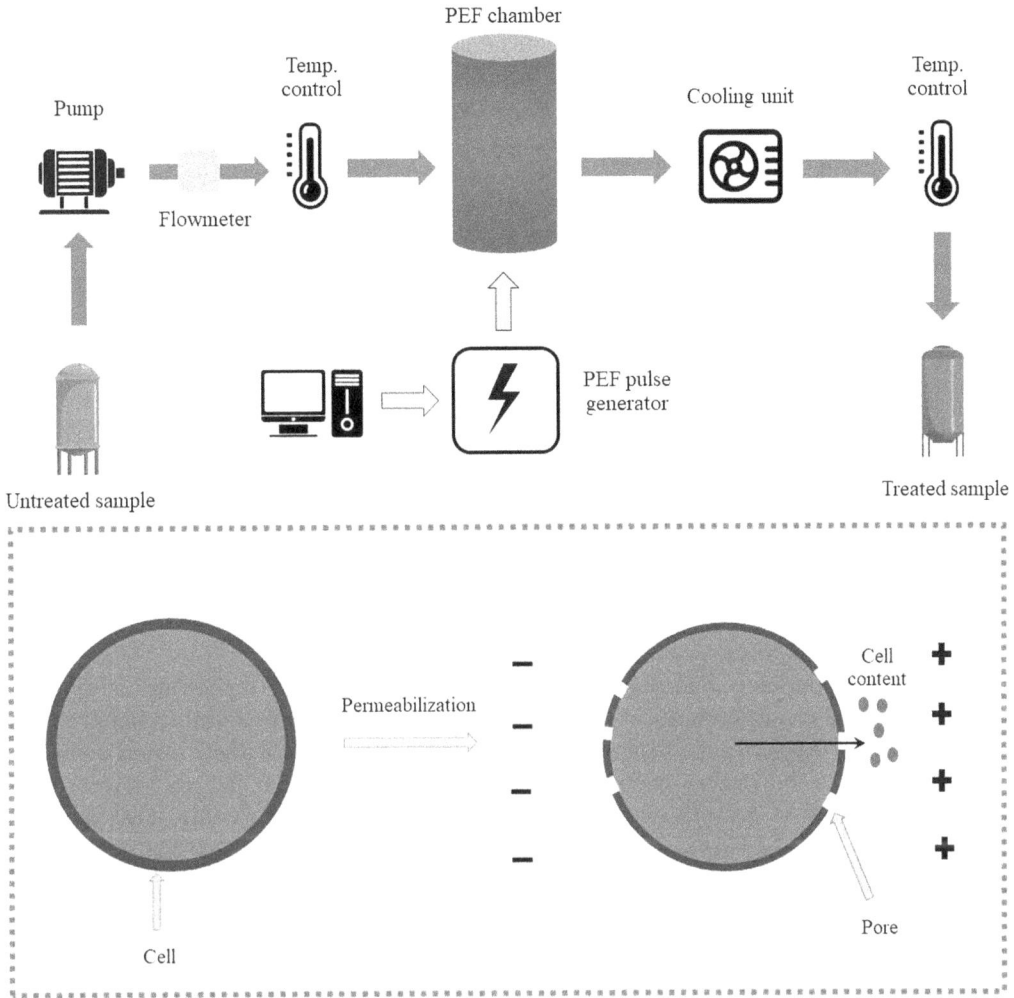

FIGURE 7.6 Schematic representation of a typical PEF unit. The bottom panel illustrates the permeabilization of plant cells as a result of PEF treatment.

The term "electroporation" describes the development of temporary or permanent holes in the membranes by this mechanism, allowing the intracellular content to be released (Figure 7.6). As a result, the potential exists for PEF to assist in the preservation of heat-sensitive compounds (Alirezalu et al., 2020).

7.4.4 Enzyme-Assisted Extraction

Enzymatic extraction is highly beneficial in a variety of technologies; it can also be used to extract some specific bioactive substances by hydrolyzing specific polysaccharides and lipids located in the cell wall; thus, it has been called an innovative technique for recovering surface-bound substances (Rosenthal, Pyle, and Niranjan, 1996; Student and Ravindra, 2020). Several components are inaccessible with conventional extraction techniques due to their retention in plant tissues via hydrophobic or hydrogen bonding (Gligor et al., 2019). Enzymes such as α-amylase, pectinase, and cellulase are used to break down the cell wall, enabling compounds that are bound to the substrate to be released, thereby increasing recovery and yield. Since EAE employs water as the solvent, it is a well-known ecologically safe and effective technique for extracting oil and bioactive chemicals (Puri, Sharma, and Barrow, 2012). The additional advantage of EAE being employed as a raw material pre-treatment is the reduction in extraction time, the smaller amount of solvent needed, and greater yield and better product quality (Nadar, Rao, and Rathod, 2018; Alirezalu et al., 2020).

Complex structural polysaccharides comprised of cellulose, hemicellulose, pectin, lignin, and proteins form up the plant cell wall. This structure is responsible for cell stability and resistance to intracellular component extraction. In order to get access to bioactive compounds that are located inside the cytosolic spaces and even those that are bound to the cellular wall, specific hydrolytic enzymes are used (Álvarez García, 2018; Panja, 2018). These mechanisms are depicted schematically in Figure 7.7.

7.4.5 Sub- and Supercritical Fluid Extraction

The condition of materials when pressure and temperature are less than the critical point is called subcritical fluid extraction (S_bFE). In regard to the environmental benefits of low basicity, protic, and high- to medium-polarity organic solvents, water can be a more ecologically benign alternative as a subcritical solvent. Heating water to above its boiling point and keeping it in a liquid condition under sufficient pressure is used in this technique. Water's dielectric constant is reduced, making it possible to extract many compounds with medium polarity under these conditions. Moreover, n-butane is frequently used as a subcritical fluid in the extraction of oils due to its excellent solvation power at lower critical pressures and temperatures (Zhang et al., 2019).

Supercritical extraction is characterized by the occurrence of changes in temperature and pressure such that gas turns into a supercritical fluid, resulting in the formation of a distinct liquid and gas phase. The solvent supercritical fluid flow takes place with a transfer mass where solvent convection is the dominant transport mechanism (Soquetta, Terra, and Bastos, 2018). While other new extraction techniques are still in the research stages, S_pFE has already left the laboratory since it has previously been used in the industry. Furthermore, owing to its unique properties, CO_2 is the most commonly utilized solvent in S_pFE. This is due to its low critical temperature and pressure (31.1°C and 7.38 MPa, respectively), which is crucial for the preservation of bioactive compounds in extracts, along with being an inert solvent, as reported by Chemat et al. (2019). Carbon dioxide is non-toxic, cheap, non-explosive, widely available, easily extracted, and well suited for separation purposes (Freitas et al., 2021). Figure 7.8 demonstrates the extraction of a selected sample using S_pFE.

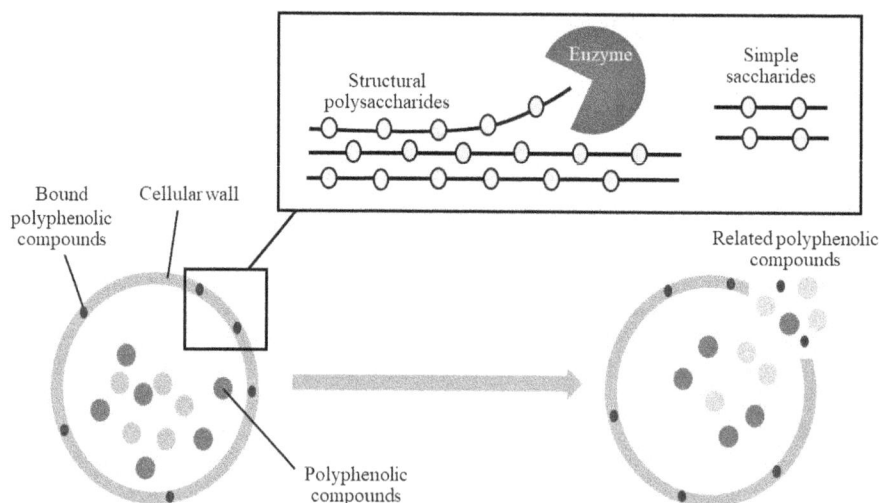

FIGURE 7.7   Enzymes break down the plant cell wall, releasing bioactive compounds.

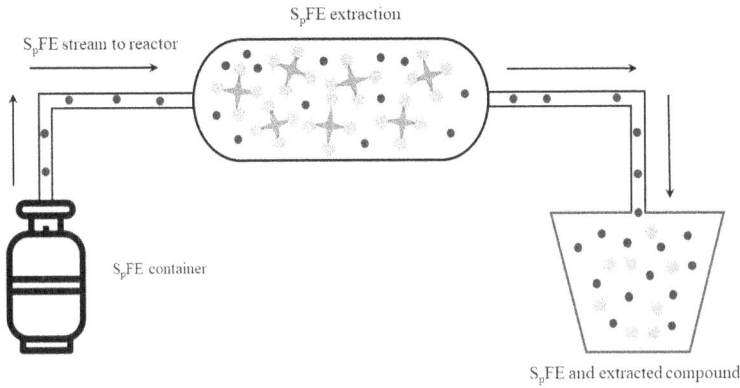

FIGURE 7.8 The schematic representation of S_pFE.

TABLE 7.3
Commercially Available BP Enriched Foods

Source	Product	Peptide	Bioactivity	Manufacturer
Milk	Lactium®	Peptide (YLGYLEQLL)	Relaxing	Ingredia, Arras Cedex, France
Whey	Myprotein™	Whole hydrolysate	Sport nutrition	The Hut Ltd., UK
Sardine	Sato Marine Super P	Peptide (VY)	Antihypertensive	Sato Pharmaceutical Co., Ltd., Tokyo, Japan
Whey or casein	Hyvital®	Whole hydrolysate	Infant nutrition	FrieslandCampina, Netherlands
Plant protein (soy, cotton, seed, wheat, pea)	Proyield®	Whole hydrolysate	Biopharmaceutical cell culture media	FrieslandCampina, Netherlands
Protein from casein, soy, malt, gelatine, and cotton	Stedygro®	Whole hydrolysate	Microbial culture media	FrieslandCampina, Netherlands
Protein from casein and whey	Lacprodan®	Whole hydrolysate	Sports nutrition and beverage	Arla Foods Ingredients, Denmark
Milk casein	Ameal S	Peptides (IPP and VPP)	ACE inhibition	Calpis, Japan
Bonito	Vasotensin®	Peptide (LKPNM)	Antihypertension	Metagenics, US
Fish	Stabilium® 200	Whole hydrolysate	Relaxing	Yalacta, France
Whey	BioZate	Peptides	Antihypertension	Davisco Foods, US
Whey	NOW®	Whole hydrolysate	Sport nutrition	NOWfoods, US

Note: A: alanine; R: arginine; N: asparagine; D: aspartic acid; C: cysteine; Q: glutamine; E: glutamic acid; G: glycine; H: histidine; I: isoleucine; L: leucine; K: lysine; M: methionine; F: phenylalanine; P: proline; S: serine; T: threonine; W: tryptophane; Y: tyrosine; V: valine; Hyp: 4-hydroxyproline.

7.5 INDUSTRIAL APPLICATION OF FOOD-DERIVED BIOACTIVE PEPTIDES

The commercial interest in BPs from food or agro-industrial waste is limited by the complications in methodology for quality assurance, evidence of bioavailability, lack of clinical trials and bioefficacy, allergenicity and toxicity, scientific data, and bitterness of peptides. Additionally, when health benefits are claimed for these peptides, a range of demands and guidelines must be met for them to reach the market. Despite the challenges associated with the industrial application of BPs, some manufacturers have already rolled out products to the market (Table 7.3).

7.6 CONCLUSION AND FUTURE PERSPECTIVES

Agro-industrial waste is an opportunity, an opportunity to utilize waste and reduce environmental hazards. Each sector should make efforts in food production, processing, marketing, and final consumption to minimize generating food waste. Governing bodies should monitor industries and get them to adopt novel technologies that reduce food waste and make consumers aware of the importance of using value-added products developed from food waste. Studies aimed at sustainably reusing waste have advanced significantly on a laboratory scale and have begun to be implemented on an industrial scale. Bioactive proteins and peptides remain an abundant source of functional foods with health benefits. Some of the important bioactivity include antioxidants, ACE inhibition, and antimicrobials, which need further elucidation in future research. In order to extract peptides with bioactive properties, novel extraction methods are required for maximum yield with less investment. This has to be expedited so that food-processing resources and energy consumption can be significantly reduced, which can be done with a strong alliance between the scientific community, industry, and the relevant stakeholders. More studies with the potential to scale up are needed to elucidate the effectiveness of UAE, MAE, EAE, and SFE for sustainable application on an industrial scale that encourages bioeconomy through sustainable production alternatives and improves society's quality of life.

REFERENCES

Acquah, Caleb, Elisa Di Stefano, and Chibuike C. Udenigwe. 2018. "Role of Hydrophobicity in Food Peptide Functionality and Bioactivity." *Journal of Food Bioactives* 4: 88–98. https://doi.org/10.31665/jfb.2018 .4164.

Agyei, Dominic, Clarence M. Ongkudon, Chan Yi Wei, Alan S. Chan, and Michael K. Danquah. 2016. "Bioprocess Challenges to the Isolation and Purification of Bioactive Peptides." *Food and Bioproducts Processing* 98: 244–56. https://doi.org/10.1016/j.fbp.2016.02.003.

Ahmed, Raju, and Byung Soo Chun. 2018. "Subcritical Water Hydrolysis for the Production of Bioactive Peptides from Tuna Skin Collagen." *Journal of Supercritical Fluids* 141(November): 88–96. https://doi .org/10.1016/j.supflu.2018.03.006.

Ahn, Chang Bum, Young Sook Cho, and Jae Young Je. 2015. "Purification and Anti-Inflammatory Action of Tripeptide from Salmon Pectoral Fin Byproduct Protein Hydrolysate." *Food Chemistry* 168: 151–56. https://doi.org/10.1016/j.foodchem.2014.05.112.

Ahn, Chang Bum, You Jin Jeon, Yong Tae Kim, and Jae Young Je. 2012. "Angiotensin i Converting Enzyme (ACE) Inhibitory Peptides from Salmon Byproduct Protein Hydrolysate by Alcalase Hydrolysis." *Process Biochemistry* 47(12): 2240–45. https://doi.org/10.1016/j.procbio.2012.08.019.

Ahn, Chang Bum, Jeong Gyun Kim, and Jae Young Je. 2014. "Purification and Antioxidant Properties of Octapeptide from Salmon Byproduct Protein Hydrolysate by Gastrointestinal Digestion." *Food Chemistry* 147: 78–83. https://doi.org/10.1016/j.foodchem.2013.09.136.

Ajibola, Comfort F., Joseph B. Fashakin, Tayo N. Fagbemi, and Rotimi E. Aluko. 2011. "Effect of Peptide Size on Antioxidant Properties of African Yam Bean Seed (Sphenostylis stenocarpa) Protein Hydrolysate Fractions." *International Journal of Molecular Sciences* 12(10): 6685–702. https://doi.org/10.3390/ ijms12106685.

Akyüz, Ayça, and Seda Ersus. 2021. "Optimization of Enzyme Assisted Extraction of Protein from the Sugar Beet (Beta vulgaris L.) Leaves for Alternative Plant Protein Concentrate Production." *Food Chemistry* 335(January): 127673. https://doi.org/10.1016/j.foodchem.2020.127673.

Alemán, A., M. C. Gómez-Guillén, and P. Montero. 2013. "Identification of Ace-Inhibitory Peptides from Squid Skin Collagen after In Vitro Gastrointestinal Digestion." *Food Research International* 54(1): 790–95. https://doi.org/10.1016/j.foodres.2013.08.027.

Alirezalu, Kazem, Mirian Pateiro, Milad Yaghoubi, Abolfazl Alirezalu, Seyed Hadi Peighambardoust, and Jose M. Lorenzo. 2020. "Phytochemical Constituents, Advanced Extraction Technologies and Techno-Functional Properties of Selected Mediterranean Plants for Use in Meat Products. A Comprehensive Review." *Trends in Food Science and Technology* 100(June): 292–306. https://doi.org/10.1016/j.tifs.2020.04.010.

Álvarez García, Carlos 2018. "Application of Enzymes for Fruit Juice Processing." In: *Fruit Juices: Extraction, Composition, Quality and Analysis*, 201–16. Elsevier Inc.. https://doi.org/10.1016/B978-0-12-802230-6.00011-4.

Ambigaipalan, Priyatharini, and Fereidoon Shahidi. 2017. "Bioactive Peptides from Shrimp Shell Processing Discards: Antioxidant and Biological Activities." *Journal of Functional Foods* 34: 7–17. https://doi.org/10.1016/j.jff.2017.04.013.

Asaduzzaman, A. K. M., and Byung Soo Chun. 2015. "Recovery of Functional Materials with Thermally Stable Antioxidative Properties in Squid Muscle Hydrolyzates by Subcritical Water." *Journal of Food Science and Technology* 52(2): 793–802. https://doi.org/10.1007/s13197-013-1107-7.

Balti, Rafik, Ali Bougatef, Assaâd Sila, Didier Guillochon, Pascal Dhulster, and Naima Nedjar-Arroume. 2015. "Nine Novel Angiotensin I-Converting Enzyme (ACE) Inhibitory Peptides from Cuttlefish (Sepia officinalis) Muscle Protein Hydrolysates and Antihypertensive Effect of the Potent Active Peptide in Spontaneously Hypertensive Rats." *Food Chemistry* 170: 519–25. https://doi.org/10.1016/j.foodchem.2013.03.091.

Barbosa-Canovas, G. V., M. D. Pierson, Q. H. Zhang, and D. W. Schaffner. 2000. "Pulsed Electric Fields." *Journal of Food Science* 65(8 SPEC. SUPPL.): 65–79. https://doi.org/10.1111/j.1750-3841.2000.tb00619.x.

Bayram, B., G. Ozkan, T. Kostka, E. Capanoglu, and T. Esatbeyoglu. 2021. "Valorization and Application of Fruit and Vegetable Wastes and By-Products for Food Packaging Materials." *Molecules* 26(13): 4031. https://doi.org/10.3390/ molecules26134031.

Bekhit, Alaa El Din, V. Suwandy, A. Carne, R. van de Ven, and D. L. Hopkins A. 2016. "Effect of Repeated Pulsed Electric Field Treatment on the Quality of Hot-Boned Beef Loins and Topsides." *Meat Science* 111(January): 139–46. https://doi.org/10.1016/j.meatsci.2015.09.001.

Belwal, Tarun, Farid Chemat, Petras Rimantas Venskutonis, Giancarlo Cravotto, Durgesh Kumar Jaiswal, Indra Dutt Bhatt, Hari Prasad Devkota, and Zisheng Luo. 2020. "Recent Advances in Scaling-Up of Non-Conventional Extraction Techniques: Learning from Successes and Failures." *TrAC - Trends in Analytical Chemistry.* Elsevier B.V. https://doi.org/10.1016/j.trac.2020.115895.

Brandelli, Adriano, Luisa Sala, and Susana Juliano Kalil. 2015. "Microbial Enzymes for Bioconversion of Poultry Waste into Added-Value Products." *Food Research International* 73: 3–12. https://doi.org/10.1016/j.foodres.2015.01.015.

Bruno, Siewe Fabrice, Franck Junior Anta Akouan Ekorong, Sandesh S. Karkal, M. S. B. Cathrine, and Tanaji G. Kudre. 2019. "Green and Innovative Techniques for Recovery of Valuable Compounds from Seafood By-Products and Discards: A Review." *Trends in Food Science and Technology* 85(March): 10–22. https://doi.org/10.1016/j.tifs.2018.12.004.

Chakraborty, Sushma, Ramagopal Uppaluri, and Chandan Das. 2020. "Optimization of Ultrasound-Assisted Extraction (UAE) Process for the Recovery of Bioactive Compounds from Bitter Gourd Using Response Surface Methodology (RSM)." *Food and Bioproducts Processing* 120(March): 114–22. https://doi.org/10.1016/j.fbp.2020.01.003.

Chalamaiah, M., B. Dinesh Kumar, R. Hemalatha, and T. Jyothirmayi. 2012. "Fish Protein Hydrolysates: Proximate Composition, Amino Acid Composition, Antioxidant Activities and Applications: A Review." *Food Chemistry* 135(4): 3020–38. https://doi.org/10.1016/j.foodchem.2012.06.100.

Chalamaiah, Meram, Sule Keskin Ulug, Hui Hong, and Jianping Wu. 2019. "Regulatory Requirements of Bioactive Peptides (Protein Hydrolysates) from Food Proteins." *Journal of Functional Foods* 58(January): 123–29. https://doi.org/10.1016/j.jff.2019.04.050.

Chalamaiah, Meram, Yu Wenlin, and Jianping Wu. 2018. "Immunomodulatory and Anticancer Protein Hydrolysates (Peptides) from Food Proteins: A Review." *Food Chemistry* 245: 205–22. https://doi.org/10.1016/j.foodchem.2017.10.087.

Chemat, Farid, Maryline Abert-Vian, Anne Sylvie Fabiano-Tixier, Jochen Strube, Lukas Uhlenbrock, Veronika Gunjevic, and Giancarlo Cravotto. 2019. "Green Extraction of Natural Products. Origins, Current Status, and Future Challenges." *TrAC - Trends in Analytical Chemistry.* Elsevier B.V. https://doi.org/10.1016/j.trac.2019.05.037.

Chen, Chun, Bin Zhang, Qiang Huang, Xiong Fu, and Rui Hai Liu. 2017. "Microwave-Assisted Extraction of Polysaccharides from Moringa oleifera Lam. Leaves: Characterization and Hypoglycemic Activity." *Industrial Crops and Products* 100(June): 1–11. https://doi.org/10.1016/j.indcrop.2017.01.042.

Chi, Chun Hua, and Seong Jun Cho. 2016. "Improvement of Bioactivity of Soybean Meal by Solid-State Fermentation with Bacillus amyloliquefaciens versus Lactobacillus spp. and Saccharomyces cerevisiae." *LWT - Food Science and Technology* 68: 619–25. https://doi.org/10.1016/j.lwt.2015.12.002.

Daamen, W. F., T. Hafmans, J. H. Veerkamp, and T. H. Van Kuppevelt. 2001. "Comparison of Five Procedures for the Purification of Insoluble Elastin." *Biomaterials* 22(14): 1997–2005. https://doi.org/10.1016/S0142-9612(00)00383-5.

Dai, Chunhua, Haile Ma, Ronghai He, Liurong Huang, Shuyun Zhu, Qingzhi Ding, and Lin Luo. 2017. "Improvement of Nutritional Value and Bioactivity of Soybean Meal by Solid-State Fermentation with Bacillus subtilis." *LWT - Food Science and Technology* 86: 1–7. https://doi.org/10.1016/j.lwt.2017.07.041.

Duarte, Jairo, Gabriel Vinderola, Barry Ritz, Gabriela Perdigón, and Chantal Matar. 2006. "Immunomodulating Capacity of Commercial Fish Protein Hydrolysate for Diet Supplementation." *Immunobiology* 211(5): 341–50. https://doi.org/10.1016/j.imbio.2005.12.002.

Esfandi, Ramak, Mallory E. Walters, and Apollinaire Tsopmo. 2019. "Antioxidant Properties and Potential Mechanisms of Hydrolyzed Proteins and Peptides from Cereals." *Heliyon* 5(4): e01538. https://doi.org/10.1016/j.heliyon.2019.e01538.

Esteve, C., M. L. Marina, and M. C. García. 2015. "Novel Strategy for the Revalorization of Olive (Olea europaea) Residues Based on the Extraction of Bioactive Peptides." *Food Chemistry* 167: 272–80. https://doi.org/10.1016/j.foodchem.2014.06.090.

Fan, Hongbing, and Jianping Wu. 2021. "Purification and Identification of Novel ACE Inhibitory and ACE2 Upregulating Peptides from Spent Hen Muscle Proteins." *Food Chemistry* 345(December 2020): 128867. https://doi.org/10.1016/j.foodchem.2020.128867.

Fan, Jian, Jintang He, Yongliang Zhuang, and Liping Sun. 2012. "Purification and Identification of Antioxidant Peptides from Enzymatic Hydrolysates of Tilapia (Oreochromis niloticus) Frame Protein." *Molecules* 17(11): 12836–50. https://doi.org/10.3390/molecules171112836.

Fathi, Parisa, Marzieh Moosavi-Nasab, Armin Mirzapour-Kouhdasht, and Mohammadreza Khalesi. 2021. "Generation of Hydrolysates from Rice Bran Proteins Using a Combined Ultrasonication-Alcalase Hydrolysis Treatment." *Food Bioscience* 42(February): 101110. https://doi.org/10.1016/j.fbio.2021.101110.

Fincan, Mustafa, and P. Dejmek. 2002. "In Situ Visualization of the Effect of a Pulsed Electric Field on Plant Tissue." *Journal of Food Engineering* 55(3): 223–30. https://doi.org/10.1016/S0260-8774(02)00079-1.

Fjell, Christopher D., Jan A. Hiss, Robert E. W. Hancock, and Gisbert Schneider. 2012. "Designing Antimicrobial Peptides : Form Follows Function." 11(January). https://doi.org/10.1038/nrd3591.

Franco, Daniel, Paulo E. S. Munekata, Rubén Agregán, Roberto Bermúdez, María López-Pedrouso, Mirian Pateiro, and José M. Lorenzo. 2020. "Application of Pulsed Electric Fields for Obtaining Antioxidant Extracts from Fish Residues." *Antioxidants* 9(2): 90. https://doi.org/10.3390/antiox9020090.

Freitas, Cyntia Silva, Mauricio Afonso, Manuela Leal, Giovani Carlo, Patricia Ribeiro, Vania Margaret, Flosi Paschoalin, Eduardo Mere, and Del Aguila. 2019. "Encrypted Antimicrobial and Antitumoral Peptides Recovered from a Protein- Rich Soybean (Glycine Max) by-Product." *Journal of Functional Foods* 54(September 2018): 187–98. https://doi.org/10.1016/j.jff.2019.01.024.

Freitas, Lucas Cantão, Jhonatas Rodrigues Barbosa, Ana Laura Caldas da Costa, Fernanda Wariss Figueiredo Bezerra, Rafael Henrique Holanda Pinto, and Raul Nunes de Carvalho Junior. 2021. "From Waste to Sustainable Industry: How Can Agro-Industrial Wastes Help in the Development of New Products?." *Resources, Conservation and Recycling.* Elsevier B.V. https://doi.org/10.1016/j.resconrec.2021.105466.

Gallo, Monica, Lydia Ferrara, and Daniele Naviglio. 2018. "Application of Ultrasound in Food Science and Technology: A Perspective." *Foods.* MDPI Multidisciplinary Digital Publishing Institute. https://doi.org/10.3390/foods7100164.

Gholizadeh, Ashraf, and Samira Baghban Kohnehrouz. 2013. "DUF538 Protein Super Family Is Predicted to Be the Potential Homologue of Bactericidal/Permeability-Increasing Protein in Plant System." *Protein Journal* 32(3): 163–71. https://doi.org/10.1007/s10930-013-9473-6.

Ghosh, Supratim, Amichai Gillis, Julia Sheviryov, Klimentiy Levkov, and Alexander Golberg. 2019. "Towards Waste Meat Biorefinery: Extraction of Proteins from Waste Chicken Meat with Non-Thermal Pulsed Electric Fields and Mechanical Pressing." *Journal of Cleaner Production* 208(January): 220–31. https://doi.org/10.1016/j.jclepro.2018.10.037.

Gildberg, Asbjørn, Jan Arne Arnesen, Bjørn Steinar Sæther, Jaran Rauø, and Even Stenberg. 2011. "Angiotensin I-Converting Enzyme Inhibitory Activity in a Hydrolysate of Proteins from Northern Shrimp (Pandalus borealis) and Identification of Two Novel Inhibitory Tri-Peptides." *Process Biochemistry* 46(11): 2205–9. https://doi.org/10.1016/j.procbio.2011.08.003.

Gligor, Octavia, Andrei Mocan, Cadmiel Moldovan, Marcello Locatelli, Gianina Crişan, and Isabel C. F. R. Ferreira. 2019. "Enzyme-Assisted Extractions of Polyphenols – A Comprehensive Review." *Trends in Food Science and Technology*. Elsevier Ltd. https://doi.org/10.1016/j.tifs.2019.03.029.

Gómez, Belen, Paulo E. S. Munekata, Mohsen Gavahian, Francisco J. Barba, Francisco J. Martí-Quijal, Tomás Bolumar, Paulo Cezar Bastianello Campagnol, Igor Tomasevic, and Jose M. Lorenzo. 2019. "Application of Pulsed Electric Fields in Meat and Fish Processing Industries: An Overview." *Food Research International*. Elsevier Ltd. https://doi.org/10.1016/j.foodres.2019.04.047.

Görgüç, Ahmet, Cavit Bircan, and Fatih Mehmet Yılmaz. 2019. "Sesame Bran as an Unexploited By-Product: Effect of Enzyme and Ultrasound-Assisted Extraction on the Recovery of Protein and Antioxidant Compounds." *Food Chemistry* 283(June): 637–45. https://doi.org/10.1016/j.foodchem.2019.01.077.

Görgüç, Ahmet, Esra Gençdağ, and Fatih Mehmet Yılmaz. 2020. "Bioactive Peptides Derived from Plant Origin By-Products: Biological Activities and Techno-Functional Utilizations in Food Developments – A Review." *Food Research International* 136(January): 109504. https://doi.org/10.1016/j.foodres.2020.109504.

Górska-Warsewicz, Hanna, Wacław Laskowski, Olena Kulykovets, Anna Kudlińska-Chylak, Maksymilian Czeczotko, and Krystyna Rejman. 2018. "Food Products as Sources of Protein and Amino Acids—The Case of Poland." *Nutrients* 10(12). https://doi.org/10.3390/nu10121977.

Guang, Cuie, and Robert D. Phillips. 2009. "Plant Food-Derived Angiotensin i Converting Enzyme Inhibitory Peptides." *Journal of Agricultural and Food Chemistry* 57(12): 5113–20. https://doi.org/10.1021/jf900494d.

Guo, Bingfeng, Boda Yang, Aude Silve, S. Akaberi, Daniel Scherer, Ioannis Papachristou, Wolfgang Frey, U. Hornung, and Nicolaus Dahmen. 2019. "Hydrothermal Liquefaction of Residual Microalgae Biomass after Pulsed Electric Field-Assisted Valuables Extraction." *Algal Research* 43(November): 101650. https://doi.org/10.1016/j.algal.2019.101650.

Han, Z., A. Park, and W. W. Su. 2018. "Valorization of Papaya Fruit Waste through Low-Cost Fractionation and Microbial Conversion of Both Juice and Seed Lipids." *RSC Advances* 8(49): 27963–72. https://doi.org/10.1039/C8RA05539D.

Haney, Evan F., and Robert Bob E. W. Hancock. n.d. "Peptide Design Applications for Antimicrobial and Immunomodulatory." https://doi.org/10.1002/bip.22250.

Harnedy, Pádraigín A., and Richard J. FitzGerald. 2012. "Bioactive Peptides from Marine Processing Waste and Shellfish: A Review." *Journal of Functional Foods* 4(1): 6–24. https://doi.org/10.1016/j.jff.2011.09.001.

Himaya, S. W. A., Dai Hung Ngo, Bomi Ryu, and Se Kwon Kim. 2012. "An Active Peptide Purified from Gastrointestinal Enzyme Hydrolysate of Pacific Cod Skin Gelatin Attenuates Angiotensin-1 Converting Enzyme (ACE) Activity and Cellular Oxidative Stress." *Food Chemistry* 132(4): 1872–82. https://doi.org/10.1016/j.foodchem.2011.12.020.

Hou, Hu, Yan Fan, Bafang Li, Changhu Xue, Yu Guangli, Zhaohui Zhang, and Xue Zhao. 2012. "Purification and Identification of Immunomodulating Peptides from Enzymatic Hydrolysates of Alaska Pollock Frame." *Food Chemistry* 134(2): 821–28. https://doi.org/10.1016/j.foodchem.2012.02.186.

Huang, Wen Hao, Jie Sun, Hui He, Hua Wei Dong, and Jiang Tao Li. 2011. "Antihypertensive Effect of Corn Peptides, Produced by a Continuous Production in Enzymatic Membrane Reactor, in Spontaneously Hypertensive Rats." *Food Chemistry* 128(4): 968–73. https://doi.org/10.1016/j.foodchem.2011.03.127.

Jain, Surangna, and Anil Kumar Anal. 2017. "Production and Characterization of Functional Properties of Protein Hydrolysates from Egg Shell Membranes by Lactic Acid Bacteria Fermentation." *Journal of Food Science and Technology* 54(5): 1062–72. https://doi.org/10.1007/s13197-017-2530-y.

Jain, Surangna, and Anil Kumar Anal. 2018. "Preparation of Eggshell Membrane Protein Hydrolysates and Culled Banana Resistant Starch-Based Emulsions and Evaluation of Their Stability and Behavior in Simulated Gastrointestinal Fluids." *Food Research International* 103: 234–42. https://doi.org/10.1016/j.foodres.2017.10.042.

Jayathilakan, K., Khudsia Sultana, K. Radhakrishna, and A. S. Bawa. 2012. "Utilization of Byproducts and Waste Materials from Meat, Poultry and Fish Processing Industries: A Review." *Journal of Food Science and Technology* 49(3): 278–93. https://doi.org/10.1007/s13197-011-0290-7.

Je, Jae Young, Jae Young Cha, Young Sook Cho, Hee Young Ahn, Jae Hong Lee, Young Su Cho, and Chang Bum Ahn. 2013. "Hepatoprotective Effect of Peptic Hydrolysate from Salmon Pectoral Fin Protein Byproducts on Ethanol-Induced Oxidative Stress in Sprague-Dawley Rats." *Food Research International* 51(2): 648–53. https://doi.org/10.1016/j.foodres.2013.01.045.

Joshi, Ila, Janagaraj K, Peer Mohamed Noorani K, and Rasool Abdul Nazeer. 2020. "Isolation and Characterization of Angiotensin I-Converting Enzyme (ACE-I) Inhibition and Antioxidant Peptide from by-Catch Shrimp (Oratosquilla Woodmasoni) Waste." *Biocatalysis and Agricultural Biotechnology* 29(December 2019): 101770. https://doi.org/10.1016/j.bcab.2020.101770.

Kannan, Arvind, Navam S. Hettiarachchy, Maurice Marshall, Sivakumar Raghavan, and Hordur Kristinsson. 2011. "Shrimp Shell Peptide Hydrolysates Inhibit Human Cancer Cell Proliferation." *Journal of the Science of Food and Agriculture* 91(10): 1920–24. https://doi.org/10.1002/jsfa.4464.

Kaur, Arshdeep, Bababode Adesegun Kehinde, Poorva Sharma, Deepansh Sharma, and Sawinder Kaur. 2021. "Recently Isolated Food-Derived Antihypertensive Hydrolysates and Peptides: A Review." *Food Chemistry* 346(November 2020): 128719. https://doi.org/10.1016/j.foodchem.2020.128719.

Khan, Asmatullah, Halima Nazar, Syed Mubashar Sabir, Muhammad Irshad, Shahid Iqbal Awan, Rizwan Abbas, Muhammad Akram, et al. 2014. "Antioxidant Activity and Inhibitory Effect of Some Commonly Used Medicinal Plants against Lipid Per-Oxidation in Mice Brain." *African Journal of Traditional, Complementary and Alternative Medicines* 11(5): 83–90. https://doi.org/10.4314/ajtcam.v11i5.14.

Khawli, Fadila Al, Mirian Pateiro, Rubén Domínguez, José M. Lorenzo, Patricia Gullón, Katerina Kousoulaki, Emilia Ferrer, Houda Berrada, and Francisco J. Barba. 2019. "Innovative Green Technologies of Intensification for Valorization of Seafood and Their By-Products." *Marine Drugs*. MDPI AG. https://doi.org/10.3390/md17120689.

Koubaa, Mohamed, Elena Roselló-Soto, Jana Šic Žlabur, Anet Režek Jambrak, Mladen Brnčić, Nabil Grimi, Nadia Boussetta, and Francisco J. Barba. 2015. "Current and New Insights in the Sustainable and Green Recovery of Nutritionally Valuable Compounds from Stevia Rebaudiana Bertoni." *Journal of Agricultural and Food Chemistry* 63(31): 6835–46. https://doi.org/10.1021/acs.jafc.5b01994.

Kshetri, Pintubala, Subhra Saikat Roy, Shamjetshabam Babeeta Chanu, Thangjam Surchandra Singh, K. Tamreihao, Susheel Kumar Sharma, Meraj Alam Ansari, and Narendra Prakash. 2020. "Valorization of Chicken Feather Waste into Bioactive Keratin Hydrolysate by a Newly Purified Keratinase from Bacillus sp. RCM-SSR-102." *Journal of Environmental Management* 273(August): 111195. https://doi.org/10.1016/j.jenvman.2020.111195.

Kumar, S., J. V. Nishanth, Bala Nambisan Siji, and C. Mohandas. 2012. "Activity and Synergistic Antimicrobial Activity between Diketopiperazines against Bacteria In Vitro." *Applied Biochemistry and Biotechnology* 168(8): 2285–96. https://doi.org/10.1007/s12010-012-9937-8.

Lasekan, Adeseye, Fatimah Abu Bakar, and Dzulkifly Hashim. 2013. "Potential of Chicken By-Products as Sources of Useful Biological Resources." *Waste Management* 33(3): 552–65. https://doi.org/10.1016/j.wasman.2012.08.001.

Lee, Jung Kwon, Jeon Joong Kyun, and Hee Guk Byun. 2014. "Antihypertensive Effect of Novel Angiotensin I Converting Enzyme Inhibitory Peptide from Chum Salmon (Oncorhynchus keta) Skin in Spontaneously Hypertensive Rats." *Journal of Functional Foods* 7(1): 381–89. https://doi.org/10.1016/j.jff.2014.01.021.

Lemes, Ailton Cesar, Luisa Sala, Joana Da Costa Ores, Anna Rafaela Cavalcante Braga, Mariana Buranelo Egea, and Kátia Flávia Fernandes. 2016. "A Review of the Latest Advances in Encrypted Bioactive Peptides from Protein-Richwaste." *International Journal of Molecular Sciences* 17(6). https://doi.org/10.3390/ijms17060950.

Li, Ying, and Jianmei Yu. 2015. "Research Progress in Structure-Activity Relationship of Bioactive Peptides." *Journal of Medicinal Food* 18(2): 147–56. https://doi.org/10.1089/jmf.2014.0028.

Li, Ying, Jianzhong Zhou, Xiaoxiong Zeng, and Jianmei Yu. 2016. "A Novel ACE Inhibitory Peptide Ala-His-Leu-Leu Lowering Blood Pressure in Spontaneously Hypertensive Rats." *Journal of Medicinal Food* 19(2): 181–86. https://doi.org/10.1089/jmf.2015.3483.

Liang, Qiufang, Meram Chalamaiah, Wang Liao, Xiaofeng Ren, Haile Ma, and Jianping Wu. 2020. "Zein Hydrolysate and Its Peptides Exert Anti-Inflammatory Activity on Endothelial Cells by Preventing TNF-α-Induced NF-KB Activation." *Journal of Functional Foods* 64(October 2019): 103598. https://doi.org/10.1016/j.jff.2019.103598.

Liu, Mingyuan. 2012. Optimization of Extraction Parameters for Protein from Beer Waste Brewing Yeast Treated by Pulsed Electric Fields (PEF). *African Journal of Microbiology Research* 6(22): 4739–46. https://doi.org/10.5897/ajmr12.117.

Liu, Rui, Lujuan Xing, Qingquan Fu, Guang Hong Zhou, and Wan Gang Zhang. 2016. "A Review of Antioxidant Peptides Derived from Meat Muscle and By-Products." *Antioxidants* 5(3). https://doi.org/10.3390/antiox5030032.

Mala, Thatchajaree, Muhammad Bilal Sadiq, and Anil Kumar Anal. 2021. "Comparative Extraction of Bromelain and Bioactive Peptides from Pineapple Byproducts by Ultrasonic- and Microwave-Assisted Extractions." *Journal of Food Process Engineering* 44(6): e13709. https://doi.org/10.1111/jfpe.13709.

Malison, Arichaya, Pornlert Arpanutud, and Suwimon Keeratipibul. 2021. "Chicken Foot Broth Byproduct: A New Source for Highly Effective Peptide-Calcium Chelate." *Food Chemistry* 345(December 2020): 128713. https://doi.org/10.1016/j.foodchem.2020.128713.

Mellinas, A. C., A. Jiménez, and M. C. Garrigós. 2020. "Optimization of Microwave-Assisted Extraction of Cocoa Bean Shell Waste and Evaluation of Its Antioxidant, Physicochemical and Functional Properties." *LWT* 127(June): 109361. https://doi.org/10.1016/j.lwt.2020.109361.

Mena-García, A., A. I. Ruiz-Matute, A. C. Soria, and M. L. Sanz. 2019. "Green Techniques for Extraction of Bioactive Carbohydrates." *TrAC - Trends in Analytical Chemistry.* Elsevier B.V. https://doi.org/10.1016/j.trac.2019.07.023.

Nadalian, Mehdi, Nurkhuzaiah Kamaruzaman, Mohd Shakir Mohamad Yusop, Abdul Salam Babji, and Salma Mohamad Yusop. 2019. "Isolation, Purification and Characterization of Antioxidative Bioactive Elastin Peptides from Poultry Skin." *Food Science of Animal Resources* 39(6): 966–79. https://doi.org/10.5851/kosfa.2019.e90.

Nadar, Shamraja S., Priyanka Rao, and Virendra K. Rathod. 2018. "Enzyme Assisted Extraction of Biomolecules as an Approach to Novel Extraction Technology: A Review." *Food Research International.* Elsevier Ltd. https://doi.org/10.1016/j.foodres.2018.03.006.

Náthia-Neves, Grazielle, and Esther Alonso. 2021. "Valorization of Sunflower By-Product Using Microwave-Assisted Extraction to Obtain a Rich Protein Flour: Recovery of Chlorogenic Acid, Phenolic Content and Antioxidant Capacity." *Food and Bioproducts Processing* 125(January): 57–67. https://doi.org/10.1016/j.fbp.2020.10.008.

Ng, H. S., P. E. Kee, H. S. Yim, P. T. Chen, Y. H. Wei, and J. C. W. Lan. 2020. "Recent Advances on the Sustainable Approaches for Conversion and Reutilization of Food Wastes to Valuable Bioproducts." *Bioresource Technology* 302:122889. https://doi.org/10.1016/j.biortech.2020.122889.

Ngoh, Ying Yuan, and Chee Yuen Gan. 2016. "Enzyme-Assisted Extraction and Identification of Antioxidative and α-Amylase Inhibitory Peptides from Pinto Beans (Phaseolus vulgaris Cv. Pinto)." *Food Chemistry* 190: 331–37. https://doi.org/10.1016/j.foodchem.2015.05.120.

Nirmal, Nilesh Prakash, Chalat Santivarangkna, Mithun Singh Rajput, and Soottawat Benjakul. 2020. "Trends in Shrimp Processing Waste Utilization: An Industrial Prospective." *Trends in Food Science and Technology* 103(May): 20–35. https://doi.org/10.1016/j.tifs.2020.07.001.

Niu, Yan Xing, Wenlin Li, Jun Zhu, Qingde Huang, Mulan Jiang, and Fenghong Huang. 2012. "Aqueous Enzymatic Extraction of Rapeseed Oil and Protein from Dehulled Cold-Pressed Double-Low Rapeseed Cake." *International Journal of Food Engineering* 8(3). https://doi.org/10.1515/1556-3758.2530.

Nwachukwu, Ifeanyi D., and Rotimi E. Aluko. 2019. "Structural and Functional Properties of Food Protein-Derived Antioxidant Peptides." *Journal of Food Biochemistry* 43(1): 1–13. https://doi.org/10.1111/jfbc.12761.

Osorio, L. L. D. R., E. Flórez-López, and C. D. Grande-Tovar. 2021. "The Potential of Selected Agri-Food Loss and Waste to Contribute to a Circular Economy: Applications in the Food, Cosmetic and Pharmaceutical Industries." *Molecules* 26(2): 515. https://doi.org/10.3390/molecules26020515.

Panja, Palash. 2018. "Green Extraction Methods of Food Polyphenols from Vegetable Materials." *Current Opinion in Food Science.* Elsevier Ltd. https://doi.org/10.1016/j.cofs.2017.11.012.

Piovesana, Susy, Anna Laura Capriotti, Chiara Cavaliere, Giorgia La Barbera, Carmela Maria Montone, Riccardo Zenezini Chiozzi, and Aldo Laganà. 2018. "Recent Trends and Analytical Challenges in Plant Bioactive Peptide Separation, Identification and Validation." *Analytical and Bioanalytical Chemistry* 410(15): 3425–44. https://doi.org/10.1007/s00216-018-0852-x.

Prajapati, Saugat, Sushil Koirala, and Anil Kumar Anal. 2021. "Bioutilization of Chicken Feather Waste by Newly Isolated Keratinolytic Bacteria and Conversion into Protein Hydrolysates with Improved Functionalities." *Applied Biochemistry and Biotechnology.* https://doi.org/10.1007/s12010-021-03554-4.

Puri, Munish, Deepika Sharma, and Colin J. Barrow. 2012. "Enzyme-Assisted Extraction of Bioactives from Plants." *Trends in Biotechnology.* Elsevier Current Trends. https://doi.org/10.1016/j.tibtech.2011.06.014.

Ravindran, R., S. S. Hassan, G. A. Williams, and A. K. Jaiswal. 2018. "A Review on Bioconversion of Agro-Industrial Wastes to Industrially Important Enzymes." *Bioengineering* 5(4): 93. https://doi.org/10.3390/bioengineering5040093.

Rosenthal, A., D. L. Pyle, and K. Niranjan. 1996. "Aqueous and Enzymatic Processes for Edible Oil Extraction." *Enzyme and Microbial Technology.* Elsevier Inc.. https://doi.org/10.1016/S0141-0229(96)80004-F.

Sadh, P. K., P. Saharan, S. Duhan, and J. S. Duhan. 2017. "Bio-Enrichment of Phenolics and Antioxidant Activity of Combination of Oryza sativa and Lablab purpureus Fermented with GRAS Filamentous Fungi." *Resource-Efficient Technologies* 3(3): 347–52. https://doi.org/10.1016/j.reffit.2017.02.008.

Salve Student, Ravindra R. 2020. "Comprehensive Study of Different Extraction Methods of Extracting Bioactive Compounds from Pineapple Waste-A Review Ravindra R Salve and Subhajit Ray." *The Pharma Innovation Journal* 9(6): 327–40.

Sarmadi, Bahareh H., and Amin Ismail. 2010. "Antioxidative Peptides from Food Proteins: A Review." *Peptides* 31(10): 1949–56. https://doi.org/10.1016/j.peptides.2010.06.020.

Senphan, Theeraphol, and Soottawat Benjakul. 2012. "Compositions and Yield of Lipids Extracted from Hepatopancreas of Pacific White Shrimp (Litopenaeus Vannamei) as Affected by Prior Autolysis." *Food Chemistry* 134(2): 829–35. https://doi.org/10.1016/j.foodchem.2012.02.188.

Shrestha, P., M. B. Sadiq, and A. K. Anal. 2021. "Development of Antibacterial Biocomposites Reinforced with Cellulose Nanocrystals Derived from Banana Pseudostem." *Carbohydrate Polymer Technologies and Applications* 2: 100112. https://doi.org/10.1016/j.carpta.2021.100112

Shrestha, S., M. B. Sadiq, and A. K. Anal. 2018. "Culled Banana Resistant Starch-Soy Protein Isolate Conjugate Based Emulsion Enriched with Astaxanthin to Enhance Its Stability." *International Journal of Biological Macromolecules* 120(A): 449–59. https://doi.org/10.1016/j.ijbiomac.2018.08.066.

Sik, B., E. L. Hanczné, V. Kapcsándi, and Z. Ajtony. 2020. "Conventional and Nonconventional Extraction Techniques for Optimal Extraction Processes of Rosmarinic Acid from Six Lamiaceae Plants as Determined by HPLC-DAD Measurement." *Journal of Pharmaceutical and Biomedical Analysis* 184: 113173. https://doi.org/10.1016/j.jpba.2020.113173.

Sila, Assaad, and Ali Bougatef. 2016. "Antioxidant Peptides from Marine By-Products : Isolation, Identification and Application in Food Systems. A Review." *Journal of Functional Foods* 21: 10–26. https://doi.org/10.1016/j.jff.2015.11.007.

Song, Yihang, Yousi Fu, Shiyang Huang, Langxing Liao, Qian Wu, Yali Wang, Fuchun Ge, and Baishan Fang. 2021. "Identification and Antioxidant Activity of Bovine Bone Collagen-Derived Novel Peptides Prepared by Recombinant Collagenase from Bacillus cereus." *Food Chemistry* 349(January): 129143. https://doi.org/10.1016/j.foodchem.2021.129143.

Soquetta, Marcela Bromberger, Lisiane de Marsillac Terra, and Caroline Peixoto Bastos. 2018. "Green Technologies for the Extraction of Bioactive Compounds in Fruits and Vegetables." *CYTA - Journal of Food*. Taylor and Francis Ltd. https://doi.org/10.1080/19476337.2017.1411978.

Tacias-Pascacio, Veymar G., Daniel Castañeda-Valbuena, Roberto Morellon-Sterling, Olga Tavano, Ángel Berenguer-Murcia, Gilber Vela-Gutiérrez, Irfan A. Rather, and Roberto Fernandez-Lafuente. 2021. "Bioactive Peptides from Fisheries Residues: A Review of Use of Papain in Proteolysis Reactions." *International Journal of Biological Macromolecules* 184(April): 415–28. https://doi.org/10.1016/j.ijbiomac.2021.06.076.

Taniguchi, Masayuki, Mitsuhiro Kameda, Toshiki Namae, Akihito Ochiai, and Eiichi Saitoh. 2017. "Identification and Characterization of Multifunctional Cationic Peptides Derived from Peptic Hydrolysates of Rice Bran Protein." *Journal of Functional Foods* 34: 287–96. https://doi.org/10.1016/j.jff.2017.04.046.

Tavakoli, Omid, and Hiroyuki Yoshida. 2006. "Conversion of Scallop Viscera Wastes to Valuable Compounds Using Sub-Critical Water." *Green Chemistry* 8(1): 100–6. https://doi.org/10.1039/b507441j.

Toldrá, Fidel, Milagro Reig, M. Concepción Aristoy, and Leticia Mora. 2018. "Generation of Bioactive Peptides during Food Processing." *Food Chemistry* 267: 395–404. https://doi.org/10.1016/j.foodchem.2017.06.119.

Treffers, Charlotte, Liying Chen, Rachel C. Anderson, and Pak Lam Yu. 2005. "Isolation and Characterisation of Antimicrobial Peptides from Deer Neutrophils." *International Journal of Antimicrobial Agents* 26(2): 165–69. https://doi.org/10.1016/j.ijantimicag.2005.05.001.

Tsai, W. T., J. M. Yang, C. W. Lai, Y. H. Cheng, C. C. Lin, and C. W. Yeh. 2006. "Characterization and Adsorption Properties of Eggshells and Eggshell Membrane." *Bioresource Technology* 97(3): 488–93. https://doi.org/10.1016/j.biortech.2005.02.050.

Udenigwe, Chibuike C., and Rotimi E. Aluko. 2012. "Multifunctional Cationic Peptide Fractions from Flaxseed Protein Hydrolysates." *Plant Foods for Human Nutrition* 67(1): 1–9. https://doi.org/10.1007/s11130-012-0275-3.

Ulug, Sule Keskin, Forough Jahandideh, and Jianping Wu. 2021. "Novel Technologies for the Production of Bioactive Peptides." *Trends in Food Science and Technology* 108(March 2020): 27–39. https://doi.org/10.1016/j.tifs.2020.12.002.

Varaee, Mona, Masoud Honarvar, Mohammad H. Eikani, Mohammad R. Omidkhah, and Narges Moraki. 2019. "Supercritical Fluid Extraction of Free Amino Acids from Sugar Beet and Sugar Cane Molasses." *Journal of Supercritical Fluids* 144(February): 48–55. https://doi.org/10.1016/j.supflu.2018.10.007.

Vinatoru, M., T. J. Mason, and I. Calinescu. 2017. "Ultrasonically Assisted Extraction (UAE) and Microwave Assisted Extraction (MAE) of Functional Compounds from Plant Materials." *TrAC - Trends in Analytical Chemistry*. Elsevier B.V. https://doi.org/10.1016/j.trac.2017.09.002.

Vorobiev, Eugene, Abdel Baset Jemai, Hazem Bouzrara, Nikolai Lebovka, and Maksym Bazhal. 2004. "Pulsed Electric Field-Assisted Extraction of Juice from Food Plants." In: *Novel Food Processing Technologies*, 105–30. CRC Press. https://doi.org/10.1201/9780203997277.ch5.

Vorobiev, Eugene, and Nikolai Lebovka. 2010. "Enhanced Extraction from Solid Foods and Biosuspensions by Pulsed Electrical Energy." *Food Engineering Reviews* 2(2): 95–108. https://doi.org/10.1007/s12393-010-9021-5.

Walters, Mallory E., Ramak Esfandi, and Apollinaire Tsopmo. 2018. "Potential of Food Hydrolyzed Proteins and Peptides to Chelate Iron or Calcium and Enhance Their Absorption." *Foods* 7(10). https://doi.org/10.3390/foods7100172.

Wang, Kai, Lihua Han, Hui Hong, Jing Pan, Huaigao Liu, and Yongkang Luo. 2021. "Purification and Identification of Novel Antioxidant Peptides from Silver Carp Muscle Hydrolysate after Simulated Gastrointestinal Digestion and Transepithelial Transport." *Food Chemistry* 342(September 2020): 128275. https://doi.org/10.1016/j.foodchem.2020.128275.

Wang, Xiuming, Haixia Chen, Xuegang Fu, Shuqin Li, and Jing Wei. 2017. "A Novel Antioxidant and ACE Inhibitory Peptide from Rice Bran Protein: Biochemical Characterization and Molecular Docking Study." *LWT - Food Science and Technology* 75: 93–9. https://doi.org/10.1016/j.lwt.2016.08.047.

Wattanasiritham, Ladda, Chockchai Theerakulkait, Samanthi Wickramasekara, Claudia S. Maier, and Jan F. Stevens. 2016. "Isolation and Identification of Antioxidant Peptides from Enzymatically Hydrolyzed Rice Bran Protein." *Food Chemistry* 192: 156–62. https://doi.org/10.1016/j.foodchem.2015.06.057.

Wen, Peng, Teng Gen Hu, Robert J. Linhardt, Sen Tai Liao, Hong Wu, and Yu Xiao Zou. 2019. "Mulberry: A Review of Bioactive Compounds and Advanced Processing Technology." *Trends in Food Science and Technology*. Elsevier Ltd. https://doi.org/10.1016/j.tifs.2018.11.017.

Wong, Fai Chu, Jianbo Xiao, G. Michelle Ling Ong, Mei Jing Pang, Shao Jun Wong, Lai Kuan Teh, and Tsun Thai Chai. 2019. "Identification and Characterization of Antioxidant Peptides from Hydrolysate of Blue-Spotted Stingray and Their Stability against Thermal, PH and Simulated Gastrointestinal Digestion Treatments." *Food Chemistry* 271(February 2018): 614–22. https://doi.org/10.1016/j.foodchem.2018.07.206.

Xiong, Y. L. 2018. *Muscle Proteins. Proteins in Food Processing*. 2nd ed. Elsevier Ltd. https://doi.org/10.1016/B978-0-08-100722-8.00006-1.

Yu, X., O. Bals, N. Grimi, and E. Vorobiev. 2015. "A New Way for the Oil Plant Biomass Valorization: Polyphenols and Proteins Extraction from Rapeseed Stems and Leaves Assisted by Pulsed Electric Fields." *Industrial Crops and Products* 74(November): 309–18. https://doi.org/10.1016/j.indcrop.2015.03.045.

Yusop, Salma Mohamad, Mehdi Nadalian, Abdul Salam Babji, Wan Aida, Wan Mustapha, and Bita Forghani. 2016. "Production of Antihypertensive Elastin Peptides from Waste Poultry Skin." 2(1): 21–5. https://doi.org/10.18178/ijfe.2.1.21-25.

Zarei, Mohammad, Afshin Ebrahimpour, Azizah Abdul-Hamid, Farooq Anwar, Fatimah Abu Bakar, Robin Philip, and Nazamid Saari. 2014. "Identification and Characterization of Papain-Generated Antioxidant Peptides from Palm Kernel Cake Proteins." *Food Research International* 62: 726–34. https://doi.org/10.1016/j.foodres.2014.04.041.

Zhan, Junqi, Gaoshang Li, Yali Dang, and Daodong Pan. 2021. "Study on the Antioxidant Activity of Peptide Isolated from Porcine Plasma during In Vitro Digestion." *Food Bioscience* 42(January): 101069. https://doi.org/10.1016/j.fbio.2021.101069.

Zhang, Rui, S. Li, Zhenzhou Zhu, and Jingren He. 2019. "Recent Advances in Valorization of Chaenomeles Fruit: A Review of Botanical Profile, Phytochemistry, Advanced Extraction Technologies and Bioactivities." *Trends in Food Science and Technology* 91(September): 467–82. https://doi.org/10.1016/j.tifs.2019.07.012.

Zhao, Jing, Lihua Guo, Hongmei Zeng, Xiufen Yang, Jingjing Yuan, Huaixing Shi, Yehui Xiong, Mingjia Chen, Lei Han, and Dewen Qiu. 2012. "Purification and Characterization of a Novel Antimicrobial Peptide from Brevibacillus laterosporus Strain A60." *Peptides* 33(2): 206–11. https://doi.org/10.1016/j.peptides.2012.01.001.

Zhao, Qian Cheng, Jie Yuan Zhao, Dong Uk Ahn, Yong Guo Jin, and Xi Huang. 2019. "Separation and Identification of Highly Efficient Antioxidant Peptides from Eggshell Membrane." *Antioxidants* 8(10): 1–16. https://doi.org/10.3390/antiox8100495.

Zhu, Guang Yong, Xian Zhu, Xue Liang Wan, Qi Fan, Yan Hua Ma, Jing Qian, Xiao Lei Liu, Yi Jing Shen, and Jun Hui Jiang. 2010. "Hydrolysis Technology and Kinetics of Poultry Waste to Produce Amino Acids in Subcritical Water." *Journal of Analytical and Applied Pyrolysis* 88(2): 187–91. https://doi.org /10.1016/j.jaap.2010.04.005.

Zhu, Lijuan, Chen Jie, Xueyan Tang, and Youling L. Xiong. 2008. "Reducing, Radical Scavenging, and Chelation Properties of In Vitro Digests of Alcalase-Treated Zein Hydrolysate." *Journal of Agricultural and Food Chemistry* 56(8): 2714–21. https://doi.org/10.1021/jf703697e.

8 Production of Biopigments from Agro-Industrial Waste

Neegam Nain, Gunjan K. Katoch, Sawinder Kaur,
Sushma Gurumayum, Prasad Rasane, and Parmjit S. Panesar

CONTENTS

8.1 INTRODUCTION

Colour is an important factor in processing material and is often considered vital to the commercial success of the product. In the case of food, this is considered an important factor for consumer acceptance. Any unnatural colour is associated with spoilage or defect (Sen et al., 2019). The colour of any commodity or product is exhibited due to the presence of certain molecules and structures that are referred to as pigments, and these are capable of absorbing light in the visible range of spectra (De Carvalho et al., 2014). Colours utilized in the food-processing industry are produced either synthetically or extracted from natural sources. However, the use of synthetic colours has steadily declined in recent years, owing to side effects such as allergic reactions caused by certain dyes, including azorubin and tartrazin; tartrazin has a toxic effect on human lymphocyte cells and can bind with DNA (Durán et al., 2002; Babitha et al., 2006; Wang et al., 2013); brilliant blue and tartrazine FD&C Yellow No. 5 cause a hyperactivity disorder; and cancer, DNA damage, neurotoxicity, and hyperactive behaviour in children caused by Allura red (Sabnis, 2010). Most synthetic colourants are derived from a non-renewable source, i.e., petroleum (Kumar et al., 2015).

DOI: 10.1201/9781003125679-8

Conventionally, natural pigments are isolated from plants or insect tissues (Downham and Collins, 2000) that are dependent upon raw material availability; moreover, the extracted pigments show a variable profile influenced by extraction conditions, and they are sensitive to heat, light, and pH conditions. Compared to plant and animal sources, microbial pigments have gathered more consideration for utilization as a natural pigment (Mapari et al., 2005). Pigment synthesis using the microbial fermentation process has a number of benefits, such as relatively simpler extraction, increased product yields, no dearth of substrates, no environmental barrier, and short production time (Mortensen, 2006; Poorniammal et al., 2011).

The processing of agricultural produce has always resulted in the generation of wastes of various forms (Figure 8.1). These waste products create an additional cost to the industry in its collection, disposal, and treatment. The development of value-added products is an effective method of waste management and environmental pollution control, such as the use of agricultural wastes for pigment production, and it is now the preferred technology over the use of synthetic pigments because of the carcinogenic properties of the latter (Panesar et al., 2015). Microbial pigments are environmentally friendly, cost-effective, as well as a potential nutrient source due to the utilization of agro-waste for producing pigments. The production is easily scalable from the laboratory scale to industrial fermenters.

The microbial strains that are commonly used for pigment production (e.g., *Bacillus subtilis*, *Corynebacterium glutamicum*, *Escherichia coli*, and *Pseudomonas putida*) can also be effortlessly genetically modified according to preferences (Venil et al., 2013). The microbial pigments possess many pharmacological properties such as antioxidant, antitumour, immunosuppressive, and antiproliferative functions.

For the production of pigments from micro-organisms, the microbes should be able to utilize an extensive range of wastes, be temperature and pH tolerant, and also give a good colour yield (Kumar et al., 2015). Wastes that are mostly utilized for pigment production are obtained from the dairy industry, cereal processing, fruit and vegetable processing, and agricultural byproducts (Figure 8.1). The main challenge in microbial pigment production is its extraction, as these are either produced intracellularly or extracellularly. If the pigment is intracellular, different methods of cell disruption are used, thereby increasing the steps of downstream processing. Various microbial pigments have not been approved by regulatory authorities for the use as food dyes, such as *Monascus*, which is a mixture of azaphilone pigments and are poisonous in their mature form (Kim and Ku, 2018).

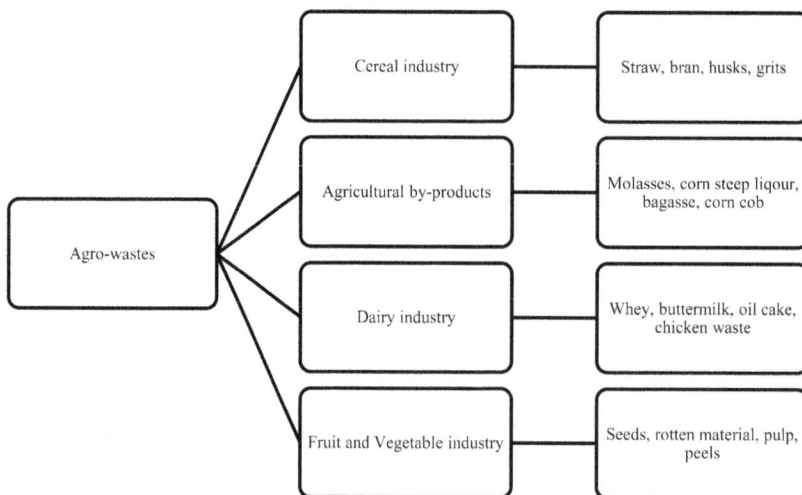

FIGURE 8.1 Different types of agro-wastes.

8.2 NATURAL AND SYNTHETIC PIGMENTS USED IN FOOD AND PHARMACEUTICAL INDUSTRIES

These include carotenoids, chlorophyll, riboflavin, turmeric, curcumin caramel, anthocyanin, betanin, etc. Despite the benefits of pigments, there are various disadvantages associated with pigments derived from natural sources, including their dependence on the availability of raw materials, pigment profile variations from batch to batch, compositional variations depending upon cultivar and climatic conditions, thermal sensitivity, light and alterations in pH, and difficulty in determining the degree of purity. Synthetic colours are classified under those colours that require certification for classification as per the Food and Drug Administration (FDA) (Purba et al., 2015).

8.2.1 SYNTHETIC PIGMENTS

Synthetic dyes were introduced by Perkins in 1856, and by the 20th century, synthetic pigments such as phthalocyanines, arylides, and quinacridones replaced natural dyes because of their cheap cost (Venil et al., 2013). They are widely used in the textile industry, agricultural and food technology research, paper production, photo-electrochemical cells, and the cosmetics industry. As a food colourant, they have been extensively exploited to enhance the appearance of food products to make them appealing to consumers. In the food industry, colourants are mostly used in sugary food products and beverages (Martins et al., 2016).

Synthetic food colourants consist of azo functional groups along with an aromatic ring structure; their use above the acceptable daily intake (ADI) value shows detrimental effects on human health (Andrade et al., 2014). These pigments are carcinogenic and teratogenic, and their precursors also have carcinogenic effects on people working with these materials . The synthetic pigments and their wastes are non-biodegradable and have various detrimental effects on the environment (Kumar et al., 2015). The FDA and European Food Safety Authority, which are responsible for monitoring the quality and safety of food commodities, have determined the ADI value of various pigments (Martins et al., 2016).

In non-alcoholic carbonated beverages, synthetic dyes such as tartrazine (E102), sunset yellow (E110), amaranth (E123), and brilliant blue (E133) are extensively used (Figure 8.2, 8.3) (Andrade et al., 2014). Titanium dioxide (E171) is largely used in confectionery goods. Erythrosine (E127), red (E128), and amaranth (E123) give a red to orange colour (Figure 8.4) to food items, and their ADI values are 0.1, 0.1, and 0.8 mg/kg body weight, respectively. These colourants are incorporated into daily consumed food such as beverages, candies, and meat products. Numerous reports have shown that the consumption of food dyes or colourants significantly affects the behaviour of children (Gostner et al., 2015; Masone and Chanforan, 2015). Tartrazine (E102), quinolone yellow (E104), sunset yellow (E110), carmoisine (E122), ponceau 4R (E124), and Allura red AC (E129) are the six main food dyes that have been proven to cause negative effects on concentration activity in children with attention deficit hyperactivity disorder. Copper chlorophyll (E121 ii) is used in fresh olives, and Green S (E142) is extensively used in canned peas, cake mixes, sauces, and mint jelly (Martins et al., 2016). Commonly used synthetic dyes are shown in Figure 8.5.

8.2.2 EFFECTS OF SYNTHETIC FOOD COLOURANTS ON HEALTH

Synthetic colours present in food are found to instigate various health problems in children. A research study was conducted in order to examine the implication of food constituting synthetic colours on rats. The high administration of tomato ketchup potato chips (30% concentration) showed an upsurge in total cholesterol and triacylglycerol levels. Administration of chocolate colour irrespective of its concentration, i.e., low (15%) and high (30%), resulted in an increase in triacylglycerol level (Sahar and Manal, 2012). The consumption of synthetic colours such as tartrazine, carmoisine, fast green, and sunset yellow in high concentrations (Figure 8.6) exhibited a significant

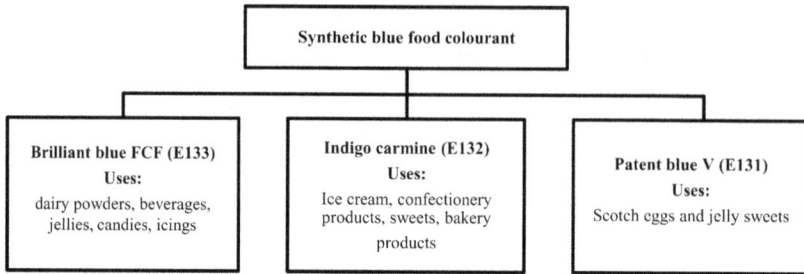

FIGURE 8.2 Synthetic blue food colourant and its uses (Bonan et al., 2013; Ma et al., 2012; Martins et al., 2016).

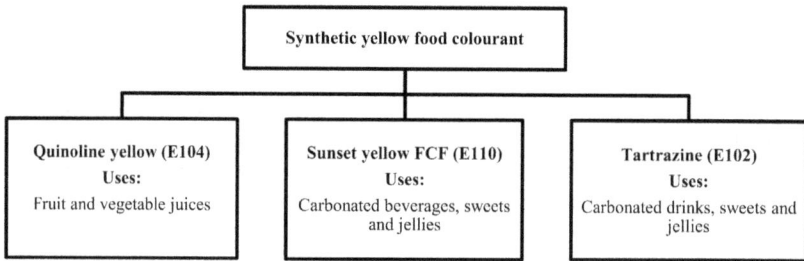

FIGURE 8.3 Synthetic yellow food colourant and its uses (Bonan et al., 2013; Martins et al., 2016).

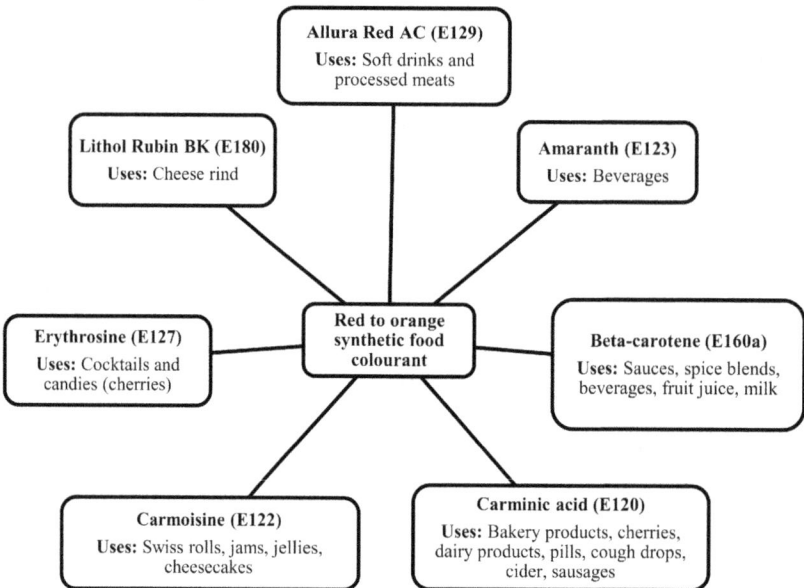

FIGURE 8.4 Red to orange synthetic food colourant and its uses (Bonan et al., 2013; Martins et al., 2016).

upsurge in serum alanine aminotransferase (ALT) and aspartate aminotransferase (AST), which led to a change in liver function. Synthetic colours were found to show unfavourable effects on essential organs (Okafor et al., 2016). One of the studies concluded that the administration of tartrazine and sulfanilic acid at higher doses in rats induced the development of free radicals causing oxidative stress. It also caused morphological changes in red blood cells, i.e., from discoid to echinocytic shape. A remarkable increase was observed in the levels of Trigylceride (TG), AST, cholesterol, creatinine, and glucose levels (Himri et al., 2011).

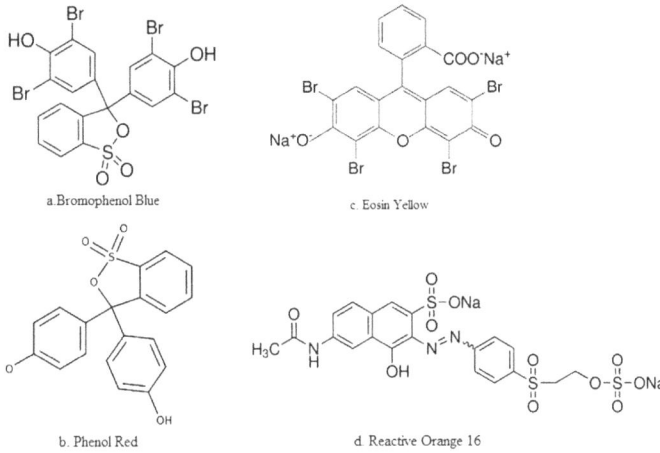

FIGURE 8.5 Commonly used synthetic dyes (Baldrian et al., 2006).

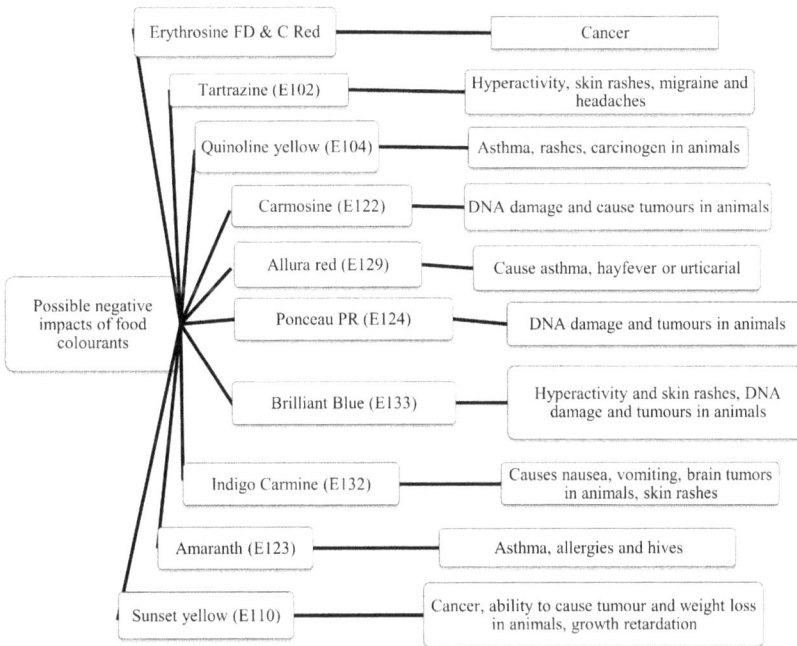

FIGURE 8.6 Possible negative impacts of food colourants (Okafor et al., 2016).

8.2.3 Natural Food Pigments

Nowadays, increased consumer awareness regarding the harmful effect of synthetic food colourants has focused on the need for the replacement of synthetic food grade colours with natural pigments. Before the development of synthetic food colourants, food colourants were obtained from naturally occurring sources like saffron, turmeric, and different types of flowers (Sigurdson et al., 2017). Pigments derived from natural sources (Figure 8.7) are expensive compared to the development of synthetic colourants (Rodriguez-Amaya, 2016). Anthocyanins, betalains, carotenoids, and porphyrins are various classes of plant pigments. Red to purple to blue (E163) can be obtained from grapes, berries, carrots, and concord grapes and used in beverages and other food products. Pink to red (E162) is acquired from beetroot. Brown (caramel) colour (E150) is derived by heating sugars, and from cottonseeds, whereas yellow to orange to red pigments are obtained from *Paracoccus*

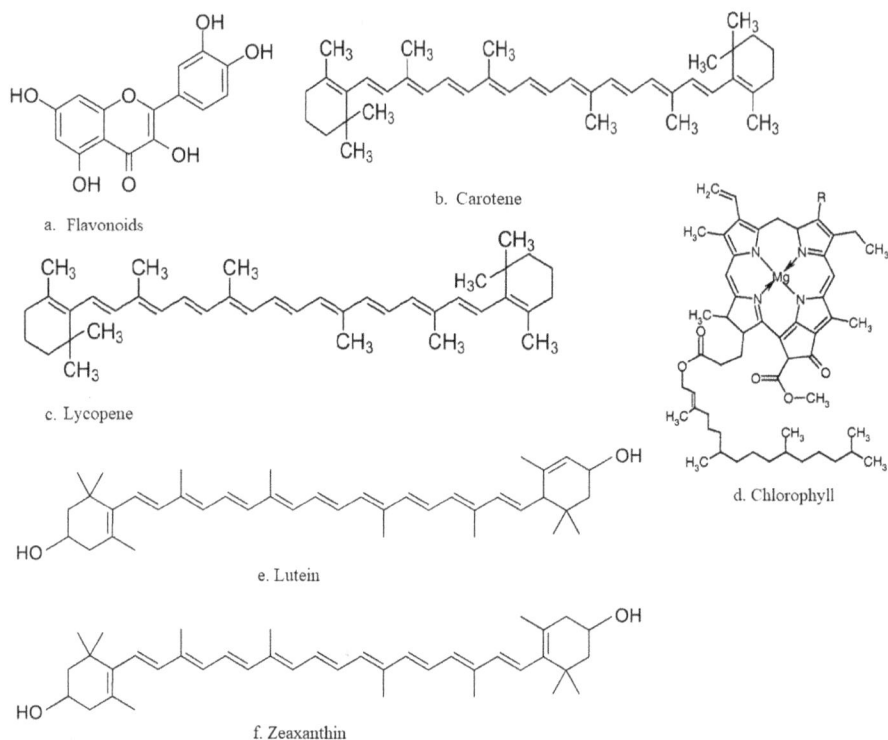

FIGURE 8.7   Commonly used natural food dyes (Shahid et al., 2013).

carotinifaciens bacterium, *Phaffia rhodozyma* yeast, *Haematococcus pluvialis* algae, carrots, yellow corn, paprika, tomato, and marigold. Green colourant (E141) can be obtained from alfalfa and blue-green colour from *Arthrospira platensis* algae (Sigurdson et al., 2017). Corn cob powder is used for producing *Monascus purpureus* KACC 42430 culture by the process of solid-state fermentation (SSF). The resulting pigment was found to be steady at a high pH (acidic conditions), high temperature, as well as in salt solutions (alkaline conditions); in addition to that, it has other industrial applications (Velmurugan et al., 2011).

8.3   MICROBIAL PIGMENTS

Various types of pigments are obtained from micro-organisms such as bacteria, mould, and fungi. Some well-studied strains of micro-organisms that have the capacity for producing biopigments from wastes belong to the genera *Monascus*, *Rhodotorula*, *Penicillium*, and *Aspergillus*. Some of the pigment-producing bacterial species are *Alteromonas rubra*, *Pseudomonas magneslorubra*, *Rugamonas rubra*, *Serratia marcescens*, *Streptoverticillium rubrireticuli*, *Streptomyces longisporus*, *Vibrio psychroerythrous*, *Vibrio gazogenes*, and *S. rubidaea* (Méndez et al., 2011; Panesar et al., 2015). Microbial pigments possess various advantages, including quick propagation of the producer strains, a wide variety of producer strains, and flexible choice of cheap substrates. There are also genetically and metabolically engineered producer strains that can be used in continuous bioreactor processes and operations (Babitha, 2009).

The parameters for pigment production are the carbon and nitrogen ratio, pH, aeration rate, and temperature. To overcome the high cost of synthetic media, several efforts have been made to produce microbial pigments by exploiting waste generated from agricultural and industrial byproducts. The first step is the selection of a suitable strain, preferably that which consistently gives a high yield in the least fermentation time (Nielsen and Olsson, 2002; Nigam and Pandey, 2009).

Strain improvement for increased yield is made by metabolic engineering and mutagenesis with UV, 1-methyl-3-nitro-1-nitrosoguanidine, and ethyl methane sulfonate (Bloor and Cranenburgh, 2006; Venil et al., 2013).

There are two types of fermentation process: SSF and submerged fermentation (SmF). In SSF, there is slower use of the substrates, hence a longer fermentation period, and it is suitable for microbes that require a low moisture content. The substrates commonly used in this type of fermentation are rice, wheat bran, straw, hay, paper pulp, fruit and vegetable remains as well as bagasse. In SmF, there is rapid substrate utilization, and it is best for microbes that demand a high moisture content. This type of fermentation commonly employs soluble sugars, molasses, and wastewater as its raw material (Subramaniyam and Vimala, 2012).

8.3.1 Fungal Pigments

Fungi are metabolically active and competent producers of several metabolites, pigments, and industrial enzymes (dos Reis Celestino et al., 2014). Numerous naturally occurring fungal species are known to produce pigments as secondary metabolites, but only a few of them are investigated for the synthesis of food colours (Mapari et al., 2006). Much work has been reported on the fungus *Monascus purpureus* as it has been traditionally used for centuries as a food colourant in China, Taiwan, and Japan (Prajapati et al., 2014). Besides *Monascus* pigments, filamentous fungi produce several non-carotenoid pigments that exhibit an extraordinary range of colours (Durán et al., 2002). The diversity of microbial pigment is not only related to chemical structures but also to the range of hues added to the colour palette of existing natural colours (Mapari et al., 2008). The non-carotenoid pigments (Table 8.1) are secondary metabolites produced by fungi, many of which have a chemical structure similar to polyketides. These polyketide pigments exhibit distinctive ultraviolet-visible (UV-vis) spectra, and polyketide synthases enzymes are responsible for the partial synthesis of these polyketide pigments (Shimizu et al., 2005). Therefore, it is assumed that these polyketides might serve as a source of new colours that possess a positive role in their use in food. Micro-organisms are recognized to secrete different types of polyketide pigments as secondary metabolites, which include carotenoids, naphthoquinones, fenazine, pyrone, acilphenol, sclertiorine, and anthraquinone (Poorniammal et al., 2011).

Penicillium spp. has also secreted pigments of a polyketide nature similar to those of *Monascus* spp. (Ogihara et al., 2001; Jiang et al., 2005; Mapari et al., 2006). In similar terms, *Epicoccum* sp. has also been reported to produce a variety of secondary metabolites, including pigments of both carotenoids (Kaur et al., 2021) and/or polyketide origin (Kaur et al., 2019b) with colour hues in the red-orange-yellow spectra; they also exhibit antioxidant properties (Stricker et al., 1981). However, the production of mycotoxin citrinin, along with the pigment in *Monascus*-fermented food, has

TABLE 8.1
List of Promising Non-carotenoid Fungal Pigment Producers

Fungal Source	Colour	Reference
Aspergillus sp.	Orange, red	Malik et al., 2012
A. glaucus	Dark red	
Blakeslea trispora	Cream	
Epicoccum nigrum	Yellow-orange	Echavarri-Erasun and Johnson, 2004; Kaur et al., 2019b
Helminthosporium catenarium	Red	Malik et al., 2012
Monascus sp.	Yellow-orange	Malik et al., 2012
Paecilomyces sinclairii	Red	Cho et al., 2002
Penicillium oxalicum	Dark red	Sardaryan, 2002; Sardaryan et al., 2004
Penicillium purpurogenum	Yellow-orange, red	Méndez et al., 2011

limited the use of these genera for natural colourant production (Mapari et al., 2005). Some species of *Aspergillus* sp. produce hydroxyanthraquinoid (HAQN) pigments along with several mycotoxins, including oxaline, secalonic acid, cyclochlorotine, citrinin, and rugulosin (Goyal et al., 2016). The pigment and toxin concentration produced varies with the cultivation conditions.

8.3.2 BACTERIAL PIGMENT

Bacteria can be isolated from the environment and purified for the production of pigments. Although extraction of bacterial pigment poses technological challenges, alternative separation techniques can be explored to get purified and concentrated forms of pigments (Venil et al., 2013). Canthaxanthin pigment produced by *D. natronolimnaea* HS-1 using cheese whey was optimized by Khodaiyan et al. (2008). Cheese whey and yeast extract were shown to improve the production of canthaxanthin (2871 ± 76 g/l) by this bacterium. The use of sucrose and glycine in the medium improved the prodigiosin production by *S. marcescens* (Su et al., 2011). Violacein production by *Dunganella* sp. B2 was found to be dependent on concentrations of L-tryptophan, potassium nitrate, and beef extract and the yield was amplified approximately 4.8-fold under optimized conditions (Wang et al., 2009).

8.3.3 YEAST PIGMENT

Yeasts are unicellular and are ubiquitous in the environment. Yeasts have been shown to have growing capabilities on agro-industrial waste and produce pigments, especially carotenoids. The different strains capable of producing pigments such as β-carotene, astaxanthin, torulene, etc. are *Blakeslea trispora* (Papaioannou and Liakopolou-Kyriakides, 2010); *Phycomyces blakesleeanus* (Almeida and Cerdá-Olmedo, 2008); *Rhodotorula* spp. (Tinoi et al., 2005); *Xanthophyllomyces dendrorhous* (Mata-Gómez et al., 2014); *Phaffia rhodozyma*; *Yarrowia lipolytica*; *Sporobolomyces* spp. The factors affecting the production of pigments are yeast strains, nutrient media, biotechnological methods, ease of separation, type of pigments, as well as the degree of purity (Libkind et al., 2005).

8.4 MICROBIAL PIGMENT PRODUCTION UTILIZING AGRO-WASTE

8.4.1 DAIRY INDUSTRY WASTES

Whey is the chief waste and inexpensive by-product of the cheese industry (Panesar et al., 2015). Around 800–850 kg of whey is produced from 1000 kg of milk, and nearly 180–190 million tonnes of whey is generated annually worldwide, creating environmental hazards (Yadav et al., 2015). Whey protein is used for the production of the pigment melanin from the actinomycetes *Dietzia schimae* NH3. Optimization of 5% whey (w/v), 2.5 g/l l-tyrosine, 0.013 g/l $CuSO_4$, pH 10.5, and 32°C temperature was adjusted to obtain a maximum yield of 790 mg/l melanin pigment. *D. schimae* NM3 melanin pigment exhibited antimicrobial activity against certain gram-positive bacteria (Eskandari and Etemadifar, 2021). *M. purpureus* fungi produced red pigments on cheese whey substrate (Lopes et al., 2013). As per the study conducted by Marova et al. (2012), *S. roseus* showed substantial variations in the biomass:carotene ratio based on the addition of whey. A significant decrease in biomass was observed in the presence of lyophilized whey that was complemented by very high beta-carotene production. *R. mucilaginosa*, a red yeast, produced a β-carotene content of 29.4 g/l in the whey medium.

8.4.2 FRUITS AND VEGETABLE PROCESSING INDUSTRY

The fruit and vegetable processing industry generates a huge amount of waste material in the form of seeds, peel, and wastewater that pose a great threat to the environment if discarded as such.

These waste materials are a potential source of macronutrients and micronutrients such as carbohydrates, starch, cellulose, vitamins, and minerals and serve as a suitable medium for microbial growth (Kaur et al., 2019a). Agro-food wastes like the peels of the orange, papaya, and carrots have been used as a substrate for producing beta-carotene pigment from *Blakeslea trispora*. They provided nitrogen and carbon to the fermentation medium (Kaur et al., 2019a). Jackfruit seed supplemented with fructose was reported to yield the highest amount of red pigment from *Monascus* (Subhasree et al., 2011). A study revealed that pomegranate pulp shows high potential for the production of yellow pigment from *Aspergillus carbonarius* (Bezirhan et al., 2020). Orange processing waste was used as the raw material for red pigment from *Monascus purpureus* ATCC 16365 and *Penicillium purpurogenum* CBS 113139 by employing SSF, SmF, and semi-solid fermentation (Kantifedaki et al., 2018). Kinnow peel powder supplemented with pea pod and magnesium sulphate was used to produce red-coloured pigment derived from *Monascus purpureus* MTCC 369 in SmF (Panesar, 2014).

8.4.3 CEREAL PROCESSING INDUSTRY

Cereal-processing industry waste is mainly comprised of germ and bran from the milling industry, spent grains from the brewing industry, and other waste generated by the starch and bread industry. These generated waste streams provide necessary nutrients that can be put to transformations using micro-organisms. *Monascus* pigments are the chief secondary metabolites of *Monascus* spp. comprising of yellow, orange, and red pigments. These pigments are known for their therapeutic importance (Liu et al., 2020). Brewer's spent grain was used as the substrate for natural red pigment production by *Monascus purpureus* CMU001 in a SmF system (Silbir and Goksungur, 2019). Hamano and Kilikian (2006) have reported the utilization of corn steep liquor for producing red pigment from *Monascus ruber*. It is also believed that corn steep liquor is a good source of various salts that are required for the production of red pigment. Tinoi et al. (2005) studied the production of carotenoid pigment by *Rhodotorula glutinis* using hydrolyzed mung bean flour waste procured from a glass noodle production factory (Panesar et al., 2015). Rice straw hydrolysate was found to be a cost-effective substrate for the production of *Monascus* pigments in SmF (Liu et al., 2020). Red pigment from *Talaromyces purpureogenus* CFRM02 pigment was produced using Bengal gram (*Cicer arietinum*) as the substrate for fermentation. This pigment was found to be non-toxic and has been proven to exhibit antioxidant and antibacterial activities as well as non-toxic effects (Pandit et al., 2019). Glycerol waste and deproteinized wastewater from potato processing can also be used as culture media by the *Rhodotorula* yeast strains, *Rhodotorula glutinis*, *Rhodotorula mucilaginosa*, and *Rhodotorula gracilis* for the synthesis of carotenoids (Kot et al., 2019).

8.4.4 AGRICULTURAL BYPRODUCTS AND CROP RESIDUES

Post-harvest activities lead to the generation of several byproducts like husk, corn cobs, bran, and bagasse (Panesar et al., 2015). Low-cost waste can be a potential substrate to produce natural pigments (Manimala and Murugesan, 2017). Solid coffee waste was used for carotenoid pigment production from the yeast *Rhodotorula mucilaginosa*. The carotenoid pigment displayed antioxidant and antimicrobial activities against several bacteria and toxigenic fungi. It is an example of a potential natural pigment developed at a low cost to replace synthetic pigments (Moreira et al., 2018). Rifamycin B (orange to reddish-brown colour) is produced by *Amycolatopsis* sp. RSP 3 using corn husk as raw material under SSF. Corn husk is an appropriate substrate as it produces a significant amount of rifamycin B in an optimized environment, and inorganic nitrogen plays an essential role in the production (Mahalaxmi et al., 2010). Cassava waste is available at a low price, and it is a high nutrient-rich substrate for producing red pigment by *Penicillium purpurogenum* (Padmapriya and Murugesan, 2016). Cassava bagasse, rice flour, and sugarcane bagasse were used to produce carotenoid pigment from *Rhodotorula mucilaginosa*. The maximum yield was obtained from cassava bagasse (Manimala and Murugesan, 2017). Table 8.2 gives a summary of microbial pigment production using agro-industrial waste.

8.5 APPLICATION OF MICROBIAL PIGMENTS

8.5.1 PHARMACEUTICAL INDUSTRY

The pharmaceutical industry uses several microbial pigments in products as their compounds exhibit antibiotic, anticancer, antiproliferative, and immunosuppressive effects. The secondary metabolites produced by micro-organisms show potential applications in treating various diseases such as cancer, diabetes mellitus, and many others. Anthocyanin shows antioxidant activity as it is able to donate a hydrogen atom to a free radical and exhibits inhibitory effects in carcinogenesis and tumour growth (Kumar et al., 2015). Prodigiosins (red pigment) fabricated using *S marcescens* and other gram-negative bacteria display antibacterial, cytotoxic and anti-inflammatory properties. Prodigiosins produced by *Streptomyces* or *Serratia* show immunosuppressing activity and antiproliferative effects on tumour cells. It is an active component in preventing and treating diabetes mellitus (Panesar et al., 2015; Kumar et al., 2015). The pigments produced using *Monascus purpureus* are able to inhibit cholesterol synthesis, therefore lowering total cholesterol and serum triglyceride levels (Mukherjee and Singh, 2011). *Monascus* pigments also aid in upholding the good health of human eyes. Melanin pigments are used in sunscreen to protect the skin from UV radiation (Soliev et al., 2011). A mixture of rhodoxanthin, beta-carotene, and rhodotorulin is present in red pigments derived from *R. gracis*, which are commercially used as vitamin A precursors (Panesar et al., 2015). Red yeast rice is produced through fermentation using *Monascus* spp., has anticholesterol activity, and reduces blood glucose levels (Kumar et al., 2015).

8.5.2 FOOD INDUSTRY

In the food industry, the appearance of the food product affects its acceptability, especially among consumers; therefore, it is important that the food product is attractive and fresh. Alternatives to artificial food colourings have been sought for a long time, especially sunset yellow, tartrazine, and quinoline yellow (Venil et al., 2013). Microbial pigments have been a promising alternative, e.g., extracted blue pigment from cultured soil bacteria that gives a natural colour to the food product and is stable. Table 8.3 shows other commonly used microbial pigments in the food industry. The microbial food colourings are also environmentally friendly and have probiotic benefits when consumed with the food product (Nagpal et al., 2011).

8.5.3 TEXTILE INDUSTRY

The microbial pigments produced via fermentation are stable, have higher yields, and give lower residues in comparison to the pigments derived from plants and animals. These pigments also possess antibacterial activity and can be used as functional dyes in the manufacturing of coloured antimicrobial textile materials (Venil et al., 2013). The bright red pigment prodigiosin obtained from *Vibrio* spp. or *S. marcescens* is utilized in dyeing fabrics like wool, acrylics, silk, and nylon, and it is resistant to the action of acids, alkalis, and detergents (Alihosseini et al., 2008; Gulani et al., 2012). Pigments from *Monascus* sp. preserve printed characters or images only if they are stored in the dark (decolourizes under visible or UV light) (Panesar et al., 2015).

8.6 CONCLUSION AND FUTURE SCOPE

The demand for natural pigments is increasing globally as consumers find natural food colourants safer in comparison to synthetic colourants. Huge amounts of waste generated by the agro-industrial

TABLE 8.2
Microbial Pigment Production Using Agro-Industrial Wastes/Residues

Microbes	Agro-industrial Waste	Pigment	Type of Fermentation	Yield	Findings	References
Monascus ruber MTCC 2326	Broken rice	Yellow and red	SSF	0.33 U/g *Monascus*-fermented rice (MFR) for yellow and 0.20 U/g MFR for red pigment	Yellow pigment production was higher than red pigments	Vidyalakshmi et al., 2009
Penicillium resticulosum	Corn waste stream	Red	Submerged fermentation	497.03±53.13 mg/l	Optimized conditions: corn cob waste based medium at 21:1 C:N ratio, pH 5.5–6.0, with yeast extract/NH_4NO_3/$NaNO_3$ and xylose or Carboxy methyl cellulose (CMC) supplementation and 12-days incubation in dark	Sopandi et al., 2013
Monascus purpureus MTCC 369	Kinnow peel and pea pod	Red	SmF	-	Optimized conditions: kinnow peel powder and pea pod containing magnesium sulphate (0.1%) with 6.5 pH, incubated at 35°C for 9 days	Panesar, 2014
Epicoccum sp.	Broken rice	Orevactaene and flavonoid	SSF	52.7 AU/g orevactaene and 77.2 AU/g flavonoid	The optimum conditions for aqueous extraction process: extraction temperature of 55.7°C, 0.79 g of fermented matter, and extraction time of 56.6 min	Kaur et al., 2019b
Monascus purpureus	Whey	Red pigments	Submerged	38.4 UA$_{510}$ nm	Optimized conditions: 7.5 g/100 ml of lactose as source of carbon, 2.5 g/100 ml of Monosodium glutamate (MSG) as nitrogen source, pH of 7; 2% (v/v) inoculation ratio	Mehri et al., 2021
Blakeslea trispora ATCC 14271	Waste cooking oil	Carotenoids	Submerged	49.3 ± 0.2 mg/g dry biomass	Waste cooking oil (50.0 g/l) supplemented with corn steep liquor (80.0 g/l) and butylated hydroxy toluene (4.0 g/l), pH 7.5	Nanou and Roukas, 2016

(Continued)

TABLE 8.2 (CONTINUED)
Microbial Pigment Production Using Agro-Industrial Wastes/Residues

Microbes	Agro-industrial Waste	Pigment	Type of Fermentation	Yield	Findings	References
				Bacteria		
Chryseobacterium artocarpi CECT 849	Liquid pineapple waste (LPW)	Yellow-orange pigment	Submerged	152 mg/l	Optimized conditions: LPW 20% (v/v), L-tryptophan 125 g/l and KH_2PO_4 12.5 g/l	Aruldass et al., 2016
Microbacterium sp.	Olive mill waste	Carotenoid	Submerged	2 mg/l	An inverse relationship exists between *Microbacterium* sp. growth and carotene production	Borroni et al., 2017
Dietzia maris NIT-D	Sugarcane bagasse hydrolysate	Tran-canthaxanthin	Submerged	-	pH of 5.5, incubated at 25°C with shaking for 5 days, shaker speed 120 rpm and inoculum 2%	Goswami et al., 2015
Chromobacterium violaceum UTM5	Liquid pineapple waste	violacein	Submerged	16256 ± 440 mg/l^{-1}	Violet pigment can achieve better stability during the storage of pH 7, temperature 25°C–30°C for 30 days in the dark	Aruldass et al., 2015
Chromobacterium violaceum	Sugarcane bagasse, molasses, solid pineapple waste, brown sugar	violacein	Submerged	0.82 g/l^{-1}	Highest yield obtained on 3 g sugarcane bagasse containing with 10% (v/v) of L-tryptophan	Ahmad et al., 2012
Oscillatoria sp. *50 A*	Sugar industry effluent	Cyanobacterial phycocyanin	Submerged	0.072 ± 0.001 mg/ml	Enhanced yield was obtained in 30 days with optimized media containing 2.5 ml/l each of sugar industrial effluent and tannery effluent	Jawaharraj et al., 2017
Serratia marcescens CF-53	Peanut oil cake	Prodigiosin	Submerged	40 mg/ml	Maximum amount of pigment was obtained in peanut oil cake extract at 30°C for 42 h using 8% inoculum density	Naik et al., 2012

(Continued)

TABLE 8.2 (CONTINUED)
Microbial Pigment Production Using Agro-Industrial Wastes/Residues

Microbes	Agro-industrial Waste	Pigment	Type of Fermentation	Yield	Findings	References
Serratia marcescens TNU02	Demineralized crab shell powder	Prodigiosin	Submerged	5100 mg/l	Yield was significantly enhanced when the casein/demineralized crab shell powder ratio was controlled in the range of 3/7–4/6; initial pH of 6.15, at 27°C for 8 h in dark	Nguyen et al., 2020
			Yeast			
Rhodotorula glutinis	Mung bean waste flour	Carotenoid	Submerged	3.48±0.02 mg/l	Optimized conditions: 23.63 g/l of hydrolyzed mung bean waste flour, 51.76 g/l of sweet potato extract, pH 5.91, 30.38°C, agitation rate of 258 rpm and incubation time of 94.78 h	Tinoi et al., 2005
Rhodotorula mucilaginosa MTCC-1403	Onion peels, potato skin, mung bean husk, and pea pods	Carotenoids	Submerged	100 µg carotenoids per g of dry biomass	Optimum conditions were pH 6.1, 25.8°C incubation temperature and agitation 119.6 rpm	Sharma and Ghoshal, 2020

TABLE 8.3
Major Microbial Pigments Used in the Food Industry

Pigment Name	Microbes Extracted From	Uses	References
β-Carotene (yellow to orange-red)	*Blakeslea trispora, Mucor circinelloides, Phycomyces blakesleeanus*	In vegetable oils, margarine, various emulsions, orange drinks, and micro-encapsulated beads.	Nigam and Luke, 2016
Arpink Red (red colour)	*Penicillium oxalicum*	In various food as per the Codex Alimentarius Commission.	Kumar et al., 2015
Riboflavin (yellow colour)	*Eremothcium ashbyii, Candida flaleri, Ashbya gossypii, Bacillus subtilis, Saccharomyces cerevisiae*	In baby foods, breakfast cereals, sauces, milk products, pastas, and energy drinks.	Panesar et al., 2015
Monascus pigment (red, orange and yellow colour)	*M. pilosus, M. ruberand, M. purpureus, M. froridanus*	In wines, particularly red; tofu, sausages, hams, and seafoods.	Panesar et al., 2015
Lycopene (red colour)	*Fusarium sporotrichioides Blakeslea trispora*	In meat colouring.	Sen et al., 2019

sector can be a threat to the environment because of its organic matter content. The development of novel and innovative biotechnologies for the utilization of these waste streams can help to develop new and variable hues of pigment that can be used in different processing industries. Thus, the use of agro-industrial waste as a substrate for producing microbial pigments is a sustainable technique to solve the issue of environmental pollution and, at the same time, reduce the cost of its production. There is a need to explore biodiversity to search for new micro-organisms as well as carry out toxicity studies of these developed pigments for their safer use in the pharmaceutical, food, and textile industries.

REFERENCES

Ahmad, W.A., Yusof, N.Z., Nordin, N., Zakaria, Z.A. and Rezali, M.F., 2012. Production and characterization of violacein by locally isolated *Chromobacterium violaceum* grown in agricultural wastes. *Applied Biochemistry and Biotechnology, 167*(5), pp.1220–1234.

Alihosseini, F., Ju, K.S., Lango, J., Hammock, B.D. and Sun, G., 2008. Antibacterial colorants: Characterization of prodiginines and their applications on textile materials. *Biotechnology Progress, 24*(3), pp.742–747.

Almeida, E.R. and Cerdá-Olmedo, E., 2008. Gene expression in the regulation of carotene biosynthesis in phycomyces. *Current Genetics, 53*(3), pp.129–137.

Aruldass, C.A., Aziz, A., Venil, C.K., Khasim, A.R. and Ahmad, W.A., 2016. Utilization of agro-industrial waste for the production of yellowish-orange pigment from Chryseobacterium artocarpi CECT 8497. *International Biodeterioration and Biodegradation, 113*, pp.342–349.

Aruldass, C.A., Venil, C.K. and, Ahmad, W.A., 2015. Violet pigment production from liquid pineapple waste by *Chromobacterium violaceum* UTM5 and evaluation of its bioactivity. *RSC Advances, 5*(64), pp.51524–51536.

Babitha, S., 2009. Microbial pigments. In: *Biotechnology for Agro-Industrial Residues Utilisation* (pp. 147–162). Springer, Dordrecht.

Babitha, S., Soccol, C.R. and Pandey, A., 2006. Jackfruit seed-a novel substrate for the production of Monascus pigments through solid-state fermentation. *Food Technology and Biotechnology, 44*(4), pp.465–471.

Bezirhan Arikan, E., Canli, O., Caro, Y., Dufossé, L. and Dizge, N., 2020. Production of bio-based pigments from food processing industry by-products (apple, pomegranate, black carrot, red beet pulps) using Aspergillus carbonarius. *Journal of Fungi, 6*(4), p.240.

Baldrian, P., Merhautová, V., Gabriel, J., Nerud, F., Stopka, P., Hrubý, M. and Beneš, M.J., 2006. Decolorization of synthetic dyes by hydrogen peroxide with heterogeneous catalysis by mixed iron oxides. *Applied Catalysis B: Environmental, 66*(3–4), 258–264

Bloor, A.E. and Cranenburgh, R.M., 2006. An efficient method of selectable marker gene excision by Xer recombination for gene replacement in bacterial chromosomes. *Applied and Environmental Microbiology*, 72(4), pp.2520–2525.

Bonan, S., Fedrizzi, G., Menotta, S. and Elisabetta, C., 2013. Simultaneous determination of synthetic dyes in foodstuffs and beverages by high-performance liquid chromatography coupled with diode-array detector. *Dyes and Pigments*, 99(1), pp.36–40.

Borroni, V., González, M.T. and Carelli, A.A., 2017. Bioproduction of carotenoid compounds using two-phase olive mill waste as the substrate. *Process Biochemistry*, 54, pp.128–134.

Cho, Y.J., Park, J.P., Hwang, H.J., Kim, S.W., Choi, J.W. and Yun, J.W., 2002. Production of red pigment by submerged culture of Paecilomyces sinclairii. *Letters in Applied Microbiology*, 35(3), pp.195–202.

de Andrade, F.I., Guedes, M.I.F., Vieira, Í.G.P., Mendes, F.N.P., Rodrigues, P.A.S., Maia, C.S.C., Ávila, M.M.M. and de Matos Ribeiro, L., 2014. Determination of synthetic food dyes in commercial soft drinks by TLC and ion-pair HPLC. *Food Chemistry*, 157, pp.193–198.

De Carvalho, J.C., Cardoso, L.C., Ghiggi, V., Woiciechowski, A.L., de Souza Vandenberghe, L.P. and Soccol, C.R., 2014. Microbial pigments. In: *Biotransformation of Waste Biomass into High Value Biochemicals* (pp. 73–97). Springer, New York.

dos Reis Celestino, J., de Carvalho, L.E., da Paz Lima, M., Lima, A.M., Ogusku, M.M. and de Souza, J.V.B., 2014. Bioprospecting of Amazon soil fungi with the potential for pigment production. *Process Biochemistry*, 49(4), pp.569–575.

Downham, A. and Collins, P., 2000. Colouring our foods in the last and next millennium. *International Journal of Food Science and Technology*, 35(1), pp.5–22.

Durán, N., Teixeira, M.F., De Conti, R. and Esposito, E., 2002. Ecological-friendly pigments from fungi. *Critical Reviews in Food Science and Nutrition*, 42(1), pp.53–66.

Echavarri-Erasun, C. and Johnson, E.A., 2004. Stimulation of astaxanthin formation in the yeast *Xanthophyllomyces dendrorhous* by the fungus Epicoccum nigrum. *FEMS Yeast Research*, 4(4–5), pp.511–519.

Eskandari, S. and Etemadifar, Z., 2021. Biocompatibility and radioprotection by newly characterized melanin pigment and its production from Dietzia schimae NM3 in optimized whey medium by response surface methodology. *Annals of Microbiology*, 71(1): 1–13

Gostner, J.M., Becker, K., Ueberall, F. and Fuchs, D., 2015. The good and bad of antioxidant foods: An immunological perspective. *Food and Chemical Toxicology*, 80, pp.72–79.

Goswami, G., Chaudhuri, S. and Dutta, D., 2015. Studies on the stability of a carotenoid produced by a novel isolate using low cost agro-industrial residue and its application in different model systems. *LWT - Food Science and Technology*, 63(1), pp.780–790.

Goyal, S., Ramawat, K.G. and Mérillon, J.M., 2016. Different shades of fungal metabolites: An overview. *Fungal Metabolites*, 1, p.29.

Gulani, C., Bhattacharya, S. and Das, A., 2012. Assessment of process parameters influencing the enhanced production of prodigiosin from Serratia marcescens and evaluation of its antimicrobial, antioxidant and dyeing potentials. *Malaysian Journal of Microbiology*, 8(2), pp.116–122.

Hamano, P.S. and Kilikian, B.V., 2006. Production of red pigments by Monascus ruber in culture media containing corn steep liquor. *Brazilian Journal of Chemical Engineering*, 23(4), pp.443–449.

Himri, I., Bellahcen, S., Souna, F., Belmekki, F., Aziz, M., Bnouham, M., Zoheir, J., Berkia, Z., Mekhfi, H. and Saalaoui, E., 2011. A 90-day oral toxicity study of tartrazine, a synthetic food dye, in wistar rats. *International Journal of Pharmacy and Pharmaceutical Sciences*, 3, 159–169.

Jawaharraj, K., Shivani, K., Shakambari, G., Sameer Kumar, R., Ashokkumar, B. and Varalakshmi, P., 2017. Agro-Industrial Waste Based Phycocyanin Production from Oscillatoria sp. 50 A: daf-16 Modulating Effect in Caenorhabditis elegans and p53 Dependent Apoptosis in HeLa cells. *ChemistrySelect*, 2(36), 11989–11994.

Jiang, Y., Li, H.B., Chen, F. and Hyde, K.D., 2005. Production potential of water-soluble Monascus red pigment by a newly isolated Penicillium sp. *Journal of Agriculture and Technology*, 1(1), pp.113–126.

Kantifedaki, A., Kachrimanidou, V., Mallouchos, A., Papanikolaou, S. and Koutinas, A.A., 2018. Orange processing waste valorisation for the production of bio-based pigments using the fungal strains *Monascus purpureus* and *Penicillium purpurogenum*. *Journal of Cleaner Production*, 185, pp.882–890.

Kaur, P., Ghoshal, G. and Jain, A., 2019a. Bio-utilization of fruits and vegetables waste to produce β-carotene in solid-state fermentation: Characterization and antioxidant activity. *Process Biochemistry*, 76, pp.155–164.

Kaur, S., Panesar, P.S., Gurumayum, S., Rasane, P. and Kumar, V., 2019b. Optimization of aqueous extraction of orevactaene and flavanoid pigments produced by *Epicoccum nigrum*. *Pigment and Resin Technology*, 48(4), pp.301–308.

Kaur, S., Panesar, P.S., Gurumayum, S., Rasane, P. and Kumar, V., 2021. Optimization of carotenoid pigment extraction from Epicoccum nigrum fermented wheat bran. *Industrial Biotechnology*, *17*(2), pp.100–104.

Khodaiyan, F., Razavi, S.H. and Mousavi, S.M., 2008. Optimization of canthaxanthin production by Dietzia natronolimnaea HS-1 from cheese whey using statistical experimental methods. *Biochemical Engineering Journal*, *40*(3), pp.415–422.

Kim, D. and Ku, S., 2018. Beneficial effects of Monascus sp. KCCM 10093 pigments and derivatives: A mini review. *Molecules*, *23*(1), p.98.

Kot, A.M., Błażejak, S., Kieliszek, M., Gientka, I. and Bryś, J., 2019. Simultaneous production of lipids and carotenoids by the red yeast Rhodotorula from waste glycerol fraction and potato wastewater. *Applied Biochemistry and Biotechnology*, *189*(2), pp.589–607.

Kumar, A., Vishwakarma, H.S., Singh, J., Dwivedi, S. and Kumar, M., 2015. Microbial pigments: Production and their applications in various industries. *International Journal of Pharmaceutical, Chemical & Biological Sciences*, *5*(1), pp.203–212.

Libkind, D., Sommaruga, R., Zagarese, H. and van Broock, M., 2005. Mycosporines in carotenogenic yeasts. *Systematic and Applied Microbiology*, *28*(8), pp.749–754.

Liu, J., Luo, Y., Guo, T., Tang, C., Chai, X., Zhao, W., Bai, J. and Lin, Q., 2020. Cost-effective pigment production by Monascus purpureus using rice straw hydrolysate as substrate in submerged fermentation. *Journal of Bioscience and Bioengineering*, *129*(2), pp.229–236.

Lopes, F.C., Tichota, D.M., Pereira, J.Q., Segalin, J., de Oliveira Rios, A. and Brandelli, A., 2013. Pigment production by filamentous fungi on agro-industrial byproducts: An eco-friendly alternative. *Applied Biochemistry and Biotechnology*, *171*(3), pp.616–625.

Ma, K., Yang, Y.N., Jiang, X.X., Zhao, M. and Cai, Y.Q., 2012. Simultaneous determination of 20 food additives by high performance liquid chromatography with photo-diode array detector. *Chinese Chemical Letters*, *23*(4), pp.492–495.

Mahalaxmi, Y., Sathish, T., Rao, C.S. and Prakasham, R.S., 2010. Corn husk as a novel substrate for the production of rifamycin B by isolated Amycolatopsis sp. RSP 3 under SSF. *Process Biochemistry*, *45*(1), pp.47–53.

Malik, K., Tokkas, J. and Goyal, S., 2012. Microbial pigments: A review. *International Journal of Microbial Resource Technology*, *1*(4), pp.361–365.

Manimala, M.R.A. and Murugesan, R., 2017. Studies on carotenoid pigment production by yeast Rhodotorula mucilaginosa using cheap materials of agro-industrial origin. *The Pharmaceutical Innovation*, *6*(1), p.80.

Mapari, S.A., Hansen, M.E., Meyer, A.S. and Thrane, U., 2008. Computerized screening for novel producers of Monascus-like food pigments in Penicillium species. *Journal of Agricultural and Food Chemistry*, *56*(21), pp.9981–9989.

Mapari, S.A., Meyer, A.S. and Thrane, U., 2006. Colorimetric characterization for comparative analysis of fungal pigments and natural food colorants. *Journal of Agricultural and Food Chemistry*, *54*(19), pp.7027–7035.

Mapari, S.A., Nielsen, K.F., Larsen, T.O., Frisvad, J.C., Meyer, A.S. and Thrane, U., 2005. Exploring fungal biodiversity for the production of water-soluble pigments as potential natural food colorants. *Current Opinion in Biotechnology*, *16*(2), pp.231–238.

Marova, I., Carnecka, M., Halienova, A., Certik, M., Dvorakova, T. and Haronikova, A., 2012. Use of several waste substrates for carotenoid-rich yeast biomass production. *Journal of Environmental Management*, *95*, pp.S338–S342.

Martins, N., Roriz, C.L., Morales, P., Barros, L. and Ferreira, I.C., 2016. Food colorants: Challenges, opportunities and current desires of agro-industries to ensure consumer expectations and regulatory practices. *Trends in Food Science and Technology*, *52*, pp.1–15.

Masone, D. and Chanforan, C., 2015. Study on the interaction of artificial and natural food colorants with human serum albumin: A computational point of view. *Computational Biology and Chemistry*, *56*, pp.152–158.

Mata-Gómez, L.C., Montañez, J.C., Méndez-Zavala, A. and Aguilar, C.N., 2014. Biotechnological production of carotenoids by yeasts: An overview. *Microbial Cell Factories*, *13*(1), pp.1–11.

Mehri, D., Perendeci, N.A. and Goksungur, Y., 2021. Utilization of whey for red pigment production by Monascus purpureus in submerged fermentation. *Fermentation*, *7*(2), p.75.

Méndez, A., Pérez, C., Montañéz, J.C., Martínez, G. and Aguilar, C.N., 2011. Red pigment production by Penicillium purpurogenum GH2 is influenced by pH and temperature. *Journal of Zhejiang University. Science. Part B*, *12*(12), pp.961–968.

Moreira, M.D., Melo, M.M., Coimbra, J.M., Dos Reis, K.C., Schwan, R.F. and Silva, C.F., 2018. Solid coffee waste as alternative to produce carotenoids with antioxidant and antimicrobial activities. *Waste Management*, *82*, pp.93–99.

Mortensen, A., 2006. Carotenoids and other pigments as natural colorants. *Pure and Applied Chemistry*, *78*(8), pp.1477–1491.

Mukherjee, G. and Singh, S.K., 2011. Purification and characterization of a new red pigment from Monascus purpureus in submerged fermentation. *Process Biochemistry*, *46*(1), pp.188–192.

Nagpal, N., Munjal, N. and Chatterjee, S., 2011. Microbial pigments with health benefits-a mini review. *Trends in Bioscience*, *4*, pp.157–160.

Naik, C., Srisevita, J.M., Shushma, K.N., Farah, N., Shilpa, A.C., Muttanna, C.D., Darshan, N. and Sannadurgappa, D., 2012. Peanut oil cake: A novel substrate for enhanced cell growth and prodigiosin production from Serratia marcescens CF-53. *Journal of Research in Biology*, *2*(6), pp.549–557.

Nanou, K. and Roukas, T., 2016. Waste cooking oil: A new substrate for carotene production by Blakeslea trispora in submerged fermentation. *Bioresource Technology*, *203*, pp.198–203.

Nguyen, V.B., Nguyen, D.N., Nguyen, A.D., Ngo, V.A., Ton, T.Q., Doan, C.T., Pham, T.P., Tran, T.P.H. and Wang, S.L., 2020. Utilization of crab waste for cost-effective bioproduction of prodigiosin. *Marine Drugs*, *18*(11), p.523.

Nielsen, J. and Olsson, L., 2002. An expanded role for microbial physiology in metabolic engineering and functional genomics: Moving towards systems biology. *FEMS Yeast Research*, *2*(2), pp.175–181.

Nigam, P.S. and Luke, J.S., 2016. Food additives: Production of microbial pigments and their antioxidant properties. *Current Opinion in Food Science*, *7*, pp.93–100.

Nigam, P.S.N. and Pandey, A. eds., 2009. *Biotechnology for Agro-Industrial Residues Utilisation: Utilisation of Agro-Residues*. Springer Science & Business Media, Dordrecht.

Ogihara, J., Kato, J., Oishi, K. and Fujimoto, Y., 2001. PP-R, 7-(2-hydroxyethyl)-monascorubramine, a red pigment produced in the mycelia of Penicillium sp. AZ. *Journal of Bioscience and Bioengineering*, *91*(1), pp.44–47.

Okafor, S.N., Obonga, W., Ezeokonkwo, M.A., Nurudeen, J., Orovwigho, U. and Ahiabuike, J., 2016. Assessment of the health implications of synthetic and natural food Colourants–A critical review. *Pharmaceutical and Biosciences Journal*, 4(4): 01–11.

Padmapriya, C. and Murugesan, R., 2016. Optimization of SSF parameters for natural red pigment production from Penicillium purpurogenum using cassava waste by central composite design. *Journal of Applied and Natural Science*, *8*(3), pp.1663–1669.

Pandit, S.G., Ramesh, K.P.M., Puttananjaiah, M.H. and Dhale, M.A., 2019. Cicer arietinum (Bengal gram) husk as alternative for Talaromyces purpureogenus CFRM02 pigment production: Bioactivities and identification. *LWT*, *116*, p.108499.

Panesar, R., 2014. Bioutilization of kinnow waste for the production of biopigments using submerged fermentation. *International Journal of Food and Nutritional Sciences*, *3*(1), p.9.

Panesar, R., Kaur, S. and Panesar, P.S., 2015. Production of microbial pigments utilizing agro-industrial waste: A review. *Current Opinion in Food Science*, *1*, pp.70–76.

Papaioannou, E.H. and Liakopoulou-Kyriakides, M., 2010. Substrate contribution on carotenoids production in Blakeslea trispora cultivations. *Food and Bioproducts Processing*, *88*(2–3), pp.305–311.

Poorniammal, R., Gunasekaran, S. and Ariharasivakumar, G., 2011. Toxicity evaluation of fungal food colourant from thermomyces sp. in albino mice. *Journal of Science and Industrial Research*, *70*(9), pp. 773–777.

Prajapati, V.S., Soni, N., Trivedi, U.B. and Patel, K.C., 2014. An enhancement of red pigment production by submerged culture of Monascus purpureus MTCC 410 employing statistical methodology. *Biocatalysis and Agricultural Biotechnology*, *3*(2), pp.140–145.

Purba, M.K., Agrawal, N. and Shukla, S.K., 2015. Detection of non-permitted food colors in edibles. *Journal of Forensic Research*, *S4*, p.1.

Rodriguez-Amaya, D.B., 2016. Natural food pigments and colorants. *Current Opinion in Food Science*, *7*, pp.20–26.

Sabnis, R.W., 2010. *Handbook of Biological Dyes and Stains: Synthesis and Industrial Applications*. John Wiley & Sons, New Jersey.

Sahar, S.A.S. and, Manal, M.E.M.S., 2012. The effects of using color foods of children on immunity properties and liver, kidney on rats. *Food and Nutrition Sciences*, *3*(7), pp.897–904.

Sardaryan, E., 2002. Strain of the microorganism Penicillium oxalicum var. *Armeniaca and its application*. *US Patent*, *6*(340), p.586.

Sardaryan, E., Zihlova, H., Strnad, R. and Cermakova, Z., 2004. Arpink Red–meet a new natural red food colorant of microbial origin. Edited by Laurent Dufossé, Université de Bretagne Occidentale, Quimper. *Pigments in Food, More than Colours*, pp.207–208.

Sen, T., Barrow, C.J. and Deshmukh, S.K., 2019. Microbial pigments in the food industry—Challenges and the way forward. *Frontiers in Nutrition*, 6, p.7.

Shahid, M. and Mohammad, F., 2013. Recent advancements in natural dye applications: A review. *Journal of Cleaner Production*, 53, pp.310–331.

Sharma, R. and Ghoshal, G., 2020. Optimization of carotenoids production by *Rhodotorula mucilaginosa* (MTCC-1403) using agro-industrial waste in bioreactor: A statistical approach. *Biotechnology Reports*, 25, p.e00407.

Shimizu, T., Kinoshita, H., Ishihara, S., Sakai, K., Nagai, S. and Nihira, T., 2005. Polyketide synthase gene responsible for citrinin biosynthesis in Monascus purpureus. *Applied and Environmental Microbiology*, 71(7), pp.3453–3457.

Sigurdson, G.T., Tang, P. and Giusti, M.M., 2017. Natural colorants: Food colorants from natural sources. *Annual Review of Food Science and Technology*, 8, pp.261–280.

Silbir, S. and Goksungur, Y., 2019. Natural red pigment production by Monascus purpureus in submerged fermentation systems using a food industry waste: Brewer's spent grain. *Foods*, 8(5), p.161.

Soliev, A.B., Hosokawa, K. and Enomoto, K., 2011. Bioactive pigments from marine bacteria: Applications and physiological roles. *Evidence-Based Complementary and Alternative Medicine: eCAM*, 2011, 17.

Sopandi, T., Wardah, A., Surtiningsih, T., Suwandi, A. and Smith, J.J., 2013. Utilization and optimization of a waste stream cellulose culture medium for pigment production by *Penicillium* spp. *Journal of Applied Microbiology*, 114(3), pp.733–745.

Stricker, R., Romailler, G., Turian, G. and Tzanos, D., 1981. Production and food applications of a mold hydro-soluble yellow pigment (Epicoccum nigrum Link). *Lebensmittel-Wissenschaft und -Technologie – Food Science and Technology* 14(1): 18–20.

Su, W.T., Tsou, T.Y. and Liu, H.L., 2011. Response surface optimization of microbial prodigiosin production from Serratia marcescens. *Journal of the Taiwan Institute of Chemical Engineers*, 42(2), pp.217–222.

Subhasree, R.S., Babu, P.D., Vidyalakshmi, R. and Mohan, V.C., 2011. Effect of carbon and nitrogen sources on stimulation of pigment production by Monascus purpureus on jackfruit seeds. *International Journal of Microbiological Research (IJMR)*, 2(2), pp.184–187.

Subramaniyam, R. and Vimala, R., 2012. Solid state and submerged fermentation for the production of bioactive substances: A comparative study. *International Journal of Natural Sciences*, 3(3), pp.480–486.

Tinoi, J., Rakariyatham, N. and Deming, R.L., 2005. Simplex optimization of carotenoid production by Rhodotorula glutinis using hydrolyzed mung bean waste flour as substrate. *Process Biochemistry*, 40(7), pp.2551–2557.

Velmurugan, P., Hur, H., Balachandar, V., Kamala-Kannan, S., Lee, K.J., Lee, S.M., Chae, J.C., Shea, P.J. and Oh, B.T., 2011. Monascus pigment production by solid-state fermentation with corn cob substrate. *Journal of Bioscience and Bioengineering*, 112(6), pp.590–594.

Venil, C.K., Zakaria, Z.A. and Ahmad, W.A., 2013. Bacterial pigments and their applications. *Process Biochemistry*, 48(7), pp.1065–1079.

Vidyalakshmi, R., Paranthaman, R., Murugesh, S. and Singaravadivel, K., 2009. Stimulation of Monascus pigments by intervention of different nitrogen sources. *Global Journal of Biotechnology and Biochemistry*, 4(1), pp.25–28.

Wang, H., Jiang, P., Lu, Y., Ruan, Z., Jiang, R., Xing, X.H., Lou, K. and Wei, D., 2009. Optimization of culture conditions for violacein production by a new strain of Duganella sp. B2. *Biochemical Engineering Journal*, 44(2–3), pp.119–124.

Wang, Y., Zhang, B., Lu, L., Huang, Y. and Xu, G., 2013. Enhanced production of pigments by addition of surfactants in submerged fermentation of Monascus purpureus H1102. *Journal of the Science of Food and Agriculture*, 93(13), pp.3339–3344.

Yadav, J.S.S., Yan, S., Pilli, S., Kumar, L., Tyagi, R.D. and Surampalli, R.Y., 2015. Cheese whey: A potential resource to transform into bioprotein, functional/nutritional proteins and bioactive peptides. *Biotechnology Advances*, 33(6), pp.756–774.

9 Production, Characterization, and Industrial Application of Biosurfactants from Agro-Industrial Wastes

Suwan Panjanapongchai, Anushuya Guragain, and Anil Kumar Anal

CONTENTS

DOI: 10.1201/9781003125679-9

9.1 INTRODUCTION

The addition of value to agro-industrial waste exhibits a huge potential in the production of sustainable compounds such as biosurfactants. Biosurfactants are amphipathic compounds with various structures and chemical properties that are produced by a wide array of micro-organisms. Biosurfactants can be categorized into high molecular mass surfactants, such as glycolipids, and low molecular mass surfactants, such as emulsan, based on their microbiological origin and chemical composition. The presence of a higher amount of macronutrients, such as carbohydrates, proteins, and lipids, in the agro-industrial waste are ideal for use as a substrate in the cost-effective microbial production of biosurfactants (Freitas et al., 2016).

The increased popularity of biosurfactants is associated with their peculiar property of minimizing and reducing the interfacial tension of liquids. In relation to its popularity, the forecasted growth of the global biosurfactants market during the period of 2020–2024 is 6% compound annual growth rate (CAGR) (Ahuja and Singh 2020). Biosurfactants are preferred over chemical surfactants for their environmentally friendly nature, higher selectivity, non-toxicity, multifunctionality, and specificity under extreme pH and temperature (Shekhar, Sundaramanickam, and Balasubramanian, 2015). In addition to that, biosurfactants' hydrophilic and hydrophobic structure makes them a unique compound that can be used as an emulsifier, dispersal, solubilizer, mobilizer, and surface tension reducer (Olasanmi and Thring, 2019).

Moreover, biosurfactants are used in various applications such as bioremediation to clean up pollutants through their ability to increase the bioavailability of contaminants to microbes and hydrocarbon solubilization and mobilization in soils through biosurfactant-augmented washing (Luna et al., 2015; Pacwa-Płociniczak et al., 2011). Currently, high production cost, low production yield, and expensive downstream and recovery processes have restricted the commercial production of biosurfactants in bulk quantities. For growth in the industrial-scale production of biosurfactants, the production cost should be equitable with synthetic surfactants (Olasanmi and Thring, 2019). Hence, the focus has shifted towards wider and more diverse applications of biosurfactants to aid in economically viable industrial-scale production.

Agro-industrial waste products like soybean oil waste, palm oil cake, cassava flour, molasses, and whey have been utilized as cheap raw substrates for the production of biosurfactants. These cheaper substrates can exhibit the dual advantage of providing nutrients for microbial growth along with a source of isolation for potential biosurfactant-producing microbes (Lee et al., 2009). Utilization of cheap raw materials like agro-waste as a nutrient source can contribute to the cost-effective production of biosurfactants along with optimization of parameters related to growth and production (Rane et al., 2017). Also, exploration of the potential application of biosurfactants in diverse fields, including agriculture, the food industry, pharmaceuticals, cosmetics, detergents, and bioremediation are being done, which could ultimately help in economic production (Gudiña et al., 2015).

Although agro-industrial waste exhibits tremendous potential in the production of biosurfactants, the selection of raw waste should ensure sufficient nutrients for microbial growth and cost-effective production. Optimization of biosurfactant production can be done by optimizing the ratio of different elements such as C:N, C:P, C:Fe, or C:Mg. The utilization of agro-industrial waste is one of the many steps required to optimize production. Along with the advantages of using agro-industrial waste, there are many challenges that lie within the chain of production. The presence of undesired compounds, varied composition of raw materials, and tedious processing requirements are some of the major limitations associated with the use of agro-waste on an industrial scale (Banat et al., 2014).

Moreover, due to the high cost associated with the recovery and purification of microbial surfactants, large-scale production of biosurfactants has not been achieved at an adequate level. Researchers have analyzed and reviewed various alternative strategies to overcome high-cost input for downstream processing. For instance, biosurfactant producers can be isolated as a renewable substrate to maintain the properties and amount of biosurfactants (Wasoh et al., 2017).

Although recent development has presented a promising future for biosurfactants, it is still a challenge to produce biosurfactants with minimum economic input. Utilizing a low-cost substrate and optimizing process parameters with the emphasis on alternative low-cost, effective production substrate and downstream processing needs to be explored more (Banat et al., 2014). Hence, the future lies in far-sighted strategies for the economical production of biosurfactants using agro-industrial waste along with a low-cost recovery process.

9.2 SURFACTANTS

Surfactants are a group of compounds with a hydrophilic and hydrophobic part that are separated by liquid interfaces, for instance, liquid/liquid or liquid/solid or liquid/gas, through varying degrees of hydrogen bridges and polarity. By changing the surface and interfacial properties, the surfactant generates an emulsion in which water can be solubilized in oil and vice versa. A polar segment is often a hydrocarbon chain; nevertheless, it can be ionic, non-ionic, or amphoteric (Mao et al., 2015). The addition of surfactants results in an increment in the solubility of hydrophilic moieties and a decrease in both the surface and interfacial tensions between interfaces. Determining the effectiveness and efficiency is essential to assessing the surfactants. The surfactants efficiency is generally calculated by critical micelle concentration (CMC). The lowest concentration point of a surfactant where molecules assemble as micelles is the CMC. They have the lowest steady surface tension, i.e., there is no further effect with additional surfactant. Moreover, effectiveness is connected to surface and interfacial tensions, which can be studied through hydrophilic–lipophilic balance (HLB). For instance, HLB indicates the possibility of the formation of water in oil or vice versa (Santos et al., 2016).

Currently, most surfactants produced commercially are extracted chemically from petroleum, and are, in general, toxic and difficult to break down. Hence, the limitations of chemical surfactants, mainly environmental issues, have led researchers to find alternatives such as biosurfactants that can be produced through fermentation process (Usman et al., 2016). In addition, consumer awareness of environmental issues and recent environmental legislations have directed the development of natural surfactants, such as biosurfactants, as a promising alternative to existing chemical surfactants.

9.2.1 EFFECT OF SURFACTANT

The diverse and wide applicability of surfactants has contributed to the world economy, and expansion is possible with an increasing market. However, the negative consequences of surfactant production cannot be ignored. For instance, the discharge of wastewater containing surfactants into water bodies has negatively impacted the ecosystem. The main effects of surfactants on plants, animals, the environment, and related areas are discussed further in this section.

9.2.1.1 Effect on Aquatic Plants

The extent of damage caused by surfactants to aquatic plants can be linked to their concentration level in the water. As the concentration of surfactants in water increases, its negative impact on the growth of algae and other micro-organisms increases, which ultimately erodes the food chain of aquatic plants in water. Surfactants have the potential to cause acute poisoning as membrane permeability increases due to the disintegration of the material exosome and cellular structure. Moreover, enzymes such as superoxide dismutase and catalase, along with the chlorophyll content and peroxidase activity, decrease in aquatic plants (Gudiña et al., 2015).

9.2.1.2 Effects on Aquatic Animals

The concentration of surfactants has the potential to cause toxicity that can pass into aquatic animals through feeding and skin perforations. Similarly, surfactants can enter an animal through various parts such as the gills, blood, kidney, and liver when their concentration is very high in the water, ultimately producing a toxic effect (Abdel-Raouf, Al-Homaidan, and Ibraheem, 2012). In the animal kingdom, fish readily absorb surfactants from their body surface and gills, which then pass to the organs through blood circulation. The increase in serum transaminase and alkaline phosphatase levels in fish indicates a repel effect due to the exposure to surfactants. The contaminated fish then enter the human body via the food chain resulting in the production of various enzymes that affects the body's immune system. The toxicological assessment of surfactants on bacteria and algae can be represented as ECO50 which indicates the degree of suppression of surfactants on the movement of bacteria and algae within a 24-hour period (Yuan et al., 2014).

9.2.1.3 Effects on Water Environment

hen discharged into water bodies, wastewater containing surfactants can cause serious environmental impacts, including water pollution. A 0.1 mg/l surfactant concentration in water might result in persistent foam, and subsequent accumulation of foam results in the formation of an insulation layer. The insulation layer reduces the exchange of gases between the outer environment and inner water bodies resulting in the reduction of the dissolved oxygen level in the water. A high surfactant concentration level leads to a high insulation layer that results in the death of the microbial population and other living beings due to hypoxia. The excess CMC of surfactant in the water increases insoluble and soluble water pollutants. This causes the creation of an adsorption layer from molecules with no previous adsorption energy. The formation of an adsorption layer signifies indirect pollution and ultimately changes the properties of the water (Yuan et al., 2014). Furthermore, surfactants can lower the microbial population and impede the degradation of other pollutants.

9.2.1.4   Effects on the Human Body

The food cycle, application of surfactants, and subsequent transfer of surfactants to the human body affect human immunity. The surfactant has an effect on the skin and the body as a whole. The application of surfactants, such as detergent, which is commonly used by humans, can cause irritation and skin allergies after long-term use. When surfactants enter the human body, they impede the enzyme system and thus hamper the normal physiological condition. The accumulation of surfactants in the human body can cause a toxic effect as they are very difficult to break down (Yuan et al., 2014).

9.3   BIOSURFACTANTS

Biosurfactants are amphipathic compounds with various structures and chemical properties that are produced by a wide array of micro-organisms (Felix et al., 2019). The increased popularity of biosurfactants is associated with their peculiar property of minimizing and reducing the interfacial tension of liquids. Biosurfactants are preferred over chemical surfactants forir environmentally friendly nature, higher selectivity, lower toxicity, lower CMC, multifunctionality, and specificity under extreme pH and temperature (Fenibo, Douglas, and Stanley, 2019). In addition, the hydrophilic and hydrophobic structure of biosurfactants makes them a unique compound that can be used as an emulsifier, dispersal, solubilizer, mobilizer, and surface tension reducer (Olasanmi and Thring, 2019).

Adding value to agro-industrial waste has huge potential in the production of sustainable compounds such as biosurfactants. Biosurfactants can be categorized into high molecular mass surfactants and low molecular mass surfactants, such as emulsan, based on their microbiological origin and chemical composition. The presence of high amounts of macronutrients such as carbohydrates, proteins, and lipids in agro-industrial waste makes it the best substrate in terms of cost-effectiveness (Chandankere et al., 2014; Freitas et al., 2016). This section focuses on the multifunctional property of biosurfactants along with their properties, classification, and application Table 9.1.

9.3.1   Global Biosurfactant Market

The biosurfactants market was estimated at more than US$1.5 billion in 2019 (Ahuja and Singh, 2020). Personal care products, house detergents, oilfield chemicals, and industrial cleaners are expected to stimulate the demand for biosurfactants. These are natural substances that are employed in the creation of cosmetic products due to their biodegradability. They are commonly used in a variety of skincare products, ranging from heavy-duty creams to light emulsions and lotions. The increasing demand for sustainable goods in a variety of cleaning applications is predicted to increase the demand for biosurfactants throughout the forecast period. The increased demand for the diverse components of formulators will boost the growth of the cleaning market industry, decreasing the input costs of formulators by 10–15%. The product is biodegradable, non-toxic, and has an outstanding foaming action, making it ideal for use in hand soaps, face cleanser, dental care, and shampoo formulations.

Biosurfactants are used to help reduce pollution and raise consumer knowledge about the benefits of using bio-based products. Furthermore, they are biodegradable and have a longer shelf life than competing products. The high sustainability and low toxicity of petroleum-based alternatives, along with tight regulations regarding the use of bio-based goods and the phase-out of petroleum-based products, will spur industrial development.

Even if appropriate production conditions and media have been achieved, the manufacturing process remains unfinished without a cost-effective and efficient downstream processing method. The most expensive cost of biosurfactant production is the downstream process which costs around 60% of the overall process, which is excessive and influences the market price trend for biosurfactants.

TABLE 9.1

Classification of Biosurfactants According to Their Chemical Structure and the Micro-Organisms That Produce Them

Group	Type	Microbial Identity	References
Glycolipids	Rhamnolipids	*Pseudomonas aeruginosa*	(Whang et al., 2009)
	Sophorolipids	*Candida bombicola ATCC 22214*	
	Trehalolipids	*Rhodococcus* sp. *PML026., Rhodococcus actinobacteria*	
Lipopeptides	Surfactin, Iturin, Fengysin	*Bacillus subtilis*	(Hmidet et al., 2017)
	Lichenysin	*Bacillus licheniformis*	(Madslien et al., 2013)
	Viscosin	*Pseudomonas fluorescens*	(Alsohim et al., 2014)
	Serrawettin	*Serratia marcescens*	(Hage-Hülsmann et al., 2019)
	Phomafungi	*Phomasp.* S31	(Jo atilde o et al., 2016)
Fatty acid, phospholipids, and neural lipid	Spiculisporic acid	*Penicillium spiculisporum*	(Ishigami, Zhang, and Ji, 2000)
	Lichenysin	*Bacillus licheniformis*	(Saharan, Sahu, and Sharma 2011)
	Phosphatidylethanolamine	*Rhodococcus erythropolis*	(Stancu, 2015)
Polymeric biosurfactant	Emulsan	*Acinetobacter calcoaceticus*	(Shekhar, Sundaramanickam, and Balasubramanian, 2015)
	Alasan	*Acinetobacter radioresistens*	(Stancu, 2015)
	Yasan	*Yarrowia lipolytica*	(Toren et al., 2001)
	Biodispersan	*Acinetobacter calcoaceticus RAG-1*	(Rahman and Gakpe, 2009)
Particulate biosurfactant	Vesicles	*Acinetobacter calcoaceticus*	(Muthusamy et al., 2009)
	Whole cell		

9.3.2 Biosurfactant Properties

To determine the commercial validity of a biosurfactant, it is important to subject it to conservation techniques that analyze its qualities over a 120-day period. As a result, heating techniques are used alone, a chemical that prevents mould growth and is commonly used in the manufacturing and preservation of foods. Biosurfactants share certain features, and these properties provide advantages over conventional surfactants.

9.3.2.1 Interfacial and Surface Activity

A surfactant's efficiency and effectiveness are critical properties. The effectiveness and efficiency of a surfactant are determined by surface and CMC. For instance, the CMC of a biosurfactant is between 1 and 2000 mg/l. Moreover, the interfacial between oil and water is about 1 mN/m and surface tension 30 mN/m. Effective surfactants can decrease the surface tension of water to 35 mN/m (Santos et al., 2016).

9.3.2.2   Temperature, pH, and Ionic Strength

Temperature, pH, and ionic strength are natural variables that affect a significant fraction of biosurfactants and their surface activity. *Bacillus licheniformis*, for example, produces lichenysin, that can tolerate temperatures as high as 50°C, pH values ranging from 4.5 to 9.0, and sodium chloride and calcium concentrations of up to 5% and 2.5%, respectively.

9.3.2.3   Biodegradability

Compared to synthetic surfactants, microbially produced compounds are quickly degraded, making them ideal for bioremediation. We are more concerned about the environment, which has led us to explore alternative surfactants. Due to the ecological concerns associated with synthetic chemical surfactants, biosurfactants, which are environmentally friendly, derived from micro-organisms were investigated for their possible function (Gharaei, 2011).

9.3.2.4   Low Toxicant Concentration

Although there are few publications on the hazardous nature of biosurfactants, they are generally considered to be non-harmful substances that are suitable for medical, corrective, and preventative purposes. Chemically derived surfactants exhibit a 50% lethal concentration value tenfold that of rhamnolipids against *Photobacterium phosphoreum*. Due to the low toxicity of biosurfactants, *Candida bombicola* sophorolipids were found to be beneficial in nutrition efforts (Roy, 2019).

9.3.2.5   Precision

Biosurfactants are complex compounds with distinct functional groups and, as a result, may have a narrow range of activity. This is especially advantageous for detoxification of a wide variety of pollutants and de-emulsification from industrial, including for certain culinary, medicinal, and cosmetic applications.

9.3.2.6   Digestibility and Biocompatibility

These qualities enable biomolecules to be used in a variety of industries, most notably food, pharmaceuticals, and cosmetics.

9.3.2.7   Emulsion Formation/Breaking

Biosurfactants may act as emulsifiers. Emulsions are heterogeneous systems composed of an impermeable liquid spread over another liquid in the form of droplets with a diameter greater than 0.1 mm. Emulsions are classified into two types: oil in water (o/w) and water in oil (w/o). Although emulsions are inherently unstable, by adding biosurfactants, it is possible to create a stable emulsion that will last for months. Liposan, a hydrophilic emulsifier produced by Candida lipolytica, has been utilized to create stable emulsions with edible oils. Liposan is a substance that is frequently utilized in the food and cosmetic sector for producing stable o/w emulsions (Santos et al., 2016).

9.3.2.8   Anti-adhesive agent

A biofilm is a group of microbes/other organic matter that forms on any surface (Jadhav et al., 2011). The initial step in establishing a biofilm is bacterial adhesion to the surface. This is influenced by a variety of factors, including the type of micro-organism, the hydrophobicity and electrical charge of the surface, the environmental parameters, and the micro-organism's ability to produce extracellular polymers that aid cells in adhesion. Biosurfactants can be used to alter the surface's hydrophobicity. As a result, the microbes' adhesion to the surface is affected. A surfactant produced from *Streptococcus thermophilus* prevents other thermophilic *Streptococcus* strains from colonizing the steel that causes fouling. Additionally, *Listeria monocytogenes* was prevented from sticking to the steel surface by a biosurfactant produced from *Pseudomonas fluorescens* (Kurtzman et al., 2010).

9.3.3 Biosurfactant Type

Biosurfactants can be classified based on various aspects, from microbial source to the chemical composition, in accordance with the intended use. The charge is one of the major categorization aspects in which the majority need to have either an anionic or neutral charge, whereas positively-charged ions or those of cationic nature are exhibited by biosurfactants with amine groups. The hydrophobic moiety is made up of long-chain fatty acids, while the water-soluble moiety is made up of an organic compound. The molar mass of biosurfactants ranges from 500 to 1500 Da.

9.3.3.1 Rhamnolipids

Rhamnolipids are glycolipids composed of one or two rhamnose molecules connected to one or two hydroxydecanoic acid molecules. They are a well-studied biosurfactant used to synthesize most of the glycolipids generated by *P. aeruginosa* (Zhao et al., 2019).

9.3.3.2 Sophorolipids

These are glycolipids made by yeasts. They are built up of a dimeric carbohydrate called a sophorose that is glycosidically linked to a long-chain hydroxyl fatty acid. Sophorolipids, a class of hydrophobic lipids that are frequently composed of at least six to nine different hydrophobic lipids, are preferred for many applications.

9.3.3.3 Trehalolipids

This is a different type of glycolipid. Trehalose, a disaccharide connected to a mycolic corrosive at C-6 and C-6, is found in the majority of *Mycobacterium*, *Corynebacterium*, and *Nocardia* species. Mycolic acids are unsaturated fats with a long chain length and a widespread, low hydroxy content. Trehalolipids from several living organisms vary in size and structure, the number of carbon molecules, and the degree to which they are unsaturated.

9.3.3.4 Lipopeptides and Lipoproteins

Lipopeptides are a type of surfactant. Numerous cyclic lipopeptides have decapeptide antitoxins (gramicidins) and lipopeptide antitoxins (polymyxins). These are composed of a lipid attached to a polypeptide chain.

9.3.3.5 Lichenysin

Numerous biosurfactants added to *Bacillus licheniformis* work synergistically and show exceptional temperature, ionic strength, and pH resistance. Compared with surfactin in terms of additional and physio-synthetic features, lichenysin can reduce the surface strain of water from 72 to 27 mN/m in addition to the interfacial pressure of water and n-hexadecane to 0.36 mN/m.

9.3.3.6 Fatty Acids, Phospholipids, and Neutral Lipids

While growing on n-alkanes, a few bacteria, including yeast, generate significant quantities of unsaturated fats and phospholipid surfactants. *Acinetobacter* spp. 1-N generates phosphatidyl ethanolamine-rich vesicles that form optically transparent smaller scale emulsions of alkanes in water. These biosurfactants are crucial for medicinal purposes. The reason for respiratory distress in premature newborns is a lack of a phospholipid protein complex.

9.3.3.7 Biosurfactants Composed of Polymers

The most well-known polymeric biosurfactants are alasan, liposan, lipomanan, emulsan, and a few additional polysaccharide-protein structures. *Acinetobacter calcoaceticus* RAG-1 incorporates an external heteropolysaccharide bio-emulsifier composed of strong polyanionic amphipathics. Emulsan is an excellent emulsifier of non-polar carbon in water, even at concentrations as low as 0.001–0.01%.

9.3.4 ROLES AND PROPERTIES OF BIOSURFACTANTS

Biosurfactants have diverse and multifunctional roles and properties. The wide application of biosurfactants is associated with their peculiar properties (Figures 9.1 and 9.2).

9.3.4.1 Biosurfactants and Hydrophobic Moieties Interact

Biosurfactant interaction with hydrophobic molecules has been widely studied. This property of biosurfactants is applied to petroleum as microbial enhanced oil recovery (Varjani and Upasani, 2017). The ability of biosurfactants to create stable micelles has led to its expanded applications in drug delivery systems as a result of formulations of nano-emulsions that are used to treat many diseases such as thrombosis, cancer, and Alzheimer's (Calabrese et al., 2017). The microbial production of biosurfactants aids the interaction with water-insoluble compounds, which in turn are utilized as a carbon source by microbes. Bioremediation application is associated with the property of biosurfactants to aid in absorption, biodegradation, and transport of hydrocarbons and the presence of xenobiotics in micro-organisms (dos Santos et al., 2019).

9.3.4.2 Adherence and De-adherence

The presence of both a water-soluble and insoluble part in biosurfactants allows them to interact with charged surfaces as a wetting agent. In application, most of the micro-organisms are allocated at interfaces, making them ideal molecules to be used in the interfacial environment (Markande and Nerurkar, 2014). The amphipathic molecules aid in the adhesion and de-adhesion of the microbial population at the interfaces (Satpute et al., 2016).

9.3.4.3 Formation and Elimination of Biofilms

Biosurfactants are extensively studied for their adherence and de-adherence characteristics, as mentioned in the previous section. Biofilms are formed by microbial aggregates in nature. The formation

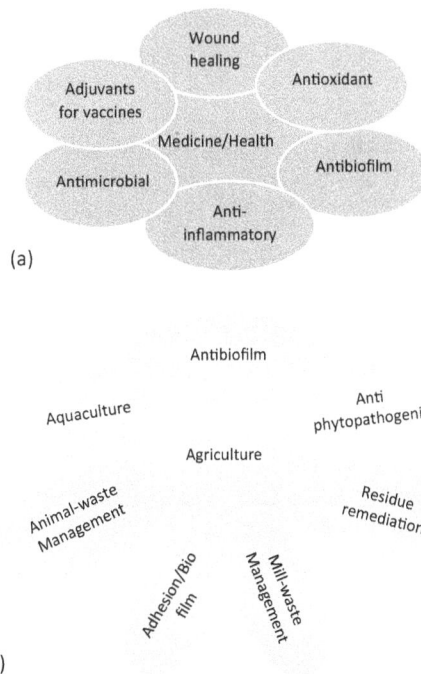

FIGURE 9.1 Biosurfactant applications and properties in (a) health sector/medicine and (b) agriculture.

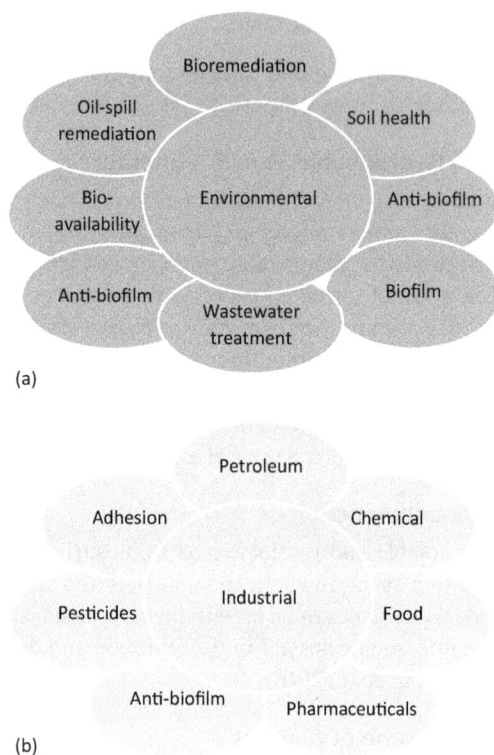

(a)

(b)

FIGURE 9.2 Biosurfactant application and properties in (a) environmental sector and (b) industrial sector.

of heterogeneous biofilms relates to interfacial bacterial wetting using surface-active agents resulting in adhesion and aggregation of micro-organisms. In microbial enhanced oil recovery (MEOR), biosurfactant-enhanced biofilms are used substantially (Geetha, Banat, and Joshi, 2019). Similarly, the amphipathic characteristics of biosurfactants can be applied as antibiofilm agents. Unwanted biofilms that appear in the environment because of industrial or even biomedical waste can be mitigated using biosurfactants. Biosurfactants in conjunction with physical methods such as mechanical abrasions and liquid flushing to the extent of thus forming biofilm. Also, pathogenic bacteria can be restricted to form a secondary biofilm through the antibiofilm activities of biosurfactants (Seghal Kiran et al., 2010).

9.3.4.4 Antimicrobial Activity

Antimicrobial activity has been demonstrated for extensively investigated biosurfactants such as surfactin, rhamnolipids, iturin, and fengysin. Antibiotic resistance has been widely documented in recent years, resulting in a rise in research on biosurfactants with antifungal, antibacterial, and antihelmintic properties (Bhawsar, Path, and Chopade, 2011). Additionally, biosurfactants have been linked to wound healing activity (Sana et al., 2019). Biosurfactants have also been reported to be used in medicine delivery, infection healing, and even as a vaccination adjuvant (Sandeep, 2017). Moreover, biosurfactants have been used against phytopathogenic fungus (Melo et al., 2021).

9.3.5 Drawbacks of Biosurfactants

The economic feasibility of large-scale biosurfactant production is a hurdle to overcome. The biotechnological procedures used to create biosurfactants are extremely inefficient, particularly for their use in petroleum and environmental applications that require large quantities of biosurfactants. Another issue associated with biosurfactants is substance purification, which is critical for pharmaceutical, cosmetic, and food applications (Reis et al., 2019). Numerous researchers have proposed solutions to these biosurfactant drawbacks. After mitigating their polluting effect, cost reductions in their manufacture might be accomplished by utilizing waste substrates (Santos et al., 2019). It is also essential to develop effective bioprocesses and optimize them successfully, which includes optimizing environmental conditions and cost-effective recovery processes to maximize biosurfactant production and recovery.

9.4 PRODUCTION OF BIOSURFACTANTS

There are various factors that affect biosurfactant production, for instance, the supply of carbon and nitrogen, the carbon:nitrogen ratio, dietary limitations, concoction, and physical factors such as temperature, pH, air circulation, divalent cations, and salinity, all of which influence the amount of biosurfactant produced (Roy, 2019).

9.4.1 SOURCES OF CARBON

Numerous studies have incorporated substantial amounts of carbon sources into the biosurfactant manufacturing process. Hydrocarbon substrate is a carbon source for the synthesis of biosurfactants. While it is evident that carbon substrates play a critical part in biosurfactant union, their importance varies according to the life form. For example, different carbon sources in the medium impacted biosurfactant synthesis during production using *Pseudomonas* sp., whereas substrates had no effect with different chain lengths and had no influence on the unsaturated fatty acid chain length (Kubicki et al., 2019).

9.4.2 SOURCE OF NITROGEN

Nitrogen is required for biosurfactant synthesis. A nitrogen-containing medium is critical for microbial development since it is required for protein and chemical synthesis. Nitrate sources were most frequently used to synthesize biosurfactants in *P. aeruginosa*. *Arthrobacter paraffineus* prefers ammonium salts and urea. While yeast separation is the most frequently used nitrogen supplier for the synthesis of biosurfactants, its major purpose in terms of application is to safeguard creatures and cultural mediums (Nurfarahin, Mohamed, and Phang, 2019).

9.4.3 ELEMENTS OF NATURE

Environmental variables have a considerable impact on the biosurfactant produced. When producing a large amount of biosurfactants, it is essential to enhance the bioprocess since the final outcome may be modified by adjusting the environmental conditions. While the majority of biosurfactant preparations are performed at temperatures ranging from 25°C to 300°C, for *Pseudomonas* sp. strain, these temperature ranges resulted in changing of the biosurfactant generated; rhamnolipid production was greatest between pH 6 and 6.5 and significantly decreased above pH 7 (Hamzah, 2016).

9.4.4 AERATION AND AGITATION

These are critical elements affecting the generation of biosurfactants because they both facilitate oxygen exchange between the gaseous and liquid phases. It might also be connected to the emulsifier's physiological capability; it has been hypothesized that the formation of bio-emulsifiers can enhance the solubility of insoluble substrates in water, therefore improving supplement transfer for the bacteria. The best surfactant production was obtained when the air current velocity was 1 vvm, and the broken-up oxygen concentration was maintained at half of the immersion (Roy, 2019).

9.4.5 SALINITY

Additionally, the salinity of a particular medium plays a vital role in biosurfactant synthesis. Moreover, opposing impressions were observed for certain biosurfactants that were unaffected by up to 10% NaCl (Nurfarahin, Mohamed, and Phang, 2019).

9.5 AGRO-INDUSTRIAL WASTE

Agro-industrial waste products like soybean oil waste, molasses, palm oil cake, cassava flour, and whey have been applied as cheap raw media for the production of biosurfactants. These cheap substrates exhibit the dual advantage of providing nutrients for microbial growth along with a source of isolation for potential biosurfactant-producing microbes (Lee et al., 2009). Utilization of cheap raw materials like agro-waste as a nutrient source can contribute to the cost-effective production of biosurfactants along with optimization of parameters related to growth and production (Banat et al., 2014). Also, exploration of the potential application of biosurfactants in diverse fields, including agriculture, the food industry, pharmaceuticals, cosmetics, detergents, and bioremediation, is being done, which could ultimately help in economic production (Gudiña et al., 2015).

Although agro-industrial waste exhibits tremendous potential in the production of biosurfactants, the selection of raw waste should ensure sufficient nutrients for microbial growth for cost-effective production. The ratio of different elements such as C:N or P or Fe or Mg can optimize biosurfactant production. The utilization of agro-industrial waste is one of the many steps required to optimize production. Along with the advantages of using agro-industrial waste, there are many challenges that lie within the chain of production. The presence of undesired compounds, the varied composition of raw material, and tedious processing requirements are some of the major limitations associated with the utilization of agro-waste on an industrial scale (Banat et al., 2014).

Moreover, due to the high cost associated with the recovery and purification of microbial surfactants, large-scale production of biosurfactants has not been achieved at an adequate level. Researchers have analyzed and reviewed various alternative strategies to overcome the high-cost input for downstream processing (Wasoh et al., 2017).

Although recent developments have presented a promising future for biosurfactants, it is still a challenge to produce biosurfactants with minimum economic input. Utilizing a low-cost substrate and optimizing process parameters with the emphasis on alternative low-cost, effective production substrates and downstream processing needs to be explored more (Banat et al., 2014). Hence, the future lies in far-sighted strategies for the economical production of biosurfactants using agro-industrial waste along with a low-cost recovery process.

9.5.1 AGRO-INDUSTRIAL WASTE AS SUBSTRATE FOR BIOSURFACTANT PRODUCTION

Agro-industrial wastes are high in carbohydrates, protein, and fats and are usually used as feed for cattle or compost before being disposed of in landfills. Instead, it can be used as an inexpensive substrate to produce microbial secondary metabolites (Chandankere et al., 2014). Numerous, less expensive substrates, such as soybean and palm oil cake, not only provide nutrients for microbial development but also serve as a key source of potential biosurfactant-generating micro-organism (Lee et al., 2009) Table 9.2. Today's civilization is defined by rising expenses, the need to recycle, and environmental concerns. As a result, attention has been placed on recovery, recycling, and reuse. Indeed, the desire to preserve the environment has resulted in the repurposing of various

TABLE 9.2

Biosurfactant Production with Agro-industry Waste as Substrate

Feedstock	Characteristics	Strain and Production	References
OOME	Olive oil mill effluent is a dark liquid made up of the water-soluble fraction of ripe olives and water used in the extraction process of olive oil. It has a significant proportion of organic matter. OOME is an excellent source of important organic chemicals such as sugars, organic acids, nitrogen compounds, and residual oil. Additionally, it includes harmful chemicals such as polyphenols.	*Pseudomonas* sp. produced rhamnolipid biosurfactants at a total conversion yield of 14 g rhamnolipids per kg OOME. *Aureobasidium thailandense* LB01 produced biosurfactant (139 mg/l) with 1.5% w/w olive oil mill wastewater.	Maneerat, 2005 Meneses et al., 2017b
Soapstock	Soapstock is a sticky, amber-coloured residue from the processing of oilseeds. Soapstock is created when edible oil is extracted and refined using an organic solvent.	*Pseudomonas aeruginosa* LBI produced 1.434 g/l of rhamnolipid after 120 hours fermentation with 2% soybean soapstock.	Lopes et al., 2021
Molasses	Molasses is a by-product of the sugar manufacturing process. Molasses is described as the runoff syrup following the crystallization process. It comprises roughly 48–56% total sugar, 9–12% organic matter, 2–4% protein, 1.5–5% potassium, 0.06% magnesium, and 0.6–2% phosphorus, 6000 mg/kg inositol, 1–3 mg/kg biotin, and 1.8 mg/kg thiamine.	*Bacillus subtilis* RSL-2 produced 12.34 g/l biosurfactant and reduced the surface tension to 24.09 mN/m with molasses as carbon sources.	Verma et al., 2020
Fats and vegetable/ frying oil	Edible oils and animal fats are widely available from the meat/food processing industry, restaurants, and households. It contains a variety of organic compounds.	*Candida tropicalis* UCP0996 produced 4.19 g/l biosurfactant with 2.5% waste frying oil, 2.5% corn steep liquor, 2.5% molasses, and 2% inoculum size.	Almeida et al., 2017
Starch-rich substrate	Starch-rich waste from agro-industry processing of potatoes, cassava, and corn steep can be obtained in enormous quantities and recognized as a promising feedstock for biotechnological operations.	*B. subtilis* DDU20161 produced 253.79 mg/l biosurfactant fermented with potato peels.	Vishal et al., 2020
Whey	Whey is a liquid by-product of the dairy industry, specifically the cheese and tofu industries. It has a high proportion of lactose (75% of dry matter) and 12–14% protein. Additionally, it contains organic acids, minerals, and vitamins. Whey disposal is a significant source of pollution, particularly in nations with a dairy economy.	*Pseudomonas fluorescens* grown on 8 g/l tofu whey produced biosurfactants that were identified as rhamnolipids, which had a CMC value of 638 mg/l and surface tension of 54 mN/m.	Kane, Mishra, and Dutta, 2016

TABLE 9.3

The Advantages and Disadvantages of Using Less Expensive Substrates in the Manufacturing of Biosurfactants

Advantages	Disadvantages
Costs associated with commercial production can be lowered.	Substrates are contaminated with undesirable chemicals.
Numerous less expensive/renewable substrates are available.	Before substrates can be developed as a source of carbon, nitrogen, or energy, they must be processed or treated.
Substrates are abundant.	The final product may acquire colour or contaminants from the substrates (e.g., molasses).
Increased biosurfactant/bio-emulsifier yield.	Purification processes must be used to obtain a pure product, which raises the cost of production.
The product's fundamental functional qualities remain constant.	Continuous supply of identical raw materials may change in composition.
Does not appear to be toxic to micro-organisms.	Raw substrates can be extremely unique to individual organisms.
Each component is environmentally friendly and safe.	A considerable quantity of raw substrates is required and obtaining a constant supply for the industrial process may be difficult.

industrial byproducts (Reike, Vermeulen, and Witjes, 2019). This is especially important to the food industry, where byproducts and waste can be recycled. The waste materials should be nutritionally balanced to enable microbial development and subsequent biosurfactant synthesis. As a substrate, industrial waste with high nutrient content, including carbohydrates and lipids, is suitable. The utilization of agro-industrial waste is the first step towards industrial-scale biosurfactant synthesis, which requires optimization of the variables involved. The following table summarizes the most often used industrial waste products for biosurfactant synthesis. However, as illustrated in Table 9.3, there are notable advantages and disadvantages to employing a low-cost substrate in the production of biosurfactants.

9.5.1.1 Olive Oil Mill Effluent

Olive oil mill effluent (OOME) is a black liquid made from ripe olives and water that is used to extract olive oil. OOME includes polyphenols that are difficult to dispose of in the environment. It does, however, include nitrogen compounds (1.2–2.4%), sugars (2–8%), residual oil (0.03–0.5%), and organic acids (0.5–1.5%). After 48 hours of fermentation with *Aureobasidium thailandense* LB01, biosurfactants of around 139 mg/l were obtained with a culture medium that included yeast extract of 0.2% and olive oil mill wastewater of 1.5% (w/w) (Meneses et al., 2017a).

9.5.1.2 Animal Fat

Different sources of fat, such as lard, have been used as a cooking medium in the food industry. Animal fat and lard are widely used for various purposes in the meat processing industry. Nevertheless, fats have recently lost a sizeable portion of the market to vegetable oils, owing to the latter's lower level of toxicity. Animal fat encourages the yeast *C. bombicola* to produce sophorolipids. The yeast *C. lipolytica* UCP 0988 produced the largest number of glycolipids with animal fat and corn steep liquor as substrate. Additionally, the substance has applications in bioremediation and oil mobilization and recovery (Santos et al., 2013).

9.5.1.3 Deep-Fried Oils

Used oil is regarded as an excellent carbon source for the creation of biosurfactants. Vegetable oils are a source of hydrophobic carbon and are primarily composed of fatty acids with carbon chains ranging from 16 to 19. Numerous oils are suitable as biosurfactant substrates. The best energy and carbon sources to produce biosurfactants are sunflower and olive oils. On residue from maize, soybean, and canola oil plants, *P. aeruginosa* strains create a biosurfactant. Canola oil residue and sodium nitrate have been shown to be sufficient for microbial growth: for 14 days, *P. aeruginosa* was cultured in 3% v/v waste canola oil and 0.4% $NaNO_3$. The yield was 0.36% rhamnolipids (Pérez-Armendáriz et al., 2019).

9.5.1.4 Molasses

Molasses is a by-product of the sugar industry. This low-cost medium contains 75% dry matter, 9–12% non-sugar organic matter, 2.5% protein, and 1.5–5.0% potassium, as well as magnesium, phosphorus, and calcium (1%). Molasses' have a thick consistency and black colour. Molasses' high sugar content (58%) is an ideal substrate for a variety of micro-organisms to synthesize biosurfactants. Molasses has been used in laboratories to synthesize a variety of microbial metabolites. *Bacillus subtilis* produces a biosurfactant in the presence of molasses as a carbon source. Utilizing molasses and other carbon sources, *Bacillus* strains were used to make biosurfactants (Rane et al., 2017).

9.5.1.5 Whey

Whey waste produced largely from the dairy industry can be used in a reuse-and-recycle approach as a suitable substrate for microbial metabolite synthesis. Various research has been done on aspects of specific growth rates and specific product formation rates that have been found to be superior to synthetic media. Numerous other nutrient-dense compounds, such as proteins, vitamins, and organic acids, are necessary for the growth of microbes and the formation of biosurfactants. (Santos et al., 2016). Additionally, whey waste needs to be disposed of, which creates an economic burden on the dairy industry, so utilizing whey as a substrate benefits both the economy and the environment.

9.5.1.6 Carbohydrate Substrates

Numerous starch-based substrates that act as renewable carbon sources are available. There are numerous potato processing plants and enterprises that generate large amounts of starch-rich waste that can be used to make biosurfactants. The starch-rich waste contains approximately 80% water, 17% carbohydrates, 2% protein, and 0.1% lipids (Burgain et al., 2011). *Bacillus subtilis* LB5a was observed to produce 2.4 g/l biosurfactants. Moreover, the medium's superficial tension was decreased from 51 to 27 mN/m, and the critical micellar concentration was observed to be 11 mg/l.

9.5.1.7 Oil Refineries

Various forms of waste from the oil processing industry, including olive oil and sunflower oil, have been found to be beneficial sources of carbon and energy, which have the potential to be an alternative source of renewable substrates for the creation of microbial surface-active substrates such as biosurfactants. Modification of oils containing short chain fatty acids (C10) integrates it into surface-active products. Palm oil mill wastewater is potentially viable as a nutrient source in the production of biosurfactants. *P. aeruginosa* SP4 has been reported to produce biosurfactants with palm oil as a nutrient medium (Pornsunthorntawee et al., 2009). *Pseudomonas fluorescence* MFS03 isolated from mangrove forest soil can utilize substrates rich in oleic acid, such as waste fried vegetable oil, glucose, and coconut oil cake, from which it can be inferred that both vegetable and coconut oils are suitable substrates for biosurfactant production (Govindammal and R., 2013).

9.5.1.8 Soap Stocks or Oil Cakes

The complex nature of soap stocks or oil cakes, which are semisolid or gummy in nature, is the result of chemical extraction or refinement of edible oils derived from seeds (Banat et al., 2014). Despite the complexity of the substrate, *Bacillus pseudomycoides* was found to successfully make soap from groundnut oil cake. Therefore, lipopeptide at a 50 g/ml concentration demonstrated the strongest antibacterial activity.

9.6   APPLICATION OF BIOSURFACTANTS

Biosurfactants are used in various applications such as bioremediation to clean up pollutants through their ability to increase the bioavailability of contaminants to microbes and hydrocarbon solubilization, and mobilization in soils through biosurfactant-augmented washing (Karlapudi et al., 2019). Currently, high production cost, low production yield, expensive downstream and recovery produces have restricted the commercial production of biosurfactants in bulk quantities. For growth in industrial-scale production of biosurfactants, the production cost should be equitable to those of synthetic surfactants (Biniarz et al., 2020). Hence, the focus has shifted towards wider and diverse applications of biosurfactants to aid in economically viable industrial-scale production.

Various research has been conducted on the characteristics, kinetics, and possible application areas such as food, medicine, and cosmetics owing to the peculiar feature of biosurfactants which are discussed further in this section.

9.6.1   Petroleum

Biosurfactants have been applied extensively in the petroleum industry (Karlapudi et al., 2019). Microbial metabolites such as biosurfactants along with biomass, biopolymers, and biosolvents can contribute to an application such as MEOR. The use of biosurfactants in MEOR applications has been found to be effective. The downstream recovery of oil from petroleum wells (MEOR) is observed to have been enhanced with biosurfactants along with transportation via sand-filled columns and metal pipes. MEOR technology can be applied in situ or ex situ. In the case of in situ MEOR application, chosen nutrients are injected into the oil well with the subsequent processes of shut-in and water-flooding phases. The shut-in phase is characterized by the enhancement of oil recovery due to the increase in the microbial load of the column. On the other hand, the production of microbial products is done outside, which is then injected into the oil well resulting in enhanced oil recovery (Geetha, Banat, and Joshi, 2019; Varjani and Upasani, 2016).

9.6.2   Pharmaceutical

Biosurfactants' ability to form a hydrophobic mixture has led to their application in the pharmaceutical industry. Numerous studies have documented its use in medication delivery systems, mixing, and formulation (Calabrese et al., 2017; Naughton et al., 2019; Dehghannoudeh et al., 2019; Sandeep, 2017). Biosurfactants have been applied as microscopic emulsions in medication delivery systems. Additionally, the heterogeneity of responses and the nature of biosurfactants must be properly understood in order to facilitate their application in enhanced drug delivery systems (Dehghannoudeh et al., 2019).

9.6.3   Food

Biosurfactants are better alternatives to chemical surfactants due to their environmentally friendly characteristics. In relation to biosurfactants' ability as a wetting agent, they can be used in the formation of a stable emulsion. It can aid in mixing food materials. Also, it can be used as an antibiofilm and cleaning agent because of its surface-active nature. The application of bio-emulsifiers in the food industry, such as the bakery industry, is associated with its ability to form an emulsion. In addition, emulsifiers in the bakery industry are used in ingredient mixing, rheological modification, improvement of water retention capacity, and maintaining dough characteristics. Moreover, the good additive and preservative characteristics of biosurfactants have been reported in the food production industry (Satpute et al., 2010; Sałek and Euston, 2019; Felix et al., 2019; Naughton et al., 2019).

9.6.4 AGRICULTURE

The application of biosurfactants in the treatment of animal and agro-industrial waste has been studied. In agricultural applications, biosurfactants have been found to work against phytopathogenic fungi (Melo et al., 2021). In a broader aspect, the use of biosurfactants is linked to the control of plant pathogens, thus protecting seeds and fertility (Naughton et al., 2019).

9.6.5 MEDICINE/HEALTH

There are several studies that have been published about the application of biosurfactants for medical purposes, such as antimicrobials, antifungals, and antivirals. Moreover, biosurfactants can increase immunomodulatory activities through their adhesive and anti-adhesive properties (Rodrigues et al., 2006; Varjani et al., 2021a). Biosurfactants are associated with antitumor and anti-inflammatory activities. Additionally, biosurfactants have been used in biomedical applications such as antibacterial and antibiofilm activity, wound healing, drug delivery systems, dermatological activity, oral activity care, and anticancer activity (Rodrigues et al., 2006; Naughton et al., 2019).

The medicinal uses of biosurfactants possess a risk as well that needs to be considered. The antimicrobial and non-cytotoxic, antitumorigenic properties of biosurfactants, such as rhamnolipids produced from *Pseudomonads* against colon cancer, have been studied, but pathogenicity is also reported, which is linked to the biosurfactants and their biosynthesis process. Also, certain types of biosurfactants from pathogens have the potential to cause allergic and related skin problems (Adu et al., 2020). Hence, a cautionary approach should be developed and followed in the case of such surfactants.

9.6.6 ANTIMICROBIAL ACTIVITY

Biosurfactants' antimicrobial properties, which include antibacterial, antifungal, mycoplasma, algicidal, anti-amoebic, zoosporicidal, and antiviral activity, have a wide range of medical applications. Actinobacteria, a popular type of eubacteria, has been extensively studied for its potential to create antibiotics and other bioactive compounds. Many amphipathic compounds have been extracted through research on actinobacteria, which have the potential for use as biosurfactants on an industrial scale (Satpute et al., 2010; Bhawsar, Path, and Chopade, 2011).

9.6.7 OTHER MEDICAL USES

Some of the biosurfactants used in this category are glycolipids, lipopeptides, and more.

9.6.7.1 Glycolipids

A biosurfactant produced from *Candida* species is known as mannosylerythritol lipid (MEL). It is a glycolipid possessing antimicrobial as well as neurological and immunological properties. Similarly, *Rhodococcus* species produced succinoyl-trehalose lipid, which could impede certain viruses. Glycolipids have been linked to growth arrest, apoptosis, and separation of murine malignant melanoma cells (Varjani and Upasani, 2019).

9.6.7.2 Lipopeptides

Lipopeptides, such as surfactin and iturin, have been reported widely for medical applications, including antitumor, antiviral, and antibacterial effects, along with inhibition of some enzymes and as immunomodulatory agents for a certain type of toxins (Meena and Kanwar, 2015).

9.6.7.3 Other Biosurfactants

Biosurfactants such as viscosin, viscosinamide, cyclic depsipeptides, massetolides A–H, etc. provide antimicrobial and antibiotic properties against *Mycobacterium* and *Streptococcus* species.

Also, biosurfactants such as pumilacidin A–G have been associated with antiviral activities (Meena and Kanwar, 2015). A glycolipoprotein biosurfactant generated by *Acinetobacter indicus* shows antiproliferative action against lung cancer cells, as well as antimicrobial and antimicrobial film-forming activity against Methicillin-resistant *Staphylococcus aureus* (Peele et al., 2019).

9.6.8 BIOFILM

The characteristic of biosurfactants as amphipathic molecules, along with their wettability property, allows them to influence biofilm production through adhesion and deadhesion of microbes, which could have a direct influence on the corrosion effect on stainless steel, carbon steel, and aluminium alloys. Lichenysin has been linked to both pre- and post-biofilm treatments associated with its anti-biofilm activity (Coronel-León et al., 2016).

9.6.9 ENVIRONMENTAL

With its amphipathic properties, biosurfactants can form stable micelles that increase the degradability of crude oil or other hydrocarbon compounds that affect soil quality and water pollution (Shindhal et al., 2021). Biosurfactants are found to contribute significantly to the xenobiotic degradation of benthic organisms (Markande et al., 2014). Moreover, biosurfactants such as rhamnolipid, sophorolipid, and surfactin from *Pseudomonas*, *Bacillus*, and *Candida*, respectively, possess potential in oil-spill bioremediation (Patel et al., 2019). Biosurfactants from diverse micro-organisms can change aerosol surface tension, which can sway the mode of precipitation. Bioremediation procedures to improve soil health has been researched extensively as a modern nano-biotechnological approach for soil and water treatment (Varjani, 2017; Felix et al., 2019; Markande, Patel, and Varjani, 2021).

Biosurfactants act as a stabilizing agent in the formation of nanoparticles, which can be utilized for general cleaning purposes. It is the essential component of an emulsion that has been applied in the field of medicine, food, and cosmetics. Surfactin is one of the most highly studied biosurfactants that has been explored in diverse fields with biotechnological improvement. Many variants of surfactin have been studied, but it still needs reliable processes to produce high yields (Geissler et al., 2019).

9.7 FUTURE PERSPECTIVE

On a positive note, recent research on optimization techniques in the production of biosurfactants has built up the idea that chemical surfactants have the potential to be replaced by high-quality biosurfactants in the near future.

Manufacturing industries are investing more resources in biosurfactants with regard to their diverse application and future prospective. Customer awareness of their environmentally friendly nature has further driven the market for biosurfactants. With the use of genetically manipulated microbes and low-cost nutritious substrates like agro-industrial waste, the production cost has been reduced at an industrial scale in recent years. Along with that, the large-scale production of biosurfactants with diverse application areas has further strengthened its future. In that regard, the expanded field for the exploration of biosurfactants includes food, pharmaceuticals, wastewater treatment, agriculture, cosmetics, petroleum, and more. Hence, future research would further decrease production costs. In the case of downstream processing, molecular methods to screen for biosurfactants need to be investigated further with the advances in genomic and proteomic methods (Shaligram et al., 2016).

9.8 CONCLUSION

Biosurfactants are preferred over chemical surfactants regarding various characteristics such as their environmentally friendly nature, higher selectivity, lower toxicity, lower CMC, multifunctionality, and specificity under extreme pH and temperature. However, biosurfactant production, which is a high production cost, has a low production yield, and expensive downstream and recovery processes have restricted the commercial production of biosurfactants in bulk quantities. For growth in industrial-scale production of biosurfactants, the production cost should be equitable with that of synthetic surfactants. Value-addition of agro-industrial waste exhibits a huge potential in the production of sustainable compounds such as biosurfactants. Agro-industrial wastes can be used as inexpensive substrates to produce microbial secondary metabolites. The utilization of agro-industrial waste is the first step towards industrial-scale biosurfactant synthesis, which requires optimization of the many variables involved. Biosurfactants are used in various applications such as bioremediation to clean up pollutants through their ability to increase the bioavailability of contaminants to microbes and hydrocarbon solubilization and mobilization in soils through biosurfactant-augmented washing. A wider and diverse application of biosurfactants will aid in economically viable industrial-scale production.

REFERENCES

Abdel-Raouf, N., A. A. Al-Homaidan, and I. B. M. Ibraheem. 2012. "Microalgae and Wastewater Treatment." *Saudi Journal of Biological Sciences* 19(3). King Saud University: 257–275.

Adu, Simms A., Patrick J. Naughton, Roger Marchant, and Ibrahim M. Banat. 2020. "Microbial Biosurfactants in Cosmetic and Personal Skincare Pharmaceutical Formulations." *Pharmaceutics* 12(11): 1–21.

Almeida, Darne G., Rita de Cássia, F.Soares da Silva, Juliana M. Luna, Raquel D. Rufino, Valdemir A. Santos, and Leonie A. Sarubbo. 2017. "Response Surface Methodology for Optimizing the Production of Biosurfactant by *Candida tropicalis* on Industrial Waste Substrates." *Frontiers in Microbiology* 8: 1–13.

Alsohim, Abdullah S., Tiffany B. Taylor, Glyn A. Barrett, Jenna Gallie, Xue-Xian Zhang, Astrid E. Altamirano-Junqueira, Louise J. Johnson, Paul B. Rainey, and Robert W. Jackson. 2014. "The Biosurfactant Viscosin Produced by *Pseudomonas fluorescens* SBW25 Aids Spreading Motility and Plant Growth Promotion." *Environmental Microbiology* 16(7): 2267–2281.

Banat, Ibrahim M., Surekha K. Satpute, Swaranjit S. Cameotra, Rajendra Patil, and Narendra V. Nyayanit. 2014. "Cost Effective Technologies and Renewable Substrates for Biosurfactants' Production." *Frontiers in Microbiology* 5: 1–19.

Bhawsar, S. D., S. D. Path, and B. A. Chopade. 2011. "Antimicrobial Activity of Purified Emulsifier of Acinetobacter Genospecies Isolated from Rhizosphere of Wheat." *Agricultural Science Digest* 31(4): 239– 246.

Biniarz, Piotr, Marius Henkel, Rudolf Hausmann, and Marcin Łukaszewicz. 2020. "Development of a Bioprocess for the Production of Cyclic Lipopeptides Pseudofactins with Efficient Purification from Collected Foam." *Frontiers in Bioengineering and Biotechnology* 8(November): 1–12.

Burgain, J., C. Gaiani, M. Linder, and J. Scher. 2011. "Encapsulation of Probiotic Living Cells: From Laboratory Scale to Industrial Applications." *Journal of Food Engineering* 104(4). Elsevier Ltd: 467–483.

Calabrese, Ilaria, Giulia Gelardi, Marcello Merli, Maria Liria Turco Liveri, and Luciana Sciascia. 2017. "Clay-Biosurfactant Materials as Functional Drug Delivery Systems: Slowing down Effect in the In Vitro Release of Cinnamic Acid." *Applied Clay Science* 135. Elsevier B.V.: 567–574.

Chandankere, Radhika, Jun Yao, Kanaji Masakorala, A. K. Jain, and Ranjan Kumar. 2014. "Enhanced Production and Characterization of Biosurfactant Produced by a Newly Isolated *Bacillus amyloliquefaciens* USTBb Using Response Surface Methodology." *International Journal of Current Microbiology and Applied Sciences* 3(2): 66–80.

Coronel-León, J., A. M. Marqués, J. Bastida, and A. Manresa. 2016. "Optimizing the Production of the Biosurfactant Lichenysin and Its Application in Biofilm Control." *Journal of Applied Microbiology* 120(1). England: 99–111.

Dehghannoudeh, Gholamreza, Kioomars Kiani, Mohammad Hassan Moshafi, Negar Dehghannoudeh, Majid Rajaee, Soodeh Salarpour, and Mandana Ohadi. 2019. "Optimizing the Immobilization of Biosurfactant-Producing Pseudomonas aeruginosa in Alginate Beads." *Journal of Pharmacy & Pharmacognosy Research* 7(6): 413–420.

Fenibo, Emmanuel Oliver, Salome Ibietela Douglas, and Herbert Okechukwu Stanley. 2019. "A Review on Microbial Surfactants: Production, Classifications, Properties and Characterization." *Journal of Advances in Microbiology* 18(3): 1–22.

Freitas, Bruno G., Juliana G. M. Brito, Pedro P. F. Brasileiro, Raquel D. Rufino, Juliana M. Luna, Valdemir A. Santos, and Leonie A. Sarubbo. 2016. "Formulation of a Commercial Biosurfactant for Application as a Dispersant of Petroleum and By-Products Spilled in Oceans." *Frontiers in Microbiology* 7: 1–9.

Geetha, S. J., Ibrahim M. Banat, and Sanket J. Joshi. 2019. "Biosurfactants: Production and Potential Applications in Microbial Enhanced Oil Recovery (MEOR)." *Biocatalysis and Agricultural Biotechnology* 14. Elsevier Ltd: 23–32.

Geissler, Mareen, Kambiz Morabbi Heravi, Marius Henkel, and Rudolf Hausmann. 2019. *Lipopeptide Biosurfactants From Bacillus species. Biobased Surfactants.* Second Edition. Elsevier Inc.

Gharaei, E. 2011. "Biosurfactants in Pharmaceutical Industry (A Mini-Review)." *American Journal of Drug Discovery and Development* 1: 58–69.

Gudiña, Eduardo J., Elisabete C. Fernandes, Ana I. Rodrigues, José A. Teixeira, and Lígia R. Rodrigues. 2015. "Biosurfactant Production by Bacillus subtilis Using Corn Steep Liquor as Culture Medium." *Frontiers in Microbiology* 6: 1–7.

Hage-Hülsmann, Jennifer, Alexander Grünberger, Stephan Thies, Beatrix Santiago-Schübel, Andreas Sebastian Klein, Jörg Pietruszka, Dennis Binder, et al. 2019. "Natural Biocide Cocktails: Combinatorial Antibiotic Effects of Prodigiosin and Biosurfactants." *PLOS ONE* 13(7): 1–23.

Hamzah, Ainon. 2016. "Determination of Optimum Conditions and Stability Study of Biosurfactant Produced by *Bacillus subtilis* UKMP-4M5." *Malaysian Journal of Analytical Science* 20(5): 986–1000.

Hmidet, Noomen, Hanen Ben Ayed, Philippe Jacques, and Moncef Nasri. 2017. "Enhancement of Surfactin and Fengycin Production by *Bacillus mojavensis* A21: Application for Diesel Biodegradation." *BioMed Research International* 2017.

Ishigami, Y., Y. J. Zhang, and F. X. Ji. 2000. "Spiculisporic Acid. Functional Development of Biosurfactant." *Chimica Oggi* 18: 32–34.

Jadhav, Mital, Satish Kalme, Dhawal Tamboli, and Sanjay Govindwar. 2011. "Rhamnolipid from Pseudomonas Desmolyticum NCIM-2112 and Its Role in the Degradation of Brown 3REL." *Journal of Basic Microbiology* 51(4): 385–396.

Jo atilde o, Marcelo Silva Lima, Odair Pereira Jos eacute, Hort ecirc ncio Batista Ieda, de Queiroz Costa Neto Pedro, Concei ccedil atilde o dos Santos Jucileuza, Pires de Ara uacute jo Solange, Correia Pantoja Mozanil, Jos eacute da Mota Adolfo, and L uacute cio de Azevedo Jo atilde o. 2016. "Potential Biosurfactant Producing Endophytic and Epiphytic Fungi, Isolated from Macrophytes in the Negro River in Manaus, Amazonas, Brazil." *African Journal of Biotechnology* 15(24): 1217–1223.

Kane, Shashank N., Ashutosh Mishra, and Anup K. Dutta. 2016. "Preface: International Conference on Recent Trends in Physics (ICRTP 2016)." *Journal of Physics: Conference Series* 755(1): 1–2.

Karlapudi, Abraham Peele, T. C. Venkateswarulu, Jahnavi Tammineedi, Lohit Kanumuri, Bharath Kumar Ravuru, Vijaya Ramu Dirisala, and Vidya Prabhakar Kodali. 2019. "Role of Biosurfactants in Bioremediation of Oil Pollution-A Review." *Petroleum* 4(3). Elsevier Taiwan LLC: 241–249.

Kubicki, Sonja, Alexander Bollinger, Nadine Katzke, Karl Erich Jaeger, Anita Loeschcke, and Stephan Thies. 2019. "Marine Biosurfactants: Biosynthesis, Structural Diversity and Biotechnological Applications." *Marine Drugs* 17(7): 1–30.

Kunal Ahuja, Sonal Singh. 2020. "Biosurfactants Market Trends 2020–2026 | Growth Projections." *Biosurfactants Market Size by Product (Sophorolipids, Rhamnolipids, Alkyl Polyglucosides [APG], Methyl Ethyl Sulfonates [MES], Sucrose Esters, Sorbitan Esters, Lipopeptides), by Application (Household Detergents, Personal Care, Industrial Cleaners, Food). Global Market Insights Inc. Retrieved April 29, 2022, from https://www.gminsights.com/industry-analysis/biosurfactants-market-report*

Kurtzman, Cletus P., Neil P. J. Price, Karen J. Ray, and Tsung Min Kuo. 2010. "Production of Sophorolipid Biosurfactants by Multiple Species of the Starmerella (Candida) Bombicola Yeast Clade." *FEMS Microbiology Letters* 311(2): 140–146.

Lee, Sang Cheol, Seung Jin Lee, Sun Hee Kim, In Hye Park, Yong Seok Lee, Soo Yeol Chung, and Yong Lark Choi. 2009. "Characterization of New Biosurfactant Produced by Klebsiella sp. Y6-1 Isolated from Waste Soybean Oil." *Bioresource Technology* 99(7): 2288–2292.

Lopes, Paulo Renato Matos, Renato Nallin Montagnolli, Jaqueline Matos Cruz, Roberta Barros Lovaglio, Carolina Rosai Mendes, Guilherme Dilarri, Jonas Contiero, and Ederio Dino Bidoia. 2021. "Production of Rhamnolipids from Soybean Soapstock: Characterization and Comparation with Synthetics Surfactants." *Waste and Biomass Valorization* 12(4). Springer Netherlands: 2013–2023.

Luna, Juliana M., Raquel D. Rufino, Alícia Maria A. T. Jara, Pedro P. F. Brasileiro, and Leonie A. Sarubbo. 2015. "Environmental Applications of the Biosurfactant Produced by Candida sphaerica Cultivated in Low-Cost Substrates." *Colloids and Surfaces. Part A: Physicochemical and Engineering Aspects* 480. Elsevier B.V.: 413–419.

Madslien, E. H., H. T. Rønning, T. Lindbäck, B. Hassel, M. A. Andersson, and P. E. Granum. 2013. "Lichenysin Is Produced by Most Bacillus licheniformis Strains. " *Journal of Applied Microbiology* 115(4): 1068–1080.

Maikudi Usman, Mohammed, Arezoo Dadrasnia, Kang Tzin Lim, Ahmad Fahim Mahmud, and Salmah Ismail. 2016. "Application of Biosurfactants in Environmental Biotechnology; Remediation of Oil and Heavy Metal." *AIMS Bioengineering* 3(3): 289–304.

Maneerat, Suppasil. 2005. "Production of Biosurfactants Using Substrates from Renewable-Resources." *Songklanakarin Journal of Science and Technology* 27(3): 675–683.

Mao, Xuhui, Rui Jiang, Wei Xiao, and Jiaguo Yu. 2015. "Use of Surfactants for the Remediation of Contaminated Soils: A Review." *Journal of Hazardous Materials* 285. Elsevier B.V.: 419–435.

Markande, Anoop, and Anuradha Nerurkar. 2014. "Biochemical Diversity of Microbial Bioemulsifiers and Their Roles in the Natural Environment." In *Microbial research: An overview.* essay, I.K. International Publishing House Pvt. Ltd.

Markande, Anoop R., Aram Mikaelyan, Binaya Bhusan Nayak, Ketan D. Patel, Niyati B. Vachharajani, Alagarsamy Vennila, Kooloth Valappil Rajendran, and Chandra S. Purushothaman. 2014. "Analysis of Midgut Bacterial Community Structure of "Neanthes Chilkaensis" from Polluted Mudflats of Gorai, Mumbai, India." *Advances in Microbiology* 04(13): 906–919.

Markande, Anoop R., Divya Patel, and Sunita Varjani. 2021. "A Review on Biosurfactants: Properties, Applications and Current Developments." *Bioresource Technology* 330 (January). Elsevier Ltd: 124963.

Meena, Khem Raj, and Shamsher S. Kanwar. 2015. "Lipopeptides as the Antifungal and Antibacterial Agents: Applications in Food Safety and Therapeutics." *BioMed Research International.* (vol. 2015, p. 9)Hindawi Publishing Corporation.

Melo, Juan J. Manjarres, Alejandro Álvarez, Cristina Ramirez, and German Bolivar. 2021. "Antagonistic Activity of Lactic Acid Bacteria against Phytopathogenic Fungi Isolated from Cherry Tomato (*Solanum Lycopersicum* Var. Cerasiforme)." *Current Microbiology* 78(4). Springer: 1399–1409.

Meneses, Dayana P., Eduardo J. Gudiña, Fabiano Fernandes, Luciana R. B. Gonçalves, Lígia R. Rodrigues, and Sueli Rodrigues. 2017a. "The Yeast-Like Fungus Aureobasidium thailandense LB01 Produces a New Biosurfactant Using Olive Oil Mill Wastewater as an Inducer." *Microbiological Research* 204 (June). Elsevier: 40–47.

Meneses, Dayana P., Eduardo J. Gudiña, Fabiano Fernandes, Luciana R. B. Gonçalves, Lígia R. Rodrigues, and Sueli Rodrigues. 2017b. "The Yeast-Like Fungus Aureobasidium thailandense LB01 Produces a New Biosurfactant Using Olive Oil Mill Wastewater as an Inducer." *Microbiological Research* 204 (June). Elsevier: 40–47.

Muthusamy, Krishnaswamy, Subbuchettiar Gopalakrishnan, Thiengungal Kochupappy Ravi, and Panchaksharam Sivachidambaram. 2009. "Biosurfactants: Properties, Commercial Production and Application." *Current Science* 94(6). Temporary Publisher: 736–747.

Naughton, P. J., R. Marchant, V. Naughton, and I. M. Banat. 2019. "Microbial Biosurfactants: Current Trends and Applications in Agricultural and Biomedical Industries." *Journal of Applied Microbiology* 127(1). England: 12–29.

Nogueira Felix, Anne Kamilly, Jeferson J. L. Martins, José Gustavo Lima Almeida, Maria Estela A. Giro, Kênia Franco Cavalcante, Vânia Maria Maciel Melo, Otília Deusdênia Loiola Pessoa, Maria Valderez Ponte Rocha, Luciana Rocha Barros Gonçalves, and Rilvia Saraiva de Santiago Aguiar. 2019. "Purification and Characterization of a Biosurfactant Produced by *Bacillus subtilis* in Cashew Apple Juice and Its Application in the Remediation of Oil-Contaminated Soil." *Colloids and Surfaces, Part B: Biointerfaces* 175 (July 2018). Elsevier: 256–263.

Nurfarahin, Abdul Hamid, Mohd Shamzi Mohamed, and Lai Yee Phang. 2019. "Culture Medium Development for Microbial-Derived Surfactants Production—An Overview." *Molecules* 23(5): 1–26.

Olasanmi, Ibukun O., and Ronald W. Thring. 2019. "The Role of Biosurfactants in the Continued Drive for Environmental Sustainability." *Sustainability (Switzerland)* 10(12): 1–12.

Pacwa-Płociniczak, Magdalena, Grazyna A. Płaza, Zofia Piotrowska-Seget, and Swaranjit Singh Cameotra. 2011. "Environmental Applications of Biosurfactants: Recent Advances." *International Journal of Molecular Sciences* 12(1): 633–654.

Patel, Seema, Ahmad Homaei, Sangram Patil, and Achlesh Daverey. 2019. "Microbial Biosurfactants for Oil Spill Remediation: Pitfalls and Potentials." *Applied Microbiology and Biotechnology* 103(1). Germany: 27–37.

Peele, Abraham, T. C. Vekateswarulu, Krupanidhi Srirama, Rohini Kota, Indira Mikkili, and Kodali Vidya Prabhakar. 2019. "Evaluation of Anti-Cancer, Anti-Microbial and Anti-Biofilm Potential of Biosurfactant Extracted from an Acinetobacter M6 Strain." *Journal of King Saud University - Science* 32: 223–227.

Pérez-Armendáriz, Beatriz, Carlos Cal-y-Mayor-Luna, Elie Girgis El-Kassis, and Luis Daniel Ortega-Martínez. 2019. "Use of Waste Canola Oil as a Low-Cost Substrate for Rhamnolipid Production Using *Pseudomonas aeruginosa*." *AMB Express* 9(1): 59–69.

Pornsunthorntawee, Orathai, Sasiwan Maksung, Onsiri Huayyai, Ratana Rujiravanit, and Sumaeth Chavadej. 2009. "Biosurfactant Production by Pseudomonas aeruginosa SP4 Using Sequencing Batch Reactors: Effects of Oil Loading Rate and Cycle Time." *Bioresource Technology* 100(2). Elsevier Ltd: 812–819.

Rahman, P. K. S. M., and Edward Gakpe. 2009. "Production, Characterisation and Applications of Biosurfactants-Review." *Biotechnology* 7(2). Asian Network for Scientific Information: 360–370.

Rane, Ashwini N., Vishakha V. Baikar, D. V. Ravi Kumar, and Rajendra L. Deopurkar. 2017. "Agro-Industrial Wastes for Production of Biosurfactant by Bacillus subtilis ANR 88 and Its Application in Synthesis of Silver and Gold Nanoparticles." *Frontiers in Microbiology* 8: 1–12.

Reike, Denise, Walter J. V. Vermeulen, and Sjors Witjes. 2019. "The Circular Economy: New or Refurbished as Ce 3.0? — Exploring Controversies in the Conceptualization of the Circular Economy through a Focus on History and Resource Value Retention Options." *Resources, Conservation and Recycling* 135 (February 2017). Elsevier: 246–264.

Reis, Carla Luzia. Borges, Emerson Yvay Almeida de Sousa, Juliana de FranÃ\Sa Serpa, Ravena Casemiro Oliveira, and JosÃ\copyright Cleiton Sousa Dos Santos. 2019. "Design of Immobilized Enzyme Biocatalysts: Drawbacks and Opportunities." *QuÃ\-Mica Nova* 42. scielo: 768–783.

Rodrigues, Lígia, Ibrahim M. Banat, José Teixeira, and Rosário Oliveira. 2006. "Biosurfactants: Potential Applications in Medicine." *The Journal of Antimicrobial Chemotherapy* 57(4). England: 609–619.

Roy, Arpita. 2019. "A Review on the Biosurfactants: Properties, Types and Its Applications." *Journal of Fundamentals of Renewable Energy and Applications* 08(01): 1–5.

Saharan, Baljeet, Ranjit Sahu, and Deepansh Sharma. 2011. "A Review on Biosurfactants: Fermentation, Current Developments and Perspectives." *Genetic Engineer and Biotechnologist* 29: 1–39.

Sałek, Karina, and Stephen R. Euston. 2019. "Sustainable Microbial Biosurfactants and Bioemulsifiers for Commercial Exploitation." *Process Biochemistry* 85(June): 143–155.

Sana, Santanu, Sriparna Datta, Dipa Biswas, Biswajit Auddy, Mradu Gupta, and Helen Chattopadhyay. 2019. "Excision Wound Healing Activity of a Common Biosurfactant Produced by Pseudomonas sp. " *Wound Medicine* 23 (January). Elsevier: 47–52.

Sandeep, Lahiry. 2017. "Biosurfactant: Pharmaceutical Perspective." *Journal of Analytical & Pharmaceutical Research* 4(3): 19–21.

Santos, Danyelle K. F., Raquel D. Rufino, Juliana M. Luna, Valdemir A. Santos, Alexandra A. Salgueiro, and Leonie A. Sarubbo. 2013. "Synthesis and Evaluation of Biosurfactant Produced by Candida lipolytica Using Animal Fat and Corn Steep Liquor." *Journal of Petroleum Science and Engineering* 105: 43–50.

Santos, Danyelle Khadydja F., Raquel D. Rufino, Juliana M. Luna, Valdemir A. Santos, and Leonie A. Sarubbo. 2016. "Biosurfactants: Multifunctional Biomolecules of the 21st Century." *International Journal of Molecular Sciences* 17(3): 1–31.

Santos, Elane Cristina Lourenço dos, Daniele Alves dos Reis Miranda, Amanda Lys dos Santos Silva, and Ana Maria Queijeiro López. 2019. "Biosurfactant Production by Bacillus Strains Isolated from Sugar Cane Mill Wastewaters." *Brazilian Archives of Biology and Technology* 62: 1–12.

Satpute, Surekha K., Ibrahim M. Banat, Prashant K. Dhakephalkar, Arun G. Banpurkar, and Balu A. Chopade. 2010. "Biosurfactants, Bioemulsifiers and Exopolysaccharides from Marine Microorganisms." *Biotechnology Advances* 28(4). Elsevier Inc.: 436–450.

Satpute, Surekha K., Gauri R. Kulkarni, Arun G. Banpurkar, Ibrahim M. Banat, Nishigandha S. Mone, Rajendra H. Patil, and Swaranjit Singh Cameotra. 2016. "Biosurfactant/S from Lactobacilli Species: Properties, Challenges and Potential Biomedical Applications." *Journal of Basic Microbiology* 56(11): 1140–1159.

Seghal Kiran, G., T. Anto Thomas, Joseph Selvin, B. Sabarathnam, and A. P. Lipton. 2010. "Optimization and Characterization of a New Lipopeptide Biosurfactant Produced by Marine Brevibacterium aureum MSA13 in Solid State Culture." *Bioresource Technology* 101(7). Elsevier Ltd: 2389–2396.

Shaligram, Shraddha, Shreyas V. Kumbhare, Dhiraj P. Dhotre, Manohar G. Muddeshwar, Atya Kapley, Neetha Joseph, Hemant P. Purohit, Yogesh S. Shouche, and Shrikant P. Pawar. 2016. "Genomic and Functional Features of the Biosurfactant Producing Bacillus sp. AM13." *Functional and Integrative Genomics* 16(5). Germany: 557–566.

Shekhar, Sudhanshu, Arumugam Sundaramanickam, and Tangavel Balasubramanian. 2015. "Biosurfactant Producing Microbes and Their Potential Applications: A Review." *Critical Reviews in Environmental Science and Technology* 45(14): 1522–1554.

Shindhal, Toral, Parita Rakholiya, Sunita Varjani, Ashok Pandey, Huu Hao Ngo, Wenshan Guo, How Yong Ng, and Mohammad J. Taherzadeh. 2021. "A Critical Review on Advances in the Practices and Perspectives for the Treatment of Dye Industry Wastewater." *Bioengineered* 12(1). Taylor & Francis: 70–87.

Stancu, Mihaela Marilena. 2015. "Response of Rhodococcus erythropolis Strain IBBPo1 to Toxic Organic Solvents." *Brazilian Journal of Microbiology* 46(4): 1009–1019.

Toren, Amir, Shiri Navon-Venezia, Eliora Z. Ron, and Eugene Rosenberg. 2001. "Emulsifying Activities of Purified Alasan ProteinS from *Acinetobacter radioresistens* KA53." *Applied and Environmental Microbiology* 67(3): 1102–1106.

Varjani, Sunita J. 2017. "Microbial Degradation of Petroleum Hydrocarbons." *Bioresource Technology* 223. Elsevier Ltd: 277–286.

Varjani, Sunita J., and Vivek N. Upasani. 2016. "Carbon Spectrum Utilization by an Indigenous Strain of Pseudomonas aeruginosa NCIM 5514: Production, Characterization and Surface Active Properties of Biosurfactant." *Bioresource Technology* 221. Elsevier Ltd: 510–516.

Varjani, Sunita J., and Vivek N. Upasani. 2017. "Critical Review on Biosurfactant Analysis, Purification and Characterization Using Rhamnolipid as a Model Biosurfactant." *Bioresource Technology* 232. Elsevier Ltd: 389–397.

Varjani, Sunita, and Vivek N. Upasani. 2019. "Influence of Abiotic Factors, Natural Attenuation, Bioaugmentation and Nutrient Supplementation on Bioremediation of Petroleum Crude Contaminated Agricultural Soil." *Journal of Environmental Management* 245 (March). Elsevier: 358–366.

Verma, Rahul, Swati Sharma, Lal Mohan Kundu, and Lalit M. Pandey. 2020. "Experimental Investigation of Molasses as a Sole Nutrient for the Production of an Alternative Metabolite Biosurfactant." *Journal of Water Process Engineering* 38 (September). Elsevier Ltd: 101632.

Vishal, Pande, Patel Vipul, Salunke Priyanka, and Patel Unnati. 2020. "Biosynthesis and Development of Novel Method for Commercial Production of Biosurfactant Utilizing Waste Potato Peels." *Indian Drugs* 57(1): 59–65.

Wasoh, Helmi, Sarinah Baharun, Murni Halim, Ahmad Firdaus B. Lajis, Arbakariya B. Ariff, and Oi-ming Lai. 2017. "Production of Rhamnolipids by Locally Isolated *Pseudomonas aeruginosa* Using Sunflower Oil as Carbon Source." *Bioremediation Science and Technology Research* 5(1): 1–6.

Whang, Liang Ming, Pao Wen G. Liu, Chih Chung Ma, and Sheng Shung Cheng. 2009. "Application of Biosurfactants, Rhamnolipid, and Surfactin, for Enhanced Biodegradation of Diesel-Contaminated Water and Soil." *Journal of Hazardous Materials* 151(1): 155–163.

Yuan, C. L., Z. Z. Xu, M. X. Fan, H. Y. Liu, Y. H. Xie, and T. Zhu. 2014. "Study on Characteristics and Harm of Surfactants." *Journal of Chemical and Pharmaceutical Research* 6(7): 2233–2237.

Zhao, Feng, Rongjiu Shi, Fang Ma, Siqin Han, and Ying Zhang. 2019. "Oxygen Effects on Rhamnolipids Production by *Pseudomonas aeruginosa.*" *Microbial Cell Factories* 17(1). BioMed Central: 1–11.

10 Carbohydrate-Based Agro-Industrial Waste
Production and Their Value Addition

Sujitta Raungrusmee

CONTENTS

10.1 CARBOHYDRATE-BASED AGRO FOOD WASTE: INTRODUCTION

Almost one-third of the food produced for consumption is lost or wasted globally. This loss and waste of food products must be reduced to combat hunger, poverty, nutritional security, and enhance food supply chain actors and zero waste challenges outlined by the sustainable development goals (SDGs) of the United Nations. Recycle, reuse, and recover, also known as 3R, has been a recently exploited principle to reduce food loss and waste and convert it to value-added products (Gunjal, 2019). Furthermore, this approach will reduce the negative environmental impacts of food loss and waste as traditional methods for disposal include incineration, dumping in landfills, and composting (Ng et al., 2020). When food waste is burned, it releases methane gas, contributing to greenhouse gas emissions and global warming. Offensive smells pollute the environment near dumping sites where food waste is deposited, making the area inhabitable, hazardous, and further losing resources (Karmee and Lin, 2014).

Carbohydrates are a major energy source for living beings on this planet. Plants use it during photosynthesis, and animals, including human beings, need carbohydrates to produce and store energy in glycogen. It is also a major structural compound in DNA and RNA (Mergenthaler et al., 2013). Chemically, carbohydrates are made up of carbon (C), hydrogen (H), and oxygen (O), which have the empirical formula $C_n(H_2O)_n$.. They are further classified into simple and complex carbohydrates based on the degree of complexity (Figure 10.1). Carbohydrate-based food waste is one of many types of agro-industrial waste. Recently, many studies have utilized carbohydrate-based food waste in creating innovative byproducts using various methods. This is mainly due to the fact that energy-rich carbohydrates can be used to produce byproducts rich in nutrients and organic matter with the possibility of its application in biofuel generation, bioplastics, bioactive oligosaccharides, polysaccharides, and carbohydrates, among others (Figure 10.2) (Torres-León et al., 2020).

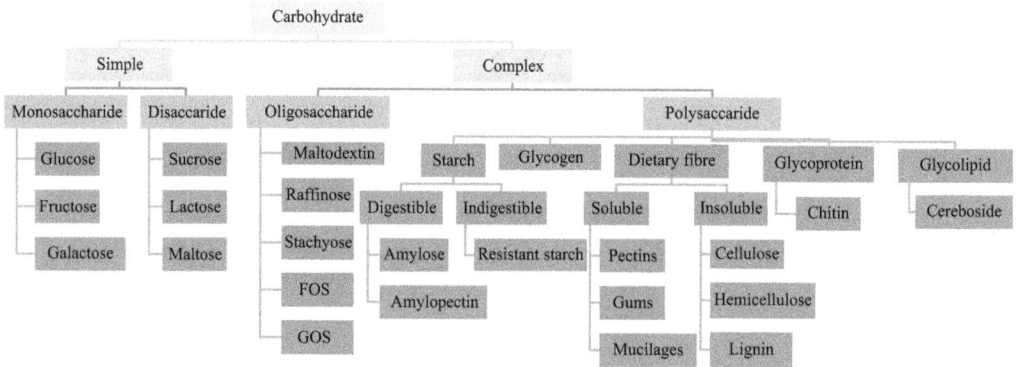

FIGURE 10.1   Classification of carbohydrates.

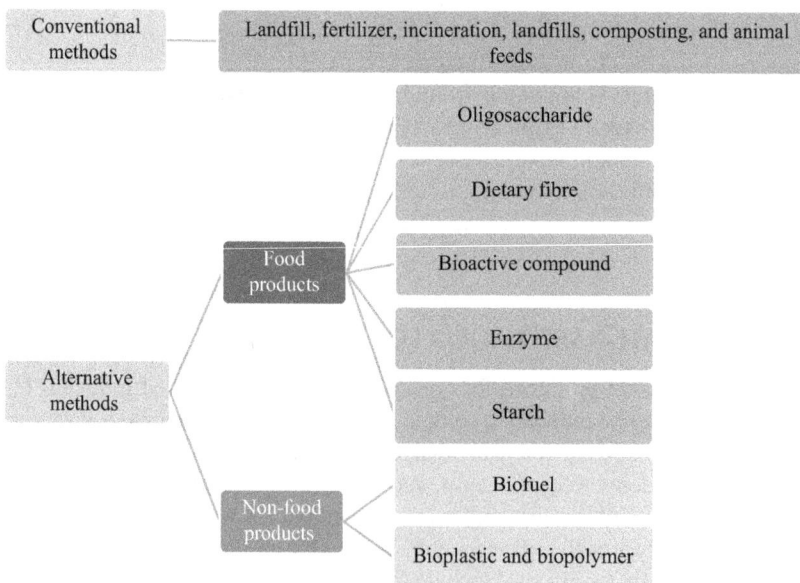

FIGURE 10.2   General diagram of the valorization of carbohydrate-based waste.

There are many examples of innovative food ingredients from carbohydrate-rich food waste, which include β-glucan from oat mill waste; polysaccharides from pomegranate peel; pectin oligosaccharides and D-limonene from citrus peel; polyphenols, glucose, fructose, and sucrose from artichoke wastewater; and others (Machado et al., 2015). Tamarin seeds, which are generally wasted, are rich in xyloglucan starch (Mohan et al., 2018).

10.2 CARBOHYDRATES FROM AGRO-INDUSTRIAL WASTE AND THEIR APPLICATIONS IN THE FOOD INDUSTRY

As discussed earlier, waste is no longer waste, and it is now seen as a resource that can be utilized to produce value-added products (Prajapati et al., 2021; Mala et al., 2021). These byproducts (peels, seeds, shells, pomace, and leaves) mainly contain bioactive compounds such as phenols, carotenoids, fatty acids, fibres, and peptides that are of huge interest to the food industry for the development of functional foods (Osorio et al., 2021).

10.2.1 STARCH

Starch, a carbohydrate material commonly found in seeds, roots, tubers, and leaves, is extracted from agricultural and industrial food byproducts. The compound can be applied in several industrial applications such as food processing, confection and beverages, paper, chemicals, binders and adhesives, fermentation, and textile industries.

Starch is commonly isolated using water, α-amylase, or alkaline compounds. Several studies examined the extraction and characterization of the extracted starch compounds in different agro-industrial products. Rinju and Harikumaran-Thampi (2021) extracted starch from pineapple plant stems, resulting in 11.08 % ± 0.77 on a wet basis, while Nakthong et al. (2017) have characterized the product which displays distinct characteristics to the ones extracted on rice, corn, and cassava starches. Agro-industrial waste products such as tamarind seeds (Mohan et al., 2018), mango kernel flour (Ferraz et al., 2019; Shahrim et al., 2018), banana peel (Lannie et al., 2017; Li et al., 2018), avocado seeds (Jiménez et al., 2022), and jackfruit seeds (Noor et al., 2014; Jiamjariyatam, 2018) were also studied and used for starch production.

10.2.2 FRUCTOOLIGOSACCHARIDES

The health benefits of fructooligosaccharides (FOS) are well documented with their role in enhancing the immune system and anti-inflammatory responses to infections. FOS is also a prebiotic substrate for probiotics in the gut to produce short-chain fatty acids. Due to their complex structure, FOS resist human digestive systems; however, once they reach the colon, probiotic microbes ferment them. Essentially, it becomes food for the survival of bacteria in the gut and maintains homeostasis with health benefits (Singh and Tingirikari, 2021). Since the calorific value of FOS is negligible, it is recommended for people suffering from lifestyle-related diseases such as diabetes (de la Rosa et al., 2019). The traditional approach to producing FOS is via sucrose as a substrate using enzymes such as β-fructofuranosidase (EC 3.2.1.26) and fructosyltransferase (EC 2.4.1.9). These enzymes convert sugar into FOS under controlled conditions. *Aspergillus niger* and *Lactobacillus bulgaricus* are the most commonly used micro-organisms for producing enzymes necessary for the biological production of FOS (Ibrahim, 2018). In this framework, agro-industrial wastes can also be seen as a potential source of FOS (de la Rosa et al., 2019) (Table 10.1).

10.2.3 DIETARY FIBRE

Certain polymers provide structural rigidity to the plant cell wall, such as cellulose, hemicellulose, lignin, and pectin; they are grouped as dietary fibres. There are two types of dietary fibre depending

TABLE 10.1

Alternative Substrate Sources and Micro-Organisms Used to Produce Fructooligosaccharides and Fructooligosaccharide-Producing Enzymes. Table adapted from de la Rosa et al. (2019)

Source	Micro-Organism	Enzyme
Molasses	*Aspergillus niger* PSSF21	β-fructofuranosidase
	Saccharomyces cerevisiae	Fructofuranosidase
	Aspergillus japonicus-FCL 119T and *Aspergillus niger* ATCC 20611	
Sugar cane bagasse	*Aspergillus flavus* NFCCI 2364	Fructosyltransferase
	Aspergillus ochraceus	β-D-fructofuranosidase
Banana peel-leaf	*A. flavus* NFCCI 2364	Fructosyltransferase
	Chrysonilia sitophila	β-fructofuranosidase
	Saccharomyces cerevisiae GVT263	β-D-fructofuranosidase
Agave mead	*Aspergillus oryzae* DIA-MF	Fructosyltransferase
Coffee byproducts	*Aspergillus japonicus*	β-fructofuranosidase
	Aspergillus japonicus	β-fructofuranosidase
Cassava wastes	*Rhizopus stolonifer* LAU 07	Fructosyltransferase
Apple pomace	*Aspergillus versicolor*	β-fructofuranosidase
Wheat bran	*Fusarium graminearum*	β-D-fructofuranosidase
	Aspergillus awamori GHRTS	
		Fructosyltransferase
Spent osmotic solutions	*Aspergillus oryzae* N74	Fructosyltransferase
Date byproducts	*Aspergillus awamori* NBRC 4033	β-D-fructofuranosidase
Cashew apple	*Lactobacillus acidophilus*	n.d.

on its water solubility and insoluble dietary fibre. Soluble dietary fibres are mainly present in fruits and vegetables, whereas cellulose and hemicellulose are present in cereals. Dietary fibres also have a range of health benefits as they have been shown to protect against cardiovascular diseases, diverticulosis, constipation, irritable bowel, colon cancer, and diabetes (Sharoba et al., 2013).

10.2.3.1   Cellulose and Hemicellulose

Almost one-third of plant material is composed of cellulose and is thus one of the most common organic compounds. Cellulose is often associated with hemicellulose and lignins in plants. Commercially available cellulose is derived from wood and other sources such as cotton, flax, hemp, jute, etc. However, food-derived cellulosic sources include soybean hulls, pea hulls, corn bran, dried beet pulp, and oat hull (Sundarraj and Ranganathan, 2018). Hemicellulose is another dietary fibre composed of anhydrides of hexose and pentose sugars that can be hydrolyzed using alkali treatments (Walia and Gupta, 2013). In agro-industrial waste, hemicellulose is present and can be used in food applications for encapsulation and emulsification (Tatar et al., 2014).

10.2.3.2   Pectin

Pectin is one of the major dietary carbohydrates present in fruits such as apples, grapes, and plums. It is found in the middle lamella of the plant cell wall and makes up about one-third of dry plant matter (Salma et al., 2012). Pectin is structurally composed of 1,4-linked α-D-galactosyluronic residues and is usually associated with cellulose, lignins, or other polyphenols (Mellinas et al., 2020). Pectin is used in food applications owing to its various properties like gelling, fruit juice preparation, and confectionery and possesses a range of health benefits such as yogurt, acid dairy drinks, bakery fillings, antioxidant, antimicrobial, and anticancer properties, respectively. Pectin extraction involves either precipitation with alcohol or with salt (Ruano et al., 2019). Pectin is commercially

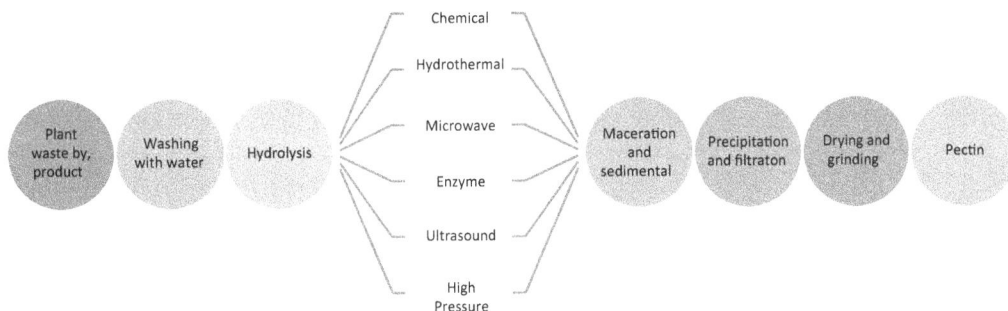

FIGURE 10.3 Production and purification of pectin from agro-waste.

produced from raw plant material, agricultural byproducts or agro-waste and purified. Production and purification of pectin from agro-waste are shown in Figure 10.3.

10.3 EXTRACTION OF MAJOR CARBOHYDRATES FROM AGRO-INDUSTRIAL WASTE AND APPLICATIONS IN THE FOOD INDUSTRY

Conventional methods for extracting carbohydrate-based, value-added compounds from agro-industrial waste include chemical extraction using alkali, solvent extraction, and drying treatments. Chemical treatment for extraction is slowly becoming obsolete, and it has been shown to damage the molecular structure of the dietary fiber and hence reduce the yield while at the same time posing environmental threats. Dry and wet milling are also quite popular as extraction processes and yields about 50–90% pure dietary fibre. Recently, enzyme assisted extraction techniques have also been explored. Other advanced extraction processes include microwave, ultrasonic, and pulsed electric field techniques (Hussain et al., 2020).

10.3.1 CHEMICAL EXTRACTION

FOS was produced from artichoke industrial waste using ultrasound assisted extraction (UAE) by direct and indirect sonication (Machado et al., 2015). The authors of this study reported the highest yield of FOS at 60°C, 360 W, and 10 min ultrasonication conditions. Further identification of the sugar demonstrated 1-kestose, nystose, fructofuranosylnystose, and raffinose (prebiotic), and fructose, glucose, and sucrose (non-prebiotic) in its structure.

Chemical methods such as alkalization, bleaching, and acid hydrolysis remain the most convenient method for extracting cellulose. Banana peels were valorized for the production of cellulose using the optimum bleaching conditions of 90% ethanol (16 h) for fat removal, followed by alkali treatment (NaOH, pH 11.6, 24 h) (Singanusong et al., 2014). Cellulose was also extracted from wheat straw using sodium chlorite (acidified) and hydrogen peroxide (alkaline) with a cellulosic yield of 81.4% and 79%, respectively (Qasim et al., 2020). Peanut shell, which is a major agro-waste, was utilized for the isolation of cellulose. Chemical treatment and extraction involved NaOH, followed by bleaching with sodium hypochlorite. Using this method, 0.39 g/g of cellulose (dry wt%) was recovered successfully (Ganguly et al., 2020). Other agro-waste such as rice husk was utilized for the extraction of cellulose with montmorillonite K-10/LiOH solution and bleaching with 2% H_2O_2. Optimum conditions of extraction were determined as 6 h, 80°C, 20% maleic acid, and 10% LiOH (in H_2O) for time, temperature, acid, and alkali, respectively, to produce 68% cellulose (Das et al., 2016). Fang et al. (2000) isolated and characterized cellulose and hemicellulose from rye straw using H_2O_2 (2%) at pH 11.5 for 12 h at 20, 30, 40, 50, 60, and 70°C, which released 44.2–71.9% of the original hemicelluloses. Sugarcane bagasse was acid hydrolyzed (H_2SO_4, 60%, 45°C) for the extraction of nanocellulose at varying times (20, 30, and 40 min).

The authors reported the increase in the hydrolysis time resulted in a lower yield of cellulosic nanofibres (Ghazy et al., 2016).

Hemicellulose was extracted from sugar beet pulp and corn waste with a 40.2% yield using alkaline treatment (NaOH, 10%, 24 h) and a 27.4% yield using acidic treatment conditions (Yılmaz Celebioglu et al., 2012). Similarly, alkaline treatments (NaOH, 4%) were applied to rice straw and *Leucaena leucocephala* with a solid:solvent ratio of (1:10 to 1:50) with yields of around 22.77% and 37.88%, respectively (Nadiha et al., 2020). Other agro-industrial wastes such as sweet sorghum bagasse and rice straw were considered to extract hemicellulose and the authors optimized the conditions of extraction using NaOH (86.1°C, 12.33% w/w, and 3.91 h) (Wei et al., 2018).

Pectin extraction by chemical means can be employed by treating a material using an acid or aqueous base at 70–90°C at an optimum time. This is done to remove the desired quality and quantity of pectin from the material. The process results in the production of low methoxyl pectin with a low degree of esterification. The most common process in the extraction of pectin is the usage of strong mineral acid that produces a high degree of esterification (50%), which contributes to a higher yield of pectin. However, strong acids used in this method are corrosive and may be dangerous to human health. In addition, the liquid generated from the process can also be a burden on the process cost and the environment (Tiwari et al., 2017). Due to this, several less acidic and more environmentally friendly compounds were considered in this process. These chemicals include sulphuric acid (Ruano et al., 2019; Abid et al., 2009), sulfurous acid, hydrochloric acid (Hwang et al., 1998; Gómez et al., 2016) or nitric acid, while citric and acetic acid has attracted considerable interest because of their environmental impact. In general, theses conventional extraction technologies require huge amounts of solvents with prolonged treatment time.

10.3.2 Hydrothermal Extraction

For the extraction of cellulose from mango waste, hydrothermal conditions were applied using low temperature, acid-free extraction that yielded around 11.63% (dry weight basis) cellulose (Matharu et al., 2016). Martin-Lara et al. (2020) investigated the efficiency of hydrothermal pre-treatment processes for the extraction of hemicellulose from Italian green pepper waste. The optimized pretreatment conditions were obtained at 180°C for 40 min and yielded 61.02% hemicellulose. The high temperature used during hydrothermal treatment was similar to the concept employed by Gallina et al. (2018).

Hydrothermal extraction of pectin is conducted in acidic conditions with water as the solvent. The compound is highly soluble in water, and a low pH environment reduces other compounds such as polyphenols. This also increases extraction yields, which helps to maintain the quality of the pectin extracted. Fishman et al. (2003) and Liu et al. (2006) have studied the optimum parameters in the extraction of pectin from albedo orange peel which showed pectin yields of 16.9% and 3.6%, respectively.

10.3.3 Microwave Assisted Extraction

Microwaves produce non-ionizing radiation with frequencies from 300 MHz up to 300 GHz (Roselló-Soto et al., 2016). In this process, the agro-wastes for pectin extraction are subjected to microwave heating. With this, the microwaves generate heat inside the plant waste material, disrupting the cell wall, and causing the release of compounds. As this microwave heat is being generated, it causes ionic conduction and dipole rotation (Wang et al., 2016). The former is based on the electrophoretic transfer of ions and electrons initiated by microwave energy, which generates an electric field responsible for particle movements, while the latter refers to the dipole rotation alternate displacement of polar molecules.

Carbohydrates were extracted from corn starch using microwave assisted extraction (MAE). The optimized conditions for extraction included a heating temperature of 176.5°C, with a time of

2 min and a solid to liquid ratio of 1/20 g/ml, which yielded 70.8% carbohydrates that were mainly composed of glucose, xylose, and arabinose (Yoshida et al., 2010).

Camani et al. (2020) applied MAE pre-treatment to extract cellulosic fibres from eucalyptus waste. The experimental design included two sets of pre-treatment conditions, one with mercerization using NaOH and the other with bleaching using H_2O_2. The optimum conditions for MAE were investigated for time (15 and 30 min) and power (100, 400, and 800 W). The authors reported the best conditions for mercerization at 30 min with 400 W and 800 W, whereas for bleaching, optimum conditions were obtained at 400 W that resulted in 99.1% cellulose purity.

Several studies have shown that when used at optimal conditions, MAE has been effective in the extraction of pectin in agro-industrial waste such as mango peel (Maran et al., 2015), *Citrullus lanatus* fruit rinds (Maran et al., 2014), and *Carica* papaya peels (Maran and Prakash, 2015). In addition, other studies on dragon fruit (Tongkham et al., 2017) and tomato process waste (Sengar et al., 2020) have shown the supremacy of MAE in comparison with the conventional method and other different extraction techniques.

10.3.4 ENZYMATIC EXTRACTION

The enzymes produced by the micro-organism *Bacillus amyloliquefaciens* were utilized to produce amylase enzymes from wheat bran, rice bran, and potato peel. Wheat bran and potato bran demonstrated the highest specific enzyme activity of 1.2 and 1.1 U/mg, respectively (Mojumdar and Deka, 2019). Banana peels were used as a substrate or nutrient source for the production of cellulase using the enzyme produced by *Trichoderma viride* GIM 3.0010 through solid-state fermentation (SSF) (Ravindran et al., 2018; Hai-Yan Sun et al., 2011). Enzyme-catalyzed production of xylitol was carried out using xylose reductase from *Candida guilliermondii* using agro-waste such as wheat straw. The substrate was pre-treated at 40 and 50°C for up to 9 h at pH 4.5 with varying xylanase concentrations (12.4 to 37.2 U). The optimal conditions determined for the production of xylitol were 30°C at pH 7 for 8 h with 7.92 U of CXR (Crude preparation of Xylose Reductase) and 10 mM NADPH with 23.9 g xylose (Walsh et al., 2018).

The enzymatic extraction of pectin has been a popular process as this is considered safe and environmentally friendly despite the process's drawback of being time-consuming and producing a lesser yield than acidic/hydrothermal methods. When pectin is extracted using enzymatic treatment, it can be combined with a physical method, such as in the study by Lim et al. (2012) to extract yuzu pomace (*Citrus junos*), where agitation and filtration were done before the enzymatic treatment in β-glucanase. Kazemi et al. (2019) have also studied pectin extraction in the processing wastes of eggplant (which includes peels and calyxes) using enzymes and MAE at optimal conditions. The results show eggplant peel pectin exhibited a higher extraction yield (29.17%) than eggplant calyx pectin, with a yield of 18.36%. At the same time, the eggplant calyx was high in rhamnogalacturonan-I (44.9%). Aside from the mechanical processes, the extraction of pectin from apple and pear pomace using recombinant polygalacturonase enzyme isolated from *Aspergillus kawachii* was also studied (Franchi et al., 2014) and was found to be more efficient at pectin extraction by a 20% higher yield than that with chemical extraction, a 50% higher yield than that with a commercial highly concentrated polygalacturonase enzyme, and an 80% higher yield than that of two enzymatic pools with high activity polygalacturonase. The treatment method resulted in a pectin yield of 26% for apple pomace and 68% for pear pomace. It has also been shown in a study by Locatelli et al. (2019) that the production of reducing compounds from the hydrolysis of pectin was more efficient using an enzymatic method, with a yield of 33% higher than that achieved using acid hydrolysis.

10.3.5 ULTRASOUND ASSISTED EXTRACTION

Similar to other compounds, ultrasound (16 Hz–16 kHz) assisted extraction (UAE) is a method that receives considerable attention in the extraction of pectin. This method produces a higher extraction yield than conventional extraction despite having a shorter extraction time and less solvent and energy. Agro-waste such as cassava (*Manihot esculenta* Crantz) was valorized for the production of starch through various optimization methods. Response surface methodology produced optimized processing conditions of 90% power, a pulse duty cycle of 1.0 S^{-1}, and solvent to sample ratio of 30:1 with an ultrasonication time of 10 min. The resulting starch from cassava had an 88.36% purity (Setyaningsih et al., 2021). Recently, sago starch was produced from sago pith waste using UAE. Processing conditions such as slurry concentration, sonication time, and temperature were optimized to 14.43%, 10 min at 38.3°C and produced 78.68% starch (Noor et al., 2022).

Cellulose and lignins were extracted from Vietnamese rice straw using a UAE alkaline treatment method that involved ultrasonication for 30 min. This pre-treatment enhanced the yield of lignin from 72.8% to 84.7% (Dinh Vu et al., 2017). Similarly, this method was further explored by Yi et al. (2019) to extract pure cellulose from sugarcane bagasse. Process parameters such as treatment temperature, treatment duration, and concentration of the alkali solution were optimized to recover pure cellulose for the production of carboxymethylcellulose.

Pectin extraction using UAE was studied by Grassino et al. (2016) in tomatoes, comparing conventional methods and UAE. It showed that the latter had the highest yield of high-quality pectin at 37 kHz in 15 min. Pectin was also successfully extracted at optimum conditions from sisal wastes (Maran and Priya, 2015) and sweet lime peels (Siddiqui et al., 2021), with yields of 29% and 36.4%, respectively.

10.3.6 High-Pressure Processing

High-pressure processing (HPP) (100–700 MPa) is a novel technique that extracts active materials from plant materials. This has been accepted as an environmentally friendly technology by the US Food and Drug Administration and is widely utilized in food industries. The high pressure modifies the cell walls in plant tissues, facilitating access to solvents, and allowing better extraction (de Oliveira et al., 2016).

Cellulosic nanocrystalline cellulose (NCC) was produced from raw rice husk biomass. The authors of this study initially used chemical extraction with delignification (80°C, HPP 5 bar, low-pressure processing (LPP) atmospheric pressure, 12 h, NaOH-4M), bleaching (NaOCl (20%) and H$_2$SO$_4$ (4M)), and hydrolysis. With the help of HPP, high-quality NCC was produced with a crystallinity of 55% and a production rate of 62%, whereas with LPP was 51% production rate only. The authors concluded that HPP is a more robust, faster, efficient, and innovative technology for the extraction of NCC and other active compounds from agro-waste (Islam et al., 2017a).

As HPP is an emerging non-thermal technology for the processing of compounds for extraction and preservation purposes, Castañón-Rodríguez et al. (2015) studied the influence of HPP and alkaline treatment on sugarcane bagasse hydrolysis for the removal of hemicellulosic compounds. The authors reported that HPP in combination with alkali treatment on pre-hydrolyzed bagasse had a significant effect on the removal of major hemicellulose.

Guo et al. (2012) studied the effect of pressure, temperature, and pressure-holding time on the extraction yield and viscosity of pectin extraction from navel orange peel. The yield of pectin (20.44% ± 0.64) was significantly higher than those extracted by traditional heating (15.47% ± 0.26) and microwave (18.13% ± 0.23). Xie et al. (2018) investigated the structural characteristics, physicochemical properties, and morphology of potato peel waste pectin treated with high hydrostatic pressure and high-pressure homogenization at 200 MPa for 5 min. The result showed that high-pressure treatments increased galacturonic acid content, the degree of esterification galactose and arabinose/rhamnose ratio, molecular weight decrease, and chemical group change, thus changing the physicochemical properties. High-pressure treatments increased the viscosity and improved the emulsifying properties.

10.4 PRODUCTION OF ENZYMES FROM AGRO-INDUSTRIAL WASTE

One of the breakthroughs in the valorization of industrial wastes is the biotransformation of wastes onto useful products such as enzymes. Most agro-industrial wastes have a variable composition of polysaccharides that support the growth of micro-organisms. Different kinds of enzymes are produced through the process of fermentation, such as SSF and submerged fermentation (SmF) (Khanal et al., 2018). The former is performed on a non-soluble material that acts both as physical support and a source of nutrients without free-flowing liquid (Bhardwaj et al., 2019). The latter is a traditional method for enzyme production from micro-organisms that has been used for a longer time. SmF involves inoculation of the microbial culture into the liquid medium to produce the desired product. Agro-industrial residues have been widely used due to their potential advantages for filamentous fungi, which can penetrate these solid substrates, aided by the presence of turgor pressure at the tip of the mycelium. This kind of production reduces costs with the development of suitable fermentation processes using agro-industrial or food waste that are easily digestible components (Kiran et al., 2014). Enzymes from agro-industrial wastes are becoming commercially and industrially recognized. Table 10.2 presents the different enzymes produced from agro-industrial waste.

10.5 BIOACTIVE CARBOHYDRATE COMPOUNDS FROM AGRO-INDUSTRIAL WASTE

The byproducts generated from the processing of fruits and vegetables are largely under-utilized and discarded as organic waste. These organic wastes that include seeds, peels, bracts, leaves, roots, bark, midribs, pulp, skin, and rinds are potential sources of bioactive compounds that have health benefits. These wastes are a potential source of many bioactive compounds (Pattnaik et al., 2021). Bioactive compounds are substances with biological activity that modulate metabolic processes when taken into the body and result in improved health conditions. The benefits of these compounds include antioxidant activity, inhibition or induction of enzymes, inhibition of receptor activities, and induction and inhibition of gene expression (Shirahigue and Ceccato-Antonini, 2020).

TABLE 10.2
Enzymes Produced from Agro-Industrial Waste

Enzyme	Substrate	Bacteria	Reference
α-amylase	Starchy potato waste Wheat bran Rice husk	*Bacillus subtilis*	Bharathiraja et al., 2017
	Starchy potato peel Wheat bran Rice husk	*Bacillus amyloliquefaciens*	Mojumdar and Deka, 2019
	Potato peel	*Bacillus licheniformis* and *Bacillus subtilis*	Shukla and Kar, 2006
	Not mentioned	*Aspergillus niger*	Wang et al., 2008
Xylanase	Wheat bran	*Aspergillus, Trichoderma, Streptomyces, Phanerochaete, Chytridiomycetes, Fibrobacterota, Clostridium, Bacillus, Pichia*	Knob et al., 2014
Pectinase	Wheat bean	*Bacillus mojavensis A21, Bacillus subtilis, Aspergillus foetidus, Bacillus pumilus*	Jahan et al., 2017
	Pineapple waste	*A. oryzae, A. flavus*, and *R. oryzae.* *A. flavus*	Thangaratham and Manimegalai, 2014

Fruits, vegetables, and whole grains are good sources of bioactive compounds, and their concentration differs with every material (Vilas-Boas et al., 2021). For example, the peels of lemons, grapes, and oranges contain more than 15% higher phenolic concentrations than those found in the fruit pulp (Sagar et al., 2018). This percentage also holds with the seeds of avocados, jackfruit, longans, and mangoes. Most of the bioactive compounds from agricultural food wastes are extracted using an organic solvent such as methanol, ethanol, or acetone (Panzella et al., 2020). The extraction process is completed by undergoing conventional techniques that require temperature and agitation on top of the sample and solvent. However, these methods contribute to low extraction yield and toxicity, and are non-renewable. Therefore, alternative methods are being researched and studied, resulting in modern green or clean techniques due to reduced use of energy and the addition of an organic solvent, which is beneficial to the environment (Soquetta et al., 2018). Alternative extracting methodologies have been reported, such as MAE, UAE, supercritical fluid extraction (SpFE) (Vilas-Boas et al., 2021), and deep eutectic solvent extraction.

10.6 PRODUCTION OF VALUE-ADDED PRODUCTS FROM CARBOHYDRATE AGRO-INDUSTRIAL WASTE

10.6.1 BIOFUEL

The concept of energy from waste is becoming popular worldwide as being capable of producing different types of fuel. Due to its high lignocellulosic material content (e.g., cellulose, hemicellulose, pectin, and non-carbohydrate which are lignin), which are widely used in the production of biofuels, agricultural byproducts are also now being researched as a potential biofuel source (Pattanaik et al., 2021; Zein et al., 2016). Some of this lignocellulosic biomass includes coconut husks, cotton waste, sugarcane bagasse, wheat, rice, and corn straw, forestry wood residues, and waste papers included in the second generation of biofuels (Machineni, 2020).

Agricultural biomass is abundant worldwide, and it can be considered an alternative source of renewable and sustainable materials that can be used as potential materials for different applications. Biomass is all organic matter, including crops, food, plants and plant wastes, forestry residues, industrial wastes, animal residues, and domestic wastes that can be used as a source of energy, generally referred to as plant or plant-based material. On the other hand, lignocellulose is the main structural constituent of plants, mainly composed of cellulose, hemicellulose, and lignin. Cellulose is a homopolymer of anhydroglucose units linked together via β-1,4-glycosidic bonds, in contrast with hemicelluloses, which are complex amorphous polymers that consist of pentose and hexose monomers and exist in the cell wall in association with lignin, a natural amorphous phenol. The hydrolysis of hemicellulose is easier than crystalline cellulose, but lignin has a higher energy density than hemicellulose and cellulose (Ribeiro et al., 2021).

Biomass is reformed in thermo-chemical catalytic and non-catalytic processes such as pyrolysis, gasification, liquefaction, SpFE, and supercritical water liquefaction to produce maximum energetic exploitable liquid and gaseous products (Arshad et al., 2018). Several commodities can be produced with biomass, one of which is biogas, and bioethanol can be used as an alternative to petroleum gas.

10.6.2 BIOPLASTIC AND BIOPOLYMER

Plastics are a highly problematic commodity as it takes more than a lifetime for them to decompose; therefore, alternatives for this compound are being found, and one of these is bioplastics. Bioplastics have less of an impact on the environment than plastics if they are made from agro-industrial wastes. Cellulose can be extracted from halum in the stems of fruits and vegetables, grains, and seeds and are a potential source of biopolymers that can be applied in food or packaging due to its highly structured network (de Moura et al., 2017). Ramesh and Radhakrishnan (2019) compared biodegradable polyvinyl alcohol (PVA) based film incorporated with fennel seed oil and cellulose nanoparticles (CNP), PVA film with chitosan nano-particle (CHNP), and a PVA based film with a combination of CNP and CHNP. Similarly, as compared to regular PVA film, PVA with fennel seed oil and CNP had considerably higher antibacterial activity and free radical scavenging activity. In addition to this, renewable agro-wastes such as bagasse, wood waste, empty fruit bunches, and starch material have been used for bioplastic production.

10.7 CONCLUSION

The waste generated from the food processing industry causes serious problems to our ecology. Though initially perceived as waste, they have the tremendous potential to be valorized for the generation of resources. These compounds have bioactive properties that can be utilized for food applications. However, current extraction and scale up techniques limit their use.

This chapter thus provides detailed information on the utilization of carbohydrate food waste and byproducts to produce valuable food functional ingredients and biofuel production as carbohydrates are organic compounds of carbon, hydrogen, and oxygen and are a major source of energy in the diet. Valorizing the carbohydrates found in agro-waste products reduces its volume, reducing its risk of causing serious environmental and economic problems and potentially transforming these wastes into value-added products. Several innovative methods are described for their potential in the extraction of value-added compounds. Traditional chemical methods such as alkali and acid hydrolysis still exist because of their convenience and result-oriented approach, whereas emerging technologies like MAE, UAE, and HPP still need more research and optimization to get better outcomes. Having said that, the world is on the right track toward the valorization of such agro-waste in order to become sustainable and, in doing so, achieving SDGs such as zero hunger (SDG 2), good health and well-being (SDG 3), and responsible consumption and production (SDG 12).

REFERENCES

Abid, H., Hussain, A., and Ali, J. 2009. Technique of optimum extraction of pectin from sour orange peels and its chemical evaluation. *Journal of the Chemical Society of Pakistan* 31(3):459–461.

Arshad, M., Zia, M. A., Shah, F. A., and Ahmad, M. 2018. An overview of biofuel. In: *Perspectives on Water Usage for Biofuels Production* (pp. 1–37), Springer.

Bharathiraja, S., Suriya, J., Krishnan, M., Manivasagan, P., and Kim, S. K. 2017. Production of enzymes from agricultural wastes and their potential industrial applications. In: *Advances in Food and Nutrition Research* (Vol. 80, pp. 125–148). Academic Press.

Bhardwaj, N., Kumar, B., and Verma, P. 2019. A detailed overview of xylanases: An emerging biomolecule for current and future prospective. *Bioresources and Bioprocessing* 6(1):1–36.

Camani, P. H., Anholon, B. F., Toder, R. R., and Rosa, D. S. 2020. Microwave-assisted pretreatment of eucalyptus waste to obtain cellulose fibers. *Cellulose* 27(7):3591–3609.

Castañón-Rodríguez, J. F., Welti-Chanes, J., Palacios, A. J., Torrestiana-Sanchez, B., Ramírez de León, J. A., Velázquez, G., and Aguilar-Uscanga, M. G. 2015. Influence of high pressure processing and alkaline treatment on sugarcane bagasse hydrolysis. *CyTA-Journal of Food* 13(4):613–620.

Das, A. M., Hazarika, M. P., Goswami, M., Yadav, A., and Khound, P. 2016. Extraction of cellulose from agricultural waste using montmorillonite K-10/LiOH and its conversion to renewable energy: Biofuel by using Myrothecium gramineum. *Carbohydrate Polymers* 141:20–27.

de la Rosa, O., Flores-Gallegos, A. C., Muñíz-Marquez, D., Nobre, C., Contreras-Esquivel, J. C., and Aguilar, C. N. 2019. Fructooligosaccharides production from agro-wastes as alternative low-cost source. *Trends in Food Science and Technology* 91:139–146.

de Moura, I. G., de Sá, A. V., Abreu, A. S. L. M., and Machado, A. V. A. 2017. Bioplastics from agro-wastes for food packaging applications. In: *Food Packaging* (pp. 223–263). Academic Press.

de Oliveira, C. F., Gurak, P. D., Cladera-Olivera, F., Marczak, L. D. F., and Karwe, M. 2016. Combined effect of high-pressure and conventional heating on pectin extraction from passion fruit peel. *Food and Bioprocess Technology* 9(6):1021–1030.

Dinh Vu, N., Thi Tran, H., Bui, N. D., Duc Vu, C., and Viet Nguyen, H. 2017. Lignin and cellulose extraction from Vietnam's rice straw using ultrasound-assisted alkaline treatment method. *International Journal of Polymer Science*, 2017: 1063695.

Fang, J. M., Sun, R. C., and Tomkinson, J. 2000. Isolation and characterization of hemicelluloses and cellulose from rye straw by alkaline peroxide extraction. *Cellulose* 7(1):87–107.

Ferraz, C. A., Fontes, R. L., Fontes-Sant'Ana, G. C., Calado, V., López, E. O., and Rocha-Leão, M. H. 2019. Extraction, modification, and chemical, thermal and morphological characterization of starch from the agro-industrial residue of Mango (*Mangifera indica* L) var. Ubá. *Starch-Stärke* 71(1–2):1800023.

Fishman, M. L., Chau, H. K., Coffin, D. R., and Hotchkiss, A. T. 2003. A comparison of lime and orange pectin which were rapidly extracted from albedo. In Advances in pectin and pectinase research (pp. 107–122). Springer, Dordrecht.

Franchi, M. L., Marzialetti, M. B., Pose, G. N., and Cavalitto, S. F. 2014. Evaluation of enzymatic pectin extraction by a recombinant polygalacturonase (PGI) from apples and pears pomace of Argentinean production and characterization of the extracted pectin. *Journal of Food Processing and Preservation* 5(8):1000352.

Gallina, G., Cabeza, Á., Grénman, H., Biasi, P., García-Serna, J., and Salmi, T. 2018. Hemicellulose extraction by hot pressurized water pretreatment at 160°C for 10 different woods: Yield and molecular weight. *The Journal of Supercritical Fluids* 133:716–725.

Ganguly, P., Sengupta, S., Das, P., and Bhowal, A. 2020. Valorization of food waste: Extraction of cellulose, lignin and their application in energy use and water treatment. *Fuel* 280:118581.

Ghazy, M. B., Esmail, F. A., El-Zawawy, W. K., Al-Maadeed, M. A., and Owda, M. E. 2016. Extraction and characterization of nanocellulose obtained from sugarcane bagasse as agro-waste. *Journal: Journal of Advances in Chemistry* 12(3):4256–4264.

Gómez, B., Yáñez, R., Parajó, J. C., and Alonso, J. L. 2016. Production of pectin-derived oligosaccharides from lemon peels by extraction, enzymatic hydrolysis and membrane filtration. *Journal of Chemical Technology & Biotechnology*, 91(1):234–247.

Grassino, A. N., Brnčić, M., Vikić-Topić, D., Roca, S., Dent, M., and Brnčić, S. R. 2016. Ultrasound assisted extraction and characterization of pectin from tomato waste. *Food Chemistry* 198:93–100.

Gunjal, B. B. 2019. Value-added products from food waste. In: *Global Initiatives for Waste Reduction and Cutting Food Loss* (pp. 20–30). IGI Global.

Guo, X., Han, D., Xi, H., Rao, L., Liao, X., Hu, X., and Wu, J. 2012. Extraction of pectin from navel orange peel assisted by ultra-high pressure, microwave or traditional heating: A comparison. *Carbohydrate Polymers* 88(2):441–448.

Hai-Yan Sun, H., Li, J., Zhao, P., and Peng, M. 2011. Banana peel: A novel substrate for cellulase production under solid-state fermentation. *African Journal of Biotechnology* 10(77):17887–17890.

Hussain, S., Jõudu, I., and Bhat, R. 2020. Dietary fiber from underutilized plant resources—A positive approach for valorization of fruit and vegetable wastes. *Sustainability* 12(13):5401.

Hwang, J. K., Kim, C. J., and Kim, C. T. 1998. Extrusion of apple pomace facilitates pectin extraction. *Journal of Food Science*, 63(5):841–844.

Ibrahim, O. O. 2018. Functional oligo-saccharides: Chemicals structure, manufacturing, health benefits, applications and regulations. *Journal of Food Chemistry and Nanotechnologyl* 4:65–76.

Islam, M. S., Kao, N., Bhattacharya, S. N., and Gupta, R. 2017a. An investigation between high and low pressure processes for nanocrystalline cellulose production from agro-waste biomass. In: *AIP Conference Proceedings* (Vol. 1914, No. 1, p. 070002), AIP Publishing LLC

Islam, S., Bhuiyan, M. R., and Islam, M. N. 2017. Chitin and chitosan: Structure, properties and applications in biomedical engineering. *Journal of Polymers and the Environment* 25(3):854–866.

Jahan, N., Shahid, F., Aman, A., Mujahid, T. Y., and Qader, S. A. U. 2017. Utilization of agro waste pectin for the production of industrially important polygalacturonase. *Heliyon* 3(6):e00330.

Jiamjariyatam, R. 2018. Effect of jackfruit seed starch (*Artocarpus heterophyllus*) microstructure on properties and characteristics of fried battered product. *Walailak Journal of Science and Technology* 15(12):879–892.

Jiménez, R., Sandoval-flores, G., Alvarado-reyna, S., Alemán-castillo, S. E., Santiago-adame, R., and Velázquez, G. 2021. Extraction of starch from Hass avocado seeds for the preparation of biofilms. *Food Science and Technology*, 42: 1–6.

Karmee, S. K., and Lin, C. S. K. 2014. Valorisation of food waste to biofuel: Current trends and technological challenges. *Sustainable Chemical Processes* 2(1):1–4.

Kazemi, M., Khodaiyan, F., and Hosseini, S. S. 2019. Utilization of food processing wastes of eggplant as a high potential pectin source and characterization of extracted pectin. *Food Chemistry* 294:339–346.

Khanal, B. K. S., Sadiq, M. B., Singh, M., and Anal, A. K. 2018. Screening of antibiotic residues in fresh milk of Kathmandu Valley, Nepal. *Journal of Environmental Science and Health. Part B* 53(1):57–86.

Kiran, E. U., Trzcinski, A. P., Ng, W. J., and Liu, Y. 2014. Enzyme production from food wastes using a biorefinery concept. *Waste and Biomass Valorization* 5(6):903–917.

Knob, A., Fortkamp, D., Prolo, T., Izidoro, S. C., and Almeida, J. M. 2014. Agro-residues as alternative for xylanase production by filamentous fungi. *BioResources* 9(3):5738–5773.

Lannie, H., Kuncoro, F., and Raymond, R. T. 2017. Isolation and characterization of Agung banana peel starch from east Java Indonesia. *International Food Research Journal* 24(3): 1324–1330.

Li, Z., Guo, K., Lin, L., He, W., Zhang, L., and Wei, C. 2018. Comparison of physicochemical properties of starches from flesh and peel of green banana fruit. *Molecules*, 23(9): 2312.

Lim, J., Yoo, J., Ko, S., and Lee, S. 2012. Extraction and characterization of pectin from Yuza (Citrus junos) pomace: A comparison of conventional-chemical and combined physical–enzymatic extractions. *Food Hydrocolloids* 29(1):160–165.

Liu, Y., Shi, J., and Langrish, T. A. G. 2006. Water-based extraction of pectin from flavedo and albedo of orange peels. *Chemical Engineering Journal*, 120(3):203–209.

Locatelli, G. O., Finkler, L., and Finkler, C. L. 2019. Comparison of acid and enzymatic hydrolysis of pectin, as inexpensive source to cell growth of Cupriavidus necator. *Anais da Academia Brasileira de Ciências* 91(2): 1–4

Machado, M. T., Eca, K. S., Vieira, G. S., Menegalli, F. C., Martinez, J., and Hubinger, M. D. 2015. Prebiotic oligosaccharides from artichoke industrial waste: Evaluation of different extraction methods. *Industrial Crops and Products* 76:141–148.

Machineni, L. 2020. Lignocellulosic biofuel production: Review of alternatives. *Biomass Conversion and Biorefinery* 10(3):779–791.

Mala, T., Sadiq, M. B., and Anal, A. K. 2021. Comparative extraction of bromelain and bioactive peptides from pineapple byproducts by ultrasonic-and microwave-assisted extractions. *Journal of Food Process Engineering* 44(6):e13709.

Maran, J. P., and Prakash, K. A. 2015. Process variables influence on microwave assisted extraction of pectin from waste Carcia papaya L. peel. *International Journal of Biological Macromolecules* 73:202–206.

Maran, J. P., and Priya, B. 2015. Ultrasound-assisted extraction of pectin from sisal waste. *Carbohydrate Polymers* 115:732–738.

Maran, J. P., Sivakumar, V., Thirugnanasambandham, K., and Sridhar, R. 2014. Microwave assisted extraction of pectin from waste Citrullus lanatus fruit rinds. *Carbohydrate Polymers* 101:786–791.

Maran, J. P., Swathi, K., Jeevitha, P., Jayalakshmi, J., and Ashvini, G. 2015. Microwave-assisted extraction of pectic polysaccharide from waste mango peel. *Carbohydrate Polymers* 123:67–71.

Martín-Lara, M. A., Chica-Redecillas, L., Pérez, A., Blázquez, G., Garcia-Garcia, G., and Calero, M. 2020. Liquid hot water pretreatment and enzymatic hydrolysis as a valorization route of Italian green pepper waste to delivery free sugars. *Foods* 9(11):1640.

Matharu, A. S., Houghton, J. A., Lucas-Torres, C., and Moreno, A. 2016. Acid-free microwave-assisted hydrothermal extraction of pectin and porous cellulose from mango peel waste–towards a zero waste mango biorefinery. *Green Chemistry* 18(19):5280–5287.

Mellinas, C., Ramos, M., Jiménez, A., and Garrigós, M. C. 2020. Recent trends in the use of pectin from agro-waste residues as a natural-based biopolymer for food packaging applications. *Materials* 13(3):673.

Mergenthaler, P., Lindauer, U., Dienel, G. A., and Meisel, A. 2013. Sugar for the brain: The role of glucose in physiological and pathological brain function. *Trends in Neurosciences* 36(10):587–597.

Mohan, C. C., Harini, K., Aafrin, B. V., Babuskin, S., Karthikeyan, S., Sudarshan, K., Renuka, V. and Sukumar, M 2018. Extraction and characterization of polysaccharides from tamarind seeds, rice mill residue, okra waste and sugarcane bagasse for its Bio-thermoplastic properties. *Carbohydrate Polymers* 186:394–401.

Mojumdar, A., and Deka, J. 2019. Recycling agro-industrial waste to produce amylase and characterizing amylase–gold nanoparticle composite. *International Journal of Recycling of Organic Waste in Agriculture* 8(1):263–269.

Nadiha, N., and Jamilah, B. 2020. Hemicellulose extraction and characterization of rice straw and Leucaena leucocephala. *KnE Social Sciences*, 2020: 46–54.

Nakthong, N., Wongsagonsup, R., and Amornsakchai, T. 2017. Characteristics and potential utilizations of starch from pineapple stem waste. *Industrial Crops and Products* 105:74–82.

Ng, H. S., Kee, P. E., Yim, H. S., Chen, P. T., Wei, Y. H., and Lan, J. C. W. 2020. Recent advances on the sustainable approaches for conversion and reutilization of food wastes to valuable bioproducts. *Bioresource Technology* 302:122889.

Noor, F., Rahman, M. J., Mahomud, M. S., Akter, M. S., Talukder, M. A. I., and Ahmed, M. 2014. Physicochemical properties of flour and extraction of starch from jackfruit seed. *International Journal of Nutrition and Food Sciences* 3(4):347.

Noor, M. H. M., Ngadi, N., Suhaidi, A. N., Inuwa, I. M., and Opotu, L. A. 2021. Response surface optimization of ultrasound-assisted extraction of sago starch From sago pith waste. *Starch-Stärke*, 2022: 1–10.

Osorio, L. L. D. R., Flórez-López, E., and Grande-Tovar, C. D. 2021. The potential of selected Agri-food loss and waste to contribute to a circular economy: Applications in the food, cosmetic and pharmaceutical industries. *Molecules* 26(2):515.

Panzella, L., Moccia, F., Nasti, R., Marzorati, S., Verotta, L., and Napolitano, A. 2020. Bioactive phenolic compounds from agri-food wastes: An update on green and sustainable extraction methodologies. *Frontiers in Nutrition* 7:60.

Pattnaik, M., Pandey, P., Martin, G. J., Mishra, H. N., and Ashokkumar, M. 2021. Innovative technologies for extraction and microencapsulation of bioactives from plant-based food waste and their applications in functional food development. *Foods*, 10(2): 279.

Prajapati, S., Koirala, S., and Anal, A. K. 2021. Bioutilization of chicken feather waste by newly isolated keratinolytic bacteria and conversion into protein hydrolysates with improved functionalities. *Applied Biochemistry and Biotechnology*, 193(8):2497–2515.

Qasim, U., Ali, Z., Nazir, M. S., Ul Hassan, S., Rafiq, S., Jamil, F., Al-Muhtaseb, A. A. H., Ali, M., Khan Niazi, M. B., Ahmad, N. M., Ullah, S., and Saqib, S 2020. Isolation of cellulose from wheat straw using alkaline hydrogen peroxide and acidified sodium chlorite treatments: Comparison of yield and properties. *Advances in Polymer Technology*, 2020: 1–7.

Ramesh, S., and Radhakrishnan, P. 2019. Cellulose nanoparticles from agro-industrial waste for the development of active packaging. *Applied Surface Science* 484:1274–1281.

Ravindran, R., Hassan, S. S., Williams, G. A., and Jaiswal, A. K. 2018. A review on bioconversion of agro-industrial wastes to industrially important enzymes. *Bioengineering* 5(4):93.

Ribeiro, L. S., de Melo Órfão, J. J., and Pereira, M. F. R. 2021. Direct catalytic conversion of agro-forestry biomass wastes into ethylene glycol over CNT supported Ru and W catalysts. *Industrial Crops and Products* 166:113461.

Rinju, R., and Harikumaran-Thampi, B. S. 2021. Characteristics of starch extracted from the stem of pineapple plant (Ananas comosus)-an agro waste from pineapple farms. *Brazilian Archives of Biology and Technology* 64: 1-14.

Roselló-Soto, E., Parniakov, O., Deng, Q., Patras, A., Koubaa, M., Grimi, N., Boussetta, N., Tiwari, B. K., Vorobiev, E., Lebovka, N., and Barba, F. J. 2016. Application of non-conventional extraction methods: Toward a sustainable and green production of valuable compounds from mushrooms. *Food Engineering Reviews* 8(2):214–234.

Ruano, P., Delgado, L. L., Picco, S., Villegas, L., Tonelli, F., Merlo, M. E. A., Merlo, Mario Eduardo Aguilera, Rigau, Javier, Diaz, Darío, and Masuelli, M. 2019. Extraction and characterization of pectins from peels of criolla oranges (Citrus sinensis): Experimental reviews. In: *Pectins-Extraction, Purification, Characterization and Applications.*(pp. 3–46) Intechopen.

Sagar, N. A., Pareek, S., Sharma, S., Yahia, E. M., and Lobo, M. G. 2018. Fruit and vegetable waste: Bioactive compounds, their extraction, and possible utilization. *Comprehensive Reviews in Food Science and Food Safety* 17(3):512–531.

Salma, M. A., Jahan, N., Islam, M. A., and Hoque, M. M. 2012. Extraction of Pectin from lemon peel: Technology development. *Journal of Chemical Engineering* 27(2):25–30.

Sengar, A. S., Rawson, A., Muthiah, M., and Kalakandan, S. K. 2020. Comparison of different ultrasound assisted extraction techniques for pectin from tomato processing waste. *Ultrasonics Sonochemistry* 61:104812.

Setyaningsih, W., Fathimah, R. N., and Cahyanto, M. N. 2021. Process optimization for ultrasound-assisted starch production from cassava (Manihot esculenta Crantz) using response surface methodology. *Agronomy* 11(1):117.

Shahrim, N. A., Sarifuddin, N., and Ismail, H. 2018. Extraction and characterization of starch From mango seeds. *Journal of Physics: Conference Series* (Vol. 1082, No. 1, p. 012019). IOP Publishing.

Sharoba, A. M., Farrag, M. A., and Abd El-Salam, A. M. 2013. Utilization of some fruits and vegetables waste as a source of dietary fiber and its effect on the cake making and its quality attributes. *Journal of Agroalimentary Processes and Technologies* 19(4):429–444.

Shirahigue, L. D., and Ceccato-Antonini, S. R. 2020. Agro-industrial wastes as sources of bioactive compounds for food and fermentation industries. *Ciência Rural* 50(4):e20190857.

Shukla, J., and Kar, R. 2006. Potato peel as a solid state substrate for thermostable α-amylase production by thermophilic Bacillus isolates. *World Journal of Microbiology and Biotechnology* 22(5):417–422.

Siddiqui, A., Chand, K., and Shahi, N. C. 2021. Effect of process parameters on extraction of pectin from sweet lime peels. *Journal of the Institution of Engineers* 102(2):469–478.

Singanusong, R., Tochampa, W., Kongbangkerd, T., and Sodchit, C. 2014. Extraction and properties of cellulose from banana peels. *Suranaree Journal of Science and Technology* 21(3):201–213.

Singh, R. P., and Tingirikari, J. M. R. 2021. Agro waste derived pectin poly and oligosaccharides: Synthesis and functional characterization. *Biocatalysis and Agricultural Biotechnology*, 31 101910.

Soquetta, M. B., Terra, L. D. M., and Bastos, C. P. 2018. Green technologies for the extraction of bioactive compounds in fruits and vegetables. *CyTA-Journal of Food* 16(1):400–412.

Sundarraj, A. A., and Ranganathan, T. V. 2018. A review on cellulose and its utilization from agro-industrial waste. *Drug Invention Today* 10(1):89–94.

Tatar, F., Tunç, M. T., Dervisoglu, M., Cekmecelioglu, D., and Kahyaoglu, T. 2014. Evaluation of hemicellulose as a coating material with gum arabic for food microencapsulation. *Food Research International* 57:168–175.

Thangaratham, T., and Manimegalai, G. 2014. Optimization and production of pectinase using agro waste by solid state and submerged fermentation. *International Journal of Current Microbiology and Applied Science* 3(9):357–365.

Tiwari, A. K., Saha, S. N., Yadav, V. P., Upadhyay, U. K., Katiyar, D., and Mishra, T. 2017. Extraction and characterization of pectin from orange peels. *International Journal of Biotechnology and Biochemistry* 13(1):39–47.

Tongkham, N., Juntasalay, B., Lasunon, P., and Sengkhamparn, N. 2017. Dragon fruit peel pectin: Microwave-assisted extraction and fuzzy assessment. *Agriculture and Natural Resources* 51(4):262–267.

Torres-León, C., Chávez-González, M. L., Hernández-Almanza, A., Martínez-Medina, G. A., Ramírez-Guzmán, N., Londoño-Hernández, L., and Aguilar, C. N. 2020. Recent advances on the microbiological and enzymatic processing for conversion of food wastes to valuable bioproducts. *Current Opinion in Food Science* 38:40–45.

Vilas-Boas, A. A., Pintado, M., and Oliveira, A. L. 2021. Natural bioactive compounds from food waste: Toxicity and safety concerns. *Foods* 10(7): 1564.

Walia, Y. K., and Gupta, D. K. 2013. Isolation, extraction and purification of hemicelluloses of Ceiba pentandra and Morus nigra. *Asian Journal of Advamced Basic Science* 2(1):29–35.

Walsh, M. K., Khliaf, H. F., and Shakir, K. A. 2018. Production of xylitol from agricultural waste by enzymatic methods. *American Journal of Agricultural and Biological Sciences* 13(1):1–8.

Wang, Q., Wang, X., Wang, X., and Ma, H. 2008. Glucoamylase production from food waste by *Aspergillus niger* under submerged fermentation. *Process Biochemistry* 43(3):280–286.

Wang, H., Ding, J., and Ren, N. 2016. Recent advances in microwave-assisted extraction of trace organic pollutants from food and environmental samples. *TrAC Trends in Analytical Chemistry*, 75:197–208.

Wei, L., Yan, T., Wu, Y., Chen, H., and Zhang, B. 2018. Optimization of alkaline extraction of hemicellulose from sweet sorghum bagasse and its direct application for the production of acidic xylooligosaccharides by *Bacillus subtilis* strain MR44. *PLOS ONE* 13(4):e0195616.

Xie, F., Zhang, W., Lan, X., Gong, S., Wu, J., and Wang, Z. 2018. Effects of high hydrostatic pressure and high pressure homogenization processing on characteristics of potato peel waste pectin. *Carbohydrate Polymers* 196:474–482.

Yi, V. C. W., Yi, T. P., Xin Xin, G., and Muei, C. L. 2019. Ultrasonic-assisted alkaline extraction process to recover cellulose from sugarcane bagasse for carboxymethylcellulose production. *AIP Conference Proceedings* 2157(1):020044.

Yılmaz Celebioglu, H., Cekmecelioglu, D., Dervisoglu, M., and Kahyaoglu, T. 2012. Effect of extraction conditions on hemicellulose yields and optimisation for industrial processes. *International Journal of Food Science and Technology* 47(12):2597–2605.

Yoshida, T., Tsubaki, S., Teramoto, Y., and Azuma, J. I. 2010. Optimization of microwave-assisted extraction of carbohydrates from industrial waste of corn starch production using response surface methodology. *Bioresource Technology* 101(20):7820–7826.

Zein, Y. M., Anal, A. K., Prasetyoko, D., and Qoniah, I. 2016. Biodiesel production from waste palm oil catalyzed by hierarchical ZSM-5 supported calcium oxide. *Indonesian Journal of Chemistry* 16(1):98–104.

11 Extraction of Fatty Acids and Micronutrients from Agro-Industrial Waste and their Application in Nutraceuticals and Cosmetics

Phan Thi Phuong Thao, Lai Phuong Phuong Thao,
Tran Quoc Toan, and Tran Thi Thu Hang

CONTENTS

11.1 INTRODUCTION

Human health has been demonstrated to benefit from polyunsaturated fatty acids, the most valuable of which are alpha-linolenic acid (ALA), eicosapentaenoic acid (EPA), and docosahexaenoic acid (DHA). Several studies have recently focused on the extraction of fatty acids from various materials, including fatty-acid-rich agro-industrial waste like rice bran, marine waste (fish, shrimp), and some oil-bearing seeds (da Silva and Jorge, 2014). Furthermore, these ingredients not only contain significant amounts of fatty acids but also micronutrients essential for humans.

DOI: 10.1201/9781003125679-11

Rice is a valuable cash crop that provides food for more than half of the world's population. Seeds are milled after harvest to remove the outer layers and reveal the white endosperm. This process generates a considerable amount of byproducts (e.g., hull, bran, cereal germ) that could be utilized as feed for livestock. The bran is a particularly interesting product, as it only accounts for 8–10% of seed weight yet contains 60–65% of the total nutrients. These nutrients include proteins (11–17%), essential amino acids, lipids (10–26%) rich in polyunsaturated fatty acids (PUFAs), and fibre (6–20%). The inorganic ash content varies from 8% to 17% and contains valuable minerals such as Ca, Fe, Mg, and Mn. Oryzanol, phytosterol, tocotrienol, squalene, polycosanol, phytic acid, ferulic acid, and inositol hexaphosphate are only a few of the beneficial components found in rice bran. Together, they confer a variety of health benefits, such as antioxidant and anti-inflammatory activity, cancer risk reduction, coronary artery disease prevention, and reduction in cholesterol levels (Khatoon and Gopalakrishna, 2004; Saikia and Deka, 2011; Sharif et al., 2014; Pengkumsri et al., 2015; Punia et al., 2021; Garofalo et al., 2020).

Despite its high nutrient value and many health benefits, a significant amount of bran is discarded or used as animal feed. This is due to the difficulty in storing it after milling, as bran spoils very easily, especially its valuable fatty acids, which are quickly hydrolyzed by lipase into free fatty acids (FFAs) and glycerol; the unsaturated bonds in PUFAs also facilitate oxidation that further degrades the nutritious value (Saunders, 1985; Dubey et al., 2019; Garofalo et al., 2020). Therefore, in order to effectively convert bran into a high value, stable product such as rice bran oil (RBO), it needs to be subjected to processing, including deactivation of the lipase, inhibition of lipid oxidation, prevention of lipid denaturation, and several more steps to ensure spoilage does not occur during storage. Several methods have been studied in order to stabilize rice bran, such as steaming, ohmic heating, sonication, parboiling, cooling and pH reduction, microwave treatment, and Infra red (IR) heating (Garba et al., 2017; Punia et al., 2021; Garofalo et al., 2020). Microwave treatment could likely be the most efficient method, as it has been shown to be cheap and quick, producing oil with low FFA content that is stable over time (Begum et al., 2015; Sharif et al., 2014; Jiang, 2019; Garofalo et al., 2020). Patil et al. (2016) discovered that microwave treatment at 700 W, 2450 MHz, at a density of 4 W/g for 5 min helps inhibit bran lipase activity for up to 90 days. Irakli et al. (2018), on the other hand, found that IR treatment at 140°C for 15 min is the most efficient approach for stabilizing rice bran. Yu et al. (2020) evaluated 11 different methods for stabilizing rice bran and ranked their effectiveness by an index based on measuring lipase activity after treatment. Their results showed that microwave treatment, pressing, steaming, low temperature (–80°C), UV treatment, and IR heating were all viable treatments. Pressing and microwave treatment are suitable for large-scale production in a short amount of time but suffer from high peroxide content and oil discolouration. Ultraviolet (UV) treatment does not affect oil quality and is energy efficient (Punia et al., 2021). Stabilized rice bran could be used directly or processed to yield high-value products.

The fishing industry generates a large quantity of solid waste and byproducts. Production of fish fillets alone creates waste that accounts for 50–70% of the total amount of fish used. Examples of waste include the skin, head, tail, and entrails, most of which are usually discarded after processing. The handling of these wastes/byproducts remains a major issue that affects both the environment and the economy. Most of them are further processed into fish powder to use for animal feed. However, there are alternatives that could yield greater economic value, such as fish oil. Fish oil is high in PUFAs, particularly omega 3 fatty acids like DHA and EPA. The organs suitable for extraction depend on the fish species and, consequently, the fatty acid content. For example, mackerel skin (usually thrown away) has a higher PUFA content compared to the internal organs and muscles (Sahena et al., 2009). Little tuny (*Euthynnus alletteratus*) liver and internal organs have a higher lipid content compared to other parts.

Fish, in general, also contain many minerals and micronutrients, as their tissue readily accumulates inorganic elements present in fresh or salt water at levels higher than terrestrial animals. Common minerals found in fish meat include sodium, potassium, calcium, magnesium, phosphorus, sulphur, iron, manganese, zinc, copper, and iodine. Fish also contain essential vitamins such as A, D, E, and B. The content varies depending on the species, organs, and extraction methods.

Dobrzański et al. (2002) analyzed the contents of oil extracted from fish byproducts. Unsaturated fatty acids accounted for 82.72%, PUFAs 35.59%, and omega-3 fatty acids 17.04%. Additionally, the oil also contains minerals and micronutrients such as: 0.23% Cl; 0.047% Ca; 24.6 mg/kg Mg; 37.57 mg/kg Na; 51.8 mg/kg Tr; 220 mg/kg Zn; 15.33 mg/kg Fe; 14.43 mg/kg Al; 7.62 mg/kg Se. The following vitamin concentrations were identified: 458 IU/g A; 240 IU/g D3; 1.21 IU/g E. Overall, the obtained fish oil is adequate for use as animal feed (Dobrzański et al., 2002). Scurria et al. (2020) evaluated the vitamin content in oil extracted from anchovy byproducts and identified only vitamin D3 at 81.5 g/kg of oil.

11.2 EXTRACTION AND PURIFICATION OF FATTY ACIDS AND MICRONUTRIENTS FROM AGRO-INDUSTRIAL WASTE

In order to obtain fatty acids and micronutrients from agro-industrial waste, many methods have been studied and used. These methods can be divided into conventional and more modern "green" methods.

11.2.1 CONVENTIONAL METHODS

11.2.1.1 Mechanical Pressing

Mechanical pressing is a conventional and widely used method to process oil-bearing seeds. It has been used to produce RBO at a small and medium scale. The method is simple, safe, and cheaper than extraction using solvents. A major advantage of mechanical pressing is that it does not rely on high temperatures or potentially toxic solvents. The obtained oil possesses good nutritional value and is free of solvent residue, making it a promising method in the eyes of consumers, who prioritize "naturalness" and "safety" (Garba et al., 2017; Punia et al., 2021). Pressing machines either use screws or hydraulics, with screw presses having higher efficiency (Garba et al., 2017).

However, the oil extraction efficiency of mechanical pressing is low. To increase the amount of retrieved oil, several pre-processing steps need to be implemented before rice bran is pressed. Low-temperature treatment (Srikaeo and Pradit, 2011) and sonication (Phan et al., 2019; Punia et al., 2021) have been shown to yield more RBO. A combination of short (cooking) time and cold pressing also helps increase efficiency (Figure 11.1).

11.2.1.2 Solvent Extraction Method

Solvent extraction is a popular method for extracting oils from various materials, despite their low lipid content (Figure 11.2, Table 11.1). RBO is usually extracted using non-polar organic solvents such as hexane, considered the most common option for RBO extraction. Hexane is cheap, has a low boiling point, is easy to recover, and has a high extraction efficiency (Garba et al., 2017; Tong and Bao, 2019; Garofalo et al., 2020). Rice bran pre-processed by ohmic heating can yield up to 92% of its oil after hexane extraction (Lakkakula et al., 2004; Garba et al., 2017).

Pengkumsri et al. (2015) looked at how the extraction process affected the composition and RBO concentration of three distinct kinds of rice. The results indicated that hexane extraction gave the highest oil yield in any rice species compared to cold pressing (40–60°C), hot pressing (80–100°C), and supercritical fluid extraction (SpFE). Hexane-extracted oils also had the highest amounts of γ-oryzanol, tocopherol, and tocotrienol. Hexane extraction also yielded oil with the best in vitro antioxidant properties (Pengkumsri et al., 2015).

Solvent extraction with the Soxhlet apparatus is also a major method for vegetable oil extraction. Liu et al. (2005) obtained 67.73% RBO by Soxhlet extraction using hexane. This method also yielded 15%–20% RBO and 18.4% RBO using heptane and kerosene ether, respectively (at

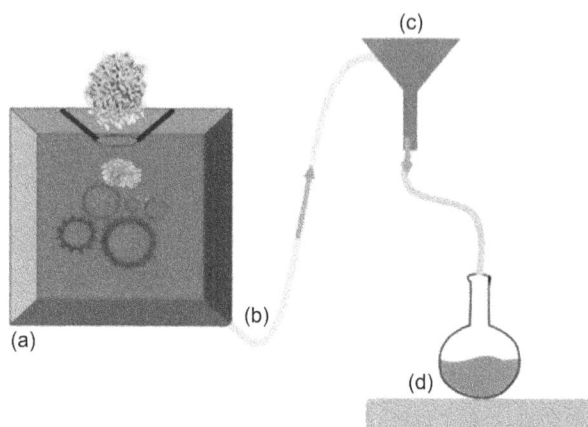

FIGURE 11.1 Method of mechanical pressing of rice bran oil (Punia et al., 2021). (a) Rice bran is fed into the press; (b) mixtures containing RBO, fibres, and other impurities; (c) filtration by a centrifuge/mechanical filters; (d) purified cold-pressed RBO.

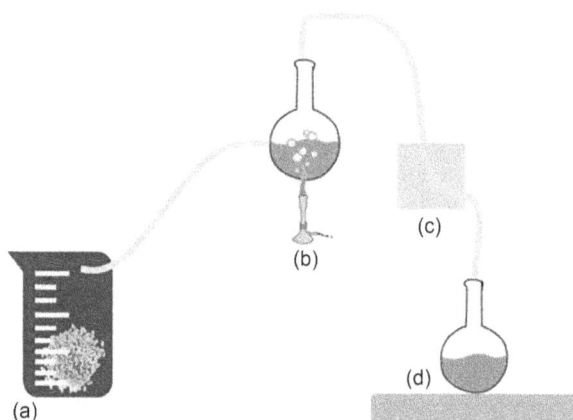

FIGURE 11.2 Solvent extraction (Punia et al., 2021). (a) Solvent and sample (rice bran); (b) vacuum distillation to remove solvents; (c) condensing chamber to obtain liquefied RBO; (d) purified RBO.

40–60°C) (Al-Okbi et al., 2014; Zaccheria et al., 2015; Garba et al., 2017). However, this method is only suitable for laboratory use.

Although providing good extraction efficiency, petroleum-based solvents such as hexane have some limitations as they are harmful to human health and they are volatile, flammable, toxic, and can cause environmental pollution (Chemat et al., 2019; Punia et al., 2021; Garofalo et al., 2020). Therefore, a suitable alternative to this type of solvent is to use green solvents, safer biological solvents such as certain short-chain alcohols (ethanol, isopropanol), ethyl acetate, and D-limonene (Zigoneanu et al., 2008; Patel and Naik, 2004; Garba et al., 2017; Toy and Bao, 2019).

RBO extraction with isopropanol and hexane at 40°C for 15 min has been studied. The amount of oil extracted with hexane was almost 40% more than that of isopropanol, but upon increasing the temperature up to 120°C, the hexane extraction yield remained unchanged, while the amount of RBO extracted with isopropanol was 25% more than that of hexane under the same conditions (Zigoneanu et al., 2008; Garba et al., 2017).

In another study, extraction with isopropanol and hexane yielded similar quantities of RBO (170.9 g/kg) and both higher yields than ethanol at 4 h using the Soxhlet apparatus. As RBO was

TABLE 11.1
Some Studies on Solvent Extraction Methods to Collect RBO

Conditions/parameters	Results/comments	Reference
Hexane, 65°C	Hexane is an effective solvent for RBO extraction	Amarasinghe and Gangodavilage (2004)
Isopropanol, hexane, ethanol Soxhlet for 4 h	RBO extraction efficiency of isopropanol is comparable to hexane (170.9 g/kg) γ-oryzanol content of extracted RBO was 2,609 mg/kg (ethanol); 2,229 mg/kg (hexane), and 2,094 mg/kg (isopropanol)	Shukla and Pratap (2017)
Ethanol (6% H_2O) Solvent/material = 4/1 (w/w); stirring speed 137.5 rpm; 82.5°C for 3 h	RBO = 20.05 g/100 g fresh bran γ-oryzanol = 4,164 mg/kg fresh bran	Oliveira et al. (2012)
Isopropanol 40°C; solvent/material = 3:1 (w/w) for 15 min	Crude oil extraction efficiency = 0.1 g/g fresh bran	Zigoneanu et al. (2008)
Acetone 30°C for 1 h; solvent/material ratio = 4:1 (w/w)	RBO extraction efficiency = 25% dry bran	Ruen-Ngam et al. (2014)
D-Limonene 176°C for 1 h, solvent/bran ratio = 5:1	RBO = 22.48% dry bran	Liu et al. (2005)
Isopropanol; ethanol (6% H_2O) 80°C for 1 h, solvent/bran = 3:1 (w/w)	RBO = 160 g/kg dry bran - oryzanol max = 1.53% RBO extracted with ethanol (6% H_2O)	Capellini et al. (2017)

Sources: Punia et al., 2021; Garofalo et al., 2020.

extracted with ethanol, however, greater quantities of γ-oryzanol (2,609 mg/kg) were found when compared to hexane (2,229 mg/kg) or isopropanol (2.094 mg/kg) (Shukla and Pratap, 2017; Punia et al., 2021).

The effects of ethanol solvent concentration, temperature, solvent/bran ratio, and stirring speed on RBO extraction were investigated by Oliveira et al. (2012). They observed the highest oil yield of 20.05 g/100 g of rice bran could be obtained utilizing ethanol containing 6% water, a solvent/bran ratio of 4:1 (w/w), temperature of 82.5°C, and a stirring speed of 137.5 rpm. The concentration of γ-oryzanol in fresh rice bran ranged from 1527 to 4164 mg/kg (Garofalo et al., 2020).

D-limonene, a citrus-processing industry agriculture waste, is also a feasible option for achieving high crude oil output with good quality (Liu and Mamidipally, 2005; Garofalo et al., 2020). However, its high boiling point (176°C) requires more energy to separate D-limonene from the oil compared to hexane (69°C).

The studies mentioned above demonstrate that it is feasible to replace hexane (toxic solvents) with biological, green solvents to extract oil from rice bran. Among them, ethanol and isopropanol are the two most promising candidates.

Various methods for extracting fish oil exist. Organic solvents such as hexane, petroleum ether, methanol, and chloroform are commonly used in traditional lipid extraction methods, which are based on the hydrophobicity of lipids. Their high extraction efficiency is hindered by the restricted/limited use of organic solvents in the food industry. Therefore, solvent extraction processes are often used at the laboratory scale for purposes of comparison or analysis.

To reduce toxicity caused by the use of organic solvents, Ciriminna et al. (2019) studied the process of extracting oil from the byproducts of anchovies (*Engraulis encrasicolus*) fillets using a green solvent called D-limonene obtained from discarded orange peels. A simple solid-liquid extraction was performed by mechanical stirring and immersion at room temperature for 21 h, followed by removal of limonene by vacuum evaporation. The measured lipid content was 1.8%, and the obtained oil was orange-yellow in colour with a pleasant smell. Oleic acid was the main monounsaturated fatty acid (MUFA) present at 23.97%, while DHA (12.38%) and EPA (5.4%) were the most predominant omega-3 PUFAs.

Tea (*Camellia sinensis*) seed oil (TSO) is considered a high vegetable oil with a fatty acid profile comparable to that of olive oil, with a high concentration of unsaturated fatty acids, particularly linoleic acid, and a low level of saturated fat. Phan et al. (2021) showed that solvent not only can obtain oil but also natural compounds from tea seed in the solvent extraction method. The study evaluated the effects of ten solvents: n-hexane (Hx), dichloromethane (DCM), ethyl acetate (EtOAc), ethanol (EtOH), EtOH:DCM (v/v, 3:1, 1:1, and 1:3), EtOH: EtOAc (v/v, 3:1, 1:1, and 1:3) on the content and antioxidant capacity of TSO (through the content of fatty acids and substances) with antioxidants such as polyphenols, carotenoids, and tocopherol present in the oil. The results showed that in terms of polyphenol content, TSO extracted from ethanol solvent gave the highest total polyphenol content, followed by the solvent mixture EtOH:EtOAc v/v 3:1, EtOH:EtOAc v/v 1:1, and EtOAc, so ethanol and ethyl acetate are solvents that extract many polyphenol components from seeds into oil. The antioxidant capacity of the oil was positively correlated with the polyphenol content. Among the studied solvents, extraction with dichloromethane solvent gave TSO the highest carotenoid content, followed by solvent mixtures of ethanol and dichloromethane. However, extraction with ethanol solvent showed that the carotenoid content in TSO did not differ from those of the above solvent systems in terms of statistical significance. Besides, extraction with ethanol solvent gives the highest content of polyunsaturated fatty acids in TSO and low MUFA content. The saturated fatty acid content of TSO when extracted with solvent EtOH:DCM v:v 3:1 was the lowest, the remaining contents when extracted in other solvents did not differ much.

Wet rendering and dry rendering are the two most widely utilized extraction processes on an industrial scale, both of which do not need the use of chemicals. Wet rendering includes steaming fish to break down the cellular structure, and then squeezing the oil out of the cooked fish. The process consists of three basic steps: high-temperature cooking (85–95°C), pressing, and centrifugation. It can yield a large amount of crude fish oil. For food consumption, further steps are required to refine the resulting crude oil.

Suseno et al. (2015) studied the lipid extraction process from byproducts (head, entrails, and skin) of tilapia (*Oreochromis niloticus*) fillet by the wet rendering method. The results showed that the optimum extraction time and temperature was 70°C for 35 min, with a crude oil yield of 6.44% (1.15% EPA and 1.03% DHA). In another study, Pudtikajorn and Soottawat Benjakul (2020) extracted lipids from skipjack tuna eyes by wet rendering at 121°C for 20 min using the autoclave. The oil recovery efficiency was 16.49%, and the obtained product had high a PUFA content (40.98%), along with a saturated fatty acid (SFA) and MUFA content of 38.08% and 20.94%, respectively. DHA (C22: 6, n3) was the predominant fatty acid (32.06%), while EPA accounted for 5.52%. The oil had a light yellow colour and a slightly fishy odour, outstanding properties by cooking oil standards, and contained high levels of DHA, making it viable as a dietary supplement or product. In 2017, Nazir et al. studied the impact of several extraction methods on the physicochemical characteristics and fatty acid content of tuna oil (*Thunnus albacares*). The three extraction methods used were precooked wet rendering extraction (samples were autoclaved at 105°C for 30 min, pressed, solid-liquid separation, and centrifugation to separate crude oil); acid silage extraction (soaked in 3% formic acid at room temperature for 4–7 days), and Soxhlet extraction using hexane solvent. Degumming (removal of phosphatides, proteins, residues, carbohydrates, water, and resins without decreasing the fatty acids in the oil), neutralization, and bleaching were used to refine the crude oil. The results showed that wet rendering was the most effective, with a 12.80% yield compared to acid silage (6.16%) and solvent extraction (8.49%). The PUFA content did not differ significantly between wet rendering (44.34%) and solvent extraction (44.89%), with both of them higher than that of acid silage (32.77%). The DHA and EPA content obtained by wet rendering was the highest (8.65% and 6.12%) compared with acid silage (5.55% and 5.21%) and solvent extraction (6.44 % and 3.28%) (Nazir et al., 2017).

11.2.2 Non-conventional (or Green Extraction) Methods

With the goal of being environmentally friendly and good for consumers, many modern green extraction techniques have also been studied, such as SpFE, subcritical water extraction (SWE), ultrasound assisted extraction (UAE), microwave assisted extraction (MAE), and enzyme assisted extraction (EAE). These novel techniques all help reduce extraction time, energy consumption, and solvent quantity. They can be integrated into existing extraction processes as improvements to cut down the number of required steps without compromising product quality (Garofalo et al., 2020; Punia et al., 2021; Garba et al., 2017).

11.2.2.1 SpFE

SpFE is a green method for extracting a wide range of bioactive compounds. This method has been extensively studied and has received much attention in research and application. Instead of using harmful organic solvents, supercritical fluids (i.e., CO_2) are used to dissolve the solutes in the material.

SpFE is an extraction process that uses a liquid at pressures and temperatures above its critical point. CO_2 is often used to replace organic solvents, especially in the food industry. It has a low critical pressure and temperature (31.06°C and 7.37 MPa), is non-flammable, non-toxic, non-corrosive, cheap, and pure, and is recoverable and recyclable. Although it is a non-polar solvent, it may be made more polar and selective for specific chemicals by adding a more polar co-solvent, such as water or ethanol, to improve solubility and extraction rates (Rozzi and Singh, 2002; Santos et al., 2017; Garofalo et al., 2020; Punia et al., 2021, Rovetto and Aieta, 2017).

A supercritical fluid system, basically, includes a pump, an extractor vessel, a pressure regulator valve, and a separator (Figure 11.3). The extractor vessel is controlled at a constant desired temperature and pressure. Besides, there is a tank of fluid, valve, and fluid cooler involved in an SpFE system. In the extraction process, the material is placed in the extractor vessel closed. Then CO_2 is pumped into the substrates (Ahmad et al., 2019; Markom et al., 2001).

SpFE leaves no solvent residue, yielding extracts of high purity that can be used directly without further processing (Straccia et al., 2012). SpFE even assists in overcoming some of the drawbacks of traditional extraction methods. Because of its superior intercellular penetration capabilities, SpFE can be quicker and more efficient than organic solvent extraction. This technique also produces a yield that is equivalent to hexane extraction (Balachandran et al., 2008).

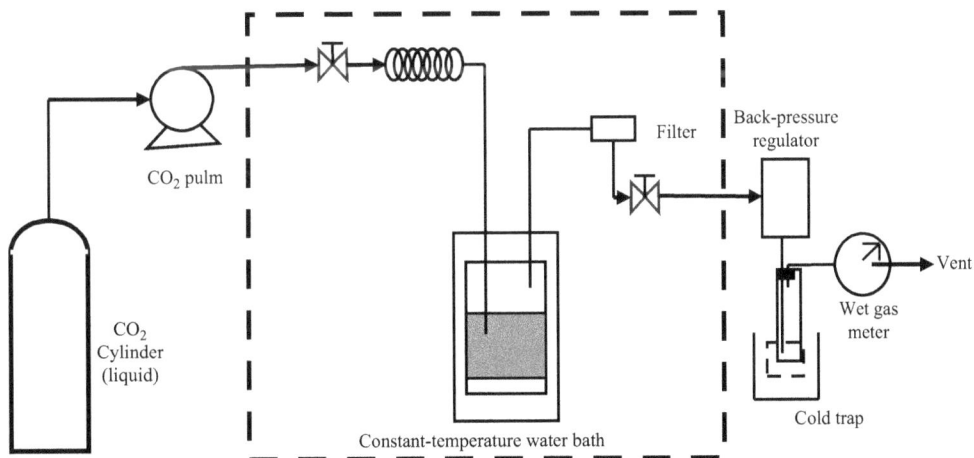

FIGURE 11.3 Schematic diagram of an SpFE system (Markom et al., 2001).

TABLE 11.2

Several Studies on Using SC-CO$_2$ to Produce RBO

Conditions	Results	Reference
45°C, 35 MPa	RBO = 22.2%	Sparks et al. (2006)
0.25 MPa, 40°C	RBO = 12.7%	Soares et al. (2016)
680 Atmospheric pressure (atm), 50°C, 25 min or 15–20 min, flow rate 250 ml/min	RBO = 5.39 mg/g fresh bran γ-oryzanol = 674.6 mg/g RBO	Xu and Godber (2000)
40 MPa, 65°C, 0.45 ml/min, 6 h	γ-oryzanol = 11,371.79 mg/kg dry bran	Imsanguan et al. (2008)
60°C, 500 bar, 1.5 h, 40 g/min	RBO = 20% γ-oryzanol = 11,100 ppm	Balachandran et al. (2008)
40°C, 200 bar, 30 min, ethanol as co-solvent	RBO = 25.48%	Juchen et al. (2019)
350 bar, 60°C, 1.5 l/min, 3 h	RBO = 15% dry bran	Al-Okbi et al. (2013)

Sources: Garofalo et al., 2020; Punia et al., 2021; Garba et al., 2017; Toy and Bao, 2019.

Many studies have been conducted to extract RBO by supercritical CO$_2$ (SC-CO$_2$) extraction (Table 11.2). The maximum RBO yield extracted with SC-CO$_2$ ranges from 20.4% to 25%, corresponding to a recovery rate of up to 99% by hexane (Kuk and Dowd, 1998; Punia et al., 2021; Garba et al., 2017). Soares et al. (2016) reported that the highest yield of RBO extracted by SC-CO$_2$ was 12.7% at 0.25 Mpa and 40°C. The obtained RBO attained a maximum antioxidant activity when extracted at 0.25 MPa and 60°C (Toy and Bao, 2019). Balachandran et al. (2008) reported a 20% RBO yield while extracting with SC-CO$_2$ at 60°C, 500 bar, for 1.5 h with a CO$_2$ flow rate of 40 g/min, compared to 22.5% when extracted with hexane (Garba et al., 2017). Imsanguan et al. (2008) surveyed different temperatures (45–65°C) and pressures at 38 and 48 MPa, with a CO$_2$ flow rate maintained at 0.45 ml/min. The highest yield of γ-oryzanol (11,371.79 mg/kg rice bran) was obtained at 48 MPa and 65°C in continuous extraction. The amount of γ-oryzanol in SpFE extraction was higher than the peak value obtained using hexane extraction, with a shorter reaction time (Garofalo et al., 2020).

RBO extracted by the SC-CO$_2$ method had a good yield, possibly with better quality, colouration, higher amount of antioxidants, and less unwanted impurities. However, this method requires investing in expensive equipment.

Using SC-CO$_2$ as a solvent is a method that has received more attention in fish oil extraction. Because CO$_2$ is non-toxic, non-corrosive, and non-flammable, it was chosen as the extraction solvent. Furthermore, because it is gaseous and leaves no trace in the product, it may be readily eliminated following extraction. This method is useful for extracting thermally unstable compounds due to its low-temperature operation. Furthermore, the oxygen-free environment aids in the extraction of fatty acids, particularly PUFAs, by reducing oxidation (Nazir et al., 2017; Alfio et al., 2021). The extraction efficiency compared with traditional methods is equivalent or maybe even better.

Ferdosh et al. (2015) evaluated the quality of lipids extracted from the head, skin, and entrails of three tuna species (*Thunnus tonggol*, *Euthynnus affinis*, and *Auxis thazard*) using the SC-CO$_2$ extraction technique (pressure condition 40MPa, 65°C) for 2 h, 2.4 ml CO$_2$ and 0.6 ml ethanol/min, v/v) and Soxhlet extraction. The results showed that the total lipid content extracted by the two methods was similar, with the head yielding the most oil (28.4–36.2%) compared to the skin and entrails. The fatty acid composition of the oil did not differ between the two methods used. SFAs (39.7–48.5%) were predominant compared to MUFAs (21.9–26.6%) and PUFA (24.1–27.9%) in all species. DHA was the main PUFA, and accounted for about 80% of total PUFAs, 17.0–19.9% in the head; 15.7–17.3% in the skin and 14.3–16.1% in the entrails.

When compared to the Soxhlet technique, SC-CO$_2$ extracted oils contain lower levels of FFA and peroxide (PV). FFA and PV ranged from 1.8% to 5.0% and 1.2% to 2.4%, respectively. As a result, SC-CO$_2$ is an excellent technique for extracting omega 3 fatty-acid-rich fish oil from tuna wastes. When SpFE is used, pre-treatment of the material is required to reduce the moisture content to less than 20%. Since water interferes with CO$_2$ diffusion and affects the solubility of oils in CO$_2$, the raw material is usually lyophilized at −40°C before extraction (Alfio et al., 2021; Ivanovs and Blumberga, 2017). Although SpFE is a promising method, it is still difficult to replace wet rendering at the industrial scale due to its significantly higher production investment costs.

11.2.2.2 Subcritical Fluid Extraction

When compared to supercritical fluid extraction, subcritical extraction has the benefit of requiring reduced pressure and temperature, as well as being less expensive. The most commonly used green solvent for subcritical extraction is water, but CO$_2$ can also be used.

Pourali et al. (2009) demonstrated that SWE may be used to stabilize rice bran and extract RBO. The findings indicated that lipase was fully destroyed at 240°C for 10 min, and the highest RBO yield was 249 mg/g dry rice bran (94% of the total oil from rice bran), which was equivalent to hexane extraction (Garofalo et al., 2020). Chia et al. (2015) compared subcritical CO$_2$ extraction to hexane extraction and found that the RBO yields were 13.0%–14.5% and 22.0%, respectively, for the two techniques. Oil extracted using subcritical CO$_2$ included roughly ten times more oryzanol compounds and tocols, as well as lower FFA levels and peroxide values than oils extracted with hexane (Punia et al., 2021).

These findings indicate that subcritical solvent extraction is a competitive extraction method that does not require the use of hazardous organic solvents and has a quick reaction time. However, there is currently a lack of study and applications for this approach to RBO extraction.

11.2.2.3 EAE

When water is employed as the reaction medium, EAE is a green technique. Cellulose, hemicellulose, pectin, lignin, and protein make up the majority of the plant cell wall. Using enzymes helps break down the cell wall, creating favourable conditions for lipid release (Figure 11.4). The enzyme reaction is also less harsh compared to conventional chemical/physical processes. However, one enzyme is usually insufficient, so to improve the oil extraction efficiency, treatment of the material with a combination of different enzymes can be used as a support step for the extraction process. Cellulase, pectinase, and protease are the most common enzymes used in RBO extraction (Garofalo et al., 2020).

Hanmoungjai et al. (2002) compared Alcalase (a serine-type endo-protease) to five additional enzymes: Celluclast 1.5L, hemicellulase, Pectinex Ultra SP-L, Viscozyme L, and papain. According to the findings, Alcalase produced 75% of the oil output in just one hour, demonstrating its relevance in the extraction process (Garofalo et al., 2020).

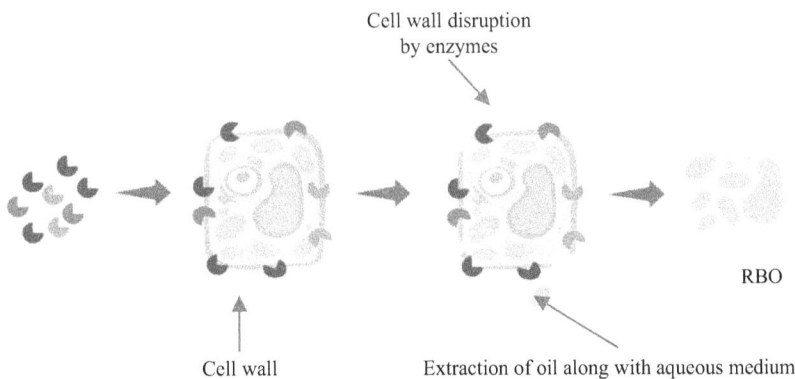

FIGURE 11.4   EAE (Punia et al., 2021).

Using 0.6L Alcalase at 50°C for 2 h, pH9, the maximum oil yield was 79% of the conventional method, and the obtained oil had a γ-oryzanol content of 1.76%. In addition, Alcalase-extracted oil had a similar composition to hexane-extracted oil but with lower FFA content and higher peroxide content, enabling a reduction in the number of refining steps (Hanmoungjai et al., 2001; Garofalo et al., 2020).

Sharma et al. (2001) used a mix of commercial enzymes: Protizyme™ (protease), Palkodex™ (α-amylase), and cellulose to try to enhance oil extraction yield. When compared to standard extraction techniques, the scientists found that 76–78% of oil could be produced at 65°C and pH 7 with continuous shaking at 80 rpm for 18 h (Garofalo et al., 2020).

Enzymatic hydrolysis is the use of enzymes derived from external sources. The addition of exogenous enzymes makes the metabolism more controlled and efficient. Enzymatic hydrolysis is, therefore, an excellent method for recovering lipids and proteins from fish and seafood processing waste. The enzymes and fish/fish byproducts utilized in this procedure must be of food-grade quality, and the enzymes must not be pathogens if they are of microbial origin. Alkaline/neutral proteases are used for hydrolysis more often than acidic proteases because they give superior outcomes. Enzymatic hydrolysis offers a higher lipid recovery than conventional thermal extraction and can compete with solvent extraction (Table 11.3) (Ramakrishnan et al., 2013; Ivanovs and Blumberga, 2017).

11.2.2.4 Microwave Assisted Extraction

MAE is a potentially green extraction technology that might ultimately replace standard solvent extractions. Microwaves rapidly heat the treated material, allowing for a quicker reaction time and less solvent and energy use. The rapid temperature rise also raises pressure inside plant cells, inducing cell wall rupture and therefore increasing the mass transfer rate between the solvent and the bioactive chemicals (Garofalo et al., 2020) (Figure 11.5).

Microwaves are non-ionizing electromagnetic waves with frequencies ranging from 300 MHz to 300 GHz (much higher than ultrasound wave frequencies). With the nature of electromagnetic radiation, microwaves have similar properties as visible light, i.e., they may be absorbed or reflected by material. There are also some mediums transparent to microwaves; in this case, microwaves may transmit through the medium without any absorption. The absorption of microwaves by a dielectric material results in the transfer of energy from the microwaves to the substance, resulting in an increase in temperature.

Heat generation in a medium put in a microwave field may occur via the mechanisms of ionic polarization (or ion conduction) and dipole rotation. In ionic polarization mechanism, when a material (e.g., food solution) containing ions is placed in an electrical field, the ions migrate through

TABLE 11.3

Fish Oil Extraction Using Enzymes

Species and organ	Pre-processing	Optimal extraction conditions	Efficiency
Farmed salmon (*Salmo salar*)	Mix thoroughly, boil at 90°C for 5 min to deactivate enzymes	240 min, 30°C, 0.5% enzyme Sea-B Zyme L200	Entrails: 13.1 g/100 g material Head: 59.9 g/100 g material
Salmon head (*Salmo salar*)	Grind finely, boil to deactivate endogenous enzymes	Enzyme/substrate (E/S)=0.05; 2 h at 55°C	Neutrase = 17.2% Flavourzyme = 17% Alcalase = 17.4%
	Grind finely, heat treat to deactivate endogenous enzymes	2 h at 50 –55°C; enzyme/substrate (E/S)=5%	Neutrase = 14.4% Protamex = 14.6% Alcalase = 19.6%

Source: Ivanovs and Blumberga, 2017.

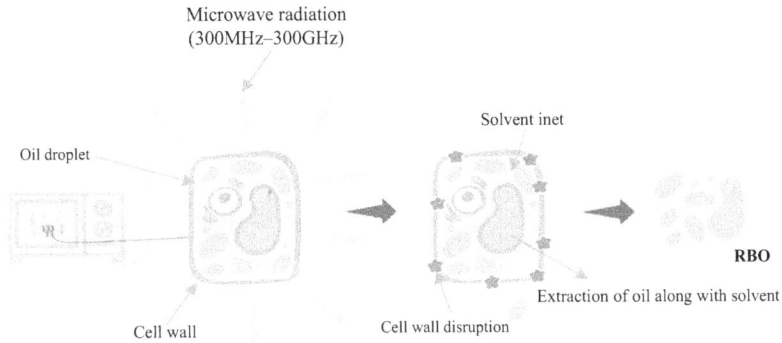

FIGURE 11.5 MAE (Punia et al., 2021).

the solution and impact each other. The collisions between ions result in the conversion of kinetic energy of moving ions into thermal energy. In a dipole rotation mechanism, the role of polar molecules in material such as water is important. When an electrical field is applied to material containing polar molecules with random orientation, based on the polarity of the field, the molecules re-orient themselves. If the microwave frequency is 2450 MHz, the change of its polarity is 2.45×10^9 (4.9×10^4) cycles/second. The rearrangement of the dipoles of a molecule with rapidly changing electric fields is called dipole rotation. The friction created when molecules rotate and contact the surrounding medium results in the production of heat (Vinatoru et al., 2017).

MEA is the method that applies microwaves during the extraction process between a solid matrix and an adequate solvent. When microwaves interact with polar solvents, the temperature of the mixture increases based on the effect of the energy generated by microwaves. Ionic polarization and dipole rotation may occur altogether or individually in MEA. Heat created inside the mixture is uniform, and a very low amount of heat energy can be lost to the surrounding environment during the process. Heat can disrupt or change the cell structure to facilitate the liberation of intercellular components. In comparison with conventional methods, MAE proved to be better from the point of view of heat and mass transfer. Heat is transferred from an outside heating medium to a combination of solvent and solid material to raise the cell temperature, while soluble components migrate from within the cell to the medium in a traditional extraction technique. Heat and mass flow in the opposite direction in this scenario. In the MAE method, heat generated inside goes out of the cell in the same direction as the solute. MAE helps to increase mass transfer coefficients and yield and lower the extraction process time.

Because of their capacity to absorb microwaves and their ecologically benign character, polar solvents such as isopropanol and ethanol are appropriate for microwave assisted oil extraction (Garofalo et al., 2020).

Zigoneanu et al. (2008) compared MAE extraction using isopropanol (polar solvent) with hexane (non-polar solvent). The solvent/rice bran ratio was 3:1 w/w, and five extraction temperatures were surveyed (40, 60, 80, 100, and 120°C). The results indicated that the extraction yield increased with temperature when isopropanol was used. At 120°C, the amount of oil extracted with isopropanol after 15 min was 50% higher than with hexane (Garofalo et al., 2020).

Kumar et al. (2016) compared the conventional extraction method with UAE and MAE, using three solvents (methanol, hexane, and petroleum ether). With methanol, about 96% of the oil in rice bran was extracted and the γ-oryzanol content was 85.0 ± 0.2 mg/kg. Both the crude oil and total phenol content of the microwave-treated methanol extract (80%) were significantly higher than those of the ultrasonic and conventional solvent extraction (Punia et al., 2021).

Terigar et al. (2011) improved the microwave extraction method in the lab and on the pilot scale. The settings in the lab were 73°C for 21 min, an ethanol/rice bran ratio of 3:1, and a flow rate of

TABLE 11.4

Several Studies on MAE

Extraction conditions	Results	References
Isopropanol, 120°C, solvent/bran = 3:1 w/w, 15 min	RBO = 0.15 g/g fresh bran	Zigoneanu et al. (2008)
Methanol, 38°C, 800 W, solvent/bran = 3:1 w/w, 1 h	Efficiency = 96.03% of recoverable RBO γ-oryzanol = 85.0 ± 0.2 mg/kg	Kumar et al. (2016)
Ethanol, 120°C, solvent/bran = 3:1 w/w, 20 min	RBO = 17.2% dry bran	Kanitkar et al. (2011)
Isopropanol, 82°C, 2.1 bar, 95 W, solvent/bran = 3:1 w/w, 30 min	RBO = 23.6 g/100 g dry bran γ-oryzanol = 2.256 ppm	Shukla and Pratap (2017)

Source: Garofalo et al., 2020.

100 ml/min. They then applied the same temperature to the pilot scale with flow rates of 1.0 and 0.6 l/min and got 90% of the oil in just 8 min (Garofalo et al., 2020) (Table 11.4).

11.2.2.5   UAE

In recent years, the application of ultrasonic waves to extract biologically active compounds has become increasingly popular because it is a cheap, environmentally friendly extraction process with high extraction yield and good reproducibility. It also uses less solvent in a shorter span of time and can be easily adapted to industrial-scale production. Therefore, this method is widely used in food and natural product processing (Punia et al., 2021; Garofalo et al., 2020).

The human ear can hear sound waves at frequencies ranging from 16 to 20 kHz. An ultrasonic wave is a form of sound that has a frequency greater than the human hearing limit. It has a frequency range of 20 Hz to 10 MHz. Ultrasonic waves require a medium to transfer their information (i.e., tissues). Ultrasonic waves are transmitted through the medium via a sequence of compression and rarefaction phases. The formation of cavitation bubbles occurs when the rarefaction cycle is higher than the attractive forces of the liquid molecules at an adequate level of power. These bubbles become bigger due to the addition of a very small amount of gas during the rarefaction or stretching phase. Afterwards, the bubbles will collapse when the compression cycle is completed. This bubble phenomenon generates mechanical energy equal to several thousand atmospheres of pressure. The localized temperature and pressure can reach 5000°C and 500 bar (Chemat et al., 2011; Vinatoru et al., 2017).

In UAE, the huge pressure released is able to break down cell membranes leading to easier penetration of the solvent into the substrates and a higher area of surface contact between the liquid and solid phase (Chemat et al., 2017).

Sound power (W), sound intensity (W/m^2), and sound power density (W/m^3) are some of the metrics used in the ultrasonic method. High-intensity ultrasound (low frequencies of 100 to 16 kHz and high power of 10 to 1000 W/cm^2) is used in UAE to improve extraction efficiency by altering the physical and chemical characteristics of the food material (Ivanovs and Blumberga, 2017).

Thanks to the effect of ultrasonic waves that cause cell wall disruption, solvent penetration is more favourable for capturing target compounds, helping to reduce extraction time and reduce solvent consumption. Besides, the extraction process is conducted at a low temperature to minimize the damage to thermal-sensitive compounds. Because it minimizes the quantity of solvents used and the extraction time, UAE is considered a green and ecologically beneficial approach (Roselló-Soto et al., 2016).

UAE is a promising method for RBO extraction. Cravotto et al. (2004) extracted RBO using ultrasound with water instead of organic solvents. At a water/bran ratio of 12/1, pH = 12, and ultrasound at 12 kHz, 300 W, and 45°C for 30 min, the oil content obtained was 21% of the dry bran. This result is similar to that of Khoei and Chekin (2016) at 25°C, 60 kHz for 70 min. Furthermore, oil extracted by UAE has a lower FFA content and less discolouration than hexane-extracted oil (Garofalo et al., 2020). These two studies showed the feasibility of UAE for low-temperature RBO extraction in aqueous media.

Ethanol, in addition to water, can be used to extract RBO with UAE. Tabaraki and Nateghi (2011) improved the extraction of RBOs, polyphenols, and antioxidants. An ultrasonic bath (35 kHz and 140 W) was used for 28 minutes at 60°C, with an ethanol concentration of 87% and a solvent-to-bran ratio of 21:1. A total of 19.83% dry bran oil was extracted (Garofalo et al., 2020) (Figures 11.6 and 11.7).

11.2.2.6 Pulsed-Electric Field Extraction

Pulsed-electric field (PEF) extraction is a method that uses high-voltage short pulses of electricity. These pulses last for a very short time, from microseconds to milliseconds. The range of pulses intensity is about 10–80 kV/cm. When a substance has electrical conductivity ability (such as a liquid food containing its ions) and is placed in an electric field, there is an electrical current run inside the substance; pulses are transmitted within the liquid based on the existence of charged molecules. In food processing and preservation, PEF extraction is a non-thermal technology used to destroy micro-organisms and keep food longer. PEF extraction causes no harmful effects to the

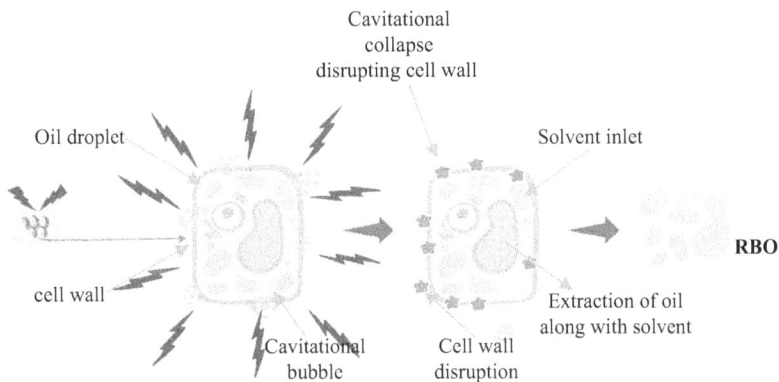

FIGURE 11.6 UAE (Punia et al., 2021).

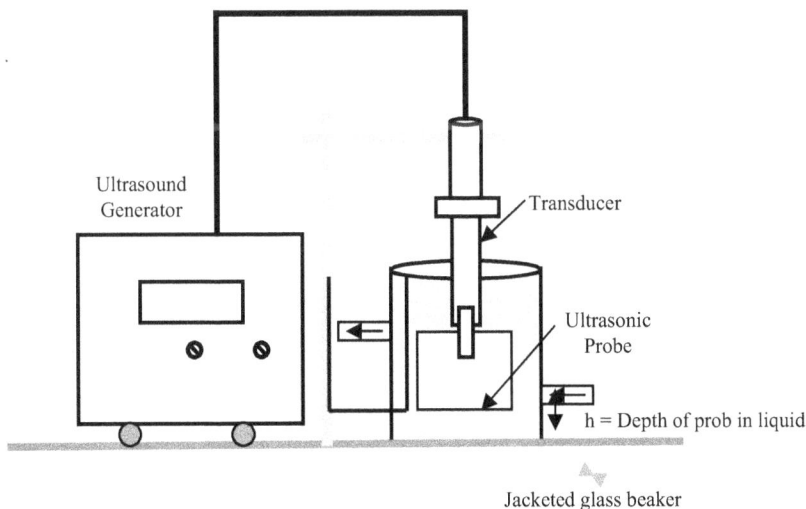

FIGURE 11.7 Schematic diagram of a UAE system.

nutritional value and sensorial quality of food and minimizes processing time. PEF treatment has been found to be highly efficient for deactivating micro-organisms, enhancing pressing efficiency, and enhancing juice extraction from food plants, as well as intensifying the dehydration and drying of food. It is considered a potential technology to provide high-quality food products to consumers (Soquetta et al., 2018).

One of the wide uses of PEF is its application in the extraction process. In PEF extraction, pulses of electricity open the pores in the cell membranes to make them softer and improve their permeability. The extraction process conducted at low temperatures leads to the protection of heat-sensitive components from deterioration.

11.2.3 FATTY ACID PURIFICATION

In order to improve quality and commercial value, as well as remove unwanted impurities that affect product quality and shelf life, the extracted crude oil needs to be refined. This process involves many physical and chemical steps. Oils extracted using modern green methods may require fewer purification steps than conventional solvent extraction methods.

The main purification methods for RBO include dewaxing, degumming, neutralization of FFA, bleaching, and steam deodorization (Garofalo et al., 2020; Toy and Bao, 2019). The refining process can change the quality of the product oil compared to crude oil.

Mezouari and Eichner (2007) identified several differences in the composition of crude and refined RBO. They include tocopherols (50.8 vs 34.3 mg/100 g), sterols (4035.6 vs 2832.0 mg/100 g), sterol ester (658.8 vs 562.0 mg/100 g) and γ-oryzanol (1.6% vs 0.5%) in crude and refined oils, respectively. Pestana-Bauer et al. (2008) discovered that the FFA composition of RBO did not change considerably during purification (Toy and Bao, 2019).

Jin et al. (2016) used multistage molecular distillation to refine the oil and achieved a yield of 80–85% for refined oil. Four coloured RBO fractions, in particular, were produced, which were light in colour and maintained about 80% of the oryzanol content of the crude oil (Toy and Bao, 2019).

Fish oil contains high levels of PUFAs, especially DHA/EPA, which have positive effects on human health. In order to be applied to food and pharmaceutical products, the crude fish oil obtained after extraction needs to be purified through steps such as removing impurities, neutralizing, bleaching, and deodorizing. Alternatively, crude oil can be enriched to obtain PUFA-rich oil by isolation of certain fractions, using techniques such as molecular distillation (MD), urea precipitation, and SpFE.

The degree of unsaturation of fatty acids can be separated via urea fractionation. FFAs are produced by hydrolyzing triglycerides, which are subsequently treated with urea and ethanol. Urea may be used to make SFAs and MUFAs, which are complex solid-phase molecules containing alkyl chains. These complexes precipitate at 20°C and are filtered out, leaving the PUFA-rich liquid fraction (Baiano, 2014; Sahena et al., 2009).

To produce omega 3 rich oils, MD, also known as short route distillation, is frequently utilized. Under high vacuum circumstances, compounds are separated based on their vapour pressure (3–10 bar), forcing the molecules to follow a route that is shorter than their normal free path. MD is widely utilized in industry; however, it has a low yield and selectivity (between 55% and 70%) compared to the market need for highly concentrated oils (85%). MD also requires temperatures in the 140–170°C range, which might damage thermally unstable materials despite the short operating time.

SC-CO$_2$ liquid extraction was also studied for the fractionation of PUFA-rich oils. This approach may be used to selectively extract low polarity lipid molecules while avoiding the extraction of polar contaminants such as certain inorganic derivatives containing heavy metals. Ferdosh et al. (2016) investigated the simultaneous extraction and fractionation of fish oil from tuna heads using SC-CO$_2$ and a co-solvent of 20% ethanol. The optimum conditions used were 40 MPa, 65°C for 120 min, and a flow rate of 3 ml/min. The total oil obtained was 35.6% (g/100 g dry matter). Simultaneous extraction and fractionation of fish oil into multiple steps (six parts) is more efficient than two or three parts for the separation of PUFA-rich fish oil. In the following fractions (4–6), PUFAs were rich in DHA, EPA, omega-3 fatty acids (27.8–43.4%), and omega-6 (5.5–7.1%). Fraction 6 was the most PUFA-rich (48.93%), with the highest amount of DHA (28.21%), EPA (6.12%), and omega-3 fatty acids (43.4%).

11.3 APPLICATION OF FATTY ACID IN NUTRACEUTICALS AND COSMETICS

11.3.1 APPLICATIONS OF RBO

Crude RBO contains 90–96% saponifiable lipids and 3–5% unsaponifiable lipids (Oliveira et al., 2011). The saponified lipid fraction is composed of 81–84% triacylglycerol, 2–3% diacylglycerol, 1–2% monoacylglycerol, 2–6% FFA, 3–4% wax, 0.8% glycolipids, and 1–2% phospholipids. Micronutrients such as tocopherols and tocotrienols, γ-oryzanol, phytosterol, and squalene are abundant in the unsaponified part (Pali, 2013; Toy and Bao, 2019; Punia et al., 2021; Singanusong and Garba, 2019).

RBO has a balanced fatty acids profile and can contain up to 47% MUFAs, 33% PUFAs, and 20% SFAs. The main unsaturated fatty acids are oleic acid (C18:1) and linoleic acid (C18:2), and the main SFA is palmitic acid (C16:0) (Garofalo et al., 2020).

One of the most important micronutrients in RBO is γ-oryzanol. It consists of ferulic acid esters combined with sterol alcohols and triterpenes. Cycloartenyl ferulate, 24-methylenecycloartenyl ferulate, campesteryl ferulate, and β-sitosteryl ferulate are the four main esters that make up roughly 80% of γ-oryzanol in RBO (Garofalo et al., 2020; Singanusong and Garba, 2019). The γ-oryzanol concentration in crude RBO can range from 1.5% to 2.9 %, according to Ghatak and Panchal (2012), and the refining process can lower this ratio (Singanusong and Garba, 2019). Furthermore, the amount of γ-oryzanol present in RBO and its composition is affected by a range of variables, including rice type, ambient conditions, farming practices, and extraction and purifying procedures.

Tocols (tocopherols and tocotrienols) are members of the vitamin E group. In rice bran lipids, all four major types of tocopherol (α, β, γ, and δ) are present. RBO includes around 0.2% tocols, with tocotrienol accounting for 70% of the total. Refined RBO is reported to contain 48–70 mg% tocopherols (Singanusong and Garba, 2019). RBO also contains vitamin K (phylloquinone) at 25.7 μg/100 g of oil (Toy and Bao, 2019).

Plant cell walls and membranes contain the majority of phytosterols (plant sterols). The concentrations of phytosterols in commercial crude and purified RBO were 1362–1376 and 858–1034 mg/100 g, respectively (Sawadikiat and Hongsprabhas, 2014) (Singanusong and Garba, 2019). Mezouari and Eichner (2007) reported phytosterol content in purified RBO at 1.5–2.0% (Toy and Bao, 2019) (Table 11.5).

Having unique properties and containing healthy ingredients, RBO is widely used in food, pharmaceuticals, and cosmetics.

11.3.1.1 Application in Food

Many Asian countries, including Japan, Korea, China, Taiwan, and Thailand, utilize RBO as a premium cooking oil. RBO is produced and consumed in large quantities in Asian countries, with annual RBO production in India, China, and Japan being approximately 472, 90, and 65 thousand tonnes, respectively (Garofalo et al., 2020). RBO has a moderate flavour and nice aroma, making it a versatile component in international cuisines. RBO is suited for high-temperature cooking methods such as frying because of its high smoke (232°C) and ignition (350°C) points (Lai et al., 2019). It takes less time to fry using RBO, and the food absorbs 15% less oil, making it more cost-effective and energy-efficient (Nayik et al., 2015). Because of its high oryzanol concentration, RBO has the greatest viscosity among cooking oils, making meals glossy and appetizing due to its lengthy stay on the food's surface (Wang, 2019; Punia et al., 2021).

TABLE 11.5

Micronutrient Composition in RBO

Micronutrient	Crude RBO (%)	Refined RBO (%)
γ-Oryzanol	1.7–2.1	1.2–1.7
Tocopherol	0.0264	0.02–0.08
Tocotrienols	0.007	0.025–0.17
Phytosterols	1.362–1.37	0.858–1.037
Squalene	0.756	0.3–0.4

Source: Singanusong and Garba, 2019.

RBO can be used in bakery products as a fat substitute with high oxidation stability. RBO has also been reported to be used as a substitute for shortening in bread making, muffins, mayonnaise, and salad dressing. Replacing 50% of shortening with RBO results in higher bread quality in terms of toasting and sensory performance (Kaur et al., 2014; Orthoefer, 2005; Garba et al., 2017).

RBO can also be used to improve dairy products. Abbas et al. (2017) improved yoghurt by combining 3% RBO and a *Bifidobacterium* strain. The obtained yoghurt had the right viscosity as well as the potential health benefits of RBO. El-Galeel and Atwaa (2017) added RBO to soft cheeses to investigate its oxidative stability and microbiological and organoleptic properties. When compared to the butylated hydroxy anisole treatment, the results revealed that the cheeses studied had a high phenolic content, superior oxidation stability, higher organoleptic characteristics, and the lowest microbiological counts (Punia et al., 2021).

RBO can also be used to replace animal fat in meat products. Pork sausage may be produced by substituting 35%, 40%, 45%, and 50% RBO for pork back fat. When compared to sausages made from pork back fat, the RBO-added sample had the lowest cooking loss (40%). RBO enhanced the quantity of unsaturated fatty acids and PUFAs in sausages while also improving tenderness and texture. Furthermore, RBO has no effect on the original sausage's taste or texture (Punia et al., 2021).

11.3.1.2 Applications in the Cosmetics Industry

RBO can also be used as an ingredient in cosmetic products. It can be added to sunscreen, nail polish, lipstick, hair conditioner, shower gel, and moisturizer. RBO contains squalene and tocotrienols, which help soften and repair the skin. RBO's γ-oryzanol can protect the skin from free radicals and reduce UV-induced lipid peroxidation, while ferulic acid and its esters stimulate hair growth and prevent skin ageing (Garba et al., 2017). RBO contains vitamin E, which nourishes skin cells, inhibits pore blockage from skin renewal, and slows the ageing process (Nayik et al., 2015). When utilized on human skin, a nano-emulsion containing 10% RBO was tested *in vivo* and shown to have promise for application in cosmetic goods due to enhanced skin moisture content, minimal irritation potential, and maintenance of normal skin pH (Bernardi et al., 2011; Garba et al., 2017). Because of these beneficial effects, RBO is a potential agent for cosmetic products.

11.3.1.3 Applications in Pharmaceuticals, Functional Foods

RBO has been found to decrease intestinal cholesterol absorption while increasing cholesterol metabolites, suggesting that it may protect against lipid abnormalities and lower the risk of heart disease. RBO can also decrease caspase activity, either indirectly or directly, preventing b-cell death and lowering diabetes risk (Liang et al., 2014). RBO's antioxidant, anti-inflammatory, anti-bacterial, anti-inflammatory, anti-diabetes, anti-cancer, anti-hypertensive, and anti-obesity characteristics can be ascribed to the fact that it includes substantial levels of micronutrients such as γ-oryzanol, tocols, and phytosterol (Toy and Bao, 2019; Garofalo et al., 2020; Punia et al., 2021).

Oryzanol is well known for its ability to reduce cholesterol by boosting bile production (Ha et al., 2005). It also lowers triglycerides and raises the good-to-bad cholesterol ratio (HDL/LDL), which is crucial for heart health (Nayik et al., 2015). Lai et al. (2012) conducted comparative research on 35 patients with type 2 diabetes and found that daily RBO consumption can reduce serum cholesterol and low-density lipoprotein cholesterol levels, which can improve blood lipid profiles (Toy and Bao, 2019).

RBO has been found to be beneficial in boosting immune system function and therefore protecting the body from sickness due to its high concentration of γ-oryzanol and phytosterols. RBO is also utilized in muscle-building supplements for bodybuilders and sportspeople (Nayik et al., 2015; Garba et al., 2017). RBO contains vitamin E, which helps to enhance nerve function and regulate endocrine hormones. γ-Oryzanol has been found to successfully decrease thyroid-stimulating hormone in hypothyroidism patients; it can also help women deal with the symptoms of menopause. According to studies, γ-oryzanol is four times more efficient than tocopherol at preventing cellular oxidation (Nayik et al., 2015).

Thanks to its bioactive ingredients and their health-promoting effects, RBO has a high potential for use in pharmaceuticals and dietary supplements.

11.3.2 APPLICATION OF OIL EXTRACTED FROM FISH PROCESSING WASTE/BYPRODUCTS

EPA and DHA, as well as omega-3 PUFAs, have been shown to improve human health by lowering the risk of cardiovascular disease, cancer, and diabetes. These lipids also have an effect on the immune system and ensure nerve development. However, DHA/EPA are essential fatty acids that the human body cannot synthesize on its own. In Europe and Japan, fish oil containing DHA/EPA is found in a wide variety of foods, including bakery products, margarine, infant formula, and dietary supplements.

However, DHA, in particular, and fish oil, in general, are quite easily oxidized under ambient conditions. Therefore, stabilization of DHA and oils to resist oxidation is necessary for commercial application in food products. Bread is considered an ideal medium for PUFA-ω-3 supplementation because the CO_2 released during baking and fermenting can help eliminate oxygen and prevent oxidation while the bread is at a high temperature. The recommended amount to be added to bread is 0.5–1.5% ω-3 oil or 1.0–2.5% ω-3 flour.

In addition, the strong scent of fish oil presents a limitation in food applications. To solve this problem, as well as ensure the stability of fish oil, fish oil encapsulation can be applied. Ghorbanzade et al. (2017) studied the packaging of fish oil with nano-liposomes and its addition to a yoghurt product, monitoring the product for 3 weeks at 4°C. The results showed that fish oil encapsulation in nanoliposomes significantly reduced the acidity, syneresis, and peroxide value of the supplemented yoghurt. Maximum amounts of DHA and EPA were retained after 21 days of storage in the nano-encapsulated fish oil yoghurt sample (57% and 12%, respectively), while the unpackaged fish oil sample had 27% DHA and 6% EPA (reduced to about 50% of its original value). Moreover, yoghurt products supplemented with nano-encapsulated fish oil had organoleptic characteristics (colour, smell, taste, texture) closer to the control samples (yoghurt samples without fish oil added).

Fish oil has also been reported as a potential supplement to reduce the severity of certain skin disorders such as photoaging, skin cancer, allergies, dermatitis, skin wounds, and melanogenesis. Fish oil and related active ingredients, such as omega-3 and omega-6 PUFAs, have been shown to be useful for maintaining skin homeostasis and ameliorating skin abnormalities. The fatty acids in fish oil can improve the skin's protective function, inhibit UV-induced inflammation and hyperpigmentation, relieve dry and itchy skin caused by dermatitis, accelerate wound healing, and prevent the development of skin cancer. The treatment of skin disorders with fish oil represents a new direction to make advances for cosmetic and dermatological use.

11.4 CONCLUSIONS

Agricultural byproducts offer certain value, including the presence of significant quantities of fatty acids and micronutrients. There are many techniques for extracting these compounds from potential waste sources, each of which has certain advantages and limitations. In terms of the quality of extracted fatty acids and micronutrients, the SC-CO$_2$ method presents the most outstanding advantages. The results of many studies show that fatty acids and micronutrients extracted from agricultural byproducts have great potential and perspectives in the field of nutraceuticals and cosmetics.

BIBLIOGRAPHY

Abbas, H. M., Shahein, N. M., Abd-Rabou, N. S., Fouad, M. T., and Zaky, W. M. 2017. Probiotic-fermented milk supplemented with rice bran oil. *International Journal of Dairy Science*, 12(3): 204–210.

Ahmad, T., Masoodi, F. A., Rather, S. A., Wani, S. M., and Gull, A. 2019. Supercritical fluid extraction: A review. Journal of Biological and Chemical Chronicles, 5: 114–122.

Al-Okbi, S. Y., Mohamed, D. A., Hamed, I. M., Agoor, F. S., Ramadan, A. M. A., El-Saed, S., and Helal, A. M. 2013. Comparative study on Egyptian rice bran extracted by solvents and supercritical CO2. *Advances in Food Sciences*, 35: 23–29.

Al-Okbi, S. Y., Mohamed, D. A., Hamed, T. E., and Esmail, R. S. H. 2014. Rice bran oil and pumpkin seed oil alleviate oxidative injury and fatty liver in rats fed high fructose diet. *Polish Journal of Food and Nutrition Sciences*, 64(2): 127–133.

Amarasinghe, B. M. W. P. K., and Gangodavilage, N. C. 2004. Rice bran oil extraction in Sri Lanka: Data for process equipment design. *Food and Bioproducts Processing*, 82(1): 54–59.

Arackal, J. J., and Parameshwari, S. 2021. A detailed evaluation of fatty acid profile and micronutrient analysis of chapattis incorporated with avocado fruit pulp. *Materials Today: Proceedings*, 45: 3268–3273. https://doi.org/10.1016/j.matpr.2020.12.468.

Arjeha, E., Akhavanb, H. R., Barzegarc, M., and Carbonell-Barrachina, Á. A. 2020. Bio-active compounds and functional properties of pistachio hull: A review. *Trends in Food Science and Technology*, 97: 55–64. https://doi.org/10.1016/j.tifs.2019.12.031.

Arnold, L. K., and Choudhury, R. B. R. 1961. The fatty acid composition of cottonseed oil at various stages of solvent extraction. *The Journal of the American Oil Chemists' Society*, 38(2): 87–89.

Ashokkumar, M. 2015. Applications of ultrasound in food and bioprocessing. *Ultrasonics Sonochemistry*, 25: 17–23.

Badal, R. 2002. Supercritical carbon dioxide extraction of lipids from raw and bioconverted rice bran. *LSU Master's Theses*. Louisiana State University and Agricultural and Mechanical College.

Balachandran, C., Mayamol, P. N., Thomas, S., Sukumar, D., Sundaresan, A., and Arumughan, C. 2008. An ecofriendly approach to process rice bran for high quality rice bran oil using supercritical carbon dioxide for nutraceutical applications. *Bioresource Technology*, 99(8): 2905–2912.

Bandar, H., Hijazi, A., Rammal, H., et al. 2013. Techniques for the extraction of bioactive compounds from Lebanese Urtica dioica. *AJPCT*, 1(6): 507–513.

Batista, A., Vetter, W., and Luckas, B. 2001. Use of focused open vessel microwave-assisted extraction as prelude for the determination of the fatty acid profile of fish – A comparison with results obtained after liquid-liquid extraction according to Bligh and Dyer. *European Food Research and Technology*, 212(3): 377–384.

Begum, A., Sarma, J., Borah, P., Moni Bhuyan, P., Saikia, R., Hussain Ahmed, T., and Rai, L. 2015. Microwave (MW) energy in enzyme deactivation: Stabilization of rice bran from few widely consumed indigenous rice cultivars (*Oryza sativa* L.) from Eastern Himalayan range. *Current Nutrition and Food Science*, 11(3): 240–245.

Belarbi El, H., Molina, E., and Chisti, Y. 2000. A process for high yield and scaleable recovery of high purity eicosapentaenoic acid esters from microalgae and fish oil. *Process Biochemistry*, 35(9): 951–969.

Bernardi, D. S., Pereira, T. A., Maciel, N. R., Bortoloto, J., Viera, G. S., Oliveira, G. C., and RochaFilho, P. A. 2011. Formation and stability of oil-in- water nanoemulsions containing rice bran oil: In vitro and in vivo assessments. *Journal of Nanobiotechnology*, 9: 44.

Canete-Rodríguez, A. M., Santos-Duenas, I. M., Jiménez-Hornero, J. E., Ehrenreich, A., Liebl, W., and García-García, I. 2016. Gluconic acid: Properties, production methods and applications—An excellent opportunity for agro-industrial by-products and waste bio-valorization. *Process Biochemistry*. https://doi.org/10.1016/j.procbio.2016.08.028.

Capellini, M. C., Giacomini, V., Cuevas, M. S., and Rodrigues, C. E. 2017. Rice bran oil extraction using alcoholic solvents: Physicochemical characterization of oil and protein fraction functionality. *Industrial Crops and Products*, 104: 133–143.

Castro-López, C., Rojas, R., Sánchez-Alejo, E. J., Niño-Medina, G., and Martínez-Ávila, G. C. G. 2016. Phenolic compounds recovery from grape fruit and by-products: An overview of extraction methods. In: *Grape and Wine Biotechnology*. https://doi.org/10.5772/64821.

Chemat, F., Fabiano-Tixier, A. S., Vian, M. A., Allaf, T., and Vorobiev, E. 2015. Solvent-free extraction of food and natural products. *Trends in Analytical Chemistry*, 71: 158–168. https://doi.org/10.1016/j.trac.2015.02.021.

Chemat, F., Rombaut, N., Meullemiestre, A., Turk, M., Perino, S., Fabiano-Tixier, A., and Abert-Vian, M. 2017. Review of green food processing techniques. Preservation, transformation, and extraction. *Innovative Food Science and Emerging Technologies*, 41: 357–377.

Chemat, F., Rombaut, N., Sicaire, A.-G., Meullemiestre, A., Fabiano-Tixier, A. S., and Abert-Vian, M. 2017. Ultrasound assisted extraction of food and natural products. Mechanisms, techniques, combinations, protocols and applications. A review. *Ultrasonics Sonochemistry*, 34: 540–560.

Chemat, F., Vian, M. A., Ravi, H. K., Khadhraoui, B., Hilali, S., Perino, S., and Tixier, A. F. 2019. Review of alternative solvents for green extraction of food and natural products: Panorama, principles, applications and prospects. *Molecules*, 24(16): 3007.

Chemat, F., Zill-e-Huma, and Khan, M. K. 2011. Applications of ultrasound in food technology: Processing, preservation and extraction. *Ultrasonics Sonochemistry*, 18(4): 813–835. https://doi.org/10.1016/j.ultsonch.2010.11.023.

Chemat, S., Aissa, A., Boumechhour, A., Arous, O., and Ait-Amar, H. 2017. Extraction mechanism of ultrasound assisted extraction and its effect on higher yielding and purity of artemisinin crystals from Artemisia annua L. leaves. *Ultrasonics Sonochemistry*, 34: 310–316.

Chia, S. L., Boo, H. C., Muhamad, K., Sulaiman, R., Umanan, F., and Chong, G. H. 2015. Effect of subcritical carbon dioxide extraction and bran stabilization methods on rice bran oil. *Journal of the American Oil Chemists' Society*, 92(3): 393–402.

Cravotto, G., Binello, A., Merizzi, G., and Avogadro, M. 2004. Improving solvent-free extraction of policosanol from rice bran by highintensity ultrasound treatment. *European Journal of Lipid Science and Technology*, 106(3): 147–151.

de Oliveira, D. A. S. B., Minozzo, M. G., Licodiedoff, S., and Waszczynskyj, N. 2016. Physicochemical and sensory characterization of refined and deodorized tuna (*Thunnus albacares*) by-product oil obtained by enzymatic hydrolysis. *Food Chemistry*, 207: 187–194. https://doi.org/10.1016/j.foodchem.2016.03.069.

Dolatowski, Z. J., Stadnik, J., and Stasiak, D. 2007. Applications of ultrasound in food technology. *Acta Scientiarum Polonorum. Technologia Alimentaria*, 6(3): 89–99.

Dovale-Rosabal, G., Rodríguez, A., Contreras, E., et al. 2019. Concentration of EPA and DHA from refined salmon oil by optimizing the urea–fatty acid addiction reaction conditions using response surface methodology. *Molecules*, 24(9): 1642. https://doi.org/10.3390/molecules24091642.

Dubey, B., Fitton, D., Nahar, S., and Howarth, M. 2019. Comparative study on the rice bran stabilization processes: A review. *Research and Development in Material Science*, 11(2). RDMS.000759.2019. DOI: 10.31031/RDMS.2019.11.000759

Eikani, M. H., Golmohammada, F., and Homami, S. S. 2012. Extraction of pomegranate (*Punica granatum* L.) seed oil using superheated hexane. *Food and Bioproducts Processing*, 90(1): 32–36. https://doi.org/10.1016/j.fbp.2011.01.002.

El-Galeel, A. A., and Atwaa, E. H. 2017. Use of rice bran oil as natural antioxidant in white soft cheese manufacturing. *Zagazig Journal of Agricultural Research*, 44(2): 605–615.

Federici, F., Fava, F., Kalogerakisc, N., and Mantzavinos, D. 2009. Valorisation of agro-industrial by-products, effluents and waste: Concept, opportunities and the case of olive mill wastewaters. *Society of Chemical Industry*, 84(6): 895–900. https://doi.org/10.1002/jctb.2165.

Ferreiraa, D. F., Barina, J. S., Binello, A., Veselov, V. V., and Cravotto, G. 2019. Highly efficient pumpkin-seed extraction with the simultaneous recovery of lipophilic and hydrophilic compounds. *Food and Bioproducts Processing*, 117: 224–330. https://doi.org/10.1016/j.fbp.2019.07.014.

Fournier, V., Destaillats, F., Juanéda, P., Dionisi, F., Lambelet, P., Sébédio, J.-L., and Berdeauxa, O. 2006. Thermal degradation of long-chain polyunsaturated fatty acids during deodorization of fish oil. *European Journal of Lipid Science and Technology*, 108(1): 33–42. https://doi.org/10.1002/ejlt.200500290.

Frieling, P. von, and Schügerl, K. 2000. Recovery of lactic acid from aqueous model solutions and fermentation broths. *Process Biochemistry*, 35: 685–696.

Garba, U., Singanusong, R., Jiamyangyuen, S., and Thongsook, T. 2017. Extraction and utilization of rice bran oil: A review. *Safety*, 17: 24.

Garofalo, S. F., Tommasi, T., and Fino, D. 2020. A short review of green extraction technologies for rice bran oil. *Biomass Conversion and Biorefinery*, 11: 569–587.

Ghatak, S. B., and Panchal, S. J. 2012. Investigation of the immunomodulatory potential of oryzanol isolated from crude rice bran oil in experimental animal models. *Phytotherapy Research*, 26(11): 1701–1708.

Gulzar, S., Raju, N., Nagarajarao, R. C., and Benjakul, S. 2020. Oil and pigments from shrimp processing by-products: Extraction, composition, bioactivities and its application-A review. *Trends in Food Science and Technology*, 100: 307–319. https://doi.org/10.1016/j.tifs.2020.04.005.

Ha, T. Y., Han, S., Kim, S. R., Kim, I. H., Lee, H. Y., and Kim, H. 2005. Bio active components in rice bran oil improve lipid profiles in rats fed a high-cholesterol diet. *Nutrition Research*, 25(6): 597–606.

Hamam, F., and Shahidi, F. 2008. Incorporation of selected long-chain fatty acids into trilinolein and trilinolenin. *Food Chemistry*, 106(1): 33–39. https://doi.org/10.1016/j.foodchem.2007.05.038.

Hanmoungjai, P., Pyle, D. L., and Niranjan, K. 2001. Enzymatic process for extracting oil and protein from rice bran. *Journal of the American Oil Chemists' Society*, 78(8): 817–821.

Hanmoungjai, P., Pyle, D. L., and Niranjan, K. 2002. Enzyme-assisted water-extraction of oil and protein from rice bran. *Journal of Chemical Technology and Biotechnology*, 77(7): 771–776.

Hasana, S. H., Ranjana, D., and Talat, M. 2010. Agro-industrial waste 'wheat bran' for the biosorptive remediation of selenium through continuous up-flow fixed-bed column. *Journal of Hazardous Materials*, 181(1–3): 1134–1142. https://doi.org/10.1016/j.jhazmat.2010.05.133.

Imsanguan, P., Roaysubtawee, A., Borirak, R., Pongamphai, S., Douglas, S., and Douglas, P. L. 2008. Extraction of α-tocopherol and γ-oryzanol from rice bran. *LWT - Food Science and Technology*, 41(8): 1417–1424.

Irakli, M., Kleisiaris, F., Mygdalia, A., and Katsantonis, D. 2018. Stabilization of rice bran and its effect on bioactive compounds content, antioxidant activity and storage stability during infrared radiation heating. *Journal of Cereal Science*, 80: 135–142.

Ivanovs, K., and Blumberga, D. 2017. Extraction of fish oil using green extraction methods: A short review. *Energy Procedia*, 128: 477–483. https://doi.org/10.1016/j.egypro.2017.09.033.

Jayakrishnan, U., Dekaa, D., and Das, G. 2018. Enhancing the volatile fatty acid production from agro-industrial waste streams through sludge pretreatment. *Environmental Science: Water Research and Technology*: 1–12. https://doi.org/10.1039/c8ew00715b.

Jiang, Y. 2019. Bioprocessing technology of rice bran oil. In: Ling-Zhi Cheong and Xuebing Xu (Eds.), *Rice Bran and Rice Bran Oil*. AOCS Press, Urbana: 97–123.

Jiao, J., Li, Z. G., Gai, Q. Y., Li, X. J., Wei, F. Y., Fu, Y. J., and Ma, W. 2014. Microwave-assisted aqueous enzymatic extraction of oil from pumpkin seeds and evaluation of its physicochemical properties, fatty acid compositions and antioxidant activities. *Food Chemistry*, 147: 17–24.

Jin, J., Xie, D., Chen, H. Q., Wang, X. S., Jin, Q. Z., and Wang, X. G. 2016. Production of rice bran oil with light color and high oryzanol content by multi stage molecular distillation. *Journal of the American Oil Chemists' Society*, 93(1): 145–153.

Juchen, P. T., Araujo, M. N., Hamerski, F., Corazza, M. L., and Voll, F. A. P. 2019. Extraction of parboiled rice bran oil with supercritical CO2 and ethanol as co-solvent: Kinetics and characterization. *Industrial Crops and Products*, 139: 111506.

Kanitkar, A., Sabliov, C. M., Balasubramanian, S., Lima, M., and Boldor, D. 2011. Microwave-assisted extraction of soybean and rice bran oil: Yield and extraction kinetics. *Transactions of the ASABE*, 54(4): 1387–1394.

Kaur, A., Jassal, V., and Bhise, S. R. 2014. Replacemnt of bakery shortening with rice bran oil in the preparation of muffins. *African Journal of Biochemistry Research*, 8(7): 141–146.

Khatoon, S., and Gopalakrishna, A. G. 2004. Fat-soluble nutraceuticals and fatty acid composition of selected Indian rice varieties. *Journal of the American Oil Chemists' Society*, 81(10): 939–943.

Khaw, K.-Y., Parat, M.-O., Shaw, P. N., and Falconer, J. R. 2017. Solvent supercritical fluid technologies to extract bioactive compounds from natural sources: A review. *Molecules*, 22(7): 1186. https://doi.org/10.3390/molecules22071186.

Khoei, M., and Chekin, F. 2016. The ultrasound-assisted aqueous extraction of rice bran oil. *Food Chemistry*, 194: 503–507.

Kuk, M. S., and Dowd, M. K. 1998. Supercritical CO2 extraction of rice bran. *Journal of the American Oil Chemists' Society*, 75(5): 623–628.

Kumar, M., Moon, U. R., and Mitra, A. 2012. Rapid separation of carotenes and evaluation of their in vitro antioxidant properties from ripened fruit waste of Areca catechu – A plantation crop of agro-industrial importance. *Industrial Crops and Products*, 40: 204–209. https://doi.org/10.1016/j.indcrop.2012.03.014.

Kumar, N., and Das, D. 2000. Erratum to "Enhancement of hydrogen production by Enterobacter cloacae IIT-BT 08". *Process Biochemistry*, 35(6): 589–593.

Kumar, P., Yadav, D., Kumar, P., Panesar, P. S., Bunkar, D. S., Mishra, D., and Chopra, H. K. 2016. Comparative study on conventional, ultrasonication and microwave assisted extraction of γ-oryzanol from rice bran. *Journal of Food Science and Technology*, 53(4): 2047–2053.

Kumar Sadh, P., Duhan, S., and Duhan, J. S. 2018. Agro-industrial wastes and their utilization using solid state fermentation: A review. *Bioresources and Bioprocessing*, 5(1): 1–15. https://doi.org/10.1186/s40643-017 -0187-z.

Kumari, P., Reddy, C. R. K., and Jha, B. 2011. Comparative evaluation and selection of a method for lipid and fatty acid extraction from macroalgae. *Analytical Biochemistry*, 415(2): 134–144. https://doi.org/10.1016 /j.ab.2011.04.010.

Lai, M. H., Chen, Y. T., Chen, Y. Y., Chang, J. H., and Cheng, H. H. 2012. Effects of rice bran oil on the blood lipids profiles and insulin resistance in type 2 diabetes patients. *Journal of Clinical Biochemistry and Nutrition*, 51(1): 15–18.

Lai, O. M., Jacoby, J. J., Leong, W. F., and Lai, W. T. 2019. Nutritional studies of rice bran oil. In: Cheong, L. Z., and Xu, X. B. (eds), *Rice Bran and Rice Bran Oil*. AOCS Press, Urbana: 19–54.

Lakkakula, N. R., Lima, M., and Walker, T. 2004. Rice bran stabilization and rice bran oil extraction using ohmic heating. *Bioresource Technology*, 92(2): 157–161.

Liang, Y., Gao, Y., Lin, Q. L., Luo, F. J., Wu, W., Lu, Q., and Liu, Y. 2014. A review of the research progress on the bioactive ingredients and physiological activities of rice bran oil. *European Food Research and Technology*, 238(2): 169–176.

Linder, M., Fanni, J., and Parmentier, M. 2005. Proteolytic extraction of salmon oil and PUFA concentration by lipases. *Marine Bioteachnology*, 15(1): 70–76. https://doi.org/10.1007/s10126-004-0149-2.

Liu, R.-L., Song, S.-H., Wu, M., He, T., and Zhang, Z.-Q. 2013. Rapid analysis of fatty acid profiles in raw nuts and seeds by microwave–ultrasonic synergistic in situ extraction–derivatisation and gas chromatography–mass spectrometry. *Food Chemistry*, 141(4): 4269–4277. https://doi.org/10.1016/j.foodchem .2013.07.011.

Liu, X., Wang, F., Liu, X., Chen, Y., and Wang, L. 2011. Fatty acid composition and physicochemical properties of ostrich fat extracted by supercritical fluid extraction. *European Journal of Lipid Science and Technology*, 113(6): 775–779. https://doi.org/10.1002/ejlt.201000434.

Liu, S. X., and Mamidipally, P. K. 2005. Quality comparison of rice bran oil extracted with d-limonene and hexane. *Cereal Chemistry Journal*, 82(2): 209–215.

Lomonaco, D., Mele, G., and Mazzetto, S. E. 2017. Cashew Nutshell Liquid (CNSL): From an agro-industrial waste to a sustainable alternative to petrochemical resources. In: Anilkumar, P. (ed.), *Cashew Nut Shell Liquid*. Springer, Cham: 19–38. https://doi.org/10.1007/978-3-319-47455-7_2.

Lordan, S., Paul Ross, R., and Stanton, C. 2011. Marine bioactives as functional food ingredients: Potential to reduce the incidence of chronic diseases. *Marine Drugs*, 9(6): 1056–1100. https://doi.org/10.3390/ md9061056.

Lü, C., Luo, X., Dong, X., Peng, J., and Cao, F. 2019. Enhanced versatile peroxidase production by *Pleurotus eryngii* using saccharide- and fatty acid-rich agro-industrial waste in solid-state fermentation. *Journal of Biobased Materials and Bioenergy*, 13(1): 134–139. https://doi.org/10.1166/jbmb.2019.1809.

Lucas, E. W. 2000. Oilseeds and oil-bearing materials. In: Loren, Z. K. J., and Kulp, K. (eds), *Handbook of Cereal Science and Technology*. Marcel Dekker Inc., New York: 297–362.

Mariana, I., Nicoleta, U., Sorin-Ştefan, B., Gheorghe, V., and Mirela, D. 2013. Actual methods for obtaining vegetable oil from oilseeds. Conference: TE-RE-RD.

Marića, B., Pavlićc, B., Čolovića, D., et al. 2020. Recovery of high-content ω–3 fatty acid oil from raspberry (*Rubus idaeus* L.) seeds: Chemical composition and functional quality. *LWT - Food Science and Technology*, 130: 109627. https://doi.org/10.1016/j.lwt.2020.109627.

Markom, M., Singh, H., and Hasan, M. 2001. Supercritical CO2 fractionation of crude palm oil. *Journal of Supercritical Fluids*, 20(1): 45–53. https://doi.org/10.1016/S0896-8446(00)00104-2.

Méndez, J. R. B., and Concha, J. L. H. Methods of extraction refining and concentration of fish oil as a source of omega-3 fatty acids. *Ciencia y Tecnología Agropecuaria*, 19(3), 645–668.

Mezouari, S., and Eichner, K. 2007. Comparative study on the stability of crude and refined rice bran oil during long-term storage at room temperature. *European Journal of Lipid Science and Technology*, 109(3): 198–205.

Mohamed, M. E. A., and Amer Eissa, A. H. 2012. Pulsed electric fields for food processing technology. In: *Structure and Function of Food Engineering*. https://doi.org/10.5772/48678.

Nayik, G. A., Majid, I., Gull, A., and Muzaffar, K. 2015. Rice bran oil, the future edible oil of India: A mini review. *Journal of Rice Research*, 3(151): 2.

Oliveira, M. S., Feddern, V., Kupski, L., Cipolatti, E. P., Badiale-Furlong, E., and SouzaSoares, L. A. 2011. Changes in lipid, fatty acids and phospholipids composition of whole rice bran after solid-state fungal fermentation. *Bioresource Technology*, 102(17): 8335–8338.

Oliveira, R., Oliveira, V., Aracava, K. K., and Costa Rodrigues, C. E. 2012. Effects of the extraction conditions on the yield and composition of rice bran oil extracted with ethanol—A response surface approach. *Food and Bioproducts Processing*, 90(1): 22–31.

Oliveira, R. C. d., Barros, S. T. D. d., and Gimenes, M. L. 2013. The extraction of passion fruit oil with green solvents. *Journal of Food Engineering*, 117(4): 458–463. https://doi.org/10.1016/j.jfoodeng.2012.12.004.

Orsavova, J., Misurcova, L., Vavra Ambrozova, J., Vicha, R., and Mlcek, J. 2015. Fatty acids composition of vegetable oils and its contribution to dietary energy intake and dependence of cardiovascular mortality on dietary intake of fatty acids. *International Journal of Molecular Sciences*, 16(6): 12871–12890. https://doi.org/10.3390/ijms160612871.

Orthoefer, F. T. 2005. Rice bran oil. *Bailey's Industrial Oil and Fat Products*, 2(7): 465–489.

Paik, M.-J., Kim, H., Lee, J., Brand, J., and Kim, K.-R. 2009. Separation of triacylglycerols and free fatty acids in microalgal lipids by solid-phase extraction for separate fatty acid profiling analysis by gas chromatography. *Journal of Chromatography. Part A*, 1216(31): 5917–5923. https://doi.org/10.1016/j.chroma .2009.06.051.

Pali, V. 2013. Rice bran oil-unique gift of nature: A review. *Agricultural Reviews*, 34(4): 288–294.

Panesar, R., Kaur, S., and Panesar, P. S. 2015. Production of microbial pigments utilizing agro-industrial waste: A review. *Current Opinion in Food Science*, 1: 70–76. https://doi.org/10.1016/j.cofs.2014.12.002.

Patel, M., and Naik, S. N. 2004. Gamma-oryzanol from rice bran oil - A review. *Journal of Scientific and Industrial Research*, 63(7): 569–578.

Patel, S. 2015. Grape seeds: Agro-industrial waste with vast functional food potential. *Emerging Bioresources with Nutraceutical and Pharmaceutical Prospects*, 6: 53–69. https://doi.org/10.1007/978-3-319-12847-4_6.

Patil, S. S., Kar, A., and Mohapatra, D. 2016. Stabilization of rice bran using microwave: Process optimization and storage studies. *Food and Bioproducts Processing*, 99: 204–211.

Pengkumsri, N., Chaiyasut, C., Sivamaruthi, B. S., Saenjum, C., Sirilun, S., Peerajan, S., Suwannalert, P., Sirisattha, S., Chaiyasut, K., and Kesika, P. 2015. The influence of extraction methods on composition and antioxidant properties of rice bran oil. *Food Science and Technology*, 35(3): 493–501.

Pernet, F., Pelletier, C. J., and Milley, J. 2006. Comparison of three solid-phase extraction methods for fatty acid analysis of lipid fractions in tissues of marine bivalves. *Journal of Chromatography A*, 1137(2): 127–137. https://doi.org/10.1016/j.chroma.2006.10.059.

Pestana-Bauer, V. R., Zambiazi, R. C., Mendonca, C. R. B., Bruscatto, M. H., Lerma-Garcia, M. J., and Ramis-Ramos, G. 2008. Quality changes and tocopherols and gamma-orizanol concentrations in rice bran oil during the refining process. *Journal of the American Oil Chemists' Society*, 85(11): 1013–1019.

Phan, V. M., Junyusen, T., Liplap, P., and Junyusen, P. 2019. Effects of ultrasonication and thermal cooking pretreatments on the extractability and quality of cold press extracted rice bran oil. *Journal of Food Process Engineering*, 42(2): e12975.

Pourali, O., Asghari, F. S., and Yoshida, H. 2009. Simultaneous rice bran oil stabilization and extraction using sub-critical water medium. *Journal of Food Engineering*, 95(3): 510–516.

Punia, S., Kumar, M., Sioha, A. K., and Purewal, S. S. 2021. Rice bran oil: Emerging trends in extraction, health benefit, and its industrial application. *Rice Science*, 28(3): 217–232. https://doi.org/10.1016/j.rsci .2021.04.002.

Quiroz, J. Q., Duran, A. M. N., Garcia, M. S., Gomez, G. L. C., and Camargo, J. J. R. 2019. Ultrasound-assisted extraction of bioactive compounds from annatto seeds, evaluation of their antimicrobial and antioxidant activity, and identification of main compounds by LC/ESI-MS analysis. *International Journal of Food Science*, 9.

Radenkovs, V., Juhnevica-Radenkova, K., Górnaś, P., and Seglina, D. 2018. Non-waste technology through the enzymatic hydrolysis of agro-industrial by-products. *Trends in Food Science and Technology*, 77: 64–76. https://doi.org/10.1016/j.tifs.2018.05.013.

Ramakrishnan, V. V., Ghaly, A. E., Brooks, M. S., and Budge, S. M. 2013. Extraction of oil from mackerel fish processing waste using alcalase enzyme. Enzyme engineering. *Enzyme Engineering*, 2: 115. https://doi .org/10.4172/2329-6674.1000115.

Raol, G. G., Raol, B. V., Prajapati, V. S., and Bhavsar, N. H. 2015. Utilization of agro-industrial waste for b-galactosidase production under solid state fermentation using halotolerant Aspergillus tubingensis GR1 isolate. *3 Biotech*, 3: 1–15. https://doi.org/10.1007/s13205-014-0236-7.

Roselló-Soto, E., Parniakov, O., Deng, Q. et al. 2016. Application of non-conventional extraction methods: Toward a sustainable and green production of valuable compounds from mushrooms. *Food Engineering Reviews*, 8(2): 214–234. https://doi.org/10.1007/s12393-015-9131-1.

Routray, W., and Orsat, V. 2011. Microwave-assisted extraction of flavonoids: A review. *Food and Bioprocess Technology*, 5: 11–68.

Rovettoa, L. J., and Aieta, N. V. 2017. Supercritical carbon dioxide extraction of cannabinoids from Cannabis sativa L. *Journal of Supercritical Fluids*, 129: 16–27. https://doi.org/10.1016/j.supflu.2017.03.014.

Rozzi, N. L., and Singh, R. K. 2002. Supercritical fluids and the food industry. *Comprehensive Reviews in Food Science and Food Safety*, 1(1): 33–44.

Ruen-Ngam, D., Thawai, C., Nokkoul, R., and Sukonthamut, S. 2014. Gamma-oryzanol extraction from upland rice bran. *International Journal of Bioscience, Biochemistry and Bioinformatics*, 4(4): 252–255.

Sahena, F., Zaidul, I. S. M., Jinap, S., Saari, N., Jahurul, H. A., Abbas, K. A., and Norulaini, N. A. 2009. PUFAs in fish: Extraction, fractionation, importance in health. *Comprehensive Reviews in Food Science and Food Safety*, 8(2): 59–74.

Sahena, F., Zaidul, I. S. M., Jinap, S., Yazid, A. M., Khatib, A., and Norulaini, N. A. N. 2009. Fatty acid compositions of fish oil extracted from different parts of Indian mackerel (Rastrelliger kanagurta) using various techniques of supercritical CO2 extraction. *Food Chemistry*, 120(3): 879–885. https://doi.org/10.1016/j.foodchem.2009.10.055.

Saikia, D., and Deka, S. C. 2011. Cereals: From staple food to nutraceuticals. *International Food Research Journal*, 18(1): 21–30.

Santos, K. A., Frohlich, P. C., Hoscheid, J., Tiuman, T. S., Gonçalves, J. E., Cardozo Filho, L., and da Silva, E. A. 2017. Candeia (Eremanthus erythroppapus) oil extraction using supercritical CO2 with ethanol and ethyl acetate cosolvents. *Journal of Supercritical Fluids*, 128: 323–330.

Saunders, R. M. 1985. Rice bran: Composition and potential food uses. *Food Reviews International*, 1(3): 465–495.

Sawadikiat, P., and Hongsprabhas, P. 2014. Phytosterols and γ-oryzanol in rice bran oils and distillates from physical refining process. *International Journal of Food Science and Technology*, 49(9): 2030–2036.

Sharif, M. K., Butt, M. S., Anjum, F. M., and Khan, S. H. 2014. Rice bran: A novel functional ingredient. *Critical Reviews in Food Science and Nutrition*, 54(6): 807–816.

Sharma, A., Khare, S. K., and Gupta, M. N. 2001. Enzyme-assisted aqueous extraction of rice bran oil. *Journal of the American Oil Chemists' Society*, 78(9): 949–951.

Shukla, H. S., and Pratap, A. 2017. Comparative studies between conventional and microwave assisted extraction for rice bran oil. *Journal of Oleo Science*, 66(9): 973–979.

Silva, A. C. d., and Jorge, N. 2014. Bioactive compounds of the lipid fractions of agro-industrial waste. *Food Research International*, 66: 493–500. https://doi.org/10.1016/j.foodres.2014.10.025.

Silvaa, M. A., Albuquerquea, T. G., Alvesb, R. C., Oliveira, M. B. P. P., and Costa, H. S. 2020. Melon (Cucumis melo L.) by-products: Potential food ingredients for novel functional foods? *Trends in Food Science and Technology*, 98: 181–189. https://doi.org/10.1016/j.tifs.2018.07.005.

Singanusong, R., and Garba, U. 2019. Micronutrients in rice bran oil. In: *Rice Bran and Rice Bran Oil*. AOCS Press: 125–158. https://www.sciencedirect.com/book/9780128128282/rice-bran-and-rice-bran-oil#book-info

Singleton, A., and Stikeleather, L. F. 1999. A solvent extractor system for the rapid extraction of lipids and trace bioactive micronutrients in oilseeds. *JAOCS Press*, 76(12) 1461–1466.

Soares, A. T., Costa, D. C., Silva, B. F., Lopes, R. G., Derner, R. B., and Antoniosi Filho, N. R. 2014. Comparative analysis of the fatty acid composition of microalgae obtained by different oil extraction methods and direct biomass transesterification. *Bioenergy Research*, 7(3): 1035–1044. https://doi.org/10.1007/s12155-014-9446-4.

Soares, I. D., Okiyama, D. C. G., and Rodrigues, C. E. d. C. 2020. Simultaneous green extraction of fat and bioactive compounds of cocoa shell and protein fraction functionalities evaluation. *Food Research International*, 137: 109622. https://doi.org/10.1016/j.foodres.2020.109622.

Soares, J. F., Dal Pra, V., de Souza, M., Lunelli, F. C., Abaide, E., da Silva, J. R. F., Kuhn, R. C., Martinez, J., and Mazutti, M. A. 2016. Extraction of rice bran oil using supercritical CO2 and compressed liquefied petroleum gas. *Journal of Food Engineering*, 170: 58–63.

Socas-Rodríguez, B., Alvarez-Rivera, G., Valdés, A., Ibañez, E., and Cifuentes, A. 2021. Food by-products and food wastes: Are they safe enough for their valorization? *Trends in Food Science and Technology*, 114: 133–147. https://doi.org/10.1016/j.tifs.2021.05.002.

Somashekar, D., Venkateshwaran, G., Srividya, C., Krishnanand, Sambaiah, K., and Lokesh, B. R. 2001. Efficacy of extraction methods for lipid and fatty acid composition from fungal cultures. *World Journal of Microbiology and Biotechnology*, 17(3): 317–320.

Sookwong, P., and Mahatheeranont, S. 2017. Supercritical CO2 extraction of rice bran oil – The technology, manufacture, and applications. *Journal of Oleo Science*, 66(6): 557–564. https://doi.org/10.5650/jos.ess17019.

Soquetta, M. B., Terra, L. d. M., and Bastos, C. P. 2017. Green technologies for the extraction of bioactive compounds in fruits and vegetables. *CyTA: Journal of Food*, 16(1): 400–412. https://doi.org/10.1080/19476337.2017.1411978.

Souza, J. R. C. L., Villanova, J. C. O., Souza, T. S., Maximino, R. C., and Menini, L. 2021. Vegetable fixed oils obtained from soursop agro-industrial waste: Extraction, characterization and preliminary evaluation of the functionality as pharmaceutical ingredients. *Environmental Technology and Innovation*, 21: 101379. https://doi.org/10.1016/j.eti.2021.101379.

Sparks, D., Hernandez, R., Zappi, M., Blackwell, D., and Fleming, T. 2006. Extraction of rice bran oil using supercritical carbon dioxide and propane. *Journal of the American Oil Chemists' Society*, 83(10): 10–16.

Srikaeo, K., and Pradit, M. 2011. Simple techniques to increase the production yield and enhance the quality of organic rice bran oils. *Journal of Oleo Science*, 60(1): 1–5.

Stark, K. D., Elswyk, M. E. V., Higgins, M. R., Weatherford, C. A., and Salem, N., Jr. 2016. Global survey of the omega-3 fatty acids, docosahexaenoic acid and eicosapentaenoic acid in the blood stream of healthy adults. *Progress in Lipid Research*, 63: 132–152. https://doi.org/10.1016/j.plipres.2016.05.001.

Stikeleather, L. F., and Singleton, J. A. 2001. Induction heated pressure vessel system and method for the rapid extraction of lipids and trace polar micronutrients from plant material. *Food and Bioproducts Processing*, 79: 169–175.

Straccia, M. C., Siano, F., Coppola, R., La Cara, F., and Volpe, M. G. 2012. Extraction and characterization of vegetable oils from cherry seed by different extraction processes. *Chemical Engineering Transactions*, 27: 391–396.

Tabaraki, R., and Nateghi, A. 2011. Optimization of ultrasonic-assisted extraction of natural antioxidants from rice bran using response surface methodology. *Ultrasonics Sonochemistry*, 18(6): 1279–1286.

Tan, C.-H., Ghazali, H. M., Kuntom, A., Tan, C.-P., and Ariffin, A. A. 2009. Extraction and physicochemical properties of low free fatty acid crude palm oil. *Food Chemistry*, 113(2): 645–650. https://doi.org/10.1016/j.foodchem.2008.07.052.

Tatke, P., and Jaiswal, Y. 2011. An overview of microwave assisted extraction and its applications in herbal drug research. *Research Journal of Medicinal Plants*, 5(1): 21–31.

Terigar, B. G., Balasubramanian, S., Sabliov, C. M., Lima, M., and Boldor, D. 2011. Soybean and rice bran oil extraction in a continuous microwave system: from laboratory-to pilot-scale. *Journal of Food Engineering*, 104(2): 208–217.

Thao, P. T. P., Anh, P. L. N., and Son, V. H. 2021. Effect of extraction solvents on quality of Vietnamese tea (*Camellia sinensis* O. Kuntze) seed oil. *Vietnam Journal of Science and Technology*, 59(2): 137–148. https://doi.org/10.15625/2525-2518/59/2/14860.

Tong, C., and Bao, J. 2019. Rice lipids and rice bran oil. In: *Rice*. AACC International Press: 131–168. https://www.sciencedirect.com/book/9780128128282/rice-bran-and-rice-bran-oil#book-info

Vinatoru, M., Mason, T. J., and Calinescu, I. 2017. Ultrasonically assisted extraction (UAE) and microwave assisted extraction (MAE) of functional compounds from plant materials. *Trends in Analytical Chemistry*, 97: 159–178. https://doi.org/10.1016/j.trac.2017.09.002.

Wang, Y. 2019. Applications of rice bran oil. In: Cheong, L. Z., and Xu, X. B. (eds), *Rice Bran and Rice Bran Oil*. AOCS Press, Urbana: 159–168.

Xu, Z., and Godber, J. S. 2000. Comparison of supercritical fluid and solvent extraction methods in extracting γ-oryzanol from rice bran. *Journal of the American Oil Chemists' Society*, 77(5): 547–551.

Yu, C. W., Hu, Q. R., Wang, H. W., and Deng, Z. Y. 2020. Comparison of 11 rice bran stabilization methods by analyzing lipase activities. *Journal of Food Processing and Preservation*, 44(4): e14370.

Zaccheria, F., Mariani, M., and Ravasio, N. 2015. The use of rice bran oil within a biorefinery concept. *Chemical and Biological Technologies in Agriculture*, 2(1): 23.

Zhang, Q.-W., Lin, L.-G., and Ye, W.-C. 2018. Techniques for extraction and isolation of natural products: A comprehensive review. *Chinese Medicine*, 13: 20. https://doi.org/10.1186/s13020-018-0177-x.

Zhou, L., Yang, F., Zhang, M, and Liu, J. 2020. A green enzymatic extraction optimization and oxidative stability of krill oil from *Euphausia superba*. *Marine Drugs*, 18(2): 82. https://doi.org/10.3390/md18020082.

Zia, S., Rafiq Khan, M., Shabbir, M. A., and Aadil, R. M. 2021. An update on functional, nutraceutical and industrial applications of watermelon by-products: A comprehensive review. *Trends in Food Science and Technology*, 114: 275–291. https://doi.org/10.1016/j.tifs.2021.05.039.

Zigoneanu, I. G., Williams, L., Xu, Z., and Sabliov, C. M. 2008. Determination of antioxidant components in rice bran oil extracted by microwave-assisted method. *Bioresource Technology*, 99(11): 4910–4918.

Zullaikah, S., Melwita, E., and Ju, Y. H. 2009. Isolation of oryzanol from crude rice bran oil. *Bioresource Technology*, 100(1): 299–302.

12 Production of Organic Acids from Agro-Industrial Waste and Their Industrial Utilization

Navneet Kaur, Parmjit S. Panesar, and Shilpi Ahluwalia

CONTENTS

DOI: 10.1201/9781003125679-12

12.1 INTRODUCTION

In the modern era, there is a tremendous surge in global energy requirements and issues related to the environment, both of which have resulted in the development of eco-friendly and sustainable ways to deal with energy requirement and pollution problems such as biorefineries (Singh et al., 2017). The motive behind these approaches is to convert biomass, i.e., organic material of plant origin formed by various activities like agriculture, aquaculture, forestry, etc., into a range of bioproducts such as bioenergy, biofuels, bio-chemicals or high value-added chemicals etc., by utilizing minimal energy expenditure (Singh nee' Nigam, 2009; Phillippini et al., 2020). Among these bioproducts are organic acids, which are high value-added products often regarded as building block chemicals and are produced mainly through microbial fermentation. Micro-organisms produce these organic acids either as end-products or as intermediates of their metabolic pathway (Sauer et al., 2008). Organic acids are also popular throughout the world owing to their part in natural ecology and the vast amount of industrial applications in the field of agriculture, cosmetics, the food industry, and the pharmaceutical sector. They have gained considerable attention currently because of their biodegradable nature and involvement in the synthesis of biodegradable polymers, thereby replacing synthetic chemicals (Khan et al., 2017).

The microbial fermentation processes for organic acid production require carbohydrates and related substrates as starting materials (Naraian and Kumari, 2017). These nutrient substrates are mainly starch and sugars, which are nowadays derived from renewable agro-substrates obtained from agricultural crops directly, e.g., cereal grains, plant tubers, plant oils, etc., or indirectly through industrial processing, which leads to the production of agro-industrial residues in the form of waste, e.g., crop residues (wheat straw, rice bran, etc.), cane and beet molasses, wood hydrolysate, etc. (Vandamme, 2009). In the past, these waste residues were released directly into the environment untreated and underutilized by burning, dumping, and landfilling, which resulted in several environmental and health problems for both humans and animals. Thus, to overcome these environmental and health issues, biotechnologists proposed a concept of integrated waste management. With this approach, these agro-industrial waste residues, being highly nutritious in composition, are getting the required attention. These were considered waste but are now regarded as raw materials for the production of high value-added chemicals. Due to their easy availability and low cost, they are utilized by micro-organisms in fermentation processes (Yusuf, 2017; Sadh et al.,2018). Hence, through this chapter, the authors will try to review various kinds of agro-industrial wastes and their utilization for the production of organic acids.

12.2 AGRO-INDUSTRIAL WASTES AND THEIR COMPOSITION

Agro-industrial residues like beet pulp, bran, corn cobs, corn stover, straw, oil cakes, and other waste materials from the food industry such as fruit and vegetable residues are getting immense attention from researchers as these residues provide abundant and cheap sources of renewable feedstocks rich in various nutrients such as sugars, proteins, oils, and other micronutrients to support the growth of micro-organisms for fermentation. These agro-industrial wastes, after their pre-treatment, are utilized for the production of various useful products (Phillippini et al., 2020). Agro-industrial waste is mainly classified into two types, i.e., agricultural residues and industrial residues.

12.2.1 Agricultural/Crop Residues

Agricultural residues are also called crop residues as these are remains left after harvesting and processing of agricultural crops and can further be categorized as field and process residues. Field residues may be defined as leftovers of crop harvest such as leaves, pods, stalks, seed pods, and stems. On the other hand, process residues are residues left after the processing of the crop into valuable products. Examples of process residues include bagasse, husks, leaves, molasses, peel, pulp, seeds, stem, straw, stalk, shell, stubble, roots, etc. The process residues are generally utilized in the manufacture of animal feed, such as oil cakes, fertilizers, and various other processes, while field residues, on the contrary, remain unused (Sadh et al., 2018).

12.2.2 Industrial Residues

With the surge in the population, the need for processed food is increasing along with the waste generated in the production of processed food. Organic residues and effluents generated by food processing industries, namely fruit and vegetable processing, snack processing, confectionery, meat, milk processing, etc., are included in industrial residues and constitute a huge amount of waste (Sadh et al., 2018).

Agro-industrial waste mainly comprises lignocellulose as the primary substance and starchy residues.

12.2.3 Lignocellulosic Waste/Biomass

Bagasse, cotton stalks, rice straw, wheat bran, and wheat straw are some of the agro-industrial residues abundant in lignocellulosic compounds. Lignocellulose is primarily composed of cellulose, hemicellulose, lignin, and other miscellaneous compounds like pectin, proteins, extractives, and ash (Michelin et al., 2014; Saini et al., 2015). Cellulose, being the main component (30–50%), is present as the skeleton holding hemicelluloses (15–35%) and lignin (10–20%). The hemicellulose and lignin fractions form the matrix and act as encrusting materials.

Cellulose, being the most abundant component and prime polysaccharide present in the plant cell wall, consists of cellobiose or (1,4)-D-glucopyranose units in a repeating fashion connected to each other via β-1,4-linkages. In nature, cellulose exists both in amorphous and crystalline forms (Michelin et al., 2014).

Followed by cellulose, another compound present in generous quantities is hemicellulose, which is heterogeneously branched. The polymer is approximately 50–200 units long and consists of pentoses, such as arabinose and xylose, and hexoses, namely galactose, glucose, and mannose, in which the acetate groups are linked via ester linkages to hydroxyl groups of sugar rings randomly. Being amorphous in nature, hemicellulose undergoes degradation quite fast (Maheshwari, 2018).

Lignin, though present in a small amount, is quite complex in nature. It forms a three-dimensional structure with phenylpropanoid units. The presence of lignin and the degree of cellulose crystallinity are responsible for a robust structure and resistance to hydrolysis by lignocellulosic biomass (Michelin et al., 2014; Maheshwari, 2018).

Although lignocellulosic waste is a rich source of carbohydrates required for microbial cultivation, the sugars present in cellulose (glucose) and hemicellulose (pentoses and hexoses) are not free to participate in microbial processes. They require pre-treatment and hydrolysis via various methods to become available for their utilization by micro-organisms. The release of sugars from cellulose is more difficult than from hemicellulose because of its recalcitrant structure provided by lignin and hemicellulose along with high crystallinity, which hampers hydrolysis by various chemical and enzymatic methods. To enhance cellulose valorization, various chemical, enzymatic, and biological pre-treatment methods are available that greatly increase the availability of sugars in pure form. The pre-treatment steps remove lignin and hemicellulose fractions and also reduce the crystallinity to a great extent, followed by chemical and enzymatic hydrolysis, which releases glucose. Enzymatic hydrolysis offers a higher yield in comparison to chemical methods such as the use of acids. However, there are several limitations associated with cellulose saccharification (Maheshwari, 2018). A few of them include the high cost of pre-treatment methods and enzymes required for enzymatic hydrolysis in addition to the strenuous purification of sugars. Due to all these limitations, along with easier and safer alternatives available in the form of sugarbeet, sugarcane, and starch for the production of edible sugars, the viability and commercial acceptance of lignocellulosic waste needs a lot of improvement (Zamani, 2015).

12.2.4 STARCHY WASTE/BIOMASS

Starch is considered the second-most abundant compound synthesized by plants, followed by cellulose. While glucose is a repeating unit in both starch and cellulose, they differ greatly in their physico-chemical properties. This is attributed to the type of linkages by which glucose monomers are linked in starch and cellulose. Starch is linked via α-1,4-linkages, while cellulose contains β-1,4-linkages. In starch, approximately 20% of units are water-insoluble amylose units linked together in a linear fashion, while water-soluble amylopectin forms the bulk amount and shows branching via α-1,6-linkages from the linear molecule. Starch has gained wide acceptance and popularity as a substrate to carry out microbial fermentation as it offers several benefits over lignocellulosic biomass. While beet and cane molasses are quite cheap as substrates and the use of lactose from whey is limited, starch is still a widely accepted substrate to carry out microbial fermentations (Rakshit, 1998). Some of the advantages offered by starchy biomass include utilization by human beings, increased sensitivity to partial and complete hydrolysis, which results in the production of useful compounds, effortless conversion to value-added products in various industries, low cost of starch converting enzymes, and high purity of starch in comparison to cellulose. The major sources of starch include cereals like barley, corn, millets, oats, rice, rye, sorghum, wheat, etc.; tubers and roots such as arrow root, cassava or tapioca, potato, and yam; fruits like annatto, apple, avocado, banana, jackfruit, kiwi fruit, lychee, longan, loquat, mango, pineapple, plantain, along with the stem, pith, and seeds (Rakhshit, 1998; Kringel et al., 2020). The composition of a major fraction of the constituents present in agro-industrial wastes is outlined in Table 12.1.

12.3 FERMENTATIVE PRODUCTION OF ORGANIC ACIDS

Organic acids may be defined as low molecular weight compounds occurring in nature as part of plant and animal tissues (Naraian and Kumari, 2017). Among the building block chemicals, they represent a rising segment that can be commercially produced by fermentation using microorganisms. They are the versatile ingredients employed in different industries such as food and beverages, pharmaceuticals, textiles, detergents, solvents, petrochemicals, dyes, adhesives, rubber, perfumes, and plastics (Singh et al., 2017). Production of organic acids at the industrial level is mainly carried out by two approaches, i.e., by chemical synthesis or by fermentation methods, with the exception of citric acid (CA), which is entirely produced by fermentation methods. However, in recent times, production has been shifting from chemical methods to fermentation methods as they give higher yields of all the organic acids belonging to the tricarboxylic acid (TCA) cycle. Among the fermentation methods also, submerged fermentation (SmF) is the method mostly preferred. Yet, the production of organic acids by solid-state fermentation (SSF) is recently gaining considerable attention due to the employment of byproducts and residues of agro-industries as low-cost substrates that not only make the process more economical but also tackle the problem of environmental pollution. Out of all the organic acids, CA ranks first in terms of production, followed by acetic acid and lactic acid in second and third position respectively (Vandenberghe et al., 2018).

TABLE 12.1

Various Types of Agro-Industrial Wastes and Their Composition

S. No.	Agro-industrial Waste Residue	Composition (% w/w)			
		Cellulose	Hemicellulose	Lignin	Starch
1.	Barley straw	33.8	21.9	13.8	–
2.	Corn stalks	61.2	19.3	6.9	–
3.	Cotton stalks	58.5	14.4	21.5	–
4.	Oat straw	39.4	27.1	17.5	–
5.	Orange peel	9.21	10.5	0.84	–
6.	Pineapple peel	18.11	–	1.37	–
7.	Potato peel waste	8.30	7.41	32.88	23.01
8.	Rice straw	39.2	23.5	36.1	16.0
9.	Sawdust	45.1	28.1	24.2	–
10.	Soya stalks	34.5	24.8	19.8	–
11.	Sugarbeet waste	26.3	18.5	2.5	–
12.	Sugarcane bagasse	30.2	56.7	13.4	–
13.	Sunflower stalks	42.1	29.7	13.4	–
14.	Wheat straw	32.9	24.0	8.9	–

S. no.	Starchy waste	% Composition	
		Starch	Amylose
15.	Annatto seeds	18.0–20.0	24.0
16.	Apple pulp	44.0–53.0	26.0–29.3
17.	Avocado seeds	27.5–29.8	15.0–16.0
18.	Banana peel	22.6	25.7
19.	Jackfruit seeds	60.0–80.0	22.1–38.3
20.	Kiwi fruit waste	51.8	23.3
21.	Lychee seeds	53.0	19.2
22.	Mango kernels	58.9–64.0	9.1–16.3
23.	Pineapple stem waste	11.0	34.0
24.	Tamarind seeds	20.0	14.2

Source: Sadh et al., 2018; Kringel et al., 2020.

12.3.1 LACTIC ACID

Lactic acid, also known as 2-hydroxypropanoic acid, is one of the most important C3 platform chemicals, generally marketed as a colourless, white, or light yellow solid or powder. It was first discovered by Carl Scheele in 1780 as the constituent of sour milk, while Lavousier named it "*acide lactique*". Then, in 1857, Pasteur discovered it as a fermentation product of milk rather than its constituent. In 1881, Fremi started fermentation production on an industrial scale. Now though, the microbial production process for lactic acid is not new, but its demand in purified form has risen recently due to the production of biodegradable and biocompatible plastic polymer, polylactide, or polylactic acid (PLA), which is widely used in manufacturing food containers and replacing petroleum-based consumables (Sauer et al., 2008; Ghaffar et al., 2014; Panesar and Kaur, 2015; Naraian and Kumari, 2017). With the extending market, the global production of lactic acid is expected to augment by 600,000 metric tonnes in the coming years. (Grewal et al., 2020).

Due to the presence of asymmetric carbon atoms, lactic acid exists in two optically active forms, i.e., D(-)-lactic acid or R-lactic acid and L(+)-lactic acid or S-lactic acid or either a racemic mixture (Holten et al., 1971). L(+)-lactic acid is more important both commercially and economically because it is consumable by human beings (Ghaffar et al., 2014; Panesar and Kaur, 2015). Lactic acid is produced commercially via both chemical and microbial processes. However, the fermentation method is preferred as chemical synthesis results in the production of a racemic mixture (±) of lactic acid while micro-organisms produce pure lactic acid (Panesar and Kaur, 2015). Several factors have a considerable role in efficient lactic acid production by fermentation methods, such as the strain used, substrate employed, and the type of fermentation (Abdel-Rahman et al., 2013).

12.3.1.1 Microbial Sources

Various microbial species can be utilized to produce lactic acid from renewable materials like bacteria, fungi, and yeast. Each class of micro-organism employed offers several advantages over each other such as the ability to utilize multiple substrates, better yield, low nutritional requirements, and greater purity of lactic acid. Similarly, mixed and genetically engineered strains are used as they offer benefits over single strains in the assimilation of complex agro-industrial residues resulting in greater productivity and yield (Abdel-Rahman et al., 2013). In bacteria, several species of the genus *Corynebacterium glutamicum*, *Escherichia coli*, *Bacillus* strains, and lactic acid bacteria (LAB) are widely employed for lactic acid production. However, the most commonly utilized strains include LAB and filamentous fungi like *Rhizopus* because they are associated with greater productivity and yield. Both of them, under aerobic conditions, consume glucose to produce lactic acid (Ghaffar et al., 2014).

LAB is a term exclusively used to represent seven genera, which are a group of gram-positive, non-motile and non-spore-forming bacteria that act on various sugars to produce lactic acid. Out of all the seven genera, *Enterococcus*, *Lactococcus*, *Lactobacillus*, and *Pediococcus* have been used widely on a commercial scale for the production of lactic acid. They have gained wide acceptance because of their GRAS (generally regarded as safe) status and generation of various immunomodulatory compounds. The LAB ferments sugars by following two pathways (Figure 12.1). In the

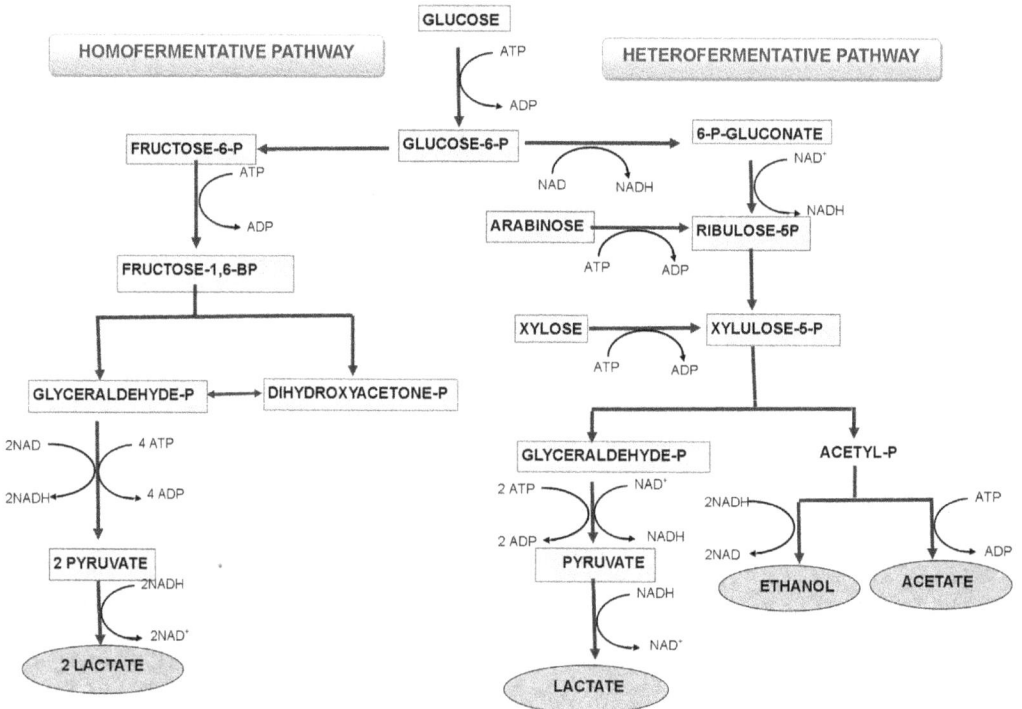

FIGURE 12.1 Homofermentative and Heterofermentative Pathway followed by LAB to produce lactic acid (Adapted from Kandler 1983).

homofermentative or Embden–Meyerhof pathway, hexoses like glucose are metabolized, resulting in the formation of two molecules of lactic acid from a single molecule of glucose. The bacteria exclusively act on glucose and do not metabolize pentoses such as xylose and gluconates. The other pathway is a heterofermentative or phosphoketolase (PK) pathway. The LAB involved in this pathway utilizes hexoses, pentoses, and certain other substances for the production of lactic acid. Moreover, as the name suggests, the pathway leads to the production of other compounds, namely acetic acid and ethanol, along with lactic acid. There is a major limitation associated with it as it not only decreases the yield but also makes downstream processing more laborious and costly (Kandler, 1983: Romo-Buchelly et al., 2019; Grewal et al., 2020).

As mentioned earlier, the filamentous fungi of the genus *Rhizopus*, particularly *R. oryzae* have become worthy of attention for producing optically pure L-lactic acid. They also provide various benefits over LAB. A few of them exhibit amylolytic properties that allow easy utilization of starch-based waste residues without any initial saccharification and little requirement for nutrients; the filamentous and pallet nature of fungi offer easy and economic downstream processing and useful byproducts in the form of fungal biomass (Abdel-Rahman et al., 2013).

12.3.1.2 Agro-Industrial Byproducts/Food Waste

Currently, around half of the total lactic acid production is mainly achieved through fermentation in one or two stages. The economic feasibility of the fermentation process mainly depends upon the cost of the substrate or raw material used. It is therefore essential to keep several factors in consideration while selecting a particular substrate: it should be inexpensive, easily available year around, abundant in quantity, contain little or no contaminant, should improve the rate of fermentation, provide high yield, and be accompanied by the low formation of byproducts (Ghaffar et al., 2014).

Agro-industrial biomass or organic waste from food industries or household waste all provide a cheap alternative to high-cost substrates, mainly pure sugars that escalate the production cost from 30% to 60%. However, the fermentation process utilizing organic waste generally requires two stages, one being pre-treatment of waste to release free sugars into the medium and the second stage comprises the utilization and fermentation of these sugars to produce lactic acid (Romo-Buchelly et al., 2019).

Lignocellulosic biomass such as sugarcane bagasse, corn cob, corn stover, wheat straw, etc., is an appealing substrate option for the production of lactic acid. However, its effective utilization is accompanied by several limitations like the failure of lactic acid micro-organisms to metabolize complex carbohydrates, namely cellulose and hemicellulose, etc. They require a multi-stage process for their complete degradation and production of lactic acid. The steps include pre-treatment, saccharification, fermentation, and downstream purification (Grewal et al., 2020) and are discussed in Figure 12.2. Starchy biomass in any form, i.e., solid or liquid, does appear to be a more appealing alternative than cellulosic biomass owing to its easy hydrolysis and presence of a wide array of micro-organisms comprising amylolytic enzymes and availability of cheap sources of enzymes. Although only a limited number of micro-organisms can directly hydrolyze starch into valuable products, numerous organisms have been reported that produce enzymes capable of converting starch into glucose and further utilize this glucose for their growth (Rakshit, 1998). Several strains of LAB have been shown to engage in the concomitant saccharification and fermentation of starch, such as *Lactobacillus amylophilus*, *Lactobacillus manihotivorans*, *Lactobacillus amylovorus*, *Lactococcus lactis*, *Lactobacillus amylolyticus*, *Lactobacillus plantarum*, *Lactobacillus paracasei*, *Lactobacillus fermentum*, *Enterococcus faecium*, *Streptococcus bovis*, *Rhizopus oryzae*, *Rhizopus arrhizus*, and *Bacillus* sp. (Panesar and Kaur, 2015; Grewal et al., 2020).

Despite this, certain substrates containing resistant starches may need pre-treatment, such as liquefaction, a process in which starchy waste is heated at 80–90°C for about 10–30 min accompanied by enzymatic hydrolysis under the action of amylolytic enzymes like α-amylase, β-amylase, and glucoamylase. Sometimes alkali (KOH) and acid (HCl) pre-treatments have also been reported in the case of kitchen waste (Grewal et al., 2020).

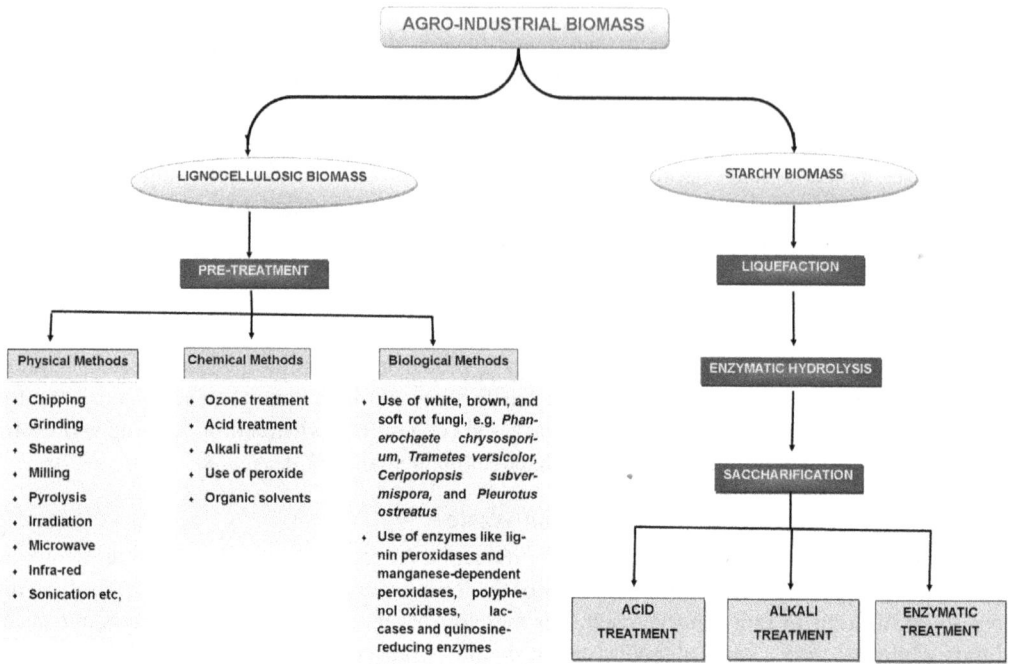

FIGURE 12.2 Pre-treatment steps involved in the valorization of lignocellulosic and starchy agro-industrial residues.

Another important substrate worth mentioning for lactic acid production is whey. It is the by-product of the cheese-manufacturing industry, and each kilogram of cheese manufactured releases about 9 l of whey. It is an important source of nutrients such as lactose (5% w/v), soluble proteins (0.8–1.0%), lipids (0.2–0.7%), and mineral salts required by micro-organisms for the cultivation and production of lactic acid (Panesar et al., 2010; Soriano-Perez et al., 2012).

12.3.1.3 Fermentation Processes

Fermentation processes for the production of lactic acid differ depending upon the substrate used, strain employed, and viscosity of the fermentation medium. Various processes used currently include batch, fed-batch, repeated, and continuous fermentation. Though studies on lactic acid are enormous, they are limited in the case of the fermentation methods used.

12.3.1.3.1 Batch Fermentation

Batch fermentation is mainly carried out at a commercial scale for the production of lactic acid, and it also presents various benefits like the low risk of contamination and higher yield. Despite these advantages, the method suffers from reduced productivity and limited availability of nutrients. The problems are mainly due to inhibition either by the product or the substrate. Thus, to overcome these difficulties and to enhance lactic acid production, several improved variants of batch fermentation have been developed like simultaneous saccharification and fermentation (SS&F), separate hydrolysis fermentation (SHF), mixed culture, or open fermentation (Abdel-Rahman et al., 2013; Beital et al., 2020).

Out of all the processes mentioned above, SS&F has been regarded as more beneficial than SHF due to several reasons such as being economical, the requirement of only a single vessel for operation, quick processing time, reduced loading of the enzyme, higher productivity, and minimize product or substrate inhibition (Abdel-Rahman et al., 2013; Kim et al., 2018). In a recent study on kimchi cabbage waste, using *Lactobacillus sakei* WIKIM31 and *L. curvatus* WIKIM38, the authors compared SHF and SS&F processes for the production of lactic acid and found that the SHF process resulted in little higher production yield, i.e., 13.2 and 13.9 g/l in the case of *L. sakei* and *L. curvatus* than SS&F, in which the yield was 12.1 and 12.7 g/l with the former and latter strains, respectively, which may be attributed to the prior hydrolysis step involved in SHF. However, lactic acid was rapidly produced (4.5% w/v) in 24 h in the case of SS&F, suggesting it to be more economically feasible than SHF. In addition, SHF appears to be a long process as it involves enzymatic hydrolysis, which reduces real productivity and also adds extra cost (Kim et al., 2018).

Similarly, batch fermentations with mixed cultures have also resulted in an improved yield of lactic acid. Mixed cultures are preferred at the industrial level because of the symbiotic associations between them that enable them to grow under nutrient-limited conditions prevailing at the industrial level, leading to an economic process by saving on the cost of the addition of expensive organic supplements. In 2005, Lee did a comparative study involving single and mixed cultures for the production of lactic acid. In the preliminary step of the study, the highest lactic-acid-producing strain, i.e., strain B1445, was selected by comparing five different strains: *Lactobacillus delbrueckii* subsp. *lactis* (ATCC 12315), *Lactobacillus casei* (NRRL-B1445), *L. delbrueckii* (NRRL-B445), *Lactobacillus helveticus* (NRRL-B1937), and *L. casei* (NRRL-B1922). Then in a later stage, lactic acid production was compared using the selected strain and the co-culture of all five strains. The rate of lactic acid production was found to be 1.01 g/(1 h) for strain B1445 and 1.48 g/(1 h) with the mixed culture, suggesting a synergistic interaction between the cultures used. In a related study, 49 g/l of lactic acid was produced from low-value date syrup via batch fermentation utilizing mixed cultures of *Streptococcus thermophiles* and *L. bulgaricus* (Bouhadi et al., 2013).

Most lactic acid fermentation is generally performed under sterile conditions using specific strains like *L. casei*, *L. planatrum*, etc. But, in comparison with standard sterile and closed fermentations, non-sterile open fermentations have also gained momentum due to several advantages like no requirement for additional sterilization equipment, which also leads to low energy requirements. Moreover, such kinds of fermentation can be carried out on-site, resulting in low transportation costs. Additionally, instead of specific micro-organisms, undefined or indigenous microflora can also be utilized. One such pilot-scale study involving the open fermentation of food waste without inoculum addition was undertaken and lactic acid production was found to be highest, i.e., 21.5 g/l at 48 h (Wang et al., 2016). In an identical study, large-scale lactic acid production was carried out in open fermentation using inedible starchy biomass and a specific strain of *Bacillus coagulans* IPE22. A one-step process involving simultaneous liquefaction, saccharification, and fermentation requiring the introduction of mesothermal amylase and glucoamylase was used, which resulted in a 0.99 g/g yield and 1.72 g/l h productivity (Wang et al., 2018).

12.3.1.3.2 Fed-Batch Fermentation

Fed-batch fermentation strategies for the production of lactic acid are attractive alternatives to batch and continuous fermentation as they result in higher biomass yield and also control the limitations of the substrate and product inhibition. There are various types of fed-batch fermentation processes depending upon the feeding methods, viz. constant, exponential, intermittent, and optimized, with or without feedback control. The main aim of these methods is to adjust the feeding rates so as to keep the substrate concentration below the inhibition level (Othman et al., 2017). This also results in a brief lag phase, high product concentration, and process productivity.

Recently, a comparative study for lactic acid production in batch, fed-batch, and continuous fermentation modes utilizing *L. casei* CCDM 198 was done, and maximum lactic acid concentration, i.e., 116.5 g/l was achieved in the case of the pulse-feeding fed-batch process (Paulova et al., 2020).

12.3.1.3.3 Repeated Fermentation

Repeated fermentation with batch or fed-batch strategies may be explained as a process in which a fraction of cells or whole cells from the previous batch are separated and inoculated repeatedly into the next fermentation. Various cell separation methods are employed for this purpose, like centrifugation in the case of bacteria, while filtration is done with fungal cultures. The method offers various benefits over conventional batch and fed-batch processes. Some of them include high yield, reduced time requirements, low labour requirements, decrease in inoculum preparation time, maximum cell concentration, and productivity with little fermentation time owing to the high volume of inoculum employed (Abdel-Rahman et al., 2013).

Several studies have reported high lactic acid yield with repeated fermentation. In the latest study on cassava waste water by amylolytic *L. plantarum* MSUL 702 utilizing repeated batch and SS&F under non-sterile conditions at room temperature, the highest lactic acid concentration (28.71 g/l) was obtained after 48 h, and the starch was mostly consumed (Tosungnoen et al., 2014).

12.3.1.3.4 Continuous Fermentation

Continuous fermentation is an attractive substitute for batch and fed-batch fermentation processes in terms of overcoming product inhibition and providing higher productivity (Abdel-Rahman et al., 2013; Thakur and Panesar, 2018). Other advantages of the method include enhanced product quality and efficiency, low operating cost, less product loss, and an environmentally friendly approach (Thakur and Panesar, 2018).

Continuous production of lactic acid was studied with immobilized *L. casei* MTCC 1423 in a two-stage process, i.e., two reactors, continuous stirred tank reactor (CSTR) and packed bed reactor (PBR) in a series and employing molasses and corn steep liquor as cheap agro-industrial substrates. The researchers found that in the first stage, i.e., the CSTR reactor stage, an increase in the dilution rate resulted in decreased lactic acid production, while in the second stage, i.e., the PBR reactor stage, the low dilution rate facilitated increased lactic acid productivity suggesting the role of the dilution rate is an important variable affecting lactic acid productivity in continuous fermentation (Thakur and Panesar, 2018). Further investigations are needed to understand the role of the dilution rate in continuous fermentations.

The studies involving lactic acid production in a continuous fermentation mode employing various agro-industrial residues, viz. alfalfa press green juice, sugar beet molasses, sugar bread, tapioca starch, and whey, indicated that the highest productivity, i.e., 10.34 g/l/h was attained in the case of molasses and it was concluded that continuous fermentation is an efficient method for the production of lactic acid regardless of the substrate used (Olszewska-Widdrat et al., 2020). A summary of lactic acid production studies performed in the last decade via various methods and by employing different agro-industrial residues is given in Table 12.2.

12.3.1.4　Recovery

Obtaining lactic acid in a pure form is both a tedious and complex process due to the presence of debris in the form of biomass and byproducts of microbial metabolism, trace amounts of other impurities, viz. residual sugars and inorganic salts, along with the presence of other organic acids. The process becomes more hectic due to the properties exhibited by lactic acid, like its high hydrophilic and low volatile nature, in addition to its ability to polymerize (Grewal et al., 2020).

Conventional methods of recovery generally begin with the removal of the biomass followed by neutralization and precipitation using an agent like $Ca(OH)_2$ and $CaCO_3$. However, this process includes various limitations such as the generation of gypsum leading to environmental issues, the high cost of the process, in addition to low purity. Thus, several other methods are getting wide acceptance these days as they are more economical, environmentally friendly, and give highly pure lactic acid. These methods include solvent extraction, ion exchange, reactive distillation, molecular distillation, and membrane-based methods such as electrodialysis, reverse osmosis, ultrafiltration, and nano-filtration. These methods are also easy to scale up (Grewal et al., 2020).

TABLE 12.2

Lactic Acid Production Utilizing Various Agro-Industrial Residues

S. no.	Strain Used	Agro-industrial Substrate employed	Mode of Fermentation	Productivity [g/(l.H)]	Yield (g/l)	References
1.	*Lactobacillus casei*	Liquid potato waste	Batch	–	16.09	Afifi, 2011
2.	*Lactobacillus casei* sub species *rhamnosus*	Pineapple waste	Repeated Batch	4.0	–	Araya-Cloutier et al., 2012
3.	*Lactobacillus bulgaricus*, *Streptococcus thermophilus*	Low-value date syrup	Batch	–	49.0	Bouhadi et al., 2013
4.	*Lactobacillus coryniformis*, *Lactobacillus paracasei*	*Curcuma longa* waste	SSCF	2.08 2.70	91.61 97.13	Nguyen et al., 2013
5.	*Lactobacillus plantarum*	Cassava starch waste water	Repeated batch and SS&F	–	28.71	Tosungnoen et al., 2014
6.	*Lactobacillus rhamnosus*	Defatted rice bran	Batch	2.56	210	Wang et al., 2014
7.	*Lactobacillus delbrueckii*	Cane molasses	Batch	3.40	81.50	Srivastava et al., 2015
8.	*Lactobacillus rhamnosus*	Cheese whey and corn steep liquor	static Fed-batch	2.25	143.7	Bernardo et al., 2016
9.	*Streptococcus thermophiles*	Carob extract	Batch	–	29.95	Bouhadi et al., 2016
10.	*Rhizopus oryzae*	Potato waste liquid	SS&F	–	15.5	Panesar and Kaur, 2016
11.	*L. casei*	Cane molasses	Batch	0.55	83	Vidra et al., 2017
12.	*Lactobacillus sakei*, *Lactobacillus curvatus*	Kimchi cabbage waste	Batch (SS&F and SHF)	–	12.1 and 12.7, 13.2 and 13.9	Kim et al., 2018
13.	*L. casei*	Molasses	Batch	2.28	130	Thakur et al., 2019
14.	*Bacillus coagulans*	Carob biomass	Batch	2.30	–	Azaizeh et al., 2020
15.	*Lactobacillus delbrueckii*	Molasses and corn steep liquor	Fed-batch	3.37	162	Beital et al., 2020
16.	*L. delbrueckii*	Soyabean meal	Fed-batch	2.4	112.3	Liang et al., 2020
17.	*Lactobacillus amylovorus*	Cassava bagasse and corn steep liquor	Fed-batch	0.46	66	Macedo et al., 2020

Note: SS&F: simultaneous saccharification and fermentation; SHF: separate hydrolysis fermentation; SSCF: simultaneous saccharification and co-fermentation.

12.3.2 CA

CA (2-hydroxy propane 1, 2, 3-TCA) is one of the most resourceful natural organic acids extensively used in food and the pharmaceutical and chemical industries. The name "citric acid" is derived from the Latin word "*citrus*", referring to the trees belonging to the genus *Citrus*. In 1784, Carl Scheele first isolated CA from lemon juice (Scheele, 1784). It is naturally found in fruits like lemons, pineapples, oranges, grapes, etc. CA is GRAS, and as a non-toxic, colourless, and water-soluble compound, it has found wide applications in industries. In the food industry, CA has been used as a preservative, antioxidant, flavour enhancer, and emulsifier, which makes it an important ingredient in various food products. It has been reported that of the total CA production, 70% utilization is only for the food industry, while the remaining 30% is used in pharmaceutical, agricultural, and chemical industries (Dhillon et al., 2013).

Many physical and chemical methods have been used for manufacturing CA, but such conventional methods are complex, expensive, and not eco-friendly (Yu et al., 2018). To fulfil the high demand for CA, a continuous process for its production is required, which can be achieved by using microbial processes. The microbial production of CA provides an economic advantage and has become popular amongst manufacturers at the industrial level. For several years, a diverse variety of micro-organisms ranging from bacteria and fungi to yeasts has been employed for CA production. The microbial production of CA from organic compounds involves a biochemical reaction that is complex and therefore requires careful control over operating mechanisms.

12.3.2.1 Microbial Sources

There are various micro-organisms (Table 12.3) that are capable of producing CA, such as those from the genus *Bacillus*, *Aspergillus*, *Candida*, *Corynebacterium* spp., *Mucor piriformis*, *Trichoderma viride*, etc. (Papagianni, 2007). However, out of numerous commercially developed strains, *Aspergillus niger* is most ideally used for the industrial production of CA (Alnassar et al.,2016; Adeoye et al., 2015; Abbas et al., 2016a). The major advantages of using *A. niger* include its ability to ferment cheaper substrates for value-added products with higher yields, genetic stability, ease of handling, tolerance for high acid, and does not create any undesirable reactions (Bakhiet and Mokhtar, 2015; Pandey, 2013).

However, the CA yield can be increased by altering the metabolism of the fungus by employing a strain improvement programme. The most widely used approach for improving strains is mutation induction in parental strains using various mutagens. Mutagenesis can be induced by physical or chemical agents. The most common physical mutagens include electromagnetic radiations like X-rays, ultraviolet rays, gamma rays, and chemical mutagens like ethyl methyl sulphate, methyl methane sulphonate, diethyl sulphonate, ethidium bromide, or a combination of both mutagens can also be used to obtain hyper-producer strains (Musilkova et al., 1983). For screening and identification of improved strain, methods like enzyme diffusion zone, the passage method, and the single spore technique are widely employed (Sawant et al., 2018).

12.3.2.2 Agro-Industrial Byproducts/Food Waste

A wide variety of substrates are utilized for the microbial production of CA, such as hydrocarbons, starchy materials, and molasses. Researchers have used various substrates using different organisms like grape pomace, apple pomace, beet molasses, cane molasses, corn starch, hydrolysate starch, olive oil, rapeseed oil, etc. The raw materials for producing CA are classified into two categories: materials with a low ash content like cane or beet sugar, dextrose syrups, etc., and materials with high non-sugar substances and ash content like cane or beet molasses, high test molasses, and crude starch hydrolysates (Rohr et al., 1983). The sugar of fruit is the main substrate required for CA fermentation. Because of the requirement for sugar, a polysaccharide can also be used with the condition that the fermenting micro-organism has the ability to secrete a hydrolytic enzyme at high acid

TABLE 12.3
Micro-Organisms Used for Producing CA

Micro-Organisms	References
Bacteria	
Corynebacterium sp.	Anastassiadis et al., 2008
Arthrobacter paraffinens	Kubicek et al., 1985
Bacillus licheniformis	Yan et al., 2013
Fungi	
Aspergillus niger	Bastos and Ribeiro, 2020; Aboyeji et al., 2020; Ajala et al., 2020; Rao and Reddy, 2013; Ramesh and Kalaiselvam, 2011;
Aspergillus flavus	Shankar and Sivakumar, 2016
Trichoderma reesei	Bastos and Ribeiro, 2020; Campanhol et al., 2019
Rhizopus stolonifera	Zafar et al., 2020
Candida zeylanoides	Bilge et al., 2018; Kim et al., 2015
Candida lypolytica	Crolla and Kennedy, 2004
Candida oleophila	Anastassiadis et al., 2002
Candida tropicalis	Hesham et al., 2020; Folake et al., 2018
Candida parapsilosis	Mamdouh et al., 2018; Thomas, 2013
C. guilliermondii	Thomas, 2013
Penicillium sp. ZE-19	Okerentugba and Anyanwu, 2014
Penicillium oxalicum	Li et al., 2016
Alternaria sp.	Shankar and Sivakumar, 2016
Yeast	
Yarrowia lipolytica	Zywicka et al., 2020, Carlos et al., 2017
Saccharomyces cerevisiae	Praveen, 2015; Acourene et al., 2011
Pichia guilliermondii	Folake et al., 2018
Pichia kluyveri	Hesham et al., 2020

concentrations. Because the process needs to be more economical and environmentally sustainable, molasses (non-crystalline effluent) from sugar refineries after the extraction of sugar can also be used. The use of molasses benefitted the industry by reducing the cost and providing a raw material with high sugar content, such as sucrose, fructose, and glucose (Dronawat et al., 1995). Diverse varieties of sugar (cane/beet), agricultural methods, and storage conditions alter the quality of the molasses, so certain pre-treatments are required before using it for fermentation (Li et al., 2014b).

Huge quantities of waste are generated throughout the world from agro-based industries, which could be used as cost-efficient resources for the production of valuable products like CA by micro-organisms (Sawant et al., 2018; Ghaffar et al., 2014). The byproducts of the fruit industry like pine-apple waste, apple pomace, banana peels, or wastes from the processing of coffee, sugarcane, and rice can be used as suitable feedstock for the CA fermentation process. However, depending upon the type of raw material used, certain pre-treatments like drying, washing, concentration, centrifugation, supplementation, etc. are required prior to their use as a substrate.

12.3.2.3 Fermentation Processes

The microbial production of CA from renewable materials is a promising and widely used approach because it is economical and uncomplicated. Around 90% of the global consumption of CA is manufactured by the fermentation processes. The fermentation process involves three different phases, which include inoculation, fermentation or incubation period, and product recovery. Over the past years, a series of developments took place in the CA fermentation process, from using *Penicillium* and *Aspergillus* on surface cultures in 1910 to the use of submerged cultures in 1948, which provided the base for other types of fermentation processes (Show et al., 2015). CA can be produced by different methods: surface fermentation, SSF, and submerged fermentation.

12.3.2.3.1 Submerged Fermentation Process

The submerged fermentation (SmF) process is the most widely used technique for CA production. It has the edge over other processes by providing higher yields and production capacities along with reducing the cost of labour and contamination risks (Max et al., 2010), besides having some limitations of the requirement for sophisticated equipment and rigorous control for large-scale production. Conventional stirred fermenters or tower fermenters with a provision to maintain the dissolved oxygen level are generally employed for carrying out the fermentation process. There are different variations of SmF process like batch, fed-batch, or continuous systems, although batch fermentation is most commonly used.

In the SmF process, sugar- or starch-based media are used, which is in liquid form. Certain raw materials require pre-treatment, supplementation of nutrients, or sterilization, which is given depending upon the substrate to be used for fermentation. Inoculation is done by adding the desired organism, and aseptic fermentation is carried out in special vessels called bioreactors for 5–10 days depending upon the process conditions. There are various parameters that affect the performance of bioreactors, which include biomass concentration, type of inoculum, and process conditions like moisture, pH, aeration, etc. (Sawant et al., 2018).

The inoculum level plays an important role in enhancing CA yield in both SmF and SS&F. Increasing the inoculum level from 7% to 11% increased the yield from 2.5 to 4.6 g/l in SS&F and 0.6–1.6 g/l in SmF. The effect of the fermentation time was observed only in SS&F, while in SmF, the increase in fermentation time decreased the production of CA (Ajala et al., 2020).

Temperature is an important factor affecting the CA fermentation process. A high temperature can lead to denaturation of the enzyme citrate synthase, accumulation of acids, or increased catabolite repression (Crolla and Kennedy, 2004). A similar trend was observed in many other studies where CA production decreased with an increase in the incubation temperature (Hesham et al., 2020; Ali et al., 2016). Using high temperatures during incubation also avoids the risk of contamination to a great extent. In CA production from sugarcane molasses, the temperature varied from 20°C to 50°C and increasing the temperature of the medium above 25°C increased the synthesis of CA, with the highest production of 3.55 g/l at 40°C after 192 h (Mamdouh et al., 2018).

Various researchers have stated different incubation temperatures for maximizing CA yield, which means that the optimal temperature is dependent upon the strain used for fermentation.

The pH is another important factor that affects enzyme activity, cell growth, and formation of the product; pH 5 was found to be optimal in many studies for maximum CA production (Hesham et al., 2020; Mamdouh et al., 2018; Sharma et al., 2018); however, a low pH during production reduces the risk of contamination and production of alcohols like oxalic acid and gluconic acid, which helps in better product recovery (Abdel-Rahman et al., 2018). Most of the studies have reported neutral to slightly acidic or slightly alkaline pH (pH 5.0–7.0) as the optimum pH for CA production (Hesham et al., 2020; Ali et al., 2016b).

Several authors have reported the addition of lower alcohols, which have less carbon atoms than that of higher alcohols and are more soluble in water, for enhancing CA production. In a particular study, the methanol concentration in the medium varied from 0% to 5% and the maximum CA production was achieved at lower methanol concentrations (3%) (Ayeni et al., 2019). In another study, the effect of different alcohols like ethanol, methanol, propanol, amyl alcohol, and ethyl acetate was studied, and 1% methanol was selected as the optimal concentration for CA production by strain NH-3 (Mamdouh et al., 2018).

12.3.2.3.2 Surface Fermentation

The surface fermentation technique was originally used for the industrial production of CA in 1919. Over the years, the SmF technique became popular, though many small- and medium-scale industries still practice the surface fermentation method (Bauweleers et al., 2014). Nutrient media like grape pomace, brewery wastes, cane molasses, and corn starch hydrolysate are commonly used (Table 12.4). In this method, micro-organisms are grown on the surface of a liquid substrate, which forms a mycelial layer on the surface of the liquid. Stackable aluminium or stainless steel trays are employed for this purpose, which are mounted over one another in fermentation rooms that can provide filtered air, controlling the oxygen supply and temperature of the fermentation process. After the completion of the fermentation time, the mycelial mats are separated from the liquid substrate (Sawant et al., 2018). The process takes place in two steps: in the first step, there is the growth of mycelia on the surface of the medium, and in the second step, there is the utilization of carbohydrates for the production of CA. This technique has some advantages like low installation and

TABLE 12.4
Fermentation Processes for the Production of CA

Fermentation Process	Substrate	Micro-Organism	Yield (g/l)	Reference
Submerged fermentation	Cassava peels	*Aspergillus niger*	1.68	Ajala et al., 2020
	Pineapple waste	*Aspergillus niger*	15.51	Ayeni et al., 2019
	Corn stalk	*Aspergillus niger*	13.20	Adudu et al., 2019
	Orange peel	*Aspergillus niger*	11.36	Vidya et al., 2018
	Synthetic medium combined with wheat bran extract	*Aspergillus niger* ATCC12846	162.7	Yu et al.,2018
	Sugar beet, molasses, and chicken feather	*Aspergillus niger*	68.8	Ozdal and Kurbanoglu, 2019
	Sugarcane molasses	*Candida parapsilosis* NH-3	4.30	Mamdouh et al., 2018
	Cassava peel-malted sorghum	*Aspergillus niger*	88.73	Adeoye et al., 2015
	Sugarcane molasses	*Candida zeylanoides*	91.0	Kim et al., 2015
	Sugarcane molasses	*Aspergillus niger*	68.7	Farooq et al., 2013
Solid-state fermentation	Cassava peels	*Aspergillus niger*	4.90	Ajala et al., 2020
	Peanut shell, orange peel	*Rhizopus stolonifera* and *Aspergillus niger*	7.65	Zafar et al., 2020
	Pomegranate peel wastes	*Aspergillus niger* B60	278.5	Roukas and Kotzekidou, 2020
	Wheat bran	*Aspergillus niger*	12.9	Hussain, 2019
	Corn cob	*Aspergillus niger*	–	Addo et al., 2016
	Sugar cane bagasse	*Aspergillus niger*	3.5	Sharan et al., 2015
	Oat bran	*Aspergillus niger*	62.0	Rao and Reddy, 2013
	Banana peel	*A. niger* UABN 210	82.0	Kareem and Rahman, 2013
	Citrus waste, brewer's spent grain, and sphagnum peat moss	*Aspergillus niger* NRRL 2001	364.5	Dhillon et al., 2013
Surface fermentation	White grape pomace	*Aspergillus niger* B60	85.0	Papadaki et al., 2019
	Apple pomace, peanut shell	*Aspergillus ornatus* and *Alternaria alternata*	2.644 ± 0.99 mg/ml)	Ali et al., 2016
	Corn starch hydrolysate	*A. niger* ATCC 9142	2.70	Nosakhare and Felix, 2012
	Brewery wastes	*Aspergillus niger*	19.0	Roukas and Kotzekidou, 2006

energy cost, as no separate instrumentation is required for aeration and agitation. But this process requires intensive manpower and is sensitive to changes in the composition of media, which adds to its limitations (Benghazi et al., 2014).

12.3.2.3.3 SSF

SSF is the most common technique adapted by several industries like food, textiles, pharmaceuticals, etc., for producing microbial metabolites using solid support in place of the liquid medium. The production of CA by SSF was first developed in Japan, where ample amounts of agro-food industry waste are easily available (Soccol et al., 2006).

The process involves the growth of micro-organisms on a moist solid substrate that provides both physical support and nutrients. The substrates used in SSF for CA production are summarized in Table 12.4. Many micro-organisms have been grown on solid substrates like *Aspergillus niger* (Ajala et al., 2020; Ali et al., 2016; Sharan et al., 2015; Rao and Reddy, 2013), *Rhizopus stolonifera* (Zafar et al., 2020), *Aspergillus ornatus*, and *Alternaria alternate* (Ali et al., 2016). Amongst all micro-organisms, filamentous fungi are best capable of growing on solid substrates because of their biochemical and physiological properties. SSF processes have certain advantages over SmF with low energy requirement and waste generation, higher productivity, easier downstream processing, less monitoring, etc. Batch experiments were conducted using cassava peel with 60% moisture for SSF and cassava peels and distilled water (20:100 w/v) in conical flasks for SmF, and a yield of 4.9 g/l was obtained by the SSF process against 1.68 g/l in SmF (Ajala et al., 2020). While in another study, SSF was carried out at a laboratory scale by *A. niger* NRRL 2001 using various substrates (citrus waste, brewer's spent grain, and sphagnum peat moss) and the highest yield of 364.4 g/kg^{-1} dry solids (DS) was obtained (Dhillon et al., 2013).

Though SSF offers several advantages, the process is mainly used for the laboratory-scale production of metabolites, hydrolytic enzymes (amylase, catalase, etc.), and spores. The scaling-up process from the lab to the industrial level is very challenging because of the technical difficulties faced during operation and control (Barragan et al., 2016). It can be concluded that the selection of fermentation method depends mainly on the final product. SSF is well suited for low-value products, while for a high-quality product that has its applications in pharmaceutical industries, SmF is preferred as it has better control over operations, takes less time, and the purification process is easy and well defined.

12.3.2.4 Recovery

After the end of the desired incubation time, the fermentation is stopped by employing heat. CA so produced may consist of mycelium or various other impurities like proteins, mineral salts, organic acids, etc., depending upon the raw materials used for fermentation. In SSF, the contents are dried in the oven at a low temperature ($\approx$50°C) for a few hours, followed by mixing in distilled water and filtration. While in SmF, coagulation of proteins is done by heating the mycelium biomass at around 70°C for 15–20 min. Precipitation is the most widely used method for the recovery of CA. Calcium hydroxide is added to the medium, which increases the pH and forms the precipitate of tri-calcium citrate and removes oxalic acid formed during fermentation by converting it to calcium oxalate. The calcium citrate precipitate is then washed with water ($\approx$90°C) to remove other impurities and treated with sulphuric acid to eliminate calcium sulphate precipitates. The solution is purified using activated carbon or ion exchange columns, which is further concentrated by a vacuum evaporation process at 40°C, which results in the formation of CA crystals. As crystallization is a temperature-dependent process, two types of crystals are formed: CA anhydrate (crystallization temperature is 20–25°C) and CA monohydrate (crystallization temperature is '36.5°C). In one such study, a temperature of 50°C for 20 min does not favour any phase transition in CA and a high purity product with 100% yield was obtained (Li et al., 2016a). The only disadvantage of the precipitation method is the generation of a large amount of waste as calcium sulphate and oxalate, micro-organism scums containing sugars, amino acids, and pigments, which, if further utilized as feed or in industries, can become an economical method (Dhillon et al., 2011). Solvent extraction has also been used as another method for recovering CA. Solvents with maximum solubility with CA, such as n-octyl alcohol, tridodecylamine, isoalkane, etc., are used, which also eliminates gypsum production during CA extraction, which is an advantage of using this method for recovery. The only drawback of this technique is that it is more expensive compared to the precipitation method.

Adsorption and ion exchange techniques have also been employed for the extraction of CA. These techniques have certain advantages of reduced energy requirements, high capacity, quick recovery, and also do not undergo phase transition. The only constraint of this process is the requirement of large quantities of de-sorbent that dilute the CA solution and results in the formation of waste liquor (Li et al., 2016a). At the laboratory level, the quantification of CA is done in a number of ways, such as spectrophotometry (Ali et al., 2016; Mamdouh et al., 2018), gas chromatography (Kumar et al., 2017), HPLC (High Performance Liquid Chromatography) (Hesham et al., 2020; Hussain, 2019; Yu et al., 2018), NMR (Nuclear Magnetic Resonance) spectroscopy (Teipel et al., 2020; Campo et al., 2006), and the easiest way is the titration method (Ajala et al., 2020; Ayeni et al., 2019; Sharma et al., 2018).

12.3.3 SUCCINIC ACID

Succinic acid, or butanedioic acid, is a white, odourless, organic acid and is diprotic in nature (Naraian and Kumari, 2017; Pais-Chanfrau et al., 2020). It is also known as amber acid, as it was first purified from amber by Agricola in 1546. Chemical synthesis of succinic acid is mainly carried out using butane as raw material through maleic anhydride and petro-based chemicals (Naraian and Kumari, 2017). However, the chemical process involves several limitations, viz. low purity and high energy requirements in terms of electricity, which not only adds to the increased cost of production but also results in environmental pollution. Thus, to cope with all these problems, alternate methods like microbial fermentation have been gaining momentum (Cao et al., 2013). It has also marked its presence as one of the 12 value-added chemicals on the list of the US Department of Energy and is extensively employed in various sectors like agriculture, chemical, food, and pharmaceuticals (Putri et al., 2020).

12.3.3.1 Microbial Sources

Succinic acid is an important intermediate produced during the TCA cycle or Krebs cycle in many biological systems. The major biosynthetic pathways for the production of succinic acid include the oxidative TCA cycle, glyoxylate shunt, and the reductive TCA cycle (Figure 12.3). Wild strains utilize the former two pathways for producing succinic acid as it is produced as an intermediate within these pathways, while the reductive pathway results in the deposition of succinic acid in a cell. Moreover, the anaerobic pathway for the production of succinic acid is more favoured as it results in the production of two succinic acid molecules from a single glucose molecule by utilizing carbon dioxide in comparison to aerobic pathways in which carbon dioxide is released and only four carbons are maintained out of the six present in glycolysis (Cao et al., 2013; Mancini et al., 2019).

Bacterial species are mostly employed to produce succinic acid, which includes *Actinobacillus succinogenes*, *Anaerobiospirillum succiniciproducens*, *Bacillus fragilis*, *Corynebacterium crenatum*, *Mannheimia succiniciproducens*, and *Ruminococcus flavefaciens*. Many fungal strains such as *Aspergillus fumigatus*, *Aspergillus niger*, *Byssochlamys nivea*, *Lentinus degener*, *Paecilomyces varioti*, and *Penicillium viniferum* have also been found to be suitable for succinate production. In yeast, *Saccharomyces cerevisiae* is used. The fungal and yeast strains can produce succinic acid under both aerobic and anaerobic conditions as they synthesize it as a by-product. Still, bacteria are more preferred than fungi and yeast due to increased productivity and ease of downstream processing (Mancini et al., 2019). Out of all the bacterial species, *Actinobacillus succinogenes* is widely employed at the industrial scale because of its high productivity, non-pathogenic status, ability to ferment various agro-industrial substrates, CO_2 fixation, resistance to high glucose and succinic acid concentrations, and potential to form biofilms (Mancini et al., 2019).

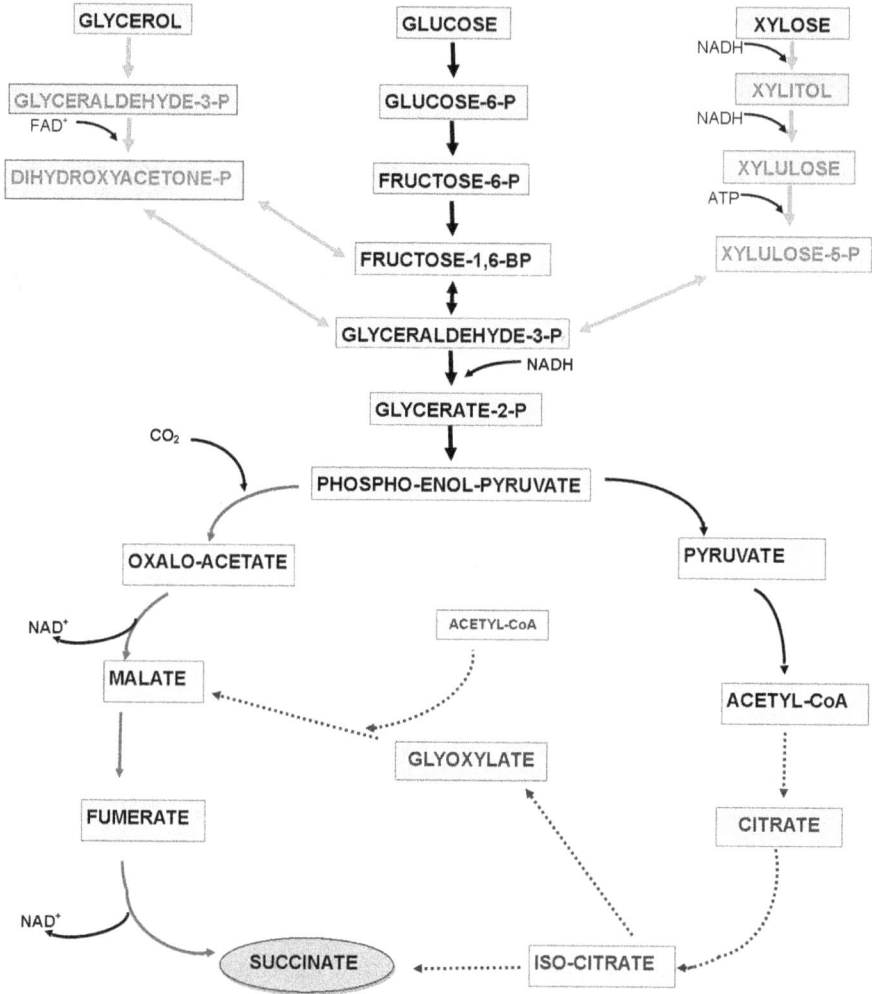

FIGURE 12.3 Common TCA cycle occurring in natural fermentative micro-organisms. The reductive pathway is indicated in black arrows. The grey arrows represent pathway occurring in *A. succinogenes* for xylose and glycerol. In engineered succinate-producing organisms such as *E. coli*, glyoxylate shunt and the oxidative branch of TCA represented with dotted arrows are exploited (Adapted from Mancini et al., 2019).

To boost the productivity and purity of the succinic acid while terminating other products, several engineered strains have also been employed that work on different strategies to improve the production of succinic acid. The strategies employed to enhance succinic acid production include diversion of pathways leading to other byproducts or acids, enhancement of succinic acid production pathways, and strategies to increase substrate transport and consumption of co-factors (Mancini et al., 2019). Recombinant *E. coli*, *C. glutamicum*, and *S. cerevisiae* have been widely used to produce succinic acid, while *A. succinogenes* is of recent interest, and the research regarding this is still in its early stage (Cao et al., 2013; Mancini et al., 2019).

12.3.3.2 Agro-Industrial Byproducts/Food Waste

Bio-based succinic acid or bio-succinic acid is mainly produced from sugars derived from biomass or other renewable sources. However, most succinic acid has been produced from biomass comprising starch-based sugars and glycerol obtained from bio-diesel production. Studies involving succinic acid production from lignocellulosic waste such as wheat straw, corn straw, molasses, corn stover, etc. are limited due to pre-treatment steps involved to release sugars and production of microbial inhibitors during the process such as furfurals, hydroxymethyl furfurals (HMF), acetic acid, and low molecular weight compounds that contribute to the reduced performance of succinic acid (Salvachua et al., 2016). The pre-treatment steps required for the lignocellulosic and starchy waste have already been discussed above in Figure 12.2.

Food waste forms one of the largest fractions of municipal waste that is detrimental to the environment. However, it contains a good amount of nutrients that can be utilized by micro-organisms for the production of various bio-based chemicals. Different types of food wastes have been utilized for the production of succinic acid like bakery and bread waste, citrus peel waste, carob pods, mixed food waste, coffee husk, cheese whey, etc. Food waste does not require any harsh pre-treatment steps like lignocellulosic waste, and only mild treatment like blending and finally enzymatic hydrolysis are sufficient.

Sometimes, in the case of bread and bakery waste for the production of succinic acid, fungal autolysis may be required, and similarly, citrus waste may require complex treatment steps (Mancini et al., 2019).

12.3.3.3 Fermentation Processes

Most wild strains employed for succinic acid production utilize TCA and glyoxylate shunt and convert phosphoenolpyruvate (PEP) to a mixture of acids like acetic, formic, and oxaloacetate first and then into malic, fumaric, and succinic, thereby leading to low productivity. However, under anaerobic fermentation conditions, PEP is converted to oxaloacetate with the help of the enzyme PEP carboxylase or PEP carboxykinase. Pyruvate can also be directly converted into oxaloacetate or malate with the addition of carbon dioxide (CO_2) by the enzyme pyruvate carboxylase. Then, under the action of the enzyme malate dehydrogenase and fumarase, oxaloacetate and malate are converted to fumarate, finally resulting in the production of two moles of succinate by fumarate reductase (Cao et al., 2013; Mancini et al., 2019). The design and operation of bioreactors play an important role in the production of succinic acid. The importance of each bioreactor is discussed below.

12.3.3.3.1 Batch Fermentation

Batch operation is one of the simplest, easiest, and most widely used approaches to producing succinic acid. It includes a temperature-controlled jacketed vessel, mechanically operated along with other equipment like a gas sparger. In this process, the substrate concentration is generally kept between 20 and 100 g/l and the operation is carried out at a temperature of 37°C–39°C. While using bacterial species, the pH optima are kept in the range 6.8–7.2 and within 4.5–6.0 in the case of a yeast culture. For succinic acid production, anaerobic conditions are maintained by sparging either oxygen-free nitrogen or carbon dioxide through the sparger, and the fermentation may take 48–100 h depending upon the operating conditions and type of strain involved (Ferone et al., 2019a).

Several batch studies have been carried out using byproducts from agro-industrial or food waste in different modes. In a novel approach involving a non-conventional substrate, i.e., the waste stream from the beverage industry was utilized for the production of succinic acid using *Actinobacillus succinogenes* in batch fermentation. The process yielded 30–40 g/l of succinic acid with a productivity of around 0.75–1.00 g/l.h (Ferone et al., 2019b).

Similarly, pre-treated chestnut shells were used as a substrate in a batch fermentation process employing *Actinobacillus succinogenes* 130Z. The process resulted in improved productivity of 0.76 g/l.h and a yield of 10 g/l (Ventrone et al., 2020).

Though several batch studies have been carried out due to the ease of operation, it still has a few limitations like low productivity. This is mainly due to the fact that instead of high biomass concentration at the end of the batch fermentation process, the product outcome is not sufficient (Ferone et al., 2019a).

12.3.3.3.2 Fed-Batch Fermentation

This type of fermentation is mainly carried out when high substrate concentration appears toxic to the strain or culture involved. In a study, the highest succinic acid concentration (25.7 g/l) and yield of 1.34 g/l were obtained by carrying out fed-batch fermentation using glycerol (a by-product of bio-diesel production) as the carbon source and *Anaerobiospirillum succiniciproducens* as the succinic acid producer. The addition of glucose in the fermentation led to a 2.5-fold increase in concentration while productivity remained constant (Bretz, 2015). Thus, it was concluded that the addition of glucose to fed-batch fermentation could lead to elevated levels of succinic acid concentration and yield, but the process did not enhance productivity at all, which is the only limitation associated with fed-batch systems (Ferone et al., 2019a). However, among the batch and fed-batch strategies, the latter results in a higher yield and productivity of succinic acid (de Barros et al., 2013).

12.3.3.3.3   Continuous Fermentation

This type of fermentation offers various benefits with free cells, like only a single inoculum is enough over longer periods, short sterilization time, and higher productivity. Several studies involving continuous fermentation with free cells have been attempted using different strains like *Basfia succiniciproducens*, *Mannheimia succiniciproducens*, genetically engineered *E. coli*, etc., and a good yield of succinic acid has been achieved. But free cells of the widely used strain of *A. succinogenes* in the continuous mode are associated with a major drawback, i.e., biofilm production (Ferone et al., 2019a).

To overcome this problem, bioreactors like CSTR, PBR, trickling bed reactor, fluidized bed reactor, airlift reactors (ALR), upflow anaerobic sludge blanket, and expanded bed reactors are designed to be used with biofilms (Ferone et al., 2019a). A study involving immobilized *A. succinogenes* in a continuous fermentation approach with xylose-rich corn stover hydrolysate resulted in an attractive succinic acid yield of 0.78 g/g and productivity of 1.77 g/l.h (Bradfield et al., 2015). Similarly, continuous succinic acid production using immobilized *A. succinogenes* in a packed bed biofilm reactor was carried out and resulted in maximum productivity of 35.0 g/l.h (Ferone et al., 2019b).

A summary of studies carried out for the production of succinic acid in the last ten years utilizing various agro-industrial residues under different fermentation conditions is given in Table 12.5.

12.3.3.4   Recovery

The recovery of succinic acid is both a costly and difficult process. Like other organic acids, succinic acid is also water-loving and has a high boiling point. Though various methods like chromatography, direct crystallization, precipitation, extraction, membrane separation, and *in situ* separation have been studied for succinic acid purification, not a single one has been found to be efficient. In the membrane separation processes, nano-filtration coupled with vapour permeation assisted esterification and dia-nano-filtration with tubular membrane module have been studied. Among both, the latter process resulted in 98% succinic acid recovery. The former method resulted in 80% protein removal but suffered from membrane fouling. Similarly, extraction associated with bipolar dialysis was employed and resulted in 90% pure succinic acid. The method has been found to be cost-effective and efficient to use on a large scale (Mancini et al., 2019).

Precipitation is another traditional method widely employed for succinic acid separation. Agents like $Ca(OH)_2$ and CaO are commonly used, which result in the precipitation of succinic acid as calcium succinate. Ammonia is also used to serve the purpose, which results in the formation of di-ammonium succinate; H_2SO_4 is added to this to obtain succinic acid and ammonium sulphate. Further purification is carried out by methanol addition followed by crystallization. Precipitation results in 81–87% pure succinic acid (Mancini et al., 2019).

Direct crystallization is another approach used for succinic acid purification. The method results in higher yield, i.e., 60–75 wt% and similar purity as precipitation. The method is generally recommended as the final step in purification (Mancini et al., 2019).

Extraction or salting out is also an efficient method. But among all the methods discussed above, membrane separation and crystallization are widely employed due to the efficient purification of succinic acid (Mancini et al., 2019).

TABLE 12.5

Production Scenario of Succinic Acid Utilizing Various Agro-Industrial Residues (data from 2011 to present)

S no.	Strain used	Agro-industrial substrate employed	Mode of fermentation	Productivity [g/(l.H)]	Yield (g/g)	References
1.	A. succinogenes	Waste bread	Batch	1.12	47.3	Joe Leung et al., 2012
2.	A. succinogenes	Sugarcane molasses and glycerol	Batch	0.155	1.30	de Barros et al., 2013
3.	A. succinogenes	Oil palm empty fruit bunch	Batch	–	23.5 g/l	Pasma et al., 2013
4.	A. succinogenes	Bakery waste	Batch	0.87	0.67	Zhang et al., 2013
5.	A. succinogenes	Carob pod waste	Batch	1.61	0.55	Carvalho et al., 2013
6.	A. succinogenes	Bakery waste	Batch	–	0.55	Lam et al., 2014
7.	A. succinogenes	Pastry	Batch	0.24	0.85	Sun et al., 2014
8.	A. succinogenes	Food waste	Batch	0.29	0.87	Sun et al., 2014
9.	Recombinant E. coli	Pastry	Batch	0.20	0.79	Sun et al., 2014
10.	Recombinant E. coli	Food waste	Batch	0.20	0.98	Sun et al., 2014
11.	A. succinogenes	Corn stover	Continuous	1.77	0.78	Bradfield et al., 2015
12.	A. succinogenes	Carob pods	Batch/ fed-batch	1.17/1.32	0.39/ 0.54	Carvalho et al., 2016
13.	A. succinogenes	Sugarcane bagasse	Batch	1.65	80.5%	Chen et al., 2016
14.	A. succinogenes	Agave tequilana bagasse	Batch	0.57	0.44	Corona-Gonzalez et al., 2016
15.	A. succinogenes	Oil palm frond bagasse	Batch	–	0.71	Luthfia et al., 2016
16.	B. succiniciproducens	Lignocellulosic hydrolysate	Batch	0.43	0.69	Salvachua et al., 2016
17.	A. succinogenes	Coffee husk	Batch	0.54	0.95	Dessie et al., 2018
18.	A. succinogenes	Waste stream from beverages	Batch	–	0.83	Ferone et al., 2019b
19.	Wild isolate	Crude glycerol	Batch	0.55	0.9	Kuenz et al., 2020
20.	A. succinogenes	Whey and lactose	Batch	0.61/ 0.9	61.1%/ 65%	Louaste and Eloutassi, 2020
21.	A. succinogenes	Organic fraction of municipal solid waste	Continuous	1.27	0.47	Stylianou et al., 2020
22.	A. succinogenes	Chestnut shells	Batch	0.76	0.62	Ventrone et al., 2020

12.3.4 Acetic Acid

Acetic acid, also known as ethanoic acid, exists as a colourless liquid. In the undiluted form, it is called glacial acetic acid, while 4–6% diluted acetic acid solution is called vinegar. It is naturally present in fruits such as pears and grapes. Acetic acid production first started with grapes in 1794 in Germany (Pal and Nayak, 2016; Naraian and Kumari, 2017). Though acetic acid mainly used in the food industry is obtained via fermentation methods, chemical methods are employed for large-scale production. At the commercial scale, it is mainly obtained from the oxidation of petroleum which derives substrates such as ethanol, acetaldehyde, methanol, butane, or ethylene. Due to exhausting oil reserves, alternative methods for acetic acid synthesis are required (Li et al., 2014b; Pal and Nayak, 2016).

12.3.4.1 Microbial Sources

A range of micro-organisms is employed for the production of acetic acid among bacteria, yeasts, and filamentous fungi. The fermentation process for the production of acetic acid is divided into two parts: first, the conversion of the substrate into ethanol by yeast or *S. cerevisiae*, and second, the production of acetic acid by acetic acid bacteria. Acetic acid bacteria represent a group of aerobic gram-negative acidophilic α-proteobacteria that oxidizes ethanol into acetic acid under acidic or neutral conditions. Two genera of this group, i.e., *Acetobacter* and *Gluconobacter*, are used for aerobic acetic acid fermentations (Xu et al., 2011). Acetobacter strains employed include *A. hansenii*, *A. xylinus*, *A. malorum*, *A. eoropheus*, *A. aceti*, *A. polyoxogenes*, *A. obodiens*, and *A. lovaniensis* (Naraian and Kumari, 2017).

Generally, vinegar is produced from one molecule of hexose on which yeast like *S. cerevisiae* acts to produce two molecules of ethanol, which are further acted upon by acetic acid bacteria to produce two molecules of acetic acid (Xu et al., 2011).

However, under anaerobic conditions, three molecules of acetic acid can be produced from one molecule of hexose by employing anaerobic acetogenic bacteria called acetogens. The most commonly used species of acetogens for acetic acid production are *Acetobacterium* and *Clostridium*. In these species, *Moorella thermoacetica*, previously called *Clostridium thermoacetica*, has been extensively studied for anaerobic acetic acid fermentation processes (Xu et al., 2011).

Among other species, *Dekkera* and *Brettanomyces* spp. of yeast also utilize glucose and ethanol as carbon sources to produce acetic acid. Similarly, filamentous fungi like *Fusarium oxysporum*, *Monilia brunnea*, and *Polyporus anceps* have also been studied as potential acetic acid producers (Naraian and Kumari, 2017).

12.3.4.2 Agro-Industrial Byproducts/Food Waste

Only a few studies have been conducted on the production of acetic acid using agro-industrial byproducts or food waste in the past few years. The reason behind this may be attributed to the fact that most of the acetic acid produced via the fermentation approach is mainly utilized in the food industry only, and the traditional routes followed already involve the use of renewable carbon sources such as apple, beer, cane, coconut, date, grape, honey, hydrolyzed starch, pears, syrup, cereals, and wine. The only limitation with the fermentation route is the availability of robust plants for commercial production (Deshmukh and Manyar, 2020). Moreover, the studies carried out in the past ten years are mainly focused on acetic acid production by membrane-based technologies.

Few studies reported have used municipal food waste and vegetable waste to obtain vinegar (Li et al., 2014; Chakraborty et al., 2017; Moestedt et al., 2019). Onion waste has also been utilized in a recent study to produce acetic acid (Kim et al., 2019).

12.3.4.3 Fermentation Processes

The fermentation approach for the production of acetic acid is quite old. Traditionally, vinegar was mainly produced via two fermentation methods, namely the Orleans method and the trickling process. However, the modern approach is based on continuous SmF (Deshmukh and Manyar, 2020).

The Orleans method is a well-known method mainly used for small-scale production of acetic acid. Generally, wooden barrels are used to carry out fermentation. The method relies on specific raw materials depending on their availability to produce exotic vinegar. However, the only limitation of this method is the slow rate of acetification. To overcome this problem, the trickling method was developed, which involves the spraying of an alcoholic substrate over the fermentation loop. The process improves the interaction between the substrate and bacteria involved. However, the formation of a biofilm limits the process by reducing the reaction rate (Deshmukh and Manyar, 2020).

Nowadays, a continuous SmF method is employed extensively for the commercial production of acetic acid. The method exhibits several advantages over traditional methods like low cost, simple fermenter design and process control, the requirement of less space, fast rate of acetification, high yield, and high quality of acetic acid (Deshmukh and Manyar, 2020).

12.3.4.4 Recovery

The recovery of acetic acid from the medium is both energy-consuming and a multi-step approach, which makes the entire procedure time consuming and expensive. The non-volatile nature of acetic acid makes it more complicated. Various methods have been investigated in the past, such as azeotropic dehydration distillation, fractional distillation, solvent extraction, a combination of the given methods, extractive distillation, and carbon adsorption. These methods, either alone or in combination, give around 95% pure acetic acid. Further, the final purification step involves the distillation of acid in the presence of $KMnO_4$ or $K_2Cr_2O_7$. The agents are capable of removing a wide range of impurities (Xu et al., 2011).

12.3.5 FUMARIC ACID

Fumaric acid, also known as (E)-2-butanedioic acid, generally occurs in the form of a white crystalline solid and possesses little acidic taste with no odour. It is a trans-C4-dicarboxylic acid that is formed naturally as an intermediate in the TCA cycle. It derives its name from the plant *Fumaria officinalis* from which it was initially extracted. The fermentation production of fumaric acid was first carried out by Ehrlich in 1911 using the fungi *Rhizopus nigricans* (Straathof and Gulic, 2012; Das et al., 2016; Naraian and Kumari, 2017). It is chemically synthesized from maleic anhydride, which is a petroleum derivative, and due to the increase in petroleum prices, the price of fumaric acid has also increased. Therefore, the fermentation production of fumaric acid employing renewable substrates has gained attention. The process is cheap and environment friendly as it results in carbon dioxide fixation (Straathof and Gulic, 2012).

12.3.5.1 Microbial Sources

Filamentous fungi belonging to the genus *Rhizopus*, namely *R. nigricans*, *R. arrhizus*, and *R. oryzae*, are considered the best natural producers of fumaric acid due to their high productivity and yield. Moreover, they can produce fumaric acid under both aerobic and anaerobic conditions. Other fungal strains from genii *Mucor*, *Aspergillus*, *Cunninghamella*, and *Cirinella* are also known to produce fumaric acid as minor fermentation products (Straathof and Gulic, 2012; Martin et al., 2018). Among the above-mentioned *Rhizopus* species, *R. oryzae* is mainly preferred over the other strains due to several technical benefits such as low nutrient requirements, high yield, and the ability to produce different morphological structures such as pellets and suspended and clumpy mycelia. These properties lead to enhanced production of fumaric acid (Das et al., 2016).

Like citric and succinic acid, fumaric acid is also produced mainly by either the TCA cycle or the reductive glyoxylate cycle. The reactions of the TCA cycle occur both in cytosol and mitochondria. Under aerobic conditions, pyruvate carboxylase results in the fixation of CO_2, which gives a high molar yield of fumarate. CO_2 fixation also results in the production of oxaloacetate; hence intermediates of the Krebs cycle are withdrawn for biosynthesis during the exponential phase. Under nitrogen-limiting conditions, the exponential phase ceases, but the process of glucose metabolism and CO_2 fixation generally commence and result in the accumulation of C4-acids. Under a non-growth situation, two molecules of fumaric acid are produced by consuming a single molecule of glucose along with fixation of two molecules of CO_2 through reductive pyruvate carboxylation, thus resulting in the maximum theoretical yield. However, reductive pyruvate carboxylation could not act as the sole pathway due to the lack of ATP production required for maintenance and transport purposes, and thus, the TCA cycle is required (Straathof and Gulik, 2012).

Pyruvate carboxylase, a cytosolic enzyme required for CO_2 carboxylation along with NAD-malate dehydrogenase and fumarase (present both in cytosol and mitochondria) results in the synthesis of fumaric acid in cytosol (Straathof and Gulik, 2012). In *Rhizopus* strains, glyoxylate bypass has also been studied as an important pathway for the fermentation production of fumaric acid, but it has been found that it cannot act as the principal pathway for the production of fumaric acid because the key enzyme of this pathway, i.e., isocitrate glyoxylate lyase, is repressed at high glucose concentrations (Strrathof and Gulik, 2012).

Genetically engineered strains of *Lactobacillus* and *E. coli* have also been employed in some processes for obtaining higher yields of fumaric acid (Martin-Dominguez et al., 2018).

12.3.5.2   Agro-Industrial Byproducts/Food Waste

The composition of the medium used mainly depends on the type of micro-organism employed. Fungi are mainly used for the biosynthesis of fumaric acid and require glucose for their growth. Several renewable raw materials are used nowadays as they are interesting both from economic and environmental points of view. Various byproducts from the agricultural and food industries have been employed for formic acid production as carbon and nitrogen sources such as apple pomace, bagasse (sugarcane and cassava), corn stover, crude glycerol, oil palm empty fruit bunch, wheat straw, potato flour, kimchi cabbage waste, dairy manure, corn starch, corn steep liquor, soya bean meal, wastewater from the brewing industry, and both solid and liquid fraction of food waste (Straathof and Gulic, 2012; Das et al., 2016; U-Thai et al., 2016; Martin-Dominguez et al., 2018; Kim et al., 2018).

12.3.5.3   Fermentation Processes

Batch bioreactors are the most common and widely employed fermentation systems for fumaric acid production. The system gives high productivities between the range of 0.18 and 0.7 g/l/h. Among other configurations, stirred tank and bubble column have been employed. Stirred tank reactors are mainly used due to ease of control and handling during the process and low operating costs due to optimum capacity and reliability. Bubble column, on the other hand, offers efficient oxygen transfer to biomass and provides good aeration rates. Moreover, they have also been associated with greater productivity and yield of fumaric acid than stirred batch reactors (Martin-Dominguez et al., 2018).

Though the above-mentioned configurations offer several benefits, due to the use of *Rhizopus* species for fumaric acid production, few operational problems pertain to their use, such as difficulty in getting proper morphology and highly viscous medium due to mycelial growth. To overcome these problems, stirred tank, fluidized bed, and bubble column reactors with immobilized biomass and rotary film contractor are employed. These reactors offer permanent morphology, prevent pH and product inhibition of micro-organisms, easy downstream processing, and enable longer production periods (Martin-Dominguez et al., 2018).

12.3.5.4   Recovery

Unlike other organic acids such as citric, lactic, acetic acid, etc., the studies related to downstream processing of fumaric acid are limited. Earlier, a combination of acids such as H_2SO_4 or HCl and base such as $CaCO_3$, $Na2CO_3$, $NaHCO_3$, $(NH_4)_2CO_3$, and $Ca(OH)_2$, also called neutralizing agents, was used, but due to the complexity and high cost of operation, new methods for recovery have been developed which include extraction with amines such as tridodecylamine (TDA), adsorption with active carbon, or ion exchange resin which result in high purity (99%) and high recovery (99%) of fumaric acid (Das et al., 2016; Martin-Dominnguez et al., 2018).

12.3.6 GLUTAMIC ACID

Glutamic acid is a non-essential amino acid used by human bodies for the biosynthesis of proteins. It plays an important role in maintaining our health and is used for the proper functioning of the brain, immune system, and digestive system. It is widely present as L-glutamic acid in nature, while D-glutamic acid is found in small quantities in fresh milk, soybeans, and some plant organs. L-glutamic acid is considered a very important commercial amino acid as it is extensively used in food, agriculture, pharmaceuticals, and chemical industries for the production of value-added products. Its sodium derivative, i.e., monosodium L-glutamate, resembles one of the five basic tastes that human senses can detect, i.e., umami, and therefore has been used as a flavour enhancer in the food industry since 1908. L-glutamic acid can be produced by chemical and microbial methods; however, the chemical method is not preferred because there is the formation of a racemic mixture of both D and L forms. Several fermentation techniques have been used for the production of glutamic acid using various strains. Many L-glutamic-acid-producing micro-organisms have been isolated so far, but the bacteria belonging to genera *Corynebacterium* and *Brevibacterium* have been used for the economical production of L-glutamic acid at industrial levels using batch and fed-batch fermentation methods. Poly-γ-glutamic acid (PGA) is a natural polymer composed of glutamic acid monomers joined by amide linkages between α-amino and γ-carboxyl groups. It can be synthesized by various micro-organisms; though, for industrial purposes, strains from the genus *Bacillus*, i.e., *B. licheniformis* and *B. subtilis*, are generally employed. It can be obtained in either L or D forms or both L and D forms, depending upon the type of strain and production media employed. As it is biodegradable, water-soluble, and a non-toxic biopolymer, PGA and its derivatives have several industrial applications such as cryoprotectants, thickeners, drug carriers, animal feed additives, heavy metal absorbers, etc. (Luo et al., 2016).

12.3.6.1 Microbial Sources and Fermentation Processes

The culture medium for glutamic acid fermentation consists of a carbon source like glucose, starch hydrolysate, molasses or their mixture, a nitrogen source, growth factors, and other salts. To make a fermentation process economical, crude nutritive sources are utilized commercially as carbon sources in the fermentation medium. Glutamic acid has been produced from various kinds of raw materials like wheat bran (Alharbi et al., 2020), cassava starch (Jyothi et al., 2005), date syrup (Ahmed et al., 2013), potato (Moosavi-Nasab et al., 2010), pineapple waste (Lawal et al., 2011), and wheat byproducts (Sari et al., 2014) using different fermentation methods. Many researchers have also investigated the use of rice straw, rice bran, sugar cane waste, cassava starch, cotton stalks, soybean, and sorghum stover for the production of PGA; Zeng et al., 2018; Mohanraj et al., 2019; Fang et al., 2020). Two fermentation approaches that have been most commonly employed for glutamic acid production are SSF and SmF. In one such study, the SmF method was adopted for the production of L-glutamic acid using *Corynebacterium glutamicum*, which under optimized conditions yielded 16.36 g/L-glutamic acid (Alharbi et al., 2019). It is very important to study the effect of various parameters on the production yield so it can be optimized. The production of acid under optimized conditions reduces the time required for designing the culture medium and also minimizes the risk of contamination. In another study, *Corynebacterium glutamicum* was used, and the effect of four variables, i.e., pH, temperature, agitation rate, and glucose concentration, was studied on the glutamic acid yield, and the process was optimized using response surface methodology (Rao et al., 2013). In a recent study, the SSF method was used for the production PGA using corn stalk and soybean meal by *Bacillus amyloliquefaciens* JX-6, and a comparison was made between sterilized and non-sterilized SSF. The results revealed that sterilized SSF resulted in a higher yield while non-sterilized SSF favoured a rapid and faster fermentation, which suggested that non-sterilized fermentation could prove to be more beneficial and economical for large-scale production of PGA (Fang et al., 2020). A low-cost fermentation process was developed (with a scale up possibility) using the SSF method with *Bacillus amyloliquefaciens* for the production of PGA from soy hull, and the process was optimized using a central composite rotary design (Jose et al., 2017).

12.3.6.2 Recovery

Downstream processing has always been crucial for developing an economical process during microbial fermentation. There are several parameters that influence the L-glutamic acid fermentation process, which in turn affect product recovery. Different methods for recovering glutamic acid from the broth culture include filtration, precipitation by complex formation, and precipitation by reducing water solubility (Luo et al., 2016). The very first step in product recovery includes the separation of biomass by centrifugation or 0.45 μm membrane filter.

In one such study, the fermented broth was precipitated at isoelectric pH (3.2) of glutamic acid resulting in a glutamic acid rich cake, which was then washed several times to remove impurities; quantitative analysis was done using the ninhydrin method. Subsequent drying in a vacuum resulted in the formation of glutamic acid crystals, which were then subjected to FTIR (Fourier-transform infrared spectroscopy) analysis (Vijayalakshmi et al., 2019). The thin layer chromatographic technique has been employed by many researchers for the qualitative detection of glutamic acid in a culture medium. For the identification of hydrolysis products, different ratios of solvent systems have been employed. A solvent system consisting of n-butanol:acetic acid:water (2:1:1) and 1-butanol, acetic acid, water in 80/20/20 (v/v) were used in two different studies for the detection of L-glutamic acid and spots were visualized using 0.02% ninhydrin (Alharbi et al., 2019; Kebede and Abate, 2016). In another study, solvents consisting of n-butanol:acetic acid:water (3:1:1) and 96% ethanol:water (63:37, v/v) were used as the primary developer for the identification of PGA (Song et al., 2019). Various researchers have made use of techniques like FTIR and NMR spectroscopies for the identification of functional groups of glutamic acid produced (Khalil et al., 2016; Jose et al., 2017; Mohanraj et al., 2019). Another technique for obtaining pure glutamic acid is the precipitation method by reducing the water solubility, which involves the addition of ethanol to the supernatant and dissolving in distilled water. After the extraction of PGA by centrifugation, ice-cold ethanol was added to the supernatant, and the crude PGA was collected after centrifugation at 10000 rpm at 4°C, which was further dissolved in water and centrifuged to remove any insoluble impurities, and finally lyophilized to get pure PGA. The KBr mode was carried out in FTIR for analysis of the PGA's structure, and the transmission spectra were recorded in the range of 400–4000cm². For determination of molecular weight, gel permeation chromatography (Jose et al., 2017) or SDS-PAGE (Sodium dodecyl sulphate-Polyacrylamide gel electrophoresis) were used (Mohanraj et al., 2019).

12.4   APPLICATIONS OF ORGANIC ACIDS

Being versatile in nature, the applications of organic acids in various industries like food, textile, chemical, and pharmaceutical are innumerable. Due to their wide range of applications, their demand is increasing each year. Some of the major applications of organic acids discussed above are given below.

12.4.1   BIODEGRADABLE POLYMERS

PLA or polylactide is a polymer of lactic acid exhibiting several important properties like high crystallinity and high melting point. It was first discovered by Carothers in 1932. It is becoming quite popular owing to its biodegradable nature and its GRAS status by the FDA. It is an emerging "green polymer" in the food packaging sector. Thermoformed PLA containers are widely used for packaging fresh produce like fruits, vegetables and salads. It is also getting attention in active and antimicrobial packaging due to several attributes like stiffness, clarity, twist and dead-fold retention, heat sealability at low temperatures and an efficient aroma and flavour barrier. It also offers several applications in the biomedical sector, such as a role in surgical sutures and implants and in drug delivery systems (Jamshidian et al., 2010).

Similarly, succinic acid is used as a starting material for the manufacture of various industrially crucial chemicals such as γ-butyrolactone (GBL), tetrahydrofuran (THF), 2-pyrrolidone (2-P), and N-methyl-2-pyrrolidone (NMP), which under appropriate conditions can generate polybutylene succinate (PBS) and poly(butylene succinate-co-butylene terephthalate), when combined with succinic acid. Both PBS and poly(butylene succinate-co-butylene terephthalate) are biodegradable plastics with several applications in various sectors (Nghiem et al., 2017).

12.4.2 FOOD INDUSTRY

Out of all the lactic acid produced worldwide, 70% is consumed by the food industry due to its several applications as an additive, preservative agent, and pH regulator in baking products, fermented meats and vegetables, confectionery and dairy industry, salad dressings, etc. Sodium and potassium salts of lactic acid are used in the meat, poultry, and fish industries to increase the shelf life of products. The sour or acidic taste imparted by lactic acid makes it a perfect acidulant in baked goods, beverages, pickled meat and vegetables and salad dressings. It is popular in the dairy industry as an acidifying agent in the manufacture of cheese and yoghurt, as a flavour enhancer, and a preservative agent. Due to its ease of handling, it is employed in chocolate and the confectionery industry as an acid regulator, flavour additive, retardant of the inversion rate, and in the production of clear candies (Castillo-Martinez et al., 2013; Krishna et al., 2019).

CA is primarily used in the food industry due to its GRAS status, pleasant taste, and high solubility in water. It is mainly used in the beverage and confectionery industry to provide a tart flavour complementing fruits and berries. It is also used to reduce sucrose inversion in candy-making besides acting as an acidulant. It functions as an emulsifier in the ice cream and processed cheese industry. It also enhances the activity of antimicrobial agents (Vandenberghe et al., 1999).

Likewise, succinic acid is used in the food industry as a sequestrant, buffer, and neutralizing agent (Saxena et al., 2017).

Acetic acid is extensively employed in the food industry as vinegar, which is used as an acidulant and also improves taste. It also acts as a preservative in the case of pickles. It improves the shelf life of cheese, and it gives it a good mouthfeel and taste. Acetic acid is also used as a flavouring agent in lambic beer, sparkling water, water kefir, and the beverage kombucha. Nowadays, it is added to edible films to give them a sour flavour. It also acts as an antimicrobial agent in the films, and acetic acid, along with chitosan, acts as an antilisteria agent in the edible films (Deshmukh and Manyar, 2020).

Due to its non-toxic nature, low-cost fumaric acid is widely utilized in the food industry for the manufacture of wheat and corn tortillas, sourdough and rye breads, refrigerated biscuit dough, fruit juice and nutraceutical drinks, gelatine desserts, gelling aids, pie fillings, and wine (Straathof and Gulik, 2012).

12.4.3 COSMETIC INDUSTRY

Lactic acid has a wide array of applications in the cosmetic and skincare industry. It acts as a skin brightening agent and lightening agent, thereby reducing dark spots on the skin. It is also used in moisturizing agents, skin rejuvenating agents, and as a humectant in various creams and lotions. It also regulates skin pH and has been proven to promote collagen production, thereby preventing wrinkles and fine lines along with maintaining skin firmness (Krishna et al., 2019). CA is used in the cosmetic industry as a buffering agent, antioxidant, and for pH adjustment (Vandenberghe et al.,

1999). Similarly, due to the antimicrobial properties exhibited, succinic acid is also used in the cosmetic industry and is known as the "elixir of youth" (Saxena et al., 2017). Acetic acid, on the other hand, is applied as a local antiseptic (Deshmukh and Manyar, 2020).

12.4.4 OTHER USES

Among its various uses, lactic acid has found a place in oral hygiene products. It also plays the role of a cleaning, descaling, and antimicrobial agent in the chemical industry. Being a powerful solvent, it is used to remove polymer and resins. It plays a significant role in enhancing animal nutrition as ammonium salt of lactic acid has been shown to improve the nutritive value of milk in cattle. Moreover, it is added to animal silage to improve its palatability and nutritive value (Castillo-Martinez et al., 2013; Krishna et al., 2019).

CA is added to powders and tablets along with bicarbonates in the pharmaceutical sector as it improves the dissolution of active ingredients. It also acts as an anticoagulating agent (Vandenberghe et al., 1999).

Sodium, potassium, and calcium salts of succinic acid are used in the medical industry to treat long-term illnesses and injuries. It also plays a role as a plant growth regulator. Succinate salts are added to ruminant feed and are a source of animal nutrition. Succinic acid has also found application in the leather industry as a water repellent and wet strength agent (Saxena et al., 2017).

Acetic acid is used in the manufacture of polyvinyl acetate, which is widely used in the paint and adhesive industry. It is also used in textiles and in silver-based photography films (Pal and Nayak, 2016).

Fumaric acid is added to animal feed as a supplement, and moreover, it enhances feed efficiency in the case of pigs and poultry (Strrathof and Gulic, 2012).

12.5 GLOBAL SCENARIO OF MICROBIAL PRODUCTION OF ORGANIC ACIDS

CA has the highest annual global production of 1.6 million tonnes by average annual growth rate (AAGR) of 3.5–4%, acetic acid and lactic acid follow thereafter with 0.19 and 0.15 million tonnes, respectively. It is estimated that the production of succinic acid will increase beyond 7 million tonnes by 2020 because it is an important acid for the production of many useful chemical compounds (Panda et al., 2019).

Various developments in the technological field have led to a rising trend of synthesizing organic acids from renewable bio-based sources such as wheat and corn. Although both microbial and chemical processes have been used for the production of all organic acids, microbial production has been favoured for many reasons. This method is not only more economically viable but also very much environmentally friendly when highly effective strains are used. Bio-based organic acids have similar chemical properties to conventional organic acids and produce comparatively far less toxic waste. These organic acids are the very basis of the manufacture of many products, not limited to food, metals, pharmaceuticals, polymers, solvents, and cosmetics.

The competitive organic acid market has a number of good players like BASF, Dow Dupont, Eastman Chemical Company, and many more. Although the organic acid market has cut-throat competition, it is still expected to grow with a 5% compound annual growth rate in the long term. This expected growth can be attributed to the vast number of industries that directly or indirectly depend on organic acids, most importantly the food, beverage, and pharmaceutical industries. These industries depend so much on the organic acid market due to the number of valuable properties of organic acids, such as their antioxidant nature, antibacterial and antifungal properties, etc.

Considering the geographical distribution, technologically advanced North America has the largest portion of the global organic acids market. The developing economies of highly populated Asian countries, particularly China and India, provide a large scope for development in this particular sector. Other regions of the world, including Africa, the Middle East, and Latin America, are also expected to see a rise in the share of the global organic acid market.

REFERENCES

Abbas, N., Safdar, W., Ali, S., Choudhry, S., and Ilahi, S. 2016. Citric acid production from *Aspergillus niger* using banana peel. *International Journal of Scientific and Engineering Research* 7(1): 1580–1583.

Abdel-Rahman, M.A., Tashiro, Y., and Sonomoto, K. 2013. Recent advances in lactic acid production by microbial fermentation processes. *Biotechnology Advances* 31(6): 877–902.

Abdelrhman, K., Abdel-Rahman, M., Desouky, S., and Al-Gamal, M. 2018. Optimization of citric acid production from sugar cane molasses using a fungal isolate, *Aspergillus fumigatus NA-1 by. Egyptian Journal of Biomedical Science* 52: 1–24.

Aboyeji, O.O., Oloke, J.K., Arinkoola, A.O., Oke, M.A., and Ishola, M.M. 2020. Optimization of media components and fermentation conditions for citric acid production from sweet potato peel starch hydrolysate by *Aspergillus niger*. *Scientific African* 10: e00554. https://doi.org/10.1016/j.sciaf.2020.e00554.

Acourene, S., Djafri, K., Ammouche, A., Djidda, A., Tama, M., and Taleb, B. 2011. Utilisation of the date wastes as substrate for the production of baker's yeast and citric acid. *Biotechnology(Faisalabad)* 10(6): 488–497.

Addo, M., Kusi, A., Ofori, L., and Obiri-Danso, K. 2016. Citric acid production by *Aspergillus niger* on a corn cob solid substrate using one-factor-at-a-time optimization method. *International Advanced Research Journal in Science, Engineering and Technology* 3(1): 95–99. https://doi.org/10.17148/iarjset.2016.3120.

Adeoye, A.O., Lateef, A., and Gueguim-Kana, E.B. 2015. Optimization of citric acid production using a mutant strain of *Aspergillus niger* on cassava peel substrate. *Biocatalysis and Agricultural Biotechnology* 4(4): 568–574. https://doi.org/10.1016/j.bcab.2015.08.004.

Adudu, J., Arekemase, S., Abdulwaliy, I., Batari, M., Raplong, H., Aronimo, B., and Sani, Y. 2019. Production of citric acid from corn stalk through submerged fermentation using *Aspergillus niger*. *Journal of Applied Sciences* 19(6): 557–564. https://doi.org/10.3923/jas.2019.557.564.

Afifi, M.M. 2011. Enhancement of lactic acid production by utilizing liquid potato wastes. *International Journal of Biological Chemistry* 5(2): 91–102.

Ahmed, Y.M., Khan, J.A., Abulnaja, K.A., and Al-Maliki, A.L. 2013. Production of glutamic acid by Corynebacterium glutamicum using dates syrup as carbon source. *African Journal of Microbiology Research* 7(19): 2072.

Ajala, A.S., Adeoye, A.O., Olaniyan, S.A., and Fasonyin, O.T. 2020. A study on effect of fermentation conditions on citric acid production from cassava peels. *Scientific African* 8: e00396. https://doi.org/10.1016/j.sciaf.2020.e00396.

Alharbi, N.S., Kadaikunnan, S., Khaled, J.M., Almanaa, T.N., Innasimuthu, G.M., Rajoo, B., Alanzi, K.F., and Rajaram, S.K. 2020. Optimization of glutamic acid production by *Corynebacterium glutamicum* using response surface methodology. *Journal of King Saud University – Science* 32(2): 1403–1408. https://doi.org/10.1016/j.jksus.2019.11.034.

Ali, S.R., Anwar, Z., Irshad, M., Mukhtar, S., and Warraich, N.T. 2016. Bio-Synthesis of citric acid from single and co-culture based fermentation technology using agro-wastes. *Journal of Radiation Research and Applied Sciences* 9(1): 57–62. https://doi.org/10.1016/j.jrras.2015.09.003.

Alnassar, M., Tayfour, A., and Afif, R.A. 2016. The study of lactose effect on citric acid production by Aspergillus niger PLA30 in Cheese whey. *Chemistry* 9: 318–322.

Anastassiadis, S., Aivasidis, A., and Wandrey, C. 2002. Citric acid production by Candida strains under intracellular nitrogen limitation. *Applied Microbiology and Biotechnology* 60(1–2): 81–87. https://doi.org/10.1007/s00253-002-1098-1.

Anastassiadis, S., Morgunov, I.G., Kamzolova, S.V., and Finogenova, T.V. 2008. Citric acid production patent review. *Recent Patents on Biotechnology* 2(2): 107–123.

Araya-Cloutier, C., Rojas-Garbanzo, C., and Velázquez-Carrillo, C. 2012. Effect of initial sugar concentration on the production of L(+) lactic acid by simultaneous enzymatic hydrolysis and fermentation of an agroindustrial waste product of pineapple (*Ananas comosus*) using *Lactobacillus casei* subspecies *rhamnosus*. *International Journal of Biotechnology for Wellness Industries* 1: 91–100.

Ayeni, A.O., Daramola, M.O., Taiwo, O., Olanrewaju, O., Oyekunle, D., Sekoai, P., and Elehinafe, F. 2019. Production of citric acid from the fermentation of pineapple waste by *Aspergillus niger*. *The Open Chemical Engineering Journal* 13(1): 88–96. https://doi.org/10.2174/1874123101913010088.

Azaizeh, H., Abu, T.H.N., Schneider, R., Klongklaew, A., and Venus, J. 2020. Production of lactic acid from carob, banana and sugarcane lignocellulose biomass. *Molecules* 25(13): 2956. https://doi.org/10.3390/molecules25132956.

Bakhiet, S.E.A., and Al-Mokhtar, E.A. 2015. Production of citric acid by *Aspergillus niger* using sugarcane molasses as substrate. *Jordan Journal of Biological Sciences* 8(3): 211–215. http://doi.org/10.12816/0026960.

Barragán, L.A.P., Figueroa, J.J.B., Rodriguez Duran, L.V., Aguilar Gonzalez, C.N., and Hennigs, C. 2016. *Fermentative Production Methods, Biotransformation of Agricultural Waste and By-Products*, pp. 189–217, Elsevier. https://doi.org/10.1016/B978-0-12-803622-8.00007-0.

Bastos, R.G., and Ribeiro, H. 2020. Citric acid production by the solid-state cultivation consortium of *Aspergillus niger* and *Trichoderma reesei* from sugarcane bagasse. *The Open Biotechnology Journal* 14(1): 32–41. https://doi.org/10.2174/1874070702014010032.

Bauweleers, H.M.K., Groeseneken, D.R., and Van, P. 2014. Genes useful for the industrial production of citric acid. In: *Google Patents*, US8637280B2

Beitel, S.M., Coelho, L.F., and Contiero, J. 2020. Efficient conversion of agroindustrial waste into D(-) lactic acid by *Lactobacillus delbrueckii* using fed-batch fermentation. *BioMed Research International* 2020: 4194052. https://doi.org/10.1155/2020/4194052.

Benghazi, L., Record, E., Suarez, A., Gomez-Vidal, J.A., Martinez, J., and de la Rubia, T. 2014. Production of the Phanerochaete flavido-alba laccase in *Aspergillus niger* for synthetic dyes decolorization and biotransformation. *World Journal of Microbiology and Biotechnology* 30(1): 201–211.

Bernardo, M.P., Coelho, L.F., Sass, D.C., and Contiero, J. 2016. l-(+)-lactic acid production by *Lactobacillus rhamnosus* B103 from dairy industry waste. *Brazilian Journal of Microbiolgy* 47(3): 640–646. https://doi.org/10.1016/j.bjm.2015.12.001.

Bilge, S.B., Mukerrem, K., and Guzin, K. 2018. Citric acid production by autochthonous *Candida zeylanoides* strains. *Journal of Biotechnology* 280: S18. https://doi.org/10.1016/j.jbiotec.2018.06.055.

Bouhadi, D., Bouziane, A., Hariri, A., Kada, I., Ouis, N., and Fatima, S. 2013. Lactic acid production by a mixed culture of lactic bacteria based on low value dates syrup and their metabolic uses. *Journal of Metabolic Syndrome*, 1(5): 1–4.

Bouhadi, D., Raho, B., Hariri, A., Benattouche, Z., Karima, Y., and Fatima, S. 2016. Utilization of the extract of pulp of carob for lactic acid production by *Streptococcus thermophiles*. *Electronic Journal of Biology* 12: 90–95.

Bradfield, M.F.A., Mohagheghi, A., Salvachua, D., Smith, H., Black, B.A., Dowe, N., Beckham, G.T., and Nicol, W. 2015. Continuous succinic acid production by *Actinobacillus succinogenes* on xylose-enriched hydrolysate. *Biotechnology for Biofuels* 8: 181. https://doi.org/10.1186/s13068-015-0363-3.

Bretz, K. 2015. Succinic acid production in Fed-Batch Fermentation of Anaerobiospirillum succiniciproducens using glycerol as carbon source. *Chemical Engineering and Technology* 38(9): 1659–1664.

Campanhol, B.S., Silveira, G.C., Castro, M.C., Ceccato-Antonini, S.R., and Bastos, R.G. 2019. Effect of the nutrient solution in the microbial production of citric acid from sugarcane bagasse and vinasse. *Biocatalysis and Agricultural Biotechnology* 19: 101147. https://doi.org/10.1016/j.bcab.2019.101147.

Campo, G., Berregi, I., Caracena, R., and Santos, J.I. 2006. Quantitative analysis of malic and citric acids in fruit juices using proton nuclear magnetic resonance spectroscopy. *Analytica Chimica Acta* 556(2): 462–468. https://doi.org/10.1016/j.aca.2005.09.039.

Cao, Y., Zhang, R., Sun, C., Cheng, T., Liu, Y., and Xian, M. 2013. Fermentative succinate production: An emerging technology to replace the traditional petrochemical processes. *BioMed Research International* 2013: 1–12.

Carlos, E.R.R., Carine, B., Stephane, E.G., Nathalie, G., Julien, C., Jean-Louis, U., Carole, M.J., Gilles, R., and Cesar, A.A.L. 2017. Dynamic metabolic modeling of lipid accumulation and citric acid production by *Yarrowia lipolytica*. *Computers and Chemical Engineering* 100: 139–152. https://doi.org/10.1016/j.compchemeng.2017.02.013.

Carpinelli Macedo, J.V., de Barros Ranke, F.F., Escaramboni, B., Campioni, T.S., Fernandez Nunez, E.G., and de Oliva Neto, P. 2020. Cost-effective lactic acid production by fermentation of agro-industrial residues. *Biocatalysis and Agricultural Biotechnology* 27: 101706. https://doi.org/10.1016/j.bcab.2020.101706

Carvalho, M., Roca, C., and Reis, M.A.M. 2013. Carob pod water extracts as feedstock for succinic acid production by *Actinobacillus succinogenes* 130Z. *Bioresource Technology* 170: 491–498.

Carvalho, M., Roca, C., and Reis, M.A.M. 2016. Improving succinic acid production by *Actinobacillus succinogenes* from raw industrial carob pods. *Bioresource Technology* 218: 491–497.

Castillo-Martinez, F.A., Balciunas, E.M., Salgado, J.M., Domínguez, J.M.G., Converti, A., and de Souza Oliveira, R.P. 2013. Lactic acid properties, applications and production: A review. *Trends in Food Science and Technology* 30(1): 70–83. https://doi.org/10.1016/j.tifs.2012.11.007.

Chakraborty, K., Kumar Saha, S., Raychaudhuri, U., and Chakraborty, R. 2017. Vinegar production from vegetable waste: Optimization of physical conditions and kinetic modeling of fermentation process. *Indian Jourrnal of Chemical Technology* 24: 508–516.

Chen, P., Tao, S., and Zheng, P. 2016. Improving succinic acid production by *Actinobacillus succinogenes* from raw industrial carob pods. Efficient and repeated production of succinic acid by turning sugarcane bagasse into sugar and support. *Bioresource Technology* 211: 406–413.

Corona-Gonzalez, R.I., Varela-Almanza, K.M., Arriola-Guevara, E., Martínez-Gómez, Áde J., Pelayo-Ortiz, C., and Toriz, G. 2016. Bagasse hydrolyzates from Agave tequilana as substrates for succinic acid production by *Actinobacillus succinogenes* in batch and repeated batch reactor. *Bioresource Technology* 205: 15–23.

Crolla, A., and Kennedy, K.J. 2004. Fed-batch production of citric acid by *Candida lipolytica* grown on n-paraffins. *Journal of Biotechnology* 110(1): 73–84. https://doi.org/10.1016/j.jbiotec.2004.01.007.

Das, R.K., Brar, S.K., and Verma, M. 2014. Enhanced fumaric acid production from brewery wastewater by immobilization technique. *Journal of Chemical Technology and Biotechnology*. http://doi.org/10.1002/jctb.4455.

Das, R.K., Brar, S.K., and Verma, M. 2016. Fumaric acid: Production and application aspects. In: *Platform Chemical Biorefinery*, pp. 133–157. Elsevier. http://doi.org/10.1016/B978-0-12-802980-0.00008-0.

de Barros, M., Freitas, S., Padilha, G.S., and Alegre, R.M. 2013. Biotechnological production of succinic acid by *Actinobacillus succinogenes Using Different Substrate*. *Chemical Engineering Transactions* 32: 985–990.

Deshmukh, G., and Manyar, H. 2020. Production pathways of acetic acid and its versatile Applications in the food industry. *Biomass*: 1–10. http://doi.org/10.5772/intechopen.92289.

Dessie, W., Zhu, J., Xin, F., Zhang, W., Jiang, Y., Wu, H., Ma, J., and Jiang, M. 2018. Bio-succinic acid production from coffee husk treated with thermochemical and fungal hydrolysis. *Bioprocess and Biosystems Engineering*. https://doi.org/10.1007/s00449-018-1974-4.

Dhillon, G.S., Brar, S.K., Kaur, S., and Verma, M. 2013. Screening of agro-industrial wastes for citric acid bioproduction by *Aspergillus niger* NRRL 2001 through solid state fermentation. *Journal of the Science of Food and Agriculture* 93(7): 1560–1567. https://doi.org/10.1002/jsfa.5920. PMID: 23108761.

Dhillon, G.S., Brar, S.K., Verma, M., and Tyagi, R.D. 2011. Utilization of different agro-industrial wastes for sustainable bioproduction of citric acid by *Aspergillus niger*. *Biochemical Engineering Journal* 54(2): 83–92.

Dronawat, S.N., Svihla, C.K., and Hanley, T.R. 1995. The effects of agitation and aeration on the production of gluconic acid by *Aspergillus niger*. *Applied Biochemistry and Biotechnology* 51–52(1): 347–354.

Fang, J., Liu, Y., Huan, C.C., Xu, L., Ji, G., and Yan, Z. 2020. Comparison of poly-γ-glutamic acid production between sterilized and non-sterilized solid-state fermentation using agricultural waste as substrates. *Journal of Cleaner Production* 255: 120248. https://doi.org/10.1016/j.jclepro.2020.120248.

Farooq, U., Faqir, M., Anjum, F., Zahoor, T., Hayat, Z., Akram, K., and Ashraf, E. 2013. Citric acid production from sugarcane molasses by *Aspergillus niger* under different fermentation conditions and substrate levels. *International Journal of Agriculture and Applied Science* 5(1): 8–16.

Ferone, M., Ercole, A., Raganati, F., Olivieri, G., Salatino, P., and Marzocchella, A. 2019b. Efficient succinic acid production from high-sugars-content beverages (HSCBs) by *Actinobacillus succinogenes*. *Biotechnology Progress* 35(5): e2863. https://doi.org/10.1002/btpr.2863.

Ferone, M., Raganati, F., Olivieri, G., and Marzocchella, A. 2019a. Bioreactors for succinic acid production processes. *Critical Reviews in Biotechnology*: 1–16. https://doi.org/10.1080/07388551.2019.1592105.

Folake, T.A., Adeyemo, S.M., and Balogun, H.O. 2018. Fermentation conditions and process optimization of citric acid production by yeasts. *The International Journal of Biotechnology, Conscientia Beam* 7(1): 51–63.

Ghaffar, T., Irshad, M., Anwar, Z., Aqil, T., Zulifqar, Z., Tariq, A., Kamran, M., Ehsan. N., and Mehmood, S. 2014. Recent trends in lactic acid biotechnology: A brief review on production to purification. *Journal of Radiation Research and Applied Sciences* 7(2): 222–229. https://doi.org/10.1016/j.jrras.2014.03.002.

Grewal, J., Sadaf, A., Yadav, N., and Khare, S. 2020. Agroindustrial waste based biorefineries for sustainable production of lactic acid. In: Thallada Bhaskar, Ashok Pandey, Eldon R. Rene, and Daniel C.W. Tsang (eds), *Waste Biorefinery*, pp. 125–153. Elsevier. https://doi.org/10.1016/B978-0-12-818228-4.00005-8.10.1016/B978-0-12-818228-4.00005-8.

Hesham, A.E., Mostafa, Y., and Alsharqi, L.E.O. 2020. Optimization of citric acid production by immobilized cells of novel yeast isolates. *Mycobiology* 48(2): 1–11. https://doi.org/10.1080/12298093.2020.1726854.

Holten, C.H., Muller, A., Rehbinder, D., and Ilra, S. 1971. *Lactic Acid: Properties and Chemistry of Lactic Acid and Derivatives*. Chemie, Weinheim, Germany.

Hussain, A.M. 2019. Citric acid production using wheat bran by *Aspergillus niger*. *Indian Journal of Public Health Research and Development* 10(6): 1213–1217. https://doi.org/10.37506/ijphrd.v10i6.8587.

Jamshidian, M., Tehrany, E.A., Imran, M., Jacquot, M., and Desobry, S. 2010. Poly-lactic acid: Production, applications, nanocomposites, and release studies. *Comprehensive Reviews in Food Science and Food Safety* 9(5): 552–571. https://doi.org/10.1111/j.1541-4337.2010.00126.x.

Joe Leung, C.C., Cheung, A.S.Y., Yan-Zhu Zhang, A., Lam, K.F., and Lin, C.S.K. 2012. Utilisation of waste bread for fermentative succinic acid production. *Biochemical Engineering Journal* 65: 10–15.

Jose, A.A., Parameswaran, B., and Ashok, P. 2017. Production and characterization of microbial poly-γ-glutamic acid from renewable resources. *International Journal of Experimental Biology* 55(7): 405–410.

Jyothi, A.N., Sasikiran, K., Nambisan, B., and Balagopalan, C. 2005. Optimization of glutamic acid production from cassava starch factory residues using *Brevibacterium divaricatum*. *Process Biochemistry* 40(11): 3576–3579. https://doi.org/10.1016/j.procbio.2005.03.046.

Kandler, O. 1983. Carbohydrate metabolism in lactic acid bacteria. *Antonie van Leeuwenhoek* 49(3): 209–224.

Kareem, S.O., and Rahman, R.A. 2013. Utilization of banana peels for citric acid production by *Aspergillus niger*. *Agriculture and Biology Journal of North America* 4(4): 384–387. https://doi.org/10.5251/abjna.2013.4.4.384.387.

Kebede, B., and Abate, D. 2016. Production of l- glutamic acid from wild bacterial isolates from soil using batch fermentation. *International Journal of Innovative Pharmaceutical Sciences and Research* 4(11): 1103–1114.

Khalil, I.R., Irorere, V.U., Radecka, I., Burns, A.T.H., Kowalczuk, M., Mason, J.L., and Khechara, M.P. 2016. Poly-γ-glutamic acid: Biodegradable polymer for potential protection of beneficial viruses. *Materials (Basel)* 9(1): 28. https://doi.org/10.3390/ma9010028.

Khan, I., Maqbool, F., Mujaddad-Ur-Rehman, Hayat, A., Munnazza, and Farooqui, S. 2017. Microbial organic acids production, biosynthetic mechanism and applications-Mini review. *Indian Journal of Geosciences* 46: 2165–2174.

Kim, H.M., Choi, I.S., Lee, S. et al. 2019. Biorefining process of carbohydrate feedstock (agricultural onion waste) to acetic acid. *ACS Omega* 4(27): 22438–22444. http://doi.org/10.1021/acsomega.9b03093.

Kim, H.M., Park, J.H., Choi, I.S. et al. 2018. Effective approach to organic acid production from agricultural kimchi cabbage waste and its potential application. *PLOS ONE* 13(11): e0207801. https://doi.org/10.1371/journal.pone.0207801.

Kim, K.H., Lee, H.Y., and Lee, C.Y. 2015. Pretreatment of sugarcane molasses and citric acid production by *Candida zeylanoides*. *Microbiology and Biotechnology Letters* 43(2): 164–168. http://doi.org/10.4014/mbl.1503.03006.

Kringel, D.H., Dias, A.R.G., da R., E., and Gandra, E.A. 2020. Fruit wastes as promising sources of starch: Extraction, properties, and applications. *Starch – Stärke* 72, 1900200. https://doi.org/10.1002/star.201900200.

Krishna, B., Saibaba, K.V.N., Gantala, S.S.N., Besetty, T., and Gopinadh, R. 2019. Industrial production of lactic acid and its applications. *International Journal of Biotechnology, (Research)* 1: 42–54.

Kubicek, C.P., Rohr, M., and Rehm, H.J. 1985. Citric acid fermentation. *Critical Reviews in Biotechnology* 3(4): 331–373.

Kuenz, A., Homann, L., Goy, K., Bromann, S., and Prüße, U. 2020. High-level production of succinic acid from crude glycerol by a wild type organism. *Catalysts* 10(5): 470. https://doi.org/10.3390/catal10050470.

Kumar, V., Sharma, A., Bhardwaj, R., and Thukral, A.K. 2017. Analysis of organic acids of tricarboxylic acid cycle in plants using GC-MS, and system modeling. *Journal of Analytical Science and Technology* 8(1): 20. https://doi.org/10.1186/s40543-017-0129-6.

Lam, K.F., Joe Leung, C.C., Lei, H.M., and Lin, C.S.K. 2014. Economic feasibility of a pilot-scale fermentative succinic acid production from bakery wastes. *Food and Bioproducts Processing* 92(3): 282–290.

Lawal, A.K., Oso, B.A., Sanni, A.I., and Olatunji, O.O. 2011. L-glutamic acid production by Bacillus spp. isolated from vegetable proteins. *African Journal of Biotechnology* 10(27): 5337– 5345.

Lee, K. 2005. Comparison of fermentative capacities of lactobacilli in single and mixed culture in industrial media. *Process Biochemistry* 40(5): 1559–1564. https://doi.org/10.1016/j.procbio.2004.04.017.

Li, C., Xu, D., Zhao, M., Sun, L., and Wang, Y. 2014a. Production optimization, purification, and characterization of a novel acid protease from a fusant by *Aspergillus oryzae* and *Aspergillus niger*. *European Food Research and Technology* 238: 1–13.

Li, Y., He, D., Niu, D., and Zhao, Y. 2014b. Acetic acid production from food wastes using yeast and acetic acid bacteria micro-aerobic fermentation. *Bioprocess and Biosystems Engineering*. http://doi.org/10.1007/s00449-014-1329-8.

Li, Q., Jiang, X., Feng, X., Wang, J., Sun, C., Zhang, H., Xian, M., and Liu, H. 2016a. Recovery processes of organic acids from fermentation broths in the biomass-based industry. *Journal of Microbiology and Biotechnology* 26(1): 1–8. https://doi.org/10.4014/jmb.1505.05049.

Li, Z., Bai, T., Dai, L. Wang, F., Tao, J., Meng, S., Hu, Y., Wang, S., and Hu, S. 2016b. A study of organic acid production in contrasts between two phosphate solubilizing fungi: *Penicillium oxalicum* and *Aspergillus niger*. *Scientific Reports* 6: 25313. https://doi.org/10.1038/srep25313.

Liang, S., Jiang, W., Song, Y., and Zhou, S.F. 2020. Improvement and metabolomics-based analysis of d-lactic acid production from agro-industrial wastes by *Lactobacillus delbrueckii* Submitted to Adaptive Laboratory Evolution. *Journal of Agriculture and Food Chemistry* 68(29): 7660–7669. https://doi.org/10.1021/acs.jafc.0c00259.

Louaste, B., and Eloutassi, N. 2020. Succinic acid production from whey and lactose by *Actinobacillus succinogenes* 130Z in batch fermentation. *Biotechnology Reports.* https://doi.org/10.1016/j.btre.2020.e00481.

Luo, Z., Guo, Y., Liu, J., Qiu, H., Zhao, M., Zou, W., and Li, S. 2016. Microbial synthesis of poly-γ -glutamic acid: Current progress, challenges, and future perspectives. *Biotechnology for Biofuels* 9: 134. https://doi.org/10.1186/s13068-016-0537-7.

Luthfia, A.A.I., Jahimb, Md., J., Harun, S., and Mohammad, A.W. 2016. Biorefinery approach towards greener succinic acid production fromoil palm frond bagasse. *Process Biochemistry* 51(10): 1527–1537.

Maheshwari, N.V. 2018. Agro-industrial lignocellulosic waste: An alternative to unravel the future bioenergy. In: A. Kumar, S. Ogita, and Y.Y. Yau (eds), *Biofuels: Greenhouse Gas Mitigation and Global Warming*, pp. 291–305. Springer, New Delhi. https://doi.org/10.1007/978-81-322-3763-1_16.

Mamdouh, E.G., Desouky, S., Mohamed, A.R., and Abdelrhman, K. 2018. High-temperature citric acid production from sugar cane molasses using a newly isolated thermotolerant yeast strain, *Candida parapsilosis* NH-3. *International Journal of Advanced Research in Biological Sciences* 5(7): 187–211. https://doi.org/10.22192/ijarbs.2018.05.07.015.

Mancini, E., Mansouri, S.S., Gernaey, K.V. et al. 2019. From second generation feed-stocks to innovative fermentation and downstream techniques for succinic acid production. *Critical Reviews in Environmental Science and Technology*: 1–45. https://doi.org/10.1080/10643389.2019.1670530.

Martin-Dominguez, V., Estevez, J., de Borja Ojembarrena, F., Santos, V., and Ladero, M. 2018. Fumaric acid production: A biorefinery perspective. *Fermentation* 4(33). http://doi.org/10.3390/fermentation4020033.

Max, B., Salgado, J.M., Rodríguez, N., Cortes, S., Converti, A., and Dominguez, J.M. 2010. Biotechnological production of citric acid. *Brazlian Journal of Microbiology* 41(4): 862–875. Https://doi.org/10.1590/S 1517-83822010000400005.

Michelin, M., Ruiz, H.A., Silva, D.P., Ruzene, D.S., Teixeira, J.A., and Polizeli, M.L.T.M. 2014. Cellulose from lignocellulosic waste. In: K. Ramawat and J.M. Mérillon (eds), *Polysaccharides*, pp. 475–511. Springer, Cham. https://doi.org/10.1007/978-3-319-03751-6_52-1.

Moestedt, J., Westerholm, M., Isaksson, S., and Schnurer, A. 2019. Inoculum source determines acetate and lactate production during anaerobic digestion of SewageSludge and food waste. *Bioengineering* 7(3): 1–19. http://doi.org/10.3390/bioengineering7010003.

Mohanraj, R., Gnanamangai, B.M., Ramesh, K., Priya, P., Srisunmathi, R., Poornima, S., Ponmurugan, P., and Robinson, J.P. 2019. Optimized production of gamma poly glutamic acid (γ-PGA) using sago. *Biocatalysis and Agricultural Biotechnology* 22: 101413. https://doi.org/10.1016/j.bcab.2019.101413.

Moosavi-Nasab, M., Izadi, M., and Hosseinpour, S. 2010. Glutamic acid production from potato by *Brevibacterium linens*. *International Journal of Biological, Biomolecular, Agricultural, Food and Biotechnological Engineering* 4(8): 610–612.

Musilkova, M., Ujcova, E., Seichert, L., and Fencl, Z. 1983. Effect of changed cultivation conditions on the morphology of *Aspergillus niger* and citric acid biosynthesis in laboratory cultivation. *Folia Microbiologica* 27: 382–332.

Naraian, R., and Kumari, S. 2017. Microbial production of organic acids. In: V.K. Gupta, H. Treichel, V. Shapaval, L. Antonio de Oliveira, and M.G. Tuohy (eds), *Microbial Functional Foods and Nutraceuticals*. https://doi.org/10.1002/9781119048961.ch5.

Nghiem, N.P., Kleff, S., and Schwegmann, S. 2017. Succinic acid: Technology Development and commercialization. *Fermentation* 3(2): 26. https://doi.org/10.3390/fermentation3020026.

Nguyen, C.M., Kim, J.S., Nguyen, T.N., Ki, K.S., Choi, G.J., hoi, Y.H., Jang, K.S., and Kim, J.C. 2013. Production of L- and D-lactic acid from waste *Curcuma longa* biomass through simultaneous saccharification and cofermentation. *Bioresource Technology* 146: 35–43. http://doi.org/10.1016/j.biortech.2013.07.035.

Okerentugba, P.O., and Anyanwu, V.E. 2014. Evaluation of citric acid Production by *Penicillium* sp. ZE-19 and its improved UV-7 strain. *Research Journal of Microbiology* 9(4): 208–215.

Olszewska-Widdrat, A., Alexandri, M., Lopez-Gomez, J.P., Roland, S., and Venus, J. 2020. Batch and continuous lactic acid fermentation based on a multi-substrate approach. *Microorganisms* 8(7): 1084. https://doi.org/10.3390/microorganisms8071084.

Othman, M., Ariff, A.B., Rios-Solis, L., and Halim, M. 2017. Extractive fermentation of lactic acid in lactic acid bacteria cultivation: A review. *Frontiers in Microbiology* 8: 2285. https://doi.org/10.3389/fmicb.2017.02285.

Ozdal, M., and Kurbanoglu, E. 2019. Use of chicken feather peptone and sugar beet molasses as low cost substrates for xanthan production by *Xanthomonas campestris* MO-03. *Fermentation* 5(1): 9. http://doi.org/10.3390/fermentation5010009.

Pais-Chanfrau, J.M., Nunez-Perez, J., Espin-Valladares, R.C. et al. 2020. Bioconversion of lactose from cheese whey to organic acids. In: *Lactose and Lactose Derivatives*, pp. 1–22. https://doi.org/10.5772/intechopen.92766.

Pal, P., and Nayak, J. 2016. Acetic acid production and purification:critical review towards process intensification. *Separation and Purification Reviews*. https://doi.org/10.1080/15422119.2016.1185017.

Panda, S.K., Sahu, L., Behera, S.K., and Ray, R.C. 2019. Research and production of organic acids and industrial potential. In: G. Molina, V. Gupta, B. Singh, and N. Gathergood (eds), *Bioprocessing for Biomolecules Production*. https://doi.org/10.1002/9781119434436.ch9.

Pandey, P. 2013. Studies on citric acid production by *Aspergillus niger* in batch fermentation. *Recent Research in Science and Technology* 5(2): 66–66.

Panesar, P.S., Kennedy, J.F., Knill, C.J., and Kosseva, M. 2010. Production of L(+) lactic acid using *Lactobacillus casei* from whey. *Brazilian Archives of Biology and Technology* 53(1): 219–226. https://doi.org/10.1590/S1516-89132010000100027.

Panesar, P., and Kaur, S. 2016. Screening of media components and process parameters for production of L(+) lactic acid from potato waste liquid using amylolytic *Rhizopus oryzae*. *Acta Alimentaria* 46: 1–11. https://doi.org/10.1556/066.2016.0013.

Panesar, P.S., and Kaur, S. 2015. Bioutilisation of agro-industrial waste for lactic acid production. *International Journal of Food Science and Technology* 50(10): 2143–2151.

Papadaki, A., Kachrimanidou, V., Papanikolaou, S., Philippoussis, A., and Diamantopoulou, P. 2019. Upgrading grape pomace through *Pleurotus* spp. cultivation for the production of enzymes and fruiting bodies. *Microorganisms* 7(7): 207. https://doi.org/10.3390/microorganisms7070207.

Papagianni, M. 2007. Advances in citric acid fermentation by *Aspergillus niger*: Biochemical aspects, membrane transport and modeling. *Biotechnology Advances* 25(3): 244–263. https://doi.org/10.1016/j.biotechadv.2007.01.002.

Pasma, S.A., Daik, R., and Maskat, M.Y. 2013. Production of succinic acid from oil palm empty fruit bunch cellulose using *Actinobacillus succinogenes*. *AIP Conference Proceedings* 1571: 753–759. https://doi.org/10.1063/1.4858745.

Paulova, L., Chmelik, J., Branska, B., Patakova, P., Drahokoupil, M., and Melzoch, K. 2020. Comparison of lactic acid production by *L. casei* in batch, fed-batch and continuous cultivation, testing the use of feather hydrolysate as a complex nitrogen source. *Brazilian Archives of Biology and Technology* 63: e20190151. https://doi.org/10.1590/1678-4324-2020190151.

Philippini, R.R., Martiniano, S.E., Ingle, A., Marcelino, P.R., Silva, G., Barbosa, F.G., Santos, J.C., and Silva, S.S. 2020. Agroindustrial byproducts for the generation of biobased products: Alternatives for sustainable biorefineries. *Frontiers in Energy Research* 8: 1–23.

Praveen, B. 2015. Production of citric acid and vinegar using normal yeast and irradiated yeast *Saccharomyces cerevisae*. *Research and Reviews: Journal of Microbiology and Biotechnology* 4(1): 1–10.

Putri, D.N., Sahlan, M., Montastruc, L., Meyer, M., Negny, S., and Hermansyah, H. 2020. Progress of fermentation methods for bio-succinic acid production using agro-industrial waste by *Actinobacillus succinogenes*. *Energy Reports* 6: 234–239.

Rakshit, S.K. 1998. Utilization of starch industry wastes. In: A.M. Martin (ed.), *Bioconversion of Waste Materials to Industrial Products*, pp. 293–315. Springer, Boston, MA. https://doi.org/10.1007/978-1-4615-5821-7_7.

Ramesh, T., and Kalaiselvam, M. 2011. An experimental study on citric acid production by *Aspergillus niger* using *Gelidiella acerosa* as a substrate. *Indian Journal of Microbiology* 51(3): 289–293. https://doi.org/10.1007/s12088-011-0066-9.

Rao, P.R., and Reddy, M.K. 2013. Production of citric acid by *Aspergillus niger* using oat bran as substrate. *International Journal of Chemistry and Chemical Engineering* 3: 181–190.

Rao, R.S.V., Sridevi, V., and Swamy, A.V.N. 2013. Statistical optimization of fermentation conditions for L-glutamic acid production by free cells of *Corynebacterium glutamicum* ATCC13032. *International Journal of Engineering Science Invention* 2(9): 23–28.

Rohr, M., Kubicek, C.P., and Kominek, J. 1983. Citric acid. In: Reed, G. and Rehm, H. J. (eds), *Biotechnology*, pp. 419–454, Vol. 3. Verlag-Chemie, Weinheim.

Romo-Buchelly, J., Rodriguez-Torres, M., and Orozco-Sanchez, F. 2019. Biotechnological valorization of agro industrial and household wastes for lactic acid production. *Revista Colombiana de Biotecnologia* 21(1): 113–127. https://doi.org/10.15446/rev.colomb.biote.v21n1.69284.

Roukas, T., and Kotzekidou, P. 2006. Production of citric acid from brewery wastes by surface fermentation using *Aspergillus niger*. *Journal of Food Science* 51(1): 225 –228. https://doi.org/10.1111/j.1365-2621.1986.tb10876.x.

Roukas, T., and Kotzekidou, P. 2020. Pomegranate peel waste: A new substrate for citric acid production by *Aspergillus niger* in solid-state fermentation under non-aseptic conditions. *Environmental Science and Pollution Research International* 27(12): 13105–13113. https://doi.org/10.1007/s11356-020-07928-9.

Sadh, P.K., Duhan, S., and Duhan, J.S. 2018. Agro-industrial wastes and their utilization using solid state fermentation: A review. *Bioresources and Bioprocessing* 5(1): 1. https://doi.org/10.1186/s40643-017-0187-z.

Saini, J.K., Saini, R., and Tewari, L. 2015. Lignocellulosic agriculture wastes as biomass feedstocks for second-generation bioethanol production: Concepts and recent developments. *3 Biotech* 5(4): 337–353. https://doi.org/10.1007/s13205-014-0246-5.

Salvachua, D., Smith, H., St. John, P.C., Mohagheghi, A., Peterson, D.J., Black, B.A., Dowe, N., and Beckham, G.T. 2016. Succinic acid production from lignocellulosic hydrolysate by *Basfia succiniciproducens*. *Bioresource Technology* 214: 558–566.

Sari, Y.W., Alting, A.C., Floris, R., Sanders, P.M., and Bruins, M.E. 2014. Glutamic acid production from wheat by-products using enzymatic and acid hydrolysis. *Biomass and Bioenergy* 67: 451–459.

Sauer, M., Porro, D., Mattanovich, D., and Branduardi, P. 2008. Microbial production of organic acids: Expanding the markets. *Trends in Biotechnology* 26(2): 100–108. https://doi.org/10.1016/j.tibtech.2007.11.006.

Sawant, O., Mahale, S., Ramchandran, V., Nagaraj, G., and Bankar, A. 2018. Fungal citric acid production using waste materials: A mini review. *Journal of Microbiology, Biotechnology and Food Sciences* 8(2): 821–828.

Saxena, R.K., Saran, S., Isar, J., and Kaushik, R. 2017. Production and applications of succinic acid. In: *Current Developments in Biotechnology and Bioengineering: Production, Isolation and Purification of Industrial Products*, pp. 601–630. Elsevier B.V. http://doi.org/10.1016/B978-0-444-63662-1.00027-0.

Scheele, C.W. 1784. Anmarkning om Citron-Saft, samt satt att crystallisera den samma (Note on lemon juice, as well as ways to crystallize the same). *Kongliga Vetenskaps Academiens Nya Handlingar* 5: 105–109.

Shankar, T., and Sivakumar, T. 2016. Screening of citric acid producing fungi from the leaf litter soil of Sathuragiri hills. *Bioengineering and Bioscience* 4(4): 56–63. https://doi.org/10.13189/bb.2016.040402.

Sharan, A., Charan, A., Bind, A., and Tiwari, S. 2015. Citric acid production from pre-treated sugarcane bagasse by *Aspergillus niger* under solid state fermentation. *Asian Journal of Biological Science* 10(2): 162–166. HTTPS://doi.org/10.15740/HAS/AJBS/10.2/162-166.

Sharma, S., Parkhey, S., Saraf, A., and Pandey, B. 2018. Production of citric acid from different agricultural waste using *Aspergillus niger*. *Indian Journal of Science and Research* 09(1): 75–81. https://doi.org/10.5958/2250-0138.2018.00014.7.

Show, P.L., Oladele, K.O., Siew, Q.Y., Zakry, F.A.A., Lan, J.C.W., and Ling, T.C. 2015. Overview of citric acid production from *Aspergillus niger*. *Frontiers in Life Science* 8(3): 271–283. https://doi.org/10.1080/21553769.2015.1033653.

Singh nee' Nigam, P. 2009. Production of organic acids from agro-industrial residues. In: *Biotechnology for Agro-Industrial Residues Utilization: Utilisation of Agro-Residues*, Pandey., A (Eds.), pp. 37–58. Springer Science and Business Media, Berlin, Germany.

Singh, R., Mittal, A., Kumar, M., and Mehta, P.K. 2017. Organic acids: An overview on microbial production. *International Journal of Advanced Biotechnology and Research* 8(1): 104–111.

Soccol, C., Vandenberghe, L., Rodrigues, C., and Pandey, A. 2006. New perspectives for citric acid production and application. *Food Technology and Biotechnology* 44(2): 141–149.

Soriano-Perez, S., Flores-Velez, L., Alonso-Davila, P., Cervantes-Cruz, G., and Arriaga, S. 2012. Production of lactic acid from cheese whey by batch cultures of *Lactobacillus helveticus* . *Annals of Microbiology* 62(1): 313–317. https://doi.org/10.1007/s13213-011-0264-z.

Song, D. Y., Reddy, L. V., Charalampopoulos, D., and Wee, Y. J. 2019. Poly-(γ-glutamic acid) Production and Optimization from Agro-Industrial Bioresources as Renewable Substrates by *Bacillus* sp. FBL-2 through Response Surface Methodology. *Biomolecules* 9(12): 754. https://doi.org/10.3390/biom9120754

Srivastava, A.K., Tripathi, A.D., Jha, A., Poonia, A., and Sharma, N. 2015. Production, optimization and characterization of lactic acid by *Lactobacillus delbrueckii* NCIM 2025 from utilizing agro-industrial byproduct (cane molasses). *Journal of Food Science and Technology* 52(6): 3571–3578. https://doi.org/10.1007/s13197-014-1423-6.

Straathof, A.J.J., and Gulik, W.M. 2012. Production of fumaric acid by fermentation. In: *Reprogramming Microbial Metabolic Pathways, Subcellular Biochemistry*, pp. 225–240. Springer Science+Business Media, Dordrecht. http://doi.org/10.1007/978-94-007-5055-5_11.

Stylianou, E., Pateraki, C., Ladakis, D., Cruz-Fernández, M., Latorre-Sánchez, M., Coll, C., and Koutinas, A. 2020. Evaluation of organic fractions of municipal solid waste as renewable feedstock for succinic acid production. *Biotechnology for Biofuels* 13: 72. https://doi.org/10.1186/s13068-020-01708-w.

Sun, Z., Li, M., Qi, Q., Gao, C., and Lin, C.S. 2014. Mixed food waste as renewable feedstock in succinic acid fermentation. *Applied Biochemistry and Biotechnology* 174(5): 1822–1833. https://doi.org/10.1007/s12010-014-1169-7.

Teipel, J.C., Hausler, T., Sommerfeld, K., Scharinger, A., Walch, S.G., Lachenmeier, D.W., and Kuballa, T. 2020. Application of 1H nuclear magnetic resonance spectroscopy as spirit drinks screener for quality and authenticity control. *Foods* 9(10): 1355. https://doi.org/10.3390/foods9101355.

Thakur, A., Panesar, P., and Saini, M.S. 2018. Continuous production of lactic acid in a two stage process using immobilized *Lactobacillus casei* MTCC 1423 cells. *ETP International Journal of Food Engineering* 4(3): 216–222. https://doi.org/10.18178/ijfe.4.3.216-222.

Thakur, A., Panesar, P.S., and Saini, M.S. 2019. L(+)-Lactic acid production by immobilized Lactobacillus casei using low cost agro-industrial waste as carbon and nitrogen sources. *Waste and Biomass Valorization* 10(5): 1119–1129. https://doi.org/10.1007/s12649-017-0129-1.

Thomas, P.W. 2013. Citric acid production by Candida species grown on a soy-based crude glycerol. *Preparative Biochemistry and Biotechnology* 43(6): 601–611. https://doi.org/10.1080/10826068.2012.762929.

Tosungnoen, S., Chookietwattana, K., and Dararat, S. 2014. Lactic acid production from repeated-batch and simultaneous saccharification and fermentation of cassava starch wastewater by amylolytic *Lactobacillus plantarum* MSUL 702. *APCBEE Procedia* 8: 204–209. https://doi.org/10.1016/j.apcbee.2014.03.028.

U-thai, P., Vaithanomsat, P., Boondaeng, A. et al. 2016. Production of fumaric acid from oil palm empty fruit bunch. *KKU Research Journal* 21(2): 221–228.

Vandamme, E.J. 2009. Agro-industrial residue utilization for industrial biotechnology products. In:Singh nee' Nigam, P., Pandey, A. (Eds.) *Biotechnology for Agro-Industrial Residues Utilization: Utilisation of Agro-Residues*, pp. 37–58. Springer Science and Business Media, Swizerland.

Vandenberghe, L.P.S., Soccol, C.R., Pandey, A., and Lebeault, J.M. 1999. Microbial production of citric acid. *Brazilian Archives of Biology and Technology* 42(3): 263–276. https://doi.org/10.1590/S1516 -89131999000300001.

Vandenberghe, L.P.S., Karp, S.G., de Oliveira, P.Z., de Carvalho, J.C., Rodrigues, C., and Soccol, C.R. 2018. Solid-state fermentation for the production of organic acids. In Pandey, A., Larroche, C., Soccol, C.R. (Eds.) *Current Developments in Biotechnology and Bioengineering*, pp. 415–434. Elsevier, Amsterdam, Netherlands. https://doi.org/10.1016/B978-0-444-63990-5.00018-9.

Ventrone, M., Schiraldi, C., Squillaci, G., Morana, A., and Cimini, D. 2020. Chestnut shells as waste material for succinic acid production from *Actinobacillus succinogenes* 130Z. *Fermentation* 6(105): 1–11. https:// doi.org/10.3390/fermentation6040105.

Vidra, A., Toth, A.J., and Nemeth, A. 2017. Lactic acid production from cane molasses. *Liquid Waste Recovery* 2(1): 13–16. https://doi.org/10.1515/lwr-2017-0003.

Vidya, P., Annapoorani, A.M., and Jalalugeen, H. 2018. Optimization and utilisation of various fruit peel as substrate for citric acid production by *Aspergillus niger* isolated from orange and carrot. *The Pharma Innovation Journal* 7(6): 141–146.

Vijayalakshmi, P., Vaddadi, U., Reddy, K., Manasa, R.V. 2019. Fermentative Scale up production of L-glutamic acid using Response Surface Methodology. *International Journal of Pharmaceutical Sciences Review and Research* 58(1): 27–33.

Wang, J., Gao, M., Wang, Q., Zhang, W., and Shirai, Y. 2016. Pilot-scale open fermentation of food waste to produce lactic acid without inoculum addition. *RSC Advances* 6(106). https://doi.org/10.1039/c6ra22760k.

Wang, Y., Cao, W., Luo, J., Qi, B., and Wan, Y. 2018. One step open fermentation for lactic acid production from inedible starchy biomass by thermophilic *Bacillus coagulans* IPE22. *Bioresource Technology* 272: 398–406. https://doi.org/10.1016/j.biortech.2018.10.043.

Wang, Y., Yang, Z., Qin, P., and Tan, T. 2014. Fermentative L-(+)-lactic acid production from defatted rice bran. *RSC Advances* 4(17): 8907–8913.

Xu, Z., Shi, Z., and Jiang, L. 2011. Acetic and propionic acids. In: Moo-Young, M. (ed.) *Comprehensive Biotechnology*, pp. 189–199. Academic Press, Burlington.

Yan, Z., Zheng, X.W., Chen, J.Y., Han, J.S., and Han, B.Z. 2013. Effect of different *Bacillus* strains on the profile of organic acids in a liquid culture of *Daqu*. *Journal of the Institute of Brewing* 119(1–2): 78–83.

Yu, B., Zhang, X., Sun, W., Xi, X., Zhao, N., Huang, Z., Ying, Z., Liu, L., Liu, D., Niu, H., Wu, J., Zhuang, W., Zhu, C., Chen, Y., and Ying, H. 2018. Continuous citric acid production in repeated-fed batch fermentation by *Aspergillus niger* immobilized on a new porous foam. *Journal of Biotechnolgy* 20: 276–277. https://doi.org/10.1016/j.jbiotec.2018.03.015.

Yusuf, M. 2017. Agro-industrial waste materials and their recycled value-added applications: Review. In: L. Martínez, O. Kharissova, and B. Kharisov (eds.), *Handbook of Ecomaterials*, pp 1–11. Springer, Cham. https://doi.org/10.1007/978-3-319-48281-1_48-1.

Zafar, M., Arshad, F., Faizi, S., Anwar, Z., Imran, M., and Mehmood, R.T. 2020. HPLC based characterization of citric acid produced from indigenous fungal strain through single and Co-Culture fermentation. *Biocatalysis and Agricultural Biotechnology* 29: 101796. https://doi.org/10.1016/j.bcab.2020.101796.

Zamani, A. 2015. Introduction to lignocellulose-based products. In: Karimi K. (ed.), *Lignocellulose-Based Bioproducts. Biofuel and Biorefinery Technologies.* Springer, Cham. https://doi.org/10.1007/978-3-319 -14033-9_1.

Zeng, W., Chen, G., Wu, Y., Dong, M., Zhang, B., and Liang, Z. 2018. Nonsterilized fermentative production of poly- γ -glutamic acid from cassava starch and corn steep powder by a thermophilic *Bacillus subtilis. Journal of Chemical Technology and Biotechnology* 93(10): 2917–2924. https://doi.org/10.1002/jctb .5646.

Zhang, A.Y., Sun, Z., Leung, C.C.J., Han, W., Lau, K.Y., Li, M., and Lin, C.S.K. 2013. Valorisation of bakery waste for succinic 1 acid production. *Green Chemistry*: 1–30. https://doi.org/10.1039/C2GC36518A.

Zywicka, A., Junka, A., Ciecholewska-Jusko, D., Migdał, P., Czajkowska, J., and Fijałkowski, K. 2020. Significant enhancement of citric acid production by *Yarrowia lipolytica* immobilized in bacterial cellulose-based carrier. *Journal of Biotechnology* 321: 13–22. https://doi.org/10.1016/j.jbiotec.2020.06.014.

13 Production of Biopolymeric Nanomaterials and Nanofibres from Agro-Industrial Waste and Their Applications

Muhammad Umar, Chaichawin Chavapradit, and Anil Kumar Anal

CONTENTS

13.1 INTRODUCTION

Agro-food byproducts are significant elements of sustainable strategies to reduce the green waste. Waste valorization and its recycling technologies have become popular and are called "second-generation food waste management" (Vasyliev and Vorobyova, 2020). A sufficient quantity of waste is produced in the fruit and vegetable industry. This waste is a precious source of raw materials as it contains many organic components, and extractions of these "green" natural components may be used to produce nanomaterials (Ravikumar et al., 2019; Ahmad and Sharma, 2012; Ahmad, Sharma, and Rai, 2012; Shankar et al., 2004). Surprisingly, some studies have authenticated that the agro-food sector may generate waste of almost 39% of the crop output every year due to improper handling, storage, packaging, and transportation (Parfitt, Barthel, and Macnaughton, 2010; Waarts et al., 2011). Making good use of food waste is a challenging task, and it can be turned into treasure if done correctly.

Only the juice-processing industry produces around 5.5 billion kg of fruit waste per year in Europe (Panouille et al., 2007). Each year, dairy waste of around 4–11 million tonnes is added to the natural environment around the globe (Ahmad et al., 2019a), and millions of tonnes of poultry, fishery, and meat industry waste are also being added to the atmosphere every year. The quantity of organic waste produced by horticulture and aquaculture is exceptionally high and contains a large number of bioactive compounds and biomolecules. The presence of this diverse and extensive food waste provides a chance to create novel food waste reprocessing strategies (Ghosh et al., 2017). Fish processing is a standard industry in coastal areas, and only the required meat is taken from the whole body; the remaining parts are discarded. The seafood industry, on average, produces around 80,000 tonnes of waste per annum. A large amount of waste material makes the degradation a slow process, resulting in waste accumulation over a long period of time (Divya and Jisha, 2018). To fabricate metal NPs, agro-industrial waste may be an environmentally friendly source and may provide a route for green chemistry (Akhtar, Panwar, and Yun, 2013).

DOI: 10.1201/9781003125679-13

Waste from the dairy processing industry is a real threat to the environment because it contains a large number of organic components, and the fat content in it produces films and grease on wastewater surface, which causes dissolved oxygen depletion that hinders oxygen transfer for aquatic life. In Europe, around 700 million tonnes of agro-waste is produced per year (Pawelczyk, 2005). The food waste created in food factories, hotels, and households was around 89 billion tonnes in 2006 in the EU-27, which causes economic and environmental problems (Morone et al., 2018). In Europe, most of the waste was from grape, pomegranate, radish, and beetroot processing industries. It is urgently necessary to develop zero-waste technologies combined with chemical methods for plant-based waste (Hoyt and Mason, 2008).

Nanotechnology is a multidisciplinary research field that involves live sciences, medical and chemical science, and engineering for the fabrication, handling, and use of ingredients at a nanometric scale (1–100 nm). On the nanometric scale, some significant characteristics are present that cannot be seen when the material is at a larger scale in size (Shah et al., 2015; Cai and Chen, 2007). Green nanotechnology has developed rapidly, emphasizing fabrication procedures that are clean and environmentally friendly instead of chemical and physical methods currently being utilized for nanoparticle manufacturing (Saratale et al., 2018).

In the last few years, nanotechnology has been intensely explored for its application in the field of chemistry (Gawande et al., 2016), pharmaceuticals (Lin, 2015; Lin, Chen, and Shi, 2018), cosmetics (Suter et al., 2016), and the food and beverage industry (McClements and Xiao, 2017). In food, the distinctive characteristics of nanomaterials lead to the enhancement of the biological, chemical, and physical properties of food (Pan and Zhong, 2016). The NPs obtained from food materials (proteins, carbohydrates, and lipids) are widely used in food to improve shelf life, sensory qualities, bioactive compounds, functional food coatings, and stabilization for emulsions (Rossi et al., 2014). Protein- and carbohydrate-based biopolymeric nanomaterials have the advantage of immunogenicity, antibacterial, biocompatible, and biodegradable properties compared to synthetic nanomaterials that cause cytotoxicity and environmental degradation problems. Moreover, they have more size distribution, and carbohydrate-based nanoparticles are mainly derived from polysaccharides such as chitosan, alginate, and pectin, and protein nanomaterials are obtained from milk, plant, and animal proteins such as sericin, albumin, whey, keratin, fibroin, lactoglobulin, gelatine, and collagen (Verma et al., 2020). It is not only the nanoparticles but the nanofibres that are considered prominent research subjects in the nanotechnology field. Compared to traditional micrometre fibres, nanofibres exhibit overall superior mechanical and structural properties (Jiang et al., 2018). The nano-sized fibres possess a high modulus, high tensile strength, and toughness that are extremely suitable as composite materials. The development of nanofibres has been widely utilized in biomedical, tissue engineering, packaging, biocatalysts, drug release, dental applications, biosensors, and molecular filtration fields (Xue et al., 2017).

13.2 EXTRACTION AND SYNTHESIS METHODS OF NANOMATERIALS FROM WASTE

Nanomaterials are fabricated by two major methods depending on the required properties: the top-down approach (sonication, acid hydrolysis, reactive extrusion, emulsion or microemulsion templating, and gamma irradiation) in which greater particles are converted to smaller (nanoscale) size particles and the bottom-up approach (nanoprecipitation and self-assembly), where small molecules (particles) are joined to fabricate the nanomaterials (Mansfield et al., 2017; Qiu et al., 2020). The acid hydrolysis technique is mainly used as it is low cost and simple, but recovery and production yield are small (Kim, Park, and Lim, 2015). Enzymatic hydrolysis is a fairly rapid method with a significantly higher amount of yield (around 85%) and low environmental impact. In enzymatic hydrolysis, compared to acid hydrolysis, the selective breakdown of specific parts of the structure occurs (Le Corre and Angellier-Coussy, 2014). Nanoprecipitation produces nanoparticles in a controlled size as no special operating conditions and equipment are needed, but its drawback is that the solvent may need to be removed after particle formation (Qiu et al., 2019). Ultrasonication consumes a high amount of energy and takes a long time to process. Being a purely mechanical process, high-pressure homogenization is clean, but the yield is not much higher as compared to other methods (Haaj et al., 2013). The second environmentally friendly mechanical method is reactive extrusion, in which extrusion is performed under semi-liquid conditions (35% moisture content), so in this technique, the yield is high (Peng, Srinivas, and Narain, 2020). Gamma radiation is a convenient and rapid modification method that converts large particles into smaller sizes (100 nm) (Singh et al., 2011).

Nanoparticles are extracted from various types of food waste, for example, chitosan nanoparticles from shellfish (Divya and Jisha, 2018) and gold nanoparticles from onion peel (Patra, Kwon, and Baek, 2016), mango peel (Yang, WeiHong, and Hao, 2014), grape waste (Xu et al., 2015), and pear fruit waste (Ghodake et al., 2010a) Table 13.1. Nanoparticles can be synthesized via physical (laser ablation, thermal decomposition, high-energy irradiation, sputtering, and ultrasonic fields) and chemical methods (lithography, photochemical reduction, and colloidal chemical method) (Ahmad et al., 2003; Sharma et al., 2007). Chemical and physical methods for metal nanoparticles are much more difficult, and the control of parameters directly affects the stability, shape, size, and physicochemical characteristics of the particles (Yang et al., 2013; Ahmed et al., 2016). The photochemical fabrication of NPs is an environmentally friendly method that involves the fabrication of NPs in different media such as surfactant micelles, cells, emulsions, and polymer films (Prasad, Kambala, and Naidu, 2013). Nanoparticle synthesis by a biological process is clean, safe, eco-friendly, cheap, and nontoxic. Bacteria, fungus, actinomycetes, plants, yeast, viruses, and algae have the potential to absorb and accumulate the metal ions in them for the fabrication of nanomaterials (Saratale et al., 2017; Kulkarni and Muddapur, 2014b).

The fabrication of metal nanoparticles without chemicals from plant waste extract using biosynthetic methods has gained much attention (Song and Kim, 2008). The biosynthesis of gold nanoparticles from the waste of grape skins, seeds, and stalks (Krishnaswamy, Vali, and Orsat, 2014), extract of pear (Ghodake et al., 2010b), mango peel (Yang, WeiHong, and Hao, 2014), and freshwater green algae and red algae have been reported (Sharma et al., 2014). In the past few years, the biological fabrication of nanoparticles has appeared as an intelligent substitute for traditional fabrication procedures. Biosynthesis uses green chemistry, which is an environmentally friendly approach utilizing multi- and uni-cellular biological individuals like actinomycetes (Ahmad et al., 2003), fungus (Mukherjee et al., 2001), bacteria (Husseiny et al., 2007), viruses (Merzlyak and Lee, 2006), plants (Philip, 2010), and yeast (Dameron et al., 1989). The biological synthesis of nanoparticles is nontoxic and an environmentally friendly method that produces particles in an extensive range of shapes, sizes, and physicochemical and compositional characteristics (Mohanpuria, Rana, and Yadav, 2008).

Plants may produce NPs using techniques based on green chemistry as they are environmentally friendly sources. For example, spherical Au NPs (20 nm) from the flower extract of night jasmine (Das, Gogoi, and Bora, 2011) and particles (7–58 nm) using coriander leaf extract (*Coriandrum sativum*) were produced (Narayanan and Sakthivel, 2008). There are several works focusing on the synthesis and use of different nanoparticles from *Azadirachta indica* (Shankar et al., 2004), *Aloe vera* (Chandran et al., 2006), *Medicago sativa* (Gardea-Torresdey et al., 2003), *Emblica officinalis* (Ankamwar et al., 2005), *Diospyros kaki* (Song and Kim, 2008), *Magnolia kobus* (Song, Jang, and Kim, 2009), *Basella alba*, *Sorghum bicolour* (Leela and Vivekanandan, 2008), and *Coriandrum* sp. (Narayanan and Sakthivel, 2008). It has been reported that the concentration and type of molecules present in food waste may affect the structure of nanomaterials and their firmness afterwards. In a study, it was found that changing the concentration of the precursor (chloroauric acid) and extract of *Cinnamomum camphora* leaf in mixed form causes changes in the shape of the nanoparticles (triangular to sphere-shaped) (Huang et al., 2007). Correspondingly, it has been reported that by altering the concentration of Aloe vera leaf extract in mixtures (chloroaurate ion), this variation not only regulates the size of the particles (50–350 nm) but it also affects the concentration of triangular

TABLE 13.1

Organic and Inorganic Nanomaterials Fabricated from Food Waste Sources

Nanoparticle	Food Waste/Source	Morphology	Size (nm)	Reference
Ag	Pineapple	Spherical	5–35	Shankar et al., 2004
Ag	Tansy fruit	Spherical	16	Poinern et al., 2013
Ag	*Citrullus colocynthis*	Spherical	31	Satyavani, Ramanathan, and Gurudeeban, 2011
Ag	Orange peel	Spherical	3–12	Konwarh et al., 2011
Ag	Orange peel	Spherical	35–2	Kaviya et al., 2011
Ag	Mandarin peel	Spherical	5–20	Basavegowda and Rok Lee, 2013
Ag	Sorghum bran	Spherical	10	Njagi et al., 2011
Ag	Gooseberry	Spherical	10–20	Ankamwar et al., 2005
Ag	*Eucalyptus macrocarpa*	Spherical and cubes	10–100	Poinern et al., 2013
Ag	*Mangifera indica*	Spherical and triangular	20	Philip, 2011
Ag	*Rhododendron dauricum*	Spherical	25–40	Mittal, Kaler, and Banerjee, 2012
Ag	Tea/coffee extracts	Spherical	60	Nadagouda and Varma, 2008
Ag	Banana peel	Nanoclusters	–	Bankar et al., 2010
Ag	Papaya	Spherical	60–80	Mude et al., 2009
Ag	Papaya	Cubic	15	Jain et al., 2009
Ag	Guava	–	0.1–0.5	Sriram and Pandidurai, 2014
Ag	Black Carrot	Spherical	4.5–17	Abubakar et al., 2014
Ag	Garlic clove	–	4–22	Ahamed et al., 2011
Ag	*Citrullus lanatus* rind	Spherical	17–0.2	Ndikau et al., 2017
Ag	*Codium capitatum*	Spherical	3–44	Kannan, Stirk, and Van Staden, 2013
Ag	*Spirogyra insignis*	Spherical	30	Ghosh et al., 2017
Ag	*Enteromorpha compressa*	Spherical	4–24	Ramkumar et al., 2017
Au	Tansy fruit	Triangular	11	Poinern et al., 2013
Au	*Eucalyptus macrocarpa*	Spherical and Hexagonal	20–100	Poinern et al., 2013
Au	*Turbinaria conoides*	Spherical and Triangular	6–10	Rajeshkumar et al., 2013
Au	*Sargassum incisifolium*	Spherical	20	Mmola et al., 2016
Au	*Stoechospermum marginatum*	Spherical	18–93	Rajathi et al., 2012
Au	*Citrullus lanatus* rind	Spherical	20–140	Patra and Baek, 2015
Au	Grape waste	Spherical	20–25	Krishnaswamy, Vali, and Orsat, 2014
Au	Rice bran	Spherical	50–100	Malhotra et al., 2014
Au	Pear	Triangular and hexagonal	200–500	Ghodake et al., 2010a
Au	Mango peel	Spherical	6–18	Yang, WeiHong, and Hao, 2014
Au	Carrot	–	432.3	Pattanayak, Muralikrishnan, and Nayak, 2014
Au	Sugar beet pulp	Spherical	10	Castro et al., 2011
Au	Gooseberry	Spherical	15–25	Ankamwar et al., 2005
Au	*Camelia sinensis*	Spherical, triangular and irregular	30–40	Vilchis-Nestor et al., 2008
Pd	Banana peel	Crystalline, irregular shape	50	Buzby, Farah-Wells, and Hyman, 2014
Pd	Tea/coffee extracts	Spherical	20	Nadagouda and Varma, 2008

(Continued)

TABLE 13.1 (CONTINUED)
Organic and Inorganic Nanomaterials Fabricated from Food Waste Sources

Nanoparticle	Food Waste/Source	Morphology	Size (nm)	Reference
Pd	Lignin	Spherical	16–20	Coccia et al., 2012
Pd	Watermelon rind	Spherical	96	Lakshmipathy et al., 2015
Fe_3O_4	Tea waste	Cubes and Pyramids	5–25	Lunge, Singh, and Sinha, 2014
Mn_3O_4	Banana peel	Spherical	20–50	Yang, WeiHong, and Hao, 2014
Au and Ag	*Aloe vera*	Spherical	50–350	Chandran et al., 2006
In_2O_3	*Aloe vera*	Spherical	5–50	Maensiri et al., 2008
Pt	*Diospyros kaki*	Crystalline	15–19	Satyavani, Ramanathan, and Gurudeeban, 2011
Complex	β-Lg and beet pectin	Spherical	300	Santipanichwong et al., 2008
Complex	β-Lg and beet pectin	Spherical	361	Chanasattru et al., 2009
Complex	β-Lg and beet pectin	Spherical	357	Jones and McClements, 2008
Complex	β-Lg and beet pectin	Spherical	199	Chanasattru et al., 2009
Complex	β-Lg and beet pectin	–	100–300	Abaee, Mohammadian, and Jafari, 2017
Particle	Waxy maize starch	–	66–320	Dai et al., 2018
Particle	Waxy potato starch	Square	50–100	Hao et al., 2018
Particle	Corn starch	–	5–100	Ma et al., 2008
Particle	Cassava starch	–	35–65	Minakawa, Faria-Tischer, and Mali, 2019

and spherical NPs generated (Chandran et al., 2006). Moreover, a low concentration of precursor (Au, 1.53 mm) in the mixture holding oil palm (*Elaese guineensis*) leaf extract generated spherical nanoparticles (27.89–14.59 nm), and a high concentration (Au, 4.055 mm,) fabricated hexagonal, triangular, spherical, and pentagonal NPs (22.88–8.21 nm) (Ahmad, Irfan, and Bhattacharjee, 2016).

In 1994, Ohya et al. developed for the first time chitosan-based nanoparticles by crosslinking and emulsifying and used these particles to deliver anticancer drugs (5-fluorouracil) (Grenha, 2012). Now there are five methods (emulsification solvent diffusion, reverse micellar, ionotropic gelation, microemulsion, and polyelectrolyte complex) available for the fabrication of chitosan-based particles (Tiyaboonchai, 2013). Numerous agri-food byproducts (rapeseed pomace, pomegranate peels, fodder radish cake, sugar beet pulp, and grape pomace) were studied for the green extraction of organic compounds; these compounds were used in the fabrication of green nanoparticles that were utilized for corrosion inhibition (Vasyliev and Vorobyova, 2020).

Grape seed extract was obtained using water as the extracting agent; using this extract as a stabilizing and reducing agent, silver NPs were synthesized that were highly stable. They were further studied using X-ray diffraction, transmission electron microscopy, and ultraviolet visible spectroscopy (Xu et al., 2015). The protein composition of rice bran includes 5% prolamin, 22% glutelins, 36% globulins, and 37% albumins (Wang et al., 1999). Rice bran albumin (RBA) has antioxidative, anticancer, anti-ageing, and antidepressant properties along with good solubility in an aqueous medium, being less allergic, biocompatible, nontoxic, low cost, and biodegradable (Zhang, Huang, and Wei, 2014). RBA was obtained from the waste of rice, with negative charges and good digestibility; it was used to prepare nanoparticles with chitosan with a self-assembly technique at a size of around 384 nm (Peng et al., 2017).

During the extraction of oil from rapeseed, the residue press cake is produced in bulk as waste material. Due to its high protein content, further processing creates a large amount of fibrous by-product; hydrothermal carbonization can convert it into carbon nanoparticles, which can be combined into micrometric spheres by acetone precipitation. These microspheres can be utilized for the encapsulation of compounds. These nanoparticles were effective biocidal agents, exhibiting cellular damage to bacteria (Das Purkayastha et al., 2014). Within a network of whey protein, nanoparticles of poly-3-hydroxybutyrate-co-hydroxyhexanoate were grafted to obtain a nano-biocomposite material. Dynamic light scattering confirms the stability of particles with a zeta-potential of −40 mV and a size of 80 nm (Corrado et al., 2021). Soy protein nanoparticles (72.42 nm) were fabricated from an isolate of defatted soy flour. These particles were added to yoghurt, which shows better antiradical scavenging activity and antioxidant properties against ferric ion. Fortified yoghurt had higher proteolytic activity and oxidative stability (Sengupta et al., 2019). In a study, milk was utilized to fabricate silver nanoparticles because it is nontoxic and environmentally friendly. Moreover, milk is readily available in large supply as a renewable resource. Without using any stabilizing or reducing agent, highly stable and bactericidal nanoparticles were produced with reasonable control over their size (Pandey et al., 2020).

Three different kinds of fibre, cellulose nanofibres (CNFs), carbon nanofibres, and synthetic polymer nanofibres, are the most explored and researched fibres in the nanomaterial field (Jiang et al., 2018). However, when related to bio-based materials, CNFs are the main focus. Plant cellulose, which is the source of CNFs, is abundant in nature; the material can be considered the most available natural polymer on Earth (Zhu et al., 2006). However, considering the material cost, availability, process efficiency, environmental impact, and sustainability, low-value agro-industrial wastes are more feasible than traditional materials such as wood sources and crops (García et al., 2016). The wastes and byproducts from agro-industrial wastes that are rich in cellulose-based byproducts are the prime candidates for the extraction and fabrication of CNFs (Kargarzadeh et al., 2018). CNFs have been extracted from various types of industrial wastes such as bagasse (Hassan et al., 2012), pineapple leaves (Cherian et al., 2011), corn cob (Wang et al., 2019; Kang et al., 2017), mandarin peel (Hiasa et al., 2014), orange peel (Hideno, Abe, and Yano, 2014), walnut shell (Harini and Mohan, 2020), banana peel (Harini, Ramya, and Sukumar, 2018), and carrot (Siqueira et al., 2016). The extraction process of nanocellulose can be conducted via different mechanical treatments such as high-pressure homogenization and ultrafine grinding (Henriksson et al., 2007). The pre-treatment steps such as ultrasonication, enzymatic treatment (Henriksson et al., 2007), and oxidation (Isogai, Saito, and Fukuzumi, 2011) are preferred to reduce the energy demand during the extraction process.

As mentioned, several agro-industrial wastes and byproducts, especially plant-based wastes, can potentially be used as sources for CNF production due to their high cellulose content. For instance, corn cobs, which are the by-product of the corn grain milling process, contain 48.1% cellulose in their structure (Pérez-Rodríguez, García-Bernet, and Domínguez, 2016). Due to the high cellulose content, the material was given different treatments to extract the CNFs (Singh et al., 2020). A combination of steam explosion, alkaline treatment, and ultrasonication method was performed to extract the CNFs from the corn cobs (Yang et al., 2017). The corn cobs were pre-treated to loosen the cell wall structure and support the removal of cementing materials, followed by mechanical treatment to remove the non-cellulosic components. As a result, CNFs with a superior crystallinity index and average diameter of 8 nm were extracted. Similarly, sugar cane bagasse, which is waste from sugar production plants, can be considered the predominant material for the extraction of CNFs due to its high cellulose content, short growth duration, and low lignin content (Pennells et al., 2020). The production of CNFs using the two-step method involving alkaline hydrogen peroxide hydrolysis and ultrasonication can effectively produce CNFs from sugar cane bagasse (Shahi et al., 2020). The CNFs obtained after ultrasonication exhibit a significantly higher crystallinity index compared to the untreated sample, and the diameter of the fibre ranged from 6 to 100 nm.

Traditionally, CNFs were extracted using a mechanical pressure process to disintegrate and break down the cellulose-based materials (Khalil et al., 2014). However, the production process requires an extremely high-energy consumption of over 25,000 kWh per tonne. Thus, the commercialization of CNFs was unviable (Klemm et al., 2011). As a consequence, the adaptation of chemical treatments such as acidic and alkaline treatments have been developed to replace the traditional method. The chemical treatment aims to cleave the glycosidic bonds linking the cellulose chains, resulting in the colloidal suspension of CNFs (Fahma et al., 2010). Even though the chemical process can efficiently fabricate the CNFs, the unfavourable surface modification and production of chemical wastes remain the issue (Tibolla, Pelissari, and Menegalli, 2014). Therefore, the current trend of CNF production has shifted toward a green technology and zero-waste approach (Nie et al., 2018). Currently, enzymatic hydrolysis has become the foremost method for CNF production to replace the acidic treatment (Tibolla, Pelissari, and Menegalli, 2014). The process is conducted using cellulase or xylanase enzyme to break down the cellulose structure with no aid from hazardous chemicals. CNFs obtained from enzymatic digestion exhibit superior structural properties such as a higher aspect ratio and a sufficient number of hydroxyl and carboxyl groups compared to chemical hydrolysis (Nie et al., 2018). Considering the reusability and processing cost of enzymatic hydrolysis, the immobilized enzyme is the preferred choice compared to free cells (Yassin et al., 2019). Immobilization of the cellulase enzyme on a polymeric gel disc was conducted (Yassin et al., 2019). Compared to a free enzyme that is used once, the immobilized enzyme can retain over 85% of enzyme activities after running for six cycles. Moreover, the CNFs acquired from the process exhibit a high aspect ratio with approximate diameter ranges from 15 to 35 nm.

13.3 CHARACTERIZATION

Nanomaterials are of versatile use due to their small size, extended surface, and quantum effect, and their physical, morphological, and chemical properties are the result of different materials used to produce these particles (Kahan et al., 2009).

The potential application of nanoparticles and nanofibres is observed using several characterization methods (Ealia and Saravanakumar, 2017). The measurement and analytical processes were performed to determine the key characteristics such as size (Quiñones et al., 2017), surface area (Sharma et al., 2007), composition (Xie et al., 2019), surface morphology (Yang et al., 2017), surface charge (Abitbol et al., 2018), rheological properties (Jiang et al., 2018), concentration, and physical, chemical, and biological properties of the developed nanomaterials (Saleh, 2020). Due to the sizes of the advanced materials, the scientific equipment or analytical tools used in the characterization must be appropriate and precise to measure the minuscule materials, specifically in nanometre scales. For instance, the size of the synthesized materials is one of the most crucial factors. The sizes must be identified to determine whether the materials fall under nano- or micro-scales (Modena et al., 2019).

Raman spectroscopy is utilized for the chemical, optical, and mechanical properties of fabricated nanomaterials (Kneipp et al., 1999; Gouadec and Colomban, 2007; Saito and Kataura, 2001). Transmission electron microscopy, field emission scanning electron microscopy, high-resolution transmission electron microscopy, and scanning electron microscopy are used to study the morphological features of nanomaterials (Denkbaş et al., 2016), and the 3D electron microscopy method has been widely used for the qualitative measurement of materials (Ersen et al., 2015). The transparent part of nanomaterials can be accessed by ultraviolet visible spectroscopy (Pavia et al., 2009), and the X-ray diffraction (XRD) technique is used to study the atomic arrangement, material structure, imperfections, and crystallite size, and crystallinity (Lee et al., 2017). Fourier Transform Infrared (FTIR) checks the linkage and functionality of nanomaterials (Wang et al., 2008), and the thermal behaviour can be examined using thermo gravimetric analysis. The cyclic voltammetry can be used to explore the electrochemical behaviour of nanomaterials (Zhou and Wu, 2013).

13.4 POTENTIAL APPLICATIONS OF NANOPARTICLES AND NANOFIBRES

Nanoparticles are being used in different chemical sensing equipment, electric devices, diagnostic tools, medicinal products, and medical treatment procedures. Gold nanoparticles have been widely used in medical applications (Sperling et al., 2008; Puvanakrishnan et al., 2012), such as protein assay, immunoassay (Hirsch et al., 2003), enhanced secondary ion mass spectrometry (Kim et al., 2006), detection of cancer cells (Kah et al., 2007), capillary electrophoresis (Tseng et al., 2005), disease diagnostics (Torres-Chavolla, Ranasinghe, and Alocilja, 2010), separation methods (Sýkora et al., 2010), and pharmaceuticals (Cai et al., 2008). Fabricated silver nanoparticles have both anti-inflammatory and antibacterial characteristics, which improve fast wound healing ability. Because of these abilities, silver nanomaterials were added to wound dressings, operation preparations, and medicinal coatings (Huang et al., 2007). NPs of platinum were extensively used in medical treatments, along with some other NPs (Hrapovic et al., 2004). Palladium NPs were utilized in catalysis and electro-catalysis applications (Gopidas, Whitesell, and Fox, 2003), optoelectronics (Pankhurst et al., 2003), chemical sensors (Coccia et al., 2012), and in many antibacterial applications. Non-noble metallic nanoparticles like iron (Pankhurst et al., 2003), selenium (Prasad et al., 2013), zinc oxide (Brayner et al., 2006), and copper (Lee et al., 2011) have been applied in cosmetic formulations, medical treatments, and antibacterial applications.

To increase the stability, bioavailability, and solubility of bioactive components susceptible to many environmental, processing, and storage factors like pH, temperature, and salt concentrations, nanoparticles have been proven to be the best aqueous-based transport (Ha et al., 2015). Nanoparticles improve the bioavailability of these compounds due to their smaller size, but their large surface areas allow interactions with gastrointestinal tract epithelial cells (Hu, McClements, and Decker, 2003). For the incorporation of these nanoparticles in food, it is essential that they are fabricated from food-grade materials such as whey protein, chitosan, pectin, and other polysaccharides obtained from food waste to meet the safety requirements (Chen, Remondetto, and Subirade, 2006). The facilitation of cellular intake of encapsulated drugs was supported by modified nanoparticles of gelatine (Balthasar et al., 2005). The specificity of a monoclonal antibody (DI17E6) in melanoma cells was improved by binding it with albumin nanoparticles (Wagner et al., 2010). Collagen, a component of organs and tissues, has an important role in the function of cells (Li et al., 2013). NPs of collagen were used to transport estradiol, which significantly improves estradiol absorption in the gastrointestinal tract (Nicklas et al., 2009). In the treatment of osteoarthritis and rheumatoid arthritis, nanoparticles of β-casein were used to orally administer the delivery of celecoxib (Bachar et al., 2012). The complex formation between pectin and whey protein also improves the encapsulation efficacy of hydrophilic compounds (Gülseren, Fang, and Corredig, 2012).

Silk proteins show gradual biodegradability, biocompatibility, good mechanical strength, temperature resistance, reduced inflammation, and immunogenicity, so this protein nanoparticle was used for bioconjugation with neutral protease, which improved the hydrolysis efficacy of the sericin molecule to its small peptides (Zhu et al., 2011). The natural protein fibrin is used in blood clotting and is also used to fabricate nanoparticles to deliver antimicrobial agents (Praveen et al., 2012). The results of fluconazole- and ciprofloxacin-loaded nanoparticles of fibrin confirmed suitable antifungal and antibacterial effects, respectively (Alphonsa et al., 2014). Nanoparticles of gliadin, an important wheat gluten protein (Duclairoir et al., 2003), were coated with pectin, which improved the stability of the particles (Joye, Nelis, and McClements, 2015). Zein (50% of maize protein) is suitable to form candidate structures to deliver drugs (daidzin and lutein) (Patel and Velikov, 2014; Shukla and Cheryan, 2001; Hu et al., 2012; Zou and Gu, 2013). NPs of zein and alginate are utilized to transport bioactive compounds with better stability and shelf life (Hu and McClements, 2015). Vitamin B12 absorption through the gut into the cells was improved by loading it on soy protein isolate nanoparticles without any toxicity on Caco-2 cells (Zhang et al., 2015).

In the case of nanofibres, the materials exhibit numerous strong points such as significantly high mechanical, biodegradability, biocompatibility, and bioactive properties (Soni et al., 2015). Due to these properties, nanofibres have been applied in several fields such as bioimaging (Liu et al., 2013), nano fillers, packaging (Gond and Gupta, 2020), biomedical (Jorfi and Foster, 2015), antimicrobial materials (Calamak et al., 2017), wound dressing (Homaeigohar and Boccaccini, 2020), drug delivery systems (Hu et al., 2014), and electrical devices (Mondal et al., 2020).

As previously mentioned, agro-industrial wastes can be the main source of bio-based nanofibres, especially CNFs. Therefore, CNFs generated from the wastes have been widely applied in several applications. For example, transparent nanopaper was synthesized from CNFs obtained from corn cob residue (Wang et al., 2019). The developed nanopaper exhibited high mechanical strength and thermal properties. In addition, the optimal transmittance of the paper was close to 90% in the visible range. Similarly, the transparent nanopaper was developed from corn cob cellulose (Kang et al., 2017). The nanopaper was developed from CNFs extracted using sonication and ultrafiltration. This resulted in a high mechanical strength nanopaper with a moderately hydrophobic surface. The extraction of nanofibre from sugarcane bagasse waste was conducted and used for packaging purposes (Gond and Gupta, 2020). Fabricated CNFs possess excellent optical properties and antibacterial activities against pathogenic micro-organisms (*E.coli* and *Bacillus*). Thus, CNFs can be used as reinforcement components in packaging applications. For wound dressing applications, apart from the development of a scaffold, CNFs extracted from sugar cane bagasse were utilized as the printing ink for a 3D printing machine for wound dressing applications (Chinga-Carrasco et al., 2019). CNF fibre shows no sign of cytotoxicity toward the L929 fibroblasts cell line. In addition, the combination of CNF ink, alginate, and Ca^{2+} can induce a dimensional change in the printing structure.

13.5   CYTOTOXICITY AND ENVIRONMENTAL CONCERNS

The production of NPs in 2004 was almost 1000 tonnes, and it is calculated that almost 800 products based on nanomaterials are being used every day (Rejeski and Lekas, 2008). The release of nanomaterials into the environment is a major side effect, which causes toxicity due to its small size, composition, and shape. The fast distribution of nanoproducts in local markets means there is an urgent necessity for new research strategies as nanoproduct toxicology is increasing (Bystrzejewska-Piotrowska, Golimowski, and Urban, 2009). Although some NPs are toxic, little is known about the mechanism of toxicology (Oberdörster, Stone, and Donaldson, 2007). Many types of research highlighted how prokaryotic and eukaryotic cells take up and accumulate nanoparticles (Liu, 2006). Since there are a variety of nanoproducts available in the market, it is important to find new ways to dispose of the end products (Bystrzejewska-Piotrowska, Golimowski, and Urban, 2009).

Physical and chemical methods of nanoparticle production add to the toxicity in the environment as they use hazardous chemicals like reducing agents, organic solvents, and stabilizers (Shankar et al., 2014). The environmental toxicity problem of nanoparticles arises from the use of these harmful materials. The toxic agents on the surface of nanoparticles and possibly the nanoparticles themselves being toxic have limited their biomedical and clinical use. Some nanoparticles are toxic because of their composition, shape, size, and surface chemistry (Kulkarni and Muddapur, 2014a).

Many *in vivo* studies have proved that ingestion of silver nanoparticles results in the abnormal composition of mucus, pigmentation of villi, and lymphocyte infiltration (Cha et al., 2008; Jeong et al., 2010). These particles may accumulate in the organs (kidneys, liver, spleen, small intestine, and stomach) (Gaillet and Rouanet, 2015; Hendrickson et al., 2016). Some lipid nanoparticles with indigestible oil phases (terpenes and hydrocarbons) are being used in foods that are not digested by the enzymes in the gut tract (McClements and Rao, 2011). The indigestible parts of carbohydrate and lipid NPs may be assimilated intact, causing an adverse effect, but no study has been found on the toxicity of these nanoparticles after ingestion. Nanoparticles of protein are also utilized in the food and pharmaceutical sectors, which may have some health-related problems by changing the bioactivity of the substance captured in it (McClements and Xiao, 2017).

REFERENCES

Abaee, Arash, Mehdi Mohammadian, and Seid Mahdi Jafari. 2017. "Whey and soy protein-based hydrogels and nano-hydrogels as bioactive delivery systems." *Trends in Food Science and Technology* 70:69–81.

Abitbol, Tiffany, Doron Kam, Yael Levi-Kalisman, Derek G. Gray, and Oded Shoseyov. 2018. "Surface charge influence on the phase separation and viscosity of cellulose nanocrystals." *Langmuir* 34(13):3925–3933.

Abubakar, Aminu Shehu, Ibrahim Bala Salisu, Shiwani Chahal, Geetika Sahni, and Ramesh Namdeo Pudake. 2014. "Biosynthesis and characterization of silver nano particles using black carrot root extract." *International Journal of Current Research and Review* 6(17):5.

Ahamed, Maqusood, M. A. Majeed Khan, M. K. J. Siddiqui, Mohamad S. AlSalhi, and Salman A. Alrokayan. 2011. "Green synthesis, characterization and evaluation of biocompatibility of silver nanoparticles." *Physica E: Low-Dimensional Systems and Nanostructures* 43(6):1266–1271. https://doi.org/10.1016/j.physe.2011.02.014.

Ahmad, Absar, Satyajyoti Senapati, M. Islam Khan, Rajiv Kumar, and Murali Sastry. 2003. "Extracellular biosynthesis of monodisperse gold nanoparticles by a novel extremophilic actinomycete, Thermomonospora sp." *Langmuir* 19(8):3550–3553.

Ahmad, Naheed, and Seema Sharma. 2012. "Biosynthesis of silver nanoparticles from biowaste pomegranate peels." *International Journal of Nanoparticles* 5(3):185–195.

Ahmad, Naheed, Seema Sharma, and Radheshyam Rai. 2012. "Rapid green synthesis of silver and gold nanoparticles using peels of *Punica granatum.*" *Advanced Materials Letters* 3(5):376–380.

Ahmad, Talha, Rana Muhammad Aadil, Haassan Ahmed, Ubaid ur Rahman, Bruna C. V. Soares, Simone L. Q. Souza, Tatiana C. Pimentel, Hugo Scudino, Jonas T. Guimarães, Erick A. Esmerino, Mônica Q. Freitas, Rafael B. Almada, Simone M. R. Vendramel, Marcia C. Silva, and Adriano G. Cruz. 2019. "Treatment and utilization of dairy industrial waste: A review." *Trends in Food Science and Technology* 88:361–372. https://doi.org/10.1016/j.tifs.2019.04.003.

Ahmad, Tausif, Muhammad Irfan, and Sekhar Bhattacharjee. 2016. "Parametric study on gold nanoparticle synthesis using aqueous Elaise guineensis (oil palm) leaf extract: Effect of precursor concentration." *Procedia Engineering* 148:1396–1401.

Ahmed, Shakeel, Mudasir Ahmad, Babu Lal Swami, and Saiqa Ikram. 2016. "A review on plants extract mediated synthesis of silver nanoparticles for antimicrobial applications: A green expertise." *Journal of Advanced Research* 7(1):17–28.

Akhtar, Mohd Sayeed, Jitendra Panwar, and Yeoung-Sang Yun. 2013. "Biogenic synthesis of metallic nanoparticles by plant extracts." *ACS Sustainable Chemistry and Engineering* 1(6):591–602.

Alphonsa, B M., P. T. Sudheesh Kumar, G. Praveen, Raja Biswas, K. P. Chennazhi, and R. Jayakumar. 2014. "Antimicrobial drugs encapsulated in fibrin nanoparticles for treating microbial infested wounds." *Pharmaceutical Research* 31(5):1338–1351.

Ankamwar, Balaprasad, Chinmay Damle, Absar Ahmad, and Murali Sastry. 2005. "Biosynthesis of gold and silver nanoparticles using Emblica officinalis fruit extract, their phase transfer and transmetallation in an organic solution." *Journal of Nanoscience and Nanotechnology* 5(10):1665–1671. https://doi.org/10.1166/jnn.2005.184.

Bachar, Michal, Amitai Mandelbaum, Irina Portnaya, Hadas Perlstein, Simcha Even-Chen, Yechezkel Barenholz, and Dganit Danino. 2012. "Development and characterization of a novel drug nanocarrier for oral delivery, based on self-assembled β-casein micelles." *Journal of Controlled Release* 160(2):164–171.

Balthasar, Sabine, Kerstin Michaelis, Norbert Dinauer, Hagen von Briesen, Jörg Kreuter, and Klaus Langer. 2005. "Preparation and characterisation of antibody modified gelatin nanoparticles as drug carrier system for uptake in lymphocytes." *Biomaterials* 26(15):2723–2732.

Bankar, Ashok, Bhagyashree Joshi, Ameeta Ravi Kumar, and Smita Zinjarde. 2010. "Banana peel extract mediated novel route for the synthesis of silver nanoparticles." *Colloids and Surfaces. Part A: Physicochemical and Engineering Aspects* 368(1):58–63. https://doi.org/10.1016/j.colsurfa.2010.07.024.

Basavegowda, Nagaraj, and Yong Rok Lee. 2013. "Synthesis of silver nanoparticles using Satsuma mandarin (Citrus unshiu) peel extract: A novel approach towards waste utilization." *Materials Letters* 109:31–33. https://doi.org/10.1016/j.matlet.2013.07.039.

Brayner, Roberta, Roselyne Ferrari-Iliou, Nicolas Brivois, Shakib Djediat, Marc F. Benedetti, and Fernand Fiévet. 2006. "Toxicological impact studies based on *Escherichia coli* bacteria in ultrafine ZnO nanoparticles colloidal medium." *Nano Letters* 6(4):866–870.

Buzby, Jean C., Hodan Farah-Wells, and Jeffrey Hyman. 2014. "The estimated amount, value, and calories of postharvest food losses at the retail and consumer levels in the United States." *USDA-ERS Economic Information Bulletin* 121.

Bystrzejewska-Piotrowska, Grazyna, Jerzy Golimowski, and Pawel L. Urban. 2009. "Nanoparticles: Their potential toxicity, waste and environmental management." *Waste Management* 29(9):2587–2595. https://doi.org/10.1016/j.wasman.2009.04.001.

Cai, Weibo, and Xiaoyuan Chen. 2007. "Nanoplatforms for targeted molecular imaging in living subjects." *Small* 3(11):1840–1854.

Cai, Weibo, Ting Gao, Hao Hong, and Jiangtao Sun. 2008. "Applications of gold nanoparticles in cancer nanotechnology." *Nanotechnology, Science and Applications* 1:17.

Calamak, Semih, Reza Shahbazi, Ipek Eroglu, Merve Gultekinoglu, and Kezban Ulubayram. 2017. "An overview of nanofiber-based antibacterial drug design." *Expert Opinion on Drug Discovery* 12(4):391–406.

Castro, Laura, M. Luisa Blázquez, Jesús A. Muñoz, Felisa González, C. García-Balboa, and A. Ballester. 2011. "Biosynthesis of gold nanowires using sugar beet pulp." *Process Biochemistry* 46(5):1076–1082.

Cha, Kyungeun, Hye-Won Hong, Yeon-Gil Choi, Min Joo Lee, Jong Hoon Park, Hee-Kwon Chae, Gyuha Ryu, and Heejoon Myung. 2008. "Comparison of acute responses of mice livers to short-term exposure to nano-sized or micro-sized silver particles." *Biotechnology Letters* 30(11):1893–1899.

Chanasattru, Wanlop, Owen Griffith Jones, Eric A. Decker, and David Julian McClements. 2009. "Impact of cosolvents on formation and properties of biopolymer nanoparticles formed by heat treatment of β-lactoglobulin–pectin complexes." *Food Hydrocolloids* 23(8):2450–2457.

Chandran, S. Prathap, Minakshi Chaudhary, Renu Pasricha, Absar Ahmad, and Murali Sastry. 2006. "Synthesis of gold nanotriangles and silver nanoparticles using Aloevera plant extract." *Biotechnology Progress* 22(2):577–583.

Chen, Lingyun, Gabriel E. Remondetto, and Muriel Subirade. 2006. "Food protein-based materials as nutraceutical delivery systems." *Trends in Food Science and Technology* 17(5):272–283.

Cherian, Bibin Mathew, Alcides Lopes Leão, Sivoney Ferreira de Souza, Ligia Maria Manzine Costa, Gabriel Molina de Olyveira, M. Kottaisamy, E. R. Nagarajan, and Sabu Thomas. 2011. "Cellulose nanocomposites with nanofibres isolated from pineapple leaf fibers for medical applications." *Carbohydrate Polymers* 86(4):1790–1798.

Chinga-Carrasco, Gary, Nanci V. Ehman, Daniel Filgueira, Jenny Johansson, María E. Vallejos, Fernando E. Felissia, Joakim Håkansson, and María C. Area. 2019. "Bagasse—A major agro-industrial residue as potential resource for nanocellulose inks for 3D printing of wound dressing devices." *Additive Manufacturing* 28:267–274.

Coccia, Francesca, Lucia Tonucci, Domenico Bosco, Mario Bressan, and Nicola d'Alessandro. 2012. "One-pot synthesis of lignin-stabilised platinum and palladium nanoparticles and their catalytic behaviour in oxidation and reduction reactions." *Green Chemistry* 14(4):1073–1078.

Corrado, Iolanda, Manar Abdalrazeq, Cinzia Pezzella, Rocco Di Girolamo, Raffaele Porta, Giovanni Sannia, C. Valeria, and L. Giosafatto. 2021. "Design and characterization of poly (3-hydroxybutyrate-co-hydroxyhexanoate) nanoparticles and their grafting in whey protein-based nanocomposites." *Food Hydrocolloids* 110:106167.

Dai, Limin, Changwei Li, Jun Zhang, and Fang Cheng. 2018. "Preparation and characterization of starch nanocrystals combining ball milling with acid hydrolysis." *Carbohydrate Polymers* 180:122–127.

Dameron, C. T., R. N. Reese, R. K. Mehra, A. R. Kortan, P. J. Carroll, M. L. Steigerwald, L. E. Brus, and D. R. Winge. 1989. "Biosynthesis of cadmium sulphide quantum semiconductor crystallites." *Nature* 338(6216):596–597. https://doi.org/10.1038/338596a0.

Das, Ratul Kumar, Nayanmoni Gogoi, and Utpal Bora. 2011. "Green synthesis of gold nanoparticles using *Nyctanthes arbortristis* flower extract." *Bioprocess and Biosystems Engineering* 34(5):615–619. https://doi.org/10.1007/s00449-010-0510-y.

Das Purkayastha, Manashi, Ajay Kumar Manhar, Manabendra Mandal, and Charu Lata Mahanta. 2014. "Industrial waste-derived nanoparticles and microspheres can be potent antimicrobial and functional ingredients." *Journal of Applied Chemistry*, 1–12..

Denkbaş, Emir Baki, Ekin Celik, Ebru Erdal, Doğa Kavaz, Öznur Akbal, Göknur Kara, and Cem Bayram. 2016. "Magnetically based nanocarriers in drug delivery." In: *Nanobiomaterials in Drug Delivery*, 285–331. Elsevier.

Divya, K., and M. S. Jisha. 2018. "Chitosan nanoparticles preparation and applications." *Environmental Chemistry Letters* 16(1):101–112.

Duclairoir, C., A.-M. Orecchioni, P. Depraetere, F. Osterstock, and E. Nakache. 2003. "Evaluation of gliadins nanoparticles as drug delivery systems: A study of three different drugs." *International Journal of Pharmaceutics* 253(1–2):133–144.

Ealia, S. Anu Mary, and M. P. Saravanakumar. 2017. "A review on the classification, characterisation, synthesis of nanoparticles and their application." IOP Conference Series. *Materials Science and Engineering*, 263(3): 1–15

Ersen, O., I. Florea, C. Hirlimann, and C. Pham-Huu. 2015. "Exploring nanomaterials with 3D electron microscopy." *Materials Today* 18(7):395–408.

Fahma, Farah, Shinichiro Iwamoto, Naruhito Hori, Tadahisa Iwata, and Akio Takemura. 2010. "Isolation, preparation, and characterization of nanofibers from oil palm empty-fruit-bunch (OPEFB)." *Cellulose* 17(5):977–985.

Gaillet, Sylvie, and Jean-Max Rouanet. 2015. "Silver nanoparticles: Their potential toxic effects after oral exposure and underlying mechanisms–a review." *Food and Chemical Toxicology* 77:58–63.

García, Araceli, Alessandro Gandini, Jalel Labidi, Naceur Belgacem, and Julien Bras. 2016. "Industrial and crop wastes: A new source for nanocellulose biorefinery." *Industrial Crops and Products* 93:26–38.

Gardea-Torresdey, Jorge L., Eduardo Gomez, Jose R. Peralta-Videa, Jason G. Parsons, Horacio Troiani, and Miguel Jose-Yacaman. 2003. "Alfalfa sprouts: A natural source for the synthesis of silver nanoparticles." *Langmuir* 19(4):1357–1361.

Gawande, Manoj B., Anandarup Goswami, François-Xavier Felpin, Tewodros Asefa, Xiaoxi Huang, Rafael Silva, Xiaoxin Zou, Radek Zboril, and Rajender S. Varma. 2016. "Cu and Cu-based nanoparticles: Synthesis and applications in catalysis." *Chemical Reviews* 116(6):3722–3811.

Ghodake, G. S., N. G. Deshpande, Y. P. Lee, and E. S. Jin. 2010. "Pear fruit extract-assisted room-temperature biosynthesis of gold nanoplates." *Colloids and Surfaces, Part B: Biointerfaces* 75(2):584–589. https://doi.org/10.1016/j.colsurfb.2009.09.040.

Ghosh, Purabi R., Derek Fawcett, Shashi B. Sharma, and Gerrard E. J. Poinern. 2017. "Production of high-value nanoparticles via biogenic processes using aquacultural and horticultural food waste." *Materials* 10(8):852.

Gond, R. K., and M. K. Gupta. 2020. "A novel approach for isolation of nanofibers from sugarcane bagasse and its characterization for packaging applications." *Polymer Composites* 41(12):5216–5226.

Gopidas, Karical R., James K. Whitesell, and Marye Anne Fox. 2003. "Synthesis, characterization, and catalytic applications of a palladium-nanoparticle-cored dendrimer." *Nano Letters* 3(12):1757–1760.

Gouadec, Gwénaël, and Philippe Colomban. 2007. "Raman spectroscopy of nanomaterials: How spectra relate to disorder, particle size and mechanical properties." *Progress in Crystal Growth and Characterization of Materials* 53(1):1–56.

Grenha, Ana. 2012. "Chitosan nanoparticles: A survey of preparation methods." *Journal of Drug Targeting* 20(4):291–300.

Gülseren, İbrahim, Yuan Fang, and Milena Corredig. 2012. "Complexation of high methoxyl pectin with ethanol desolvated whey protein nanoparticles: Physico-chemical properties and encapsulation behaviour." *Food and Function* 3(8):859–866.

Ha, Ho-Kyung, Jin Wook Kim, Mee-Ryung Lee, Woojin Jun, and Won-Jae Lee. 2015. "Cellular uptake and cytotoxicity of β-lactoglobulin nanoparticles: The effects of particle size and surface charge." *Asian-Australasian Journal of Animal Sciences* 28(3):420.

Haaj, Sihem Bel, Albert Magnin, Christian Pétrier, and Sami Boufi. 2013. "Starch nanoparticles formation via high power ultrasonication." *Carbohydrate Polymers* 92(2):1625–1632.

Hao, Yacheng, Yun Chen, Qian Li, and Qunyu Gao. 2018. "Preparation of starch nanocrystals through enzymatic pretreatment from waxy potato starch." *Carbohydrate Polymers* 184:171–177.

Harini, K., and C. Chandra Mohan. 2020. "Isolation and characterization of micro and nanocrystalline cellulose fibers from the walnut shell, corncob and sugarcane bagasse." *International Journal of Biological Macromolecules* 163:1375–1383.

Harini, K., K. Ramya, and M. Sukumar. 2018. "Extraction of nano cellulose fibers from the banana peel and bract for production of acetyl and lauroyl cellulose." *Carbohydrate Polymers* 201:329–339.

Hassan, Mohammad L., Aji P. Mathew, Enas A. Hassan, Nahla A. El-Wakil, and Kristiina Oksman. 2012. "Nanofibers from bagasse and rice straw: Process optimization and properties." *Wood Science and Technology* 46(1–3):193–205.

Hendrickson, Olga D., Sergey G. Klochkov, Oksana V. Novikova, Irina M. Bravova, Elena F. Shevtsova, Irina V. Safenkova, Anatoly V. Zherdev, Sergey O. Bachurin, and Boris B Dzantiev. 2016. "Toxicity of nanosilver in intragastric studies: Biodistribution and metabolic effects." *Toxicology Letters* 241:184–192.

Henriksson, Marielle, Gunnar Henriksson, L. A. Berglund, and Tom Lindström. 2007. "An environmentally friendly method for enzyme-assisted preparation of microfibrillated cellulose (MFC) nanofibers." *European Polymer Journal* 43(8):3434–3441.

Hiasa, Shou, Shinichiro Iwamoto, Takashi Endo, and Yusuke Edashige. 2014. "Isolation of cellulose nanofibrils from mandarin (Citrus unshiu) peel waste." *Industrial Crops and Products* 62:280–285.

Hideno, Akihiro, Kentaro Abe, and Hiroyuki Yano. 2014. "Preparation using pectinase and characterization of nanofibers from orange peel waste in juice factories." *Journal of Food Science* 79(6):N1218–N1224.

Hirsch, L. R., J. B. Jackson, A. Lee, N. J. Halas, and J. L. West. 2003. "A whole blood immunoassay using gold nanoshells." *Analytical Chemistry* 75(10):2377–2381.

Homaeigohar, Shahin, and Aldo R. Boccaccini. 2020. "Antibacterial biohybrid nanofibers for wound dressings." *Acta Biomaterialia* 107:25–49.

Hoyt, Vincent W., and Eileen Mason. 2008. "Nanotechnology: Emerging health issues." *Journal of Chemical Health and Safety* 15(2):10–15.

Hrapovic, Sabahudin, Yali Liu, Keith B. Male, and John H. T. Luong. 2004. "Electrochemical biosensing platforms using platinum nanoparticles and carbon nanotubes." *Analytical Chemistry* 76(4):1083–1088.

Hu, Daode, Changchun Lin, Liang Liu, Sining Li, and Yaping Zhao. 2012. "Preparation, characterization, and in vitro release investigation of lutein/zein nanoparticles via solution enhanced dispersion by supercritical fluids." *Journal of Food Engineering* 109(3):545–552.

Hu, Kun, and David Julian McClements. 2015. "Fabrication of biopolymer nanoparticles by antisolvent precipitation and electrostatic deposition: Zein-alginate core/shell nanoparticles." *Food Hydrocolloids* 44:101–108.

Hu, Min, D. Julian McClements, and Eric A. Decker. 2003. "Lipid oxidation in corn oil-in-water emulsions stabilized by casein, whey protein isolate, and soy protein isolate." *Journal of Agricultural and Food Chemistry* 51(6):1696–1700.

Hu, Xiuli, Shi Liu, Guangyuan Zhou, Yubin Huang, Zhigang Xie, and Xiabin Jing. 2014. "Electrospinning of polymeric nanofibers for drug delivery applications." *Journal of Controlled Release* 185:12–21.

Huang, Jiale, Qingbiao Li, Daohua Sun, Yinghua Lu, Yuanbo Su, Xin Yang, Huixuan Wang, Yuanpeng Wang, Wenyao Shao, and Ning He. 2007. "Biosynthesis of silver and gold nanoparticles by novel sundried Cinnamomum camphora leaf." *Nanotechnology* 18(10):105104.

Huang, Yuesheng, Xiaolu Li, Zhenjiang Liao, Guoan Zhang, Qun Liu, Jin Tang, Yizhi Peng, Xuesheng Liu, and Qizhi Luo. 2007. "A randomized comparative trial between Acticoat and SD-Ag in the treatment of residual burn wounds, including safety analysis." *Burns* 33(2):161–166.

Husseiny, M. I., M. Abd El-Aziz, Y. Badr, and M. A. Mahmoud. 2007. "Biosynthesis of gold nanoparticles using Pseudomonas aeruginosa." *Spectrochimica Acta. Part A: Molecular and Biomolecular Spectroscopy* 67(3):1003–1006. https://doi.org/10.1016/j.saa.2006.09.028.

Isogai, Akira, Tsuguyuki Saito, and Hayaka Fukuzumi. 2011. "TEMPO-oxidized cellulose nanofibers." *Nanoscale* 3(1):71–85.

Jain, Devendra, H. Kumar Daima, Sumita Kachhwaha, and S. L. Kothari. 2009. "Synthesis of plant-mediated silver nanoparticles using papaya fruit extract and evaluation of their anti microbial activities." *Digest Journal of Nanomaterials and Biostructures* 4(3):557–563.

Jeong, Gil Nam, Un Bock Jo, Hyeon Yeol Ryu, Yong Soon Kim, Kyung Seuk Song, and Il Je Yu. 2010. "Histochemical study of intestinal mucins after administration of silver nanoparticles in Sprague–Dawley rats." *Archives of Toxicology* 84(1):63–69.

Jiang, Shan, Baohua Shan, Jian Ouyang, Wei Zhang, Xun Yu, Peigang Li, and Baoguo Han. 2018. "Rheological properties of cementitious composites with nano/fiber fillers." *Construction and Building Materials* 158:786–800.

Jiang, Shaohua, Yiming Chen, Gaigai Duan, Changtong Mei, Andreas Greiner, and Seema Agarwal. 2018. "Electrospun nanofiber reinforced composites: A review." *Polymer Chemistry* 9(20):2685–2720.

Jones, Owen G., and D. Julian McClements. 2008. "Stability of biopolymer particles formed by heat treatment of β-lactoglobulin/beet pectin electrostatic complexes." *Food Biophysics* 3(2):191–197.

Jorfi, Mehdi, and E. Johan Foster. 2015. "Recent advances in nanocellulose for biomedical applications." *Journal of Applied Polymer Science* 132(14).

Joye, Iris J., Veronique A. Nelis, and D. Julian McClements. 2015. "Gliadin-based nanoparticles: Stabilization by post-production polysaccharide coating." *Food Hydrocolloids* 43:236–242.

Kah, JamesChen Yong, Kiang Wei Kho, Caroline Guat Leng Lee, and Colin James Richard. 2007. "Early diagnosis of oral cancer based on the surface plasmon resonance of gold nanoparticles." *International Journal of Nanomedicine* 2(4):785.

Kahan, Dan M., Donald Braman, Paul Slovic, John Gastil, and Geoffrey Cohen. 2009. "Cultural cognition of the risks and benefits of nanotechnology." *Nature Nanotechnology* 4(2):87–90.

Kang, Xingya, Peipei Sun, Shigenori Kuga, Chao Wang, Yang Zhao, Min Wu, and Yong Huang. 2017. "Thin cellulose nanofiber from corncob cellulose and its performance in transparent nanopaper." *ACS Sustainable Chemistry and Engineering* 5(3):2529–2534.

Kannan, R. R. R., W. A. Stirk, and J. Van Staden. 2013. "Synthesis of silver nanoparticles using the seaweed Codium capitatum PC Silva (Chlorophyceae)." *South African Journal of Botany* 86:1–4.

Kargarzadeh, Hanieh, Marcos Mariano, Deepu Gopakumar, Ishak Ahmad, Sabu Thomas, Alain Dufresne, Jin Huang, and Ning Lin. 2018. "Advances in cellulose nanomaterials." *Cellulose* 25(4):2151–2189.

Kaviya, S., J. Santhanalakshmi, B. Viswanathan, J. Muthumary, and K. Srinivasan. 2011. "Biosynthesis of silver nanoparticles using citrus sinensis peel extract and its antibacterial activity." *Spectrochimica Acta. Part A: Molecular and Biomolecular Spectroscopy* 79(3):594–598. https://doi.org/10.1016/j.saa .2011.03.040.

Khalil, H. P. S. Abdul, Yalda Davoudpour, Md Nazrul Islam, Asniza Mustapha, Kumar Sudesh, Rudi Dungani, and Mohammad Jawaid. 2014. "Production and modification of nanofibrillated cellulose using various mechanical processes: A review." *Carbohydrate Polymers* 99:649–665.

Kim, Hee-Young, Sung Soo Park, and Seung-Taik Lim. 2015. "Preparation, characterization and utilization of starch nanoparticles." *Colloids and Surfaces B: Biointerfaces* 126:607–620.

Kim, Young-Pil, Eunkeu Oh, Mi-Young Hong, Dohoon Lee, Min-Kyu Han, Hyun Kyong Shon, Dae Won Moon, Hak-Sung Kim, and Tae Geol Lee. 2006. "Gold nanoparticle-enhanced secondary ion mass spectrometry imaging of peptides on self-assembled monolayers." *Analytical Chemistry* 78(6):1913–1920.

Klemm, Dieter, Friederike Kramer, Sebastian Moritz, Tom Lindström, Mikael Ankerfors, Derek Gray, and Annie Dorris. 2011. "Nanocelluloses: A new family of nature-based materials." *Angewandte Chemie International Edition* 50(24):5438–5466.

Kneipp, Katrin, Harald Kneipp, Irving Itzkan, Ramachandra R. Dasari, and Michael S. Feld. 1999. "Ultrasensitive chemical analysis by Raman spectroscopy." *Chemical Reviews* 99(10):2957–2976.

Konwarh, Rocktotpal, Biswajit Gogoi, Ruby Philip, M. A. Laskar, and Niranjan Karak. 2011. "Biomimetic preparation of polymer-supported free radical scavenging, cytocompatible and antimicrobial "green" silver nanoparticles using aqueous extract of Citrus sinensis peel." *Colloids and Surfaces B: Biointerfaces* 84(2):338–345. https://doi.org/10.1016/j.colsurfb.2011.01.024.

Krishnaswamy, Kiruba, Hojatollah Vali, and Valérie Orsat. 2014. "Value-adding to grape waste: Green synthesis of gold nanoparticles." *Journal of Food Engineering* 142:210–220.

Kulkarni, Narendra, and Uday Muddapur. 2014. "Biosynthesis of metal nanoparticles: A review." *Journal of Nanotechnology* 2014:510246. https://doi.org/10.1155/2014/510246.

Lakshmipathy, R., B. Palakshi Reddy, N. C. Sarada, K. Chidambaram, and Sk Khadeer Pasha. 2015. "Watermelon rind-mediated green synthesis of noble palladium nanoparticles: Catalytic application." *Applied Nanoscience* 5(2):223–228.

Le Corre, Déborah, and Hélène Angellier-Coussy. 2014. "Preparation and application of starch nanoparticles for nanocomposites: A review." *Reactive and Functional Polymers* 85:97–120.

Lee, H. Cheun, Wei-Wen Liu, Siang-Piao Chai, Abdul Rahman Mohamed, Azizan Aziz, Cheng-Seong Khe, N. M. S. Hidayah, and U. Hashim. 2017. "Review of the synthesis, transfer, characterization and growth mechanisms of single and multilayer graphene." *RSC Advances* 7(26):15644–15693.

Lee, Hyo-Jeoung, Geummi Lee, Na Ri Jang, Jung Hyun Yun, Jae Yong Song, and Beom Soo Kim. 2011. "Biological synthesis of copper nanoparticles using plant extract." *Nanotechnology* 1(1):371–374.

Leela, Arangasamy, and Munusamy Vivekanandan. 2008. "Tapping the unexploited plant resources for the synthesis of silver nanoparticles." *African Journal of Biotechnology* 7(17): 3162–3165.

Li, Fei, Haitao Zhang, Chunhu Gu, Li Fan, Youbei Qiao, Yangchun Tao, Chong Cheng, Hong Wu, and Jun Yi. 2013. "Self-assembled nanoparticles from folate-decorated maleilated pullulan–doxorubicin conjugate for improved drug delivery to cancer cells." *Polymer International* 62(2):165–171.

Lin, Han, Yu Chen, and Jianlin Shi. 2018. "Nanoparticle-triggered in situ catalytic chemical reactions for tumour-specific therapy." *Chemical Society Reviews* 47(6):1938–1958.

Lin, Wenbin. 2015. "Introduction: Nanoparticles in medicine." *Chemical Reviews* 115(19):10407–10409.

Liu, Wanyun, Junchao Wei, Yiwang Chen, Ping Huo, and Yen Wei. 2013. "Electrospinning of poly (L-lactide) nanofibers encapsulated with water-soluble fullerenes for bioimaging application." *ACS Applied Materials and Interfaces* 5(3):680–685.

Liu, Wen-Tso. 2006. "Nanoparticles and their biological and environmental applications." *Journal of Bioscience and Bioengineering* 102(1):1–7. https://doi.org/10.1263/jbb.102.1.

Lunge, Sneha, Shripal Singh, and Amalendu Sinha. 2014. "Magnetic iron oxide (Fe3O4) nanoparticles from tea waste for arsenic removal." *Journal of Magnetism and Magnetic Materials* 356:21–31.

Ma, Xiaofei, Ruijuan Jian, Peter R. Chang, and Jiugao Yu. 2008. "Fabrication and characterization of citric acid-modified starch nanoparticles/plasticized-starch composites." *Biomacromolecules* 9(11):3314–3320.

Maensiri, S., P. Laokul, J. Klinkaewnarong, S. Phokha, V. Promarak, and Supapan Seraphin. 2008. "Indium oxide (In2O3) nanoparticles using Aloe vera plant extract: Synthesis and optical properties." *Journal of Optoelectronics and Advanced Materials* 10(3):161–165.

Malhotra, Ankit, Nishat Sharma, Navinder Kumar, Kunzes Dolma, Deepak Sharma, Hemraj S. Nandanwar, and Anirban Roy Choudhury. 2014. "Multi-analytical approach to understand biomineralization of gold using rice bran: A novel and economical route." *RSC Advances* 4(74):39484–39490.

Mansfield, Elisabeth, Debra L. Kaiser, Daisuke Fujita, and Marcel Van de Voorde. 2017. *Metrology and Standardization for Nanotechnology: Protocols and Industrial Innovations.* John Wiley & Sons.

McClements, David Julian, and Jiajia Rao. 2011. "Food-grade nanoemulsions: Formulation, fabrication, properties, performance, biological fate, and potential toxicity." *Critical Reviews in Food Science and Nutrition* 51(4):285–330.

McClements, David Julian, and Hang Xiao. 2017. "Is Nano safe in foods? Establishing the factors impacting the gastrointestinal fate and toxicity of organic and inorganic food-grade nanoparticles." *NPJ Science of Food* 1(1):1–13.

Merzlyak, Anna, and Seung-Wuk Lee. 2006. "Phage as templates for hybrid materials and mediators for nanomaterial synthesis." *Current Opinion in Chemical Biology* 10(3):246–252. https://doi.org/10.1016/j.cbpa.2006.04.008.

Minakawa, Alyne F. K., Paula C. S. Faria-Tischer, and Suzana Mali. 2019. "Simple ultrasound method to obtain starch micro-and nanoparticles from cassava, corn and yam starches." *Food Chemistry* 283:11–18.

Mittal, Amit Kumar, Abhishek Kaler, and Uttam Chand Banerjee. 2012. "Free Radical scavenging and antioxidant activity of silver nanoparticles synthesized from flower extract of *Rhododendron dauricum*." *Nano Biomedicine and Engineering* 4(3): 118–124.

Mmola, Mokone, Marilize Le Roes-Hill, Kim Durrell, John J. Bolton, Nicole Sibuyi, Mervin E. Meyer, Denzil R. Beukes, and Edith Antunes. 2016. "Enhanced antimicrobial and anticancer activity of silver and gold nanoparticles synthesised using Sargassum incisifolium aqueous extracts." *Molecules* 21(12):1633.

Modena, Mario M., Bastian Rühle, Thomas P. Burg, and Stefan Wuttke. 2019. "Nanoparticle characterization: What to measure?" *Advanced Materials* 31(32):1901556.

Mohanpuria, Prashant, Nisha K. Rana, and Sudesh Kumar Yadav. 2008. "Biosynthesis of nanoparticles: Technological concepts and future applications." *Journal of Nanoparticle Research* 10(3):507–517. https://doi.org/10.1007/s11051-007-9275-x.

Mondal, Subhadip, Revathy Ravindren, Poushali Bhawal, Beomsu Shin, Sayan Ganguly, Changwoon Nah, and Narayan Ch Das. 2020. "Combination effect of carbon nanofiber and ketjen carbon black hybrid nanofillers on mechanical, electrical, and electromagnetic interference shielding properties of chlorinated polyethylene nanocomposites." *Composites Part B Engineering* 197:108071.

Morone, Piergiuseppe, Pasquale Marcello Falcone, Enrica Imbert, and Andrea Morone. 2018. "Does food sharing lead to food waste reduction? An experimental analysis to assess challenges and opportunities of a new consumption model." *Journal of Cleaner Production* 185:749–760.

Mude, Namrata, Avinash Ingle, Aniket Gade, and Mahendra Rai. 2009. "Synthesis of silver nanoparticles using callus extract of Carica papaya — A first report." *Journal of Plant Biochemistry and Biotechnology* 18(1):83–86. https://doi.org/10.1007/BF03263300.

Mukherjee, Priyabrata, Absar Ahmad, Deendayal Mandal, Satyajyoti Senapati, Sudhakar R. Sainkar, Mohammad I. Khan, Renu Parishcha, P. V. Ajaykumar, Mansoor Alam, Rajiv Kumar, and Murali Sastry. 2001. "Fungus-mediated synthesis of silver nanoparticles and their immobilization in the mycelial matrix: A novel biological approach to nanoparticle synthesis." *Nano Letters* 1(10):515–519. https://doi.org/10.1021/nl0155274.

Nadagouda, Mallikarjuna N., and Rajender S. Varma. 2008. "Green synthesis of silver and palladium nanoparticles at room temperature using coffee and tea extract." *Green Chemistry* 10(8):859–862.

Narayanan, K. Badri, and N. Sakthivel. 2008. "Coriander leaf mediated biosynthesis of gold nanoparticles." *Materials Letters* 62(30):4588–4590. https://doi.org/10.1016/j.matlet.2008.08.044.

Ndikau, Michael, Naumih M. Noah, Dickson M. Andala, and Eric Masika. 2017. "Green synthesis and characterization of silver nanoparticles using Citrullus lanatus fruit rind extract." *International Journal of Analytical Chemistry* 2017:8108504. https://doi.org/10.1155/2017/8108504.

Nicklas, Martina, Wolfgang Schatton, Sascha Heinemann, Thomas Hanke, and Jörg Kreuter. 2009. "Preparation and characterization of marine sponge collagen nanoparticles and employment for the transdermal delivery of 17β-estradiol-hemihydrate." *Drug Development and Industrial Pharmacy* 35(9):1035–1042.

Nie, Shuangxi, Kun Zhang, Xuejiao Lin, Chenyuan Zhang, Depeng Yan, Hongming Liang, and Shuangfei Wang. 2018. "Enzymatic pretreatment for the improvement of dispersion and film properties of cellulose nanofibrils." *Carbohydrate Polymers* 181:1136–1142.

Njagi, Eric C., Hui Huang, Lisa Stafford, Homer Genuino, Hugo M. Galindo, John B. Collins, George E. Hoag, and Steven L. Suib. 2011. "Biosynthesis of iron and silver nanoparticles at room temperature using aqueous sorghum bran extracts." *Langmuir* 27(1):264–271. https://doi.org/10.1021/la103190n.

Oberdörster, Günter, Vicki Stone, and Ken Donaldson. 2007. "Toxicology of nanoparticles: A historical perspective." *Nanotoxicology* 1(1):2–25. https://doi.org/10.1080/17435390701314761.

Pan, Kang, and Qixin Zhong. 2016. "Organic nanoparticles in foods: Fabrication, characterization, and utilization." *Annual Review of Food Science and Technology* 7:245–266.

Pandey, Sadanand, Corli De Klerk, Joonwoo Kim, Misook Kang, and Elvis Fosso-Kankeu. 2020. "Eco friendly approach for synthesis, characterization and biological activities of milk protein stabilized silver nanoparticles." *Polymers* 12(6):1418.

Pankhurst, Quentin A., J Connolly, Stephen K. Jones, and J. Dobson. 2003. "Applications of magnetic nanoparticles in biomedicine." *Journal of Physics. Part D: Applied Physics* 36(13):R167.

Panouille, Maud, M.-C. Ralet, Estelle Bonnin, and J.-F. Thibault. 2007. "Recovery and reuse of trimmings and pulps from fruit and vegetable processing." In: *Handbook of Waste Management and Co-Product Recovery in Food Processing*, 417–447. Elsevier.

Parfitt, Julian, Mark Barthel, and Sarah Macnaughton. 2010. "Food waste within food supply chains: Quantification and potential for change to 2050." *Philosophical Transactions of the Royal Society of London Series B: Biological Sciences* 365(1554):3065–3081.

Patel, Ashok R., and Krassimir P. Velikov. 2014. "Zein as a source of functional colloidal Nano-and microstructures." *Current Opinion in Colloid and Interface Science* 19(5):450–458.

Patra, Jayanta Kumar, and Kwang-Hyun Baek. 2015. "Novel green synthesis of gold nanoparticles using Citrullus lanatus rind and investigation of proteasome inhibitory activity, antibacterial, and antioxidant potential." *International Journal of Nanomedicine* 10:7253–7264. https://doi.org/10.2147/IJN.S95483.

Patra, Jayanta Kumar, Yongseok Kwon, and Kwang-Hyun Baek. 2016. "Green biosynthesis of gold nanoparticles by onion peel extract: Synthesis, characterization and biological activities." *Advanced Powder Technology* 27(5):2204–2213. https://doi.org/10.1016/j.apt.2016.08.005.

Pattanayak, M., T. Muralikrishnan, and P. L. Nayak. 2014. "Green Synthesis of gold nanoparticles using Daucus carota (carrot) aqueous extract." *World Journal of Nano Science and Technology* 3:52–58.

Pavia, Donald L., Gary M. Lampman, George S. Kriz, and James R. Vyvyan. 2009. *Introduction to Spectroscopy*. BROOKS/COLE Cengage Learning.

Pawelczyk, Adam. 2005. "EU Policy and Legislation on recycling of organic wastes to agriculture." *ISAH* 1:64–71.

Peng, Hailong, Zhaodi Gan, Hua Xiong, Mei Luo, Ningxiang Yu, Tao Wen, Ronghui Wang, and Yanbin Li. 2017. "Self-assembly of protein nanoparticles from rice bran waste and their use as delivery system for curcumin." *ACS Sustainable Chemistry and Engineering* 5(8):6605–6614.

Peng, Yi-Yang, Shruti Srinivas, and Ravin Narain. 2020. "Modification of polymers." *Polymer Science and Nanotechnology: Fundamentals and Applications* 95, 104

Pennells, Jordan, Ian D. Godwin, Nasim Amiralian, and Darren J. Martin. 2020. "Trends in the production of cellulose nanofibers from non-wood sources." *Cellulose* 27(2):575–593.

Pérez-Rodríguez, Noelia, Diana García-Bernet, and J. M. Domínguez. 2016. "Effects of enzymatic hydrolysis and ultrasounds pretreatments on corn cob and vine trimming shoots for biogas production." *Bioresource Technology* 221:130–138.

Philip, Daizy. 2010. "Green synthesis of gold and silver nanoparticles using Hibiscus rosa sinensis." *Physica E: Low-Dimensional Systems and Nanostructures* 42(5):1417–1424. https://doi.org/10.1016/j.physe.2009.11.081.

Philip, Daizy. 2011. "Mangifera indica leaf-assisted biosynthesis of well-dispersed silver nanoparticles." *Spectrochimica Acta. Part A: Molecular and Biomolecular Spectroscopy* 78(1):327–331.

Poinern, Gérrard Eddy Jai, Peter Chapman, Xuan Le, and Derek Fawcett. 2013. "Green biosynthesis of gold nanometre scale plates using the leaf extracts from an Indigenous Australian plant Eucalyptus macrocarpa." *Gold Bulletin* 46(3):165–173.

Poinern, Gérrard Eddy Jai, Peter Chapman, Monaliben Shah, and Derek Fawcett. 2013. "Green biosynthesis of silver nanocubes using the leaf extracts from eucalyptus macrocarpa." *Nano Bulletin* 2(1): 130101–130107

Prasad, Kumar Suranjit, Hirnee Patel, Tirtha Patel, Khusbu Patel, and Kaliaperumal Selvaraj. 2013. "Biosynthesis of Se nanoparticles and its effect on UV-induced DNA damage." *Colloids and Surfaces B: Biointerfaces* 103:261–266.

Prasad, Tollamadugu N. V. K. V., Venkata Subba Rao Kambala, and Ravi Naidu. 2013. "Phyconanotechnology: Synthesis of silver nanoparticles using brown marine algae Cystophora moniliformis and their characterisation." *Journal of Applied Phycology* 25(1):177–182.

Praveen, G., P. R. Sreerekha, Deepthy Menon, Shantikumar V. Nair, and Krishna Prasad Chennazhi. 2012. "Fibrin nanoconstructs: A novel processing method and their use as controlled delivery agents." *Nanotechnology* 23(9):095102.

Puvanakrishnan, Priyaveena, Jaesook Park, Deyali Chatterjee, Sunil Krishnan, and James W. Tunnell. 2012. "In vivo tumor targeting of gold nanoparticles: Effect of particle type and dosing strategy." *International Journal of Nanomedicine* 7:1251.

Qiu, Chao, Yao Hu, Zhengyu Jin, David Julian McClements, Yang Qin, Xueming Xu, and Jinpeng Wang. 2019. "A review of green techniques for the synthesis of size-controlled starch-based nanoparticles and their applications as nanodelivery systems." *Trends in Food Science and Technology* 92:138–151.

Qiu, Chao, Chenxi Wang, Chen Gong, David Julian McClements, Zhengyu Jin, and Jinpeng Wang. 2020. "Advances in research on preparation, characterization, interaction with proteins, digestion and delivery systems of starch-based nanoparticles." *International Journal of Biological Macromolecules* 152:117–125.

Quiñones, Rosalynn, Deben Shoup, Grayce Behnke, Cynthia Peck, Sushant Agarwal, Rakesh K. Gupta, Jonathan W. Fagan, Karl T. Mueller, Robbie J. Iuliucci, and Qiang Wang. 2017. "Study of perfluorophosphonic acid surface modifications on zinc oxide nanoparticles." *Materials* 10(12):1363.

Rajathi, F., Arockiya Aarthi, C. Parthiban, V. Ganesh Kumar, and P. Anantharaman. 2012. "Biosynthesis of antibacterial gold nanoparticles using brown alga, Stoechospermum marginatum (kützing)." *Spectrochimica Acta. Part A: Molecular and Biomolecular Spectroscopy* 99:166–173.

Rajeshkumar, Shanmugam, Chelladurai Malarkodi, Gnanadhas Gnanajobitha, Kanniah Paulkumar, Mahendran Vanaja, Chellapandian Kannan, and Gurusamy Annadurai. 2013. "Seaweed-mediated synthesis of gold nanoparticles using Turbinaria Conoides and its characterization." *Journal of Nanostructure in Chemistry* 3(1):1–7.

Ramkumar, Vijayan Sri, Arivalagan Pugazhendhi, Kumar Gopalakrishnan, Periyasamy Sivagurunathan, Ganesh Dattatraya Saratale, Thi Ngoc Bao Dung, and Ethiraj Kannapiran. 2017. "Biofabrication and characterization of silver nanoparticles using aqueous extract of seaweed Enteromorpha compressa and its biomedical properties." *Biotechnology Reports* 14:1–7.

Ravikumar, K. V. G., Shruthi Vathaluru Sudakaran, K. Ravichandran, Mrudula Pulimi, Chandrasekaran Natarajan, and Amitava Mukherjee. 2019. "Green synthesis of NiFe Nano particles using Punica granatum peel extract for tetracycline removal." *Journal of Cleaner Production* 210:767–776.

Rejeski, David, and Deanna Lekas. 2008. "Nanotechnology field observations: Scouting the new industrial west." *Journal of Cleaner Production* 16(8):1014–1017. https://doi.org/10.1016/j.jclepro.2007.04.014.

Rossi, Marco, F. Cubadda, L. Dini, M. L. Terranova, F. Aureli, A. Sorbo, and D. Passeri. 2014. "Scientific basis of nanotechnology, implications for the food sector and future trends." *Trends in Food Science and Technology* 40(2):127–148.

Saito, Riichiro, and Hiromichi Kataura. 2001. "Optical properties and Raman spectroscopy of carbon nanotubes." *Carbon Nanotubes*:213–247.

Saleh, Tawfik A. 2020. "Nanomaterials: Classification, properties, and environmental toxicities." *Environmental Technology and Innovation* 101067.

Santipanichwong, R., M. Suphantharika, J. Weiss, and D. J. McClements. 2008. "Core-shell biopolymer nanoparticles produced by electrostatic deposition of beet pectin onto heat-denatured β -lactoglobulin aggregates." *Journal of Food Science* 73(6):N23–N30.

Saratale, Ganesh Dattatraya, Rijuta Ganesh Saratale, Giovanni Benelli, Gopalakrishnan Kumar, Arivalagan Pugazhendhi, Dong-Su Kim, and Han-Seung Shin. 2017. "Anti-diabetic potential of silver nanoparticles synthesized with Argyreia nervosa leaf extract high synergistic antibacterial activity with standard antibiotics against foodborne bacteria." *Journal of Cluster Science* 28(3):1709–1727.

Saratale, Rijuta Ganesh, Ganesh Dattatraya Saratale, Han Seung Shin, Jaya Mary Jacob, Arivalagan Pugazhendhi, Mukesh Bhaisare, and Gopalakrishanan Kumar. 2018. "New insights on the green synthesis of metallic nanoparticles using plant and waste biomaterials: Current knowledge, their agricultural and environmental applications." *Environmental Science and Pollution Research International* 25(11):10164–10183.

Satyavani, K., T. Ramanathan, and S. Gurudeeban. 2011. "Green synthesis of silver nanoparticles by using stem derived callus extract of bitter apple (Citrullus colocynthis)" Digest Journal of Nanomaterials and Biostructures 6(3):1019–1024.

Sengupta, Samadrita, Dipak K. Bhattacharyya, Riddhi Goswami, and Jayati Bhowal. 2019. "Emulsions stabilized by soy protein nanoparticles as potential functional non-dairy yogurts." *Journal of the Science of Food and Agriculture* 99(13):5808–5818.

Shah, Monaliben, Derek Fawcett, Shashi Sharma, Suraj Kumar Tripathy, and Gérrard Eddy Jai Poinern. 2015. "Green synthesis of metallic nanoparticles via biological entities." *Materials* 8(11):7278–7308.

Shahi, Naresh, Byungjin Min, Bedanga Sapkota, and Vijaya K. Rangari. 2020. "Eco-friendly cellulose nanofiber extraction from sugarcane bagasse and film fabrication." *Sustainability* 12(15):6015.

Shankar, S. Shiv, Akhilesh Rai, Absar Ahmad, and Murali Sastry. 2004. "Rapid synthesis of Au, Ag, and bimetallic Au core–Ag shell nanoparticles using Neem (Azadirachta indica) leaf broth." *Journal of Colloid and Interface Science* 275(2):496–502.

Shankar, Shiv, Lily Jaiswal, R. S. L. Aparna, and R. G. S. V. Prasad. 2014. "Synthesis, characterization, in vitro biocompatibility, and antimicrobial activity of gold, silver and gold silver alloy nanoparticles prepared from Lansium domesticum fruit peel extract." *Materials Letters* 137:75–78. https://doi.org/10.1016/j.matlet.2014.08.122.

Sharma, Banasree, Debraj Dhar Purkayastha, Subhenjit Hazra, Lohit Gogoi, Chira R. Bhattacharjee, Narendra Nath Ghosh, and Jayashree Rout. 2014. "Biosynthesis of gold nanoparticles using a freshwater green alga, Prasiola crispa." *Materials Letters* 116:94–97.

Sharma, Banasree, Debraj Dhar Purkayastha, Subhenjit Hazra, Moirangthem Thajamanbi, Chira R. Bhattacharjee, Narendra Nath Ghosh, and Jayashree Rout. 2014. "Biosynthesis of fluorescent gold nanoparticles using an edible freshwater red alga, Lemanea fluviatilis (L.) C. Ag. and antioxidant activity of biomatrix loaded nanoparticles." *Bioprocess and Biosystems Engineering* 37(12):2559–2565.

Sharma, Nilesh C, Shivendra V. Sahi, Sudip Nath, Jason G. Parsons, Jorge L. Gardea-Torresde, and Tarasankar Pal. 2007. "Synthesis of plant-mediated gold nanoparticles and catalytic role of biomatrix-embedded nanomaterials." *Environmental Science and Technology* 41(14):5137–5142.

Shukla, Rishi, and Munir Cheryan. 2001. "Zein: The industrial protein from corn." *Industrial Crops and Products* 13(3):171–192.

Singh, H. K., T. Patil, S. K. Vineeth, S. Das, A. Pramanik, and S. T. Mhaske. 2020. "Isolation of microcrystalline cellulose from corn stover with emphasis on its constituents: Corn cover and corn cob." *Materials Today: Proceedings* 27:589–594.

Singh, Sandeep, Narpinder Singh, Rajarathnam Ezekiel, and Amritpal Kaur. 2011. "Effects of gamma-irradiation on the morphological, structural, thermal and rheological properties of potato starches." *Carbohydrate Polymers* 83(4):1521–1528.

Siqueira, Gilberto, Kristiina Oksman, Sandra K. Tadokoro, and Aji P. Mathew. 2016. "Re-dispersible carrot nanofibers with high mechanical properties and reinforcing capacity for use in composite materials." *Composites Science and Technology* 123:49–56.

Song, Jae Yong, Hyeon-Kyeong Jang, and Beom Soo Kim. 2009. "Biological synthesis of gold nanoparticles using Magnolia kobus and Diopyros kaki leaf extracts." *Process Biochemistry* 44(10):1133–1138.

Song, Jae Yong, and Beom Soo Kim. 2008. "Biological synthesis of bimetallic Au/Ag nanoparticles using persimmon (Diopyros kaki) leaf extract." *Korean Journal of Chemical Engineering* 25(4):808–811.

Soni, Bhawna, and Barakat Mahmoud. 2015. "Chemical isolation and characterization of different cellulose nanofibers from cotton stalks." *Carbohydrate Polymers* 134:581–589.

Sperling, Ralph A., Pilar Rivera Gil, Feng Zhang, Marco Zanella, and Wolfgang J. Parak. 2008. "Biological applications of gold nanoparticles." *Chemical Society Reviews* 37(9):1896–1908.

Sriram, T., and V. Pandidurai. 2014. "Synthesis of silver nanoparticles from leaf extract of Psidium guajava and its antibacterial activity against pathogens." *International Journal of Current Microbiology and Applied Sciences* 3(3):146–152.

Suter, Franz, Daniel Schmid, Franziska Wandrey, and Fred Zülli. 2016. "Heptapeptide-loaded solid lipid nanoparticles for cosmetic anti-aging applications." *European Journal of Pharmaceutics and Biopharmaceutics* 108:304–309.

Sýkora, David, Václav Kašička, Ivan Mikšík, Pavel Řezanka, Kamil Záruba, Pavel Matějka, and Vladimír Král. 2010. "Application of gold nanoparticles in separation sciences." *Journal of Separation Science* 33(3):372–387.

Tibolla, Heloisa, Franciele Maria Pelissari, and Florencia Cecilia Menegalli. 2014. "Cellulose nanofibers produced from banana peel by chemical and enzymatic treatment." *LWT - Food Science and Technology* 59(2):1311–1318.

Tiyaboonchai, Waree. 2013. "Chitosan nanoparticles: A promising system for drug delivery." *Naresuan University Journal: Science and Technology (NUJST)* 11(3):51–66.

Torres-Chavolla, Edith, Romali J. Ranasinghe, and Evangelyn C. Alocilja. 2010. "Characterization and functionalization of biogenic gold nanoparticles for biosensing enhancement." *IEEE Transactions on Nanotechnology* 9(5):533–538.

Tseng, Wei-Lung, Ming-Feng Huang, Yu-Fen Huang, and Huan-Tsung Chang. 2005. "Nanoparticle-filled capillary electrophoresis for the separation of long DNA molecules in the presence of hydrodynamic and electrokinetic forces." *Electrophoresis* 26(16):3069–3075.

Vasyliev, Georgii, and Victoria Vorobyova. 2020. "Valorization of food waste to produce eco-friendly means of corrosion protection and "Green" synthesis of nanoparticles." *Advances in Materials Science and Engineering* 2020: 14

Verma, Madan L., B. Sukriti Dhanya, Varsha Rani, Meenu Thakur, J. Jeslin, and Rekha Kushwaha. 2020. "Carbohydrate and protein based biopolymeric nanoparticles: Current status and biotechnological applications." *International Journal of Biological Macromolecules* 154:390–412.

Vilchis-Nestor, Alfredo R., Victor Sánchez-Mendieta, Marco A. Camacho-López, Rosa M. Gómez-Espinosa, Miguel A. Camacho-López, and Jesús A. Arenas-Alatorre. 2008. "Solventless synthesis and optical properties of Au and Ag nanoparticles using Camellia sinensis extract." *Materials Letters* 62(17–18):3103–3105.

Waarts, Y. R., M. Eppink, E. B. Oosterkamp, S. R. C. H. Hiller, A. A. Van Der Sluis, and T. Timmermans. 2011. *Reducing Food Waste; Obstacles Experienced in Legislation and Regulations.* LEI, Part of Wageningen UR.

Wagner, Sylvia, Florian Rothweiler, Marion G. Anhorn, Daniel Sauer, Iris Riemann, Eike C. Weiss, Alisa Katsen-Globa, Martin Michaelis, Jindrich Cinatl Jr, and Daniel Schwartz. 2010. "Enhanced drug targeting by attachment of an anti αv integrin antibody to doxorubicin loaded human serum albumin nanoparticles." *Biomaterials* 31(8):2388–2398.

Wang, Guoxiu, Juan Yang, Jinsoo Park, Xinglong Gou, Bei Wang, Hao Liu, and Jane Yao. 2008. "Facile synthesis and characterization of graphene nanosheets." *The Journal of Physical Chemistry C* 112(22):8192–8195.

Wang, M., N. S. Hettiarachchy, M. Qi, W. Burks, and T. Siebenmorgen. 1999. "Preparation and functional properties of rice bran protein isolate." *Journal of Agricultural and Food Chemistry* 47(2):411–416.

Wang, Yu, Liguo Zhang, Wei Liu, Chen Cui, and Qingxi Hou. 2019. "Fabrication of optically transparent and strong nanopaper from cellulose nanofibril based on corncob residues." *Carbohydrate Polymers* 214:159–166.

Xie, Chenlu, Zhiqiang Niu, Dohyung Kim, Mufan Li, and Peidong Yang. 2019. "Surface and interface control in nanoparticle catalysis." *Chemical Reviews* 120(2):1184–1249.

Xu, Hui, Lei Wang, Hongyan Su, Liang Gu, Tingting Han, Fanbin Meng, and Chongchong Liu. 2015. "Making good use of food wastes: Green synthesis of highly stabilized silver nanoparticles from grape seed extract and their antimicrobial activity." *Food Biophysics* 10(1):12–18.

Xue, Jiajia, Jingwei Xie, Wenying Liu, and Younan Xia. 2017. "Electrospun nanofibers: New concepts, materials, and applications." *Accounts of Chemical Research* 50(8):1976–1987. https:doi.org/10.1021/acs.accounts.7b00218.

Yang, Fan, Yongfeng Li, Ting Liu, Kai Xu, Liqiang Zhang, Chunming Xu, and Jinsen Gao. 2013. "Plasma synthesis of Pd nanoparticles decorated-carbon nanotubes and its application in Suzuki reaction." *Chemical Engineering Journal* 226:52–58.

Yang, Jiaying, Hong Xu, Linping Zhang, Yi Zhong, Xiaofeng Sui, and Zhiping Mao. 2017. "Lasting superhydrophobicity and antibacterial activity of Cu nanoparticles immobilized on the surface of dopamine modified cotton fabrics." *Surface and Coatings Technology* 309:149–154.

Yang, Ning, Li WeiHong, and Lin Hao. 2014. "Biosynthesis of Au nanoparticles using agricultural waste mango peel extract and its in vitro cytotoxic effect on two normal cells." *Materials Letters* 134:67–70. https://doi.org/10.1016/j.matlet.2014.07.025.

Yang, Wengang, Tianyu Cheng, Yanhong Feng, Jinping Qu, Hezhi He, and Xingxing Yu. 2017. "Isolating cellulose nanofibers from steam-explosion pretreated corncobs using mild mechanochemical treatments." *BioResources* 12(4):9183–9197.

Yassin, Mohamed A., Abdul Aziz M. Gad, Ahmed F. Ghanem, and Mona H. Abdel Rehim. 2019. "Green synthesis of cellulose nanofibers using immobilized cellulase." *Carbohydrate Polymers* 205:255–260.

Zhang, Jing, Catherine J. Field, Donna Vine, and Lingyun Chen. 2015. "Intestinal uptake and transport of vitamin B 12-loaded soy protein nanoparticles." *Pharmaceutical Research* 32(4):1288–1303.

Zhang, Ye-Hui, Li-Hua Huang, and Zhen-Cheng Wei. 2014. "Effects of additional fibrils on structural and rheological properties of rice bran albumin solution and gel." *European Food Research and Technology* 239(6):971–978.

Zhou, Zhengping, and Xiang-Fa Wu. 2013. "Graphene-beaded carbon nanofibers for use in supercapacitor electrodes: Synthesis and electrochemical characterization." *Journal of Power Sources* 222:410–416.

Zhu, Lin, Ren-Ping Hu, Hai-Yan Wang, Yuan-Jing Wang, and Yu-Qing Zhang. 2011. "Bioconjugation of neutral protease on silk fibroin nanoparticles and application in the controllable hydrolysis of sericin." *Journal of Agricultural and Food Chemistry* 59(18):10298–10302.

Zhu, Shengdong, Yuanxin Wu, Qiming Chen, Ziniu Yu, Cunwen Wang, Shiwei Jin, Yigang Ding, and Gang Wu. 2006. "Dissolution of cellulose with ionic liquids and its application: A mini-review." *Green Chemistry* 8(4):325–327.

Zou, Tao, and Liwei Gu. 2013. "TPGS emulsified zein nanoparticles enhanced oral bioavailability of daidzin: In vitro characteristics and in vivo performance." *Molecular Pharmaceutics* 10(5):2062–2070.

14 Utilization of Agro-Industrial Wastes for Biofuel Generation

Rajeev K. Sukumaran, Meera Christopher,
AthiraRaj Sreeja-Raju, Meena Sankar,
Prajeesh Kooloth-Valappil, Valan Rebinro Gnanaraj,
Anoop Puthiyamadam, Reshma M. Mathew, and
Velayudhan Pillai-Prasannakumari Adarsh

CONTENTS

DOI: 10.1201/9781003125679-14

14.1 INTRODUCTION

Fossil fuels are regarded as the primary energy source for the world, and their over-exploitation has been highlighted as the primary cause of rising environmental pollution. Globally, countries are now conscious of replacing fossil fuels, which has shifted the focus to alternative energy sources such as biofuels (Patrizi et al., 2020), as it decreases crude oil use and dependency, therefore lowering greenhouse gas (GHG) emissions. The raw materials accessible from agricultural, forestry, and agro-industrial wastes are increasingly being used as energy resources through multiple routes of conversion. Prominent among these are the agro-industrial wastes, which can be defined as (i) residues generated in the field during harvest and preliminary processing of agricultural produce (e.g., straw, stubble); (ii) residues that are generated during processing of the agricultural harvest/ produce (e.g., husks, cane bagasse, peels), mostly at centralized facilities; and (iii) residues that are generated during factory/industry processing of the agricultural produce (e.g., fruit peels, oil cakes) (Figure 14.1). Both field residues/wastes (e.g., stems and stalks) and process residues (e.g., bagasse, husk, and molasses) can provide fermentable sugars, either directly or after processing, and can thus serve as feedstock for energy generation through a multitude of routes that include thermochemical conversion to form fuels like syngas and bio-oils, and biochemical conversion to bio-alcohols, bio-diesel, biohydrogen, biomethane, etc. The choice of a feedstock and the route for conversion is often decided by the biomass type and availability, available conversion technology, economics, demand, and a large number of other considerations.

One of the prime factors that determine the choice of biofuel to be produced and the technology for biomass conversion is the type, quantity, and availability of the feedstock/agro-residue in a geographical region. Often the largest available feedstock is crop residues generated in the field, which includes cereal straws, stover, and other lignocellulosic feedstocks. Such feedstock enjoys a universal availability, as their generation is linked to food/feed production in all cases, except dedicated energy crops. The worldwide production of lignocellulosic agro-residues was estimated at 110 billion tonnes annually (Pattanaik et al., 2019), a significant share of which is un-utilized and often burned in the field, leading to environmental issues.

From the major crops cultivated worldwide, up to 75% of the lignocellulosic biomass derived from the agricultural sector comes from three major plantations: rice, wheat, and maize (Xie and Peng, 2011). Apparently, there are advantages to using lignocellulosic biomass of such agro-residues that come from major food crops as feedstock for biofuels. Since the quantities available are enormous, there is no resource competition for land, water, and nutrients; as the crops are being grown for their food value, this makes the residues a bonus by-product, and hence there is no food versus fuel controversy. Challenges, therefore, in implementing agro-residue conversion to biofuels are mostly technical and process-related and, in some cases, on the supply chain and logistics of feedstock. Among the different fuel/energy options from the crop-residue lignocellulosic biomass, second-generation (2G) bioethanol is considered a superior option compared to non-renewable fossil

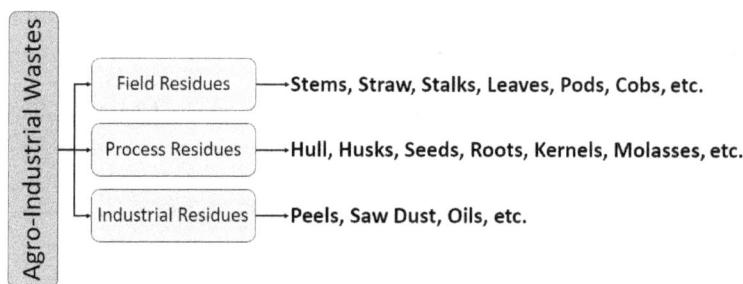

FIGURE 14.1 Classification of agro-industrial wastes.

fuels, and therefore its production and consumption are encouraged by global mandates. According to the Renewable Fuels Association's 2020 report, the worldwide production of bioethanol was estimated at 98.6 billion litres annually (Renewable Fuels Association, 2021).

The overall output of 2G bioethanol (2GE) is mostly determined by raw material availability. As per the Food and Agriculture Organization reports, worldwide rice production has risen to 760 million tonnes in recent years, which translates to approximately 1,400 million tonnes of rice straw (Satlewal et al., 2018). Similarly, the worldwide production of corn and wheat is estimated at 1.1 and 0.773 billion tonnes, respectively, per annum (Dukhnytskyi, 2019; Shahbandeh, 2021). In either case, standard conversion factors applied to these figures show that there is a huge availability of the residues coming from these crops as feedstock for biorefineries. Efficient use of these agricultural wastes might contribute to generating substantial amounts of renewable energy. Sustainable conversion of this biomass into energy/fuels requires efficient biorefineries.

Biorefineries may be driven by multiple routes of conversion ranging from direct and controlled combustion to enzymatic digestion followed by fermentation. Biomass to heat and power generation can also be considered on a small to medium scale as decentralized systems for bio-energy production (Abraham et al., 2016). Regardless of the routes, the biorefinery concept is futuristic as it has multiple deliverables from biomass conversion. This ability of biorefineries to deliver a variety of chemicals or products using different biomass conversion processes may lead to the most efficient utilization of all components of the feedstock and the most efficient operation, analogous to petroleum refineries. Biorefineries operating with multiple products are better in terms of sustainability and economics. System analyses studies by both Kaparaju et al. (2009) and Liu et al. (2011) for a wheat-straw-based biorefinery producing ethanol, succinic acid, and acetic acid as the major products and in which other byproducts (e.g., lignin) were used for steam and electricity generation, performed much better than a simple biomass-to-ethanol plant in estimated profits as well as efficiency and environmental sustainability.

This chapter discusses the appropriateness of agro-industrial waste as feedstock for biorefineries producing multiple value-added products, with a focus on 2GE production. With the world market for bioethanol projected to be 9.5 billion dollars by 2022 (Roy, 2017) and with a projected compound annual growth rate of 7.6%, this is among the more important fuel options for agro-industrial residues. This chapter also discusses, to a lesser extent, other possible fuel options and routes of conversion for agro-industrial wastes.

14.2 AGRO-RESIDUES TO BIOFUELS: TYPES OF FUELS AND THE ROUTES FOR BIOMASS CONVERSION

Agro-residues can be converted to different biofuels through appropriate processing technologies. Thermochemical conversion can yield bio-oils and syngas, whereas methane and biohydrogen can be produced using the anaerobic fermentation route; the bio-alcohols (bioethanol, biobutanol) follow the fermentation route, each having multiple steps depending on the process (Figure 14.2).

14.2.1 FUELS FROM BIOMASS THROUGH THE THERMOCHEMICAL ROUTE

Thermochemical conversion includes high-temperature chemical processes that result in bond breakage and reformation. This method is generally preferred for agro-wastes with a moisture content <30%, C/N ratio >30, and high lignin content (Pattanaik et al., 2019). Thermochemical conversions have become popular since they convert all the biomass components, thereby minimizing the cost and energy input of pre-treatment, increasing the speed of operation, and improving the availability of the infrastructure required for the process (Balagurumurthy et al., 2015). Thermochemical conversions for biofuel production generally follow the gasification, pyrolysis, or liquefaction routes, the former giving syngas and the latter two giving bio-oils.

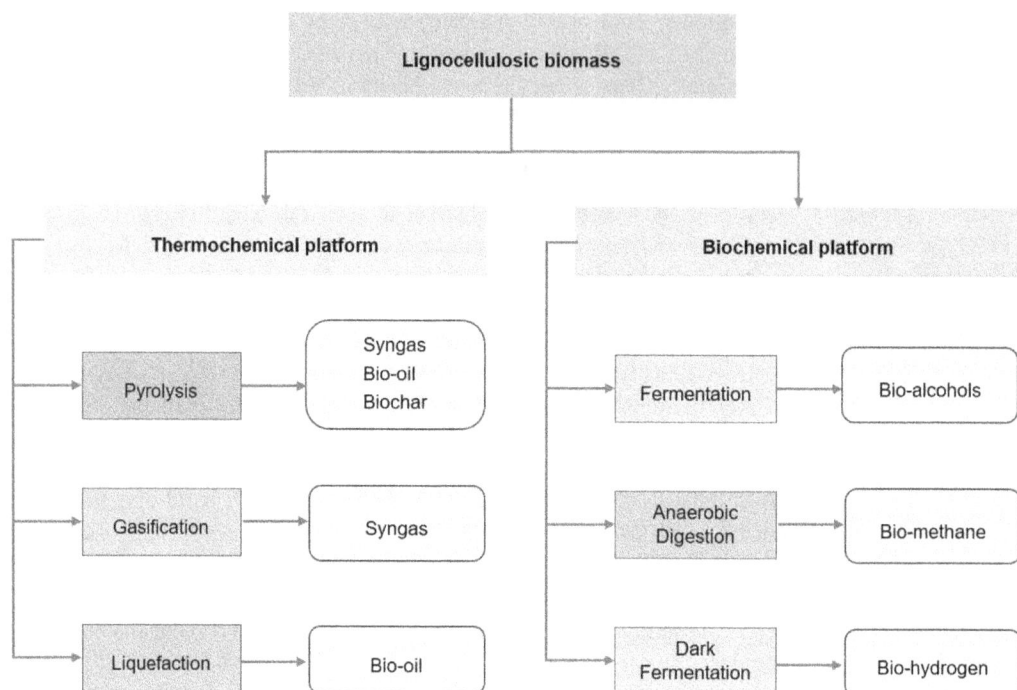

FIGURE 14.2 Main processes and their products in biomass to biofuels conversion.

14.2.1.1 Syngas

Syngas/synthesis gas, which is a mixture of carbon monoxide (30–60%), hydrogen (25–30%), carbon dioxide (5–15%), and methane (0–5%), is typically produced by gasification, in which the biomass is heated at a high temperature (800–1500°C) with steam (steam reforming), CO_2 (dry reforming), or oxygen (partial oxidation) (Ciferno and Marano, 2002). All the major agro-residues, like rice straw, wheat straw, corn straw, sugarcane bagasse, etc., can be used for generating syngas, and the composition of the gas varies according to the technology used for production. Some of the typical values for different agro-residues under different process conditions are given in Table 14.1.

Syngas can be used directly as a combustible fuel in power plants or as a replacement fuel for internal combustion engines, though it has much lesser energy density than gasoline (Boehman and Le Corre, 2008). In addition to syngas, the gasification platform can also be employed for the production of gaseous and liquid fuels (e.g., hydrogen, methane, hydrocarbons), as well as chemicals and power via a broad range of chemical and microbial processes (Molino et al., 2016). In fact, gasification is also considered the most efficient technique for the conversion of biomass to hydrogen (Lee et al., 2019). As per the International Energy Agency's data, there are 114 active biomass gasification projects worldwide and 15 idle plants, with a further 13 projects underway (IEA BioEnergy, 2018).

14.2.1.2 Bio-oils

Bio-oils, also known as pyrolysis oil or bio-crude, are synthetic fuels derived through pyrolysis, i.e., the thermal decomposition of biomass at temperatures ranging from 350°C to 550°C and in the absence of oxygen. Pyrolysis of biomass produces solid, liquid, and gaseous products. Bio-oils are dark brown liquids with high viscosity and are comprised of acids, alcohols, aldehydes, phenols, and oligomers (Rahman et al., 2018). At lower temperatures, pyrolysis results in a solid product known as biochar, and at higher temperatures, it results in gaseous products (Shafizadeh and Fu, 1973). At pyrolysis temperatures and oxygen concentrations significantly lower than that

TABLE 14.1
Composition of Syngas from Gasification of Different Agro-residues

Feedstock	Syngas Composition (% v/v)				Gasifier Type	Gasification Conditions	Reference
	H_2	CO	CO_2	CH_4			
Corn straw	48.5	33.9	12.2	5.3	N/A	GA: N/A T: 750–900°C ER: NA	Hu et al., 2019
Wheat straw	25.4	27.5	22.0	16.3	Fluidized bed gasifier	GA: Steam T: 600°C–710°C ER: N/A	Carpenter et al., 2010
Corn stover	26.9	24.7	23.7	15.3	Fluidized bed gasifier	GA: Steam T: 600–710°C ER: N/A	Carpenter et al., 2010
Hay	8.8	19.7	14.4	3.0	Fixed bed reactor	GA: Air T: 800°C ER: NA	Mikeska et al., 2020
Sugarcane bagasse	7.4–8.0	8.0–12.9	15.0–18.7	1.4–2.5	Cyclone gasifier	GA: Air T: 600–950°C ER: 0.18–0.25	Gabra et al., 2001

Note: GA: Gasifying Agent, T: Temperature, ER: Equivalence ratio, N/A: Data not available.

needed for combustion, biomass undergoes thermal decomposition to form pyrolysis oil, syngas, and biochar (Mohan et al., 2006). The ratio of these products is determined by the process conditions: at high temperatures (>500°C) and with a high heating rate (fast pyrolysis) in the absence of oxygen, the biomass vaporizes and then condenses to bio-oil, which can be up to 60% of the products (Mullen et al., 2010). In addition to fast pyrolysis, biomass may also be converted to bio-oil through liquefaction or gasification (Xu et al., 2011). Many upgradation processes are used to enhance bio-oil so that it can be used as an alternative for crude oil. The main advantage of bio-oils over gasification products is that they can be easily stored and transported (Dhyani and Bhaskar, 2018).

Bio-oils are also produced through liquefaction, which differs from gasification and pyrolysis in that it requires a liquid reaction medium during treatment at elevated temperatures and pressure (Huang et al., 2017). Liquefaction is often performed at relatively lower temperatures and high pressure in water; then, it is called hydrothermal liquefaction (HTL). HTL, when performed at lower temperatures (180°C–250°C), yields more solid products (biochar), while at temperatures above 370°C, the process yields mostly gases and is frequently referred to as hydrothermal gasification (HTG) or supercritical water gasification (SCWG) (Yang et al., 2020). HTL is typically performed at temperatures ranging from 250°C to 350°C and 5 to 35 MPa of operating pressure when bio-oils are the preferred product (Yang et al., 2020). HTL has the advantage of being able to work with high moisture content biomass, which enables the use of wet biomass as the substrate (Dimitriadis and Bezergianni, 2016) Bio-oils have unique physico-chemical properties, particularly remarkably high energy density, renewability, and sustainability, that makes it suitable as a renewable alternative fuel for burners, stationary diesel engines, turbines, and boilers (Duan et al., 2016; Tekin, 2015). However, bio-oils also have several undesirable properties, including high ash content, oxygen, high viscosity, high acidity (corrosivity), and low heating value. So, it often has to be upgraded to make it suitable as fuel. Common methods for upgradation include catalytic hydrogenation, catalytic cracking, steam reforming, emulsification, extraction, etc. More information on the upgradation processes may be obtained in Xu et al. (2011).

14.2.2 FUELS THROUGH ANAEROBIC FERMENTATION

14.2.2.1 Biomethane/Biogas

Biomethane/biogas is a mixture of gases that results from the anaerobic fermentation of organic matter. Anaerobic digestion (AD) is widely employed for the decomposition and stabilization of a range of organic wastes, including lignocellulosic biomass (LCB), and for the production of methane (Sayara and Sánchez, 2019). Methane fermentation or bio-methanation is a complex biological process composed of hydrolysis, acidogenesis, acetogenesis, and methanation. These phases of AD are performed by different microbial consortia, often interdependent and acting in succession (Chandra et al., 2012). The end product, commonly known as biogas, consists of methane (55–75%), carbon dioxide (25–45%), and small quantities of other gases (N_2, H_2, O_2, and H_2S). The leftover slurry has fertilizer value (Pattanaik et al., 2019).

Hydrolysis is the first of the phases where the complex biomass containing different polymeric components like cellulose, hemicellulose, lignin, proteins, and fats are degraded by microbial action to their simpler and often monomeric forms. This requires the concerted action of several microorganisms with lignocellulose hydrolyzing activity, besides those that produce enzymes that attack the complex polysaccharides, proteins, and lipids (Azman et al., 2015). Due to the inherent recalcitrance of LCB, hydrolysis is also the rate-limiting step of the bio-methanation process. As such, pre-treatment of the biomass prior to AD, and bioaugmentation, i.e., supplementing cellulolytic consortia/pure culture(s) is often practised for enhancing the efficiency of hydrolysis and thereby improving the yield of biogas (Mulat et al., 2018).

Acidogenesis follows the hydrolysis stage, in which the breakdown products from the hydrolysis stage (fatty acids, sugars, amino acids, etc.) are fermented by obligate or facultative anaerobes to short-chain organic acids (primarily volatile fatty acids like acetate, propionate, and butyrate), H_2, alcohols, NH_3, and CO_2 (Gerardi, 2003). In this step, the concentration of the H^+ ions formed is a critical determinant of the products of fermentation. Production of Volatile fatty acids (VFAs) is mostly accomplished by anaerobic oxidation reactions where reducing equivalents are directly transferred to hydrogen ions. The reaction is highly susceptible to H_2 partial pressures, and low ($<10^{-4}$ atmos) H_2 partial pressures are required for thermodynamic feasibility. In well-maintained anaerobic digesters, low H_2 partial pressures are maintained through hydrogenotrophic methanogens and homo-acetogens through interspecies hydrogen transfer (Pavlostathis, 2011).

The products of the acidogenic phase are oxidized to H_2, CO_2, and acetate in the acetogenic phase, which in turn form the substrates for the subsequent methanogenic phase. Acetogenesis and methanogenesis are highly interconnected through the syntrophic relationship between acetogenic and methanogenic bacteria, the latter being hydrogenotrophic. Acetate is the substrate for methane-forming bacteria, and the production of acetate by the acetogens is regulated by H_2 partial pressures, which is maintained at the required lower levels by the methanogens (Chandra et al., 2012).

Methanogenesis is the final stage in AD and is a strict anaerobic phase and exergonic reaction. Acetoclastic and hydrogenotrophic methanogens convert the acetate and the H_2 and CO_2 into methane. Most of the methane produced is derived from acetate, with the remaining from hydrogen and carbon dioxide (Angelidaki et al., 2011).

14.2.2.2 Biohydrogen

Hydrogen is a highly potential energy carrier as it has a high energy density of ~140 kJ/g and high enthalpy of combustion of 286 kJ/mol, and produces only water as the product of combustion, making it environmentally safe (Wang and Yin, 2017). Hydrogen produced through the biological route is called biohydrogen, and there are multiple routes for its production from biomass/agro-residues. This includes steam reforming of biogas/methane generated through the AD, through photofermentation or through dark fermentation. Photolytic methods employed by some algae and cyanobacteria are not discussed here as they produce hydrogen through the breakdown of water, and organic compounds are not the starting raw materials. In dark fermentation, anaerobic bacteria produce biohydrogen from carbohydrates in the absence of light, whereas in photofermentation, photosynthetic bacteria use light energy to generate biohydrogen from different organic acids (Bharathiraja et al., 2016; Hallenbeck and Benemann, 2002). Both these modes of biohydrogen production are complementary technologies and can be used sequentially in a hybrid system (Krishna, 2013).

The oxidative degradation of substrates during the heterotrophic growth of bacteria generates electrons, which are neutralized through proton reduction to generate molecular hydrogen. The production of hydrogen by dark fermentation is a universal phenomenon in anaerobic environments. The breakdown of complex carbohydrate polymers occurs when agro-residues as substrates provide the carbon sources for bacteria, which route their energy production through the generation of pyruvate. The metabolism of glucose via glycolysis generates pyruvate in a multistep reaction, which is then converted to acetyl-CoA and CO_2 by pyruvate–ferredoxin oxidoreductase in strict anaerobes and forms acetyl-CoA using pyruvate formate lyase in facultative anaerobes. In either case, the acetyl-CoA is converted finally to acetate, butyrate, or ethanol according to the biology of the microorganisms and the prevailing environmental conditions (Toledo-Alarcón et al., 2018). The reduced ferredoxin, which is generated along with acetyl-CoA in the anaerobes, is oxidized by a hydrogenase, resulting in the generation of oxidized ferredoxin and the release of electrons that reduce protons to generate molecular hydrogen. More hydrogen comes from the oxidation of NADH, which is generated in the glycolytic cycle. Most of the acetone-butanol-ethanol (ABE)-fermenting, strictly anaerobic bacteria like *Clostridium acetobutylicum*, *C. beijerinckii*, and *C. butyricum* are also efficient hydrogen producers, while the enterobacteria like *Escherichia coli*, *Enterobacter cloacae*, and *Enterobacter aerogenes* are facultative anaerobes that can produce biohydrogen (Toledo-Alarcón et al., 2018). Detailed reviews on dark fermentation can be obtained in Das and Veziroglu (2008) and Toledo-Alarcón et al. (2018).

14.2.3 Fuels from Agro-Residues Through Biochemical Route

14.2.3.1 Biodiesel from LCB

Biodiesel is another renewable fuel produced through the transesterification of fats and oils with methanol. Worldwide, biodiesel is now an accepted alternate transportation fuel, often blended with petroleum diesel. Mostly surplus/used vegetable oils/cooking oils are used for biodiesel production, but since such resources are limited, agro-residues have been targeted as renewable and cheaper feedstock for biodiesel production. Also, there have been very interesting approaches like the production of lipids in crops like sugar cane, where the biomass after sugar extraction can be used for extracting oil (Huang et al., 2017).

More common approaches towards generating oil from agro-residues include fermentation processes where oleaginous micro-organisms can be grown on biomass hydrolysates to produce oil. Oil-producing fermentative micro-organisms (known as single-cell oils) are attractive feedstock for biodiesel production because of several desirable features such as the ability to produce under all seasons and climatic conditions, a lesser requirement of space, an easiness in scale-up, minimal labour requirements, etc. (Mhlongo et al., 2021). Oleaginous micro-organisms, which can sometimes accumulate 20% or more of their weight as triacylglycerols, have multiple advantages like (i) high growth rate; (ii) high lipid accumulation; (iii) lipid composition similar to vegetable oils; (iv) having lipids with rare fatty acid structure; and (v) negligible resorption of the stored lipids for cellular metabolism (Carvalho et al., 2015; Uthandi et al., 2021). These organisms are able to attain an enhanced synthesis of fatty acids due to the presence of the enzyme ATP: citrate lyase (ACL), which provides a continuous supply of the precursor of lipid biosynthesis, acetyl-CoA (Boulton and Ratledge, 1981). In the presence of excess carbon sources, when another nutrient like nitrogen or phosphorous becomes limited, the utilization of carbohydrates through glycolysis results in an accumulation of citrate in the mitochondria due to rapid depletion of Adenosine monophosphate (AMP) as it is diverted for an alternate nitrogen supply mechanism in the organism. The excess citrate transported to the cytoplasm is converted by the ACL to acetyl-CoA, which in turn is routed to fatty acid synthesis (Ratledge and Wynn, 2002). Many microbes have been reported for the production of single-cell oils for biofuel applications, and some of them are able to grow on LCB hydrolysates and other agro-industrial residues. These include *Rhodosporidium toruloides*, *Cryptococcus* sp., *Trichosporon fermentans*, *Fusarium oxysporum*, *Rhodococcus opacus*, etc. (Patel et al., 2020). Studies on the cultivation of oleaginous organisms in biomass hydrolysates for the production of microbial oils/single-cell oils are reviewed in Diwan et al. (2018), Patel et al. (2020), Chintagunta et al. (2021), and Mhlongo et al. (2021). While the production of high-value lipids from oleaginous micro-organisms has been commercialized, the same is not true for biofuels because the high cost of production impairs commercial viability (Ochsenreither et al., 2016). Improvements in production conditions, recovery of lipids, and efficiency of organisms themselves are currently being explored for viability.

14.2.3.2 Bio-Alcohols

14.2.3.2.1 Bioethanol

Bioethanol, the 2G biofuel produced from LCB through the fermentation route, has been in focus for the past two decades due to policies and fluctuating oil prices. While the current production of ethanol is from starchy raw materials through the fermentation route (first-generation/1G ethanol), agricultural wastes, forestry residues, and any other LCB can also be used as effective feedstock for the manufacture of bioethanol when it is called 2G ethanol (Edeh, 2020). Lignocellulosic biofuels like bioethanol offer a great number of benefits, such as reduced competition with food production, carbon sequestration, and habitat improvement (Nair et al., 2017). As a fuel, bioethanol possesses several qualities like a high octane number (108), low boiling point and heat of vaporization, and reasonable energy density (~19 mj/l).

The current biochemical process for generating ethanol from LCB typically proceeds in three steps: pre-treatment, enzymatic hydrolysis, and fermentation (Figure 14.3). The lignocellulose

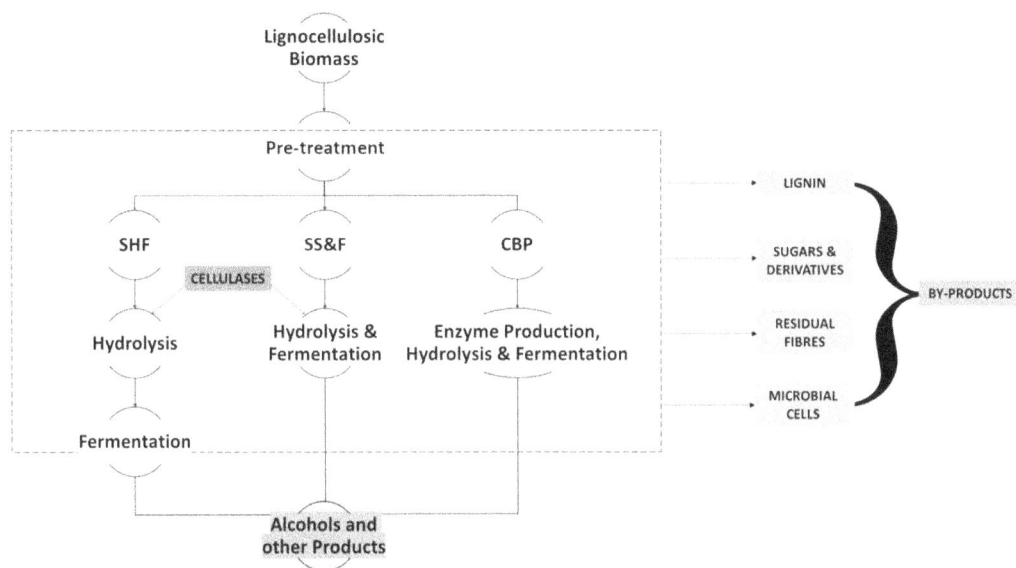

FIGURE 14.3 Schematic representation showing process schemes for alcohol production from agro-residues.

structure is highly recalcitrant, consisting of cellulose, hemicellulose, and lignin; therefore, the digestion of plant biomass to obtain fermentable sugars requires disruption of this complex structure to help enzymatic action. This is accomplished by pre-treatment methods that often target the removal of lignin or hemicellulose, exposing the cellulose to enzyme attack. Different pre-treatment methods, such as acid/alkali pre-treatment, pre-treatments using ammonia, liquid hot water, steam explosion, ozonolysis, and organosolv pre-treatments are commonly employed with newer and hybrid methods frequently being tried (Xu and Huang, 2014).

Enzymatic hydrolysis converts the polysaccharides, cellulose and hemicellulose, in the pre-treated biomass to fermentable sugars by the action of cellulases and hemicellulases, which are then fermented to ethanol by bacteria or yeasts (Sun and Cheng, 2002). From cellulose and mannan, glucose is formed, while xylans, glucans, galactans, and arabinans are produced from hemicellulose. The primary enzymes involved in cellulose degradation are endoglucanase, cellobiohydrolases, and beta-glucosidases, and for hemicellulose, these are xylanases and xylosidases (Vasić et al., 2021).

Microbes play a key role in fermentation. *Saccharomyces cerevisiae* and *Zymomonas mobilis* are the best-known organisms for ethanol fermentation, but they are limited by their ability to ferment only hexose sugars. Pentose-fermenting organisms like *Candida shehatae*, *Pichia stipitis*, and *Pachysolen tannophilus* are also commonly reported, but these organisms are inhibited by the end product and are not adapted to ferment high concentrations of sugars, unlike *Saccharomyces* (Adegboye et al., 2021). Strains of *S. cerevisiae* and *Z. mobilis* that have been genetically engineered to improve alcohol tolerance and yield are now being used for the efficient production of ethanol (Limayem and Ricke, 2012). Still, the industrial production of 2GE still has limitations with respect to fermentation, as organisms that can efficiently utilize both hexoses and pentoses simultaneously to produce practical levels of alcohol are non-existent, and the industrial performance of engineered strains still has challenges.

14.2.3.2.2 Biobutanol

Similar to bioethanol, biobutanol, which has an energy density (~29 mj/l) closer to that of gasoline, is also enjoying a renewed interest as an alternative renewable fuel (Xue and Cheng, 2019). Butanol is less hygroscopic, less corrosive, and less volatile, making this more attractive than bioethanol, at least in some cases. Even though the current industrial production of butanol is predominantly from the petrochemical route, biobutanol, i.e., butanol produced through ABE fermentation, is also steadily rising in acceptance. Biobutanol is also a drop-in fuel that can be blended in gasoline up to 85% and can be used in vehicles without any modification (Wang et al., 2017). However, ABE fermentation is limited by the high cost of feedstock and the product toxicity to *Clostridia* sp., causing inhibition of production (García et al., 2011).

Several strains of biobutanol-producing organisms have been identified, all of which belong to the genus *Clostridium*, such as *C. acetobutylicum*, *C. beijerinckii*, *C. saccharoperbutylacetonicum*, *C. sporogenes*, *C. pasteurianum*, etc. (Desai et al., 1999). The biggest challenges for the industrialization of fermentative butanol production are low yields of butanol and low butanol tolerance capability of the micro-organism. Inhibition of ABE fermentation starts in several *Clostridia* at around 10 g/l butanol or at an ABE concentration of nearly 20 g/l (Oudshoorn et al., 2009; Harvey and Meylemans, 2011; Thirmal and Dahman, 2012). Production of butanol from agro-residues has the added challenge of efficiency in utilizing pentose sugars coming from the hemicellulose fraction of lignocellulose and the major butanol-producing micro-organisms (e.g., *C. acetobutylicum* or *C. beijerinckii*) preferentially use glucose, and there is even a hierarchy in the utilization of different pentose sugars like xylose or arabinose (Aristilde et al., 2015). This results in an underutilization of the pentose fraction, as by the time the organism achieves the maximum possible glucose conversion, the ABE levels would have reached an inhibitory concentration preventing further utilization of the sugars, which leads to the pentose fraction being left behind.

14.2.4 Common Routes for Biochemical Conversion of Biomass to Fuels

Biochemical conversion processes utilize biological agents (enzymes, microbes) to break biomass into constituent polymers and then into liquid/gaseous biofuels and other byproducts (Mahalaxmi and Williford, 2016). Fuels other than those from the thermochemical route, in some form or other, depend on the breakdown of the biomass polymers and often require some conditioning and pre-treatment of biomass. Once broken down, the biomass-derived carbon can be biochemically/fermentatively converted to the respective fuels. Following discussions focused more on the biochemical route, especially on bioethanol, though the steps involved in the generation of the starting raw material, sugars are more or less similar to other biofuels like biobutanol, that is, those produced through microbial fermentation. There are generally four primary unit operations in the biochemical route: pre-processing, pre-treatment, hydrolysis, and fermentation; however, there could be fine variations in their implementation according to the type of biomass and the preferred product(s).

14.2.4.1 Pre-Processing/Conditioning

Size reduction, often known as physical pre-treatment, is the preliminary step in the biochemical conversion of biomass. This method uses physical methods to increase the surface area of the biomass, reduce the water content, and decrease the degrees of polymerization and crystallinity, which simplifies further processes (Rajendran et al., 2017). This step has been shown to improve biofuel production by 5%–25% (Hendriks and Zeeman, 2009). Commonly used methods of physical pre-treatment are milling, grinding, extrusion, microwave treatments, and ultrasonication.

14.2.4.2 Pre-Treatment

Pre-treatment alters the cellulose-hemicellulose-lignin network in the biomass for improved cellulose accessibility in the subsequent hydrolysis step by disrupting the lignin–carbohydrate interaction. It helps to remove lignin, may break down the hemicellulose, decreases crystallinity, and improves the porosity of the biomass, which increases its surface area (Chen et al., 2017). Pre-treatment methods are broadly classified as chemical, physico-chemical, and biological treatments, each employing different sets of processes. Table 14.2 lists some of the common types of pre-treatment, the conditions, method of action, and changes in biomass after pre-treatment.

TABLE 14.2

Biomass Pre-treatment Methods Used in Biochemical Conversion

Type of Pre-treatment	Common Reagents Used	Conditions	Mode of Action	Changes in Biomass	Disadvantages/Challenges	References
Chemical	Dilute acid (H_2SO_4, HNO_3, HCl, oxalic acid)	0.25–1% of acid 120–180°C 5–10 min	Hydronium ions formed during the process break glycosidic linkages	Removal of hemicellulose	Inhibitor formation. High-cost corrosion-resistant equipment	Lloyd and Wyman 2005; Yang and Tucker, 2013
	Dilute alkali (NaOH, KOH, or CaOH)	0.25–0.5% alkali 120–140°C 10–20 min	Breakage of ester bonds cross-linking lignin and xylan	Removal of lignin and some of the hemicellulose	High cost of alkali	Kim et al., 2016; Sun et al., 2016
	Organosolvent (Ethanol, methanol, acetone, ethylene glycol)	120–180°C, 45–75% of solvent, 30–90 min	Breaks internal bonds of lignin and hemicellulose	Removal of lignin and hemicellulose	High chemical costs and the need for solvent removal/recovery	Borand and Karaosmanoğlu, 2018; Zhang, Pei, and Wang, 2016
	Ionic liquid (N-methylmorpholine-n oxide monohydrate (NMMO), 1-allyl-3-methylimidazolium chloride)	70–150°C 100% of solvent 30–1440 min	Dissolving cellulose	Removal of lignin. Reduced cellulose crystallinity	High chemical cost and the need for solvent recycling	Swatloski et al., 2002; Yoo et al., 2017
Physico-chemical	Steam explosion (High-pressure steam)	180–230°C, and 10–30 bar pressure for 10–20 min	Breakdown of the glycosidic bonds in cellulose and hemicellulose and cleavage of hemicellulose–lignin bonds	Fibre splitting and expansion of inter-fibre spaces	Energy and water-intensive. Inhibitor formation	Chen and Liu, 2015; Pielhop et al., 2016
	Ammonia fibre explosion (liquid ammonia)	60–100°C under high pressure for 5–30 min	Cleaves the C–O–C bonds in lignin and ether and ester bonds between lignin–carbohydrate complex	Reduce crystallinity	Recovery and recycling of ammonia. Special reactors and safety concerns	Zhao et al., 2014; Shirkavand et al., 2016
	Supercritical CO_2	Under high pressure	CO_2 molecules penetrate into the biomass	Shatter hemicellulose and lignin	Infrastructure	Rostagno et al., 2015; Capolupo and Faraco, 2016

14.2.4.3 Enzymatic Digestion of Biomass to Sugars (Hydrolysis)

Hydrolysis is the next step in bio-alcohol production, where the polymers in the pre-treated feedstock are degraded into soluble and fermentable sugars. Glucose is produced from cellulose and mannan, and xylan, glucan, galactan, and arabinan are formed from hemicellulose (Taherzadeh and Niklasson, 2004). Two hydrolysis methods are very commonly employed: acid hydrolysis and enzymatic hydrolysis, of which the latter uses less energy, and the reaction can proceed under ambient conditions. Also, the enzymatic method is considered superior in terms of the efficiency in biomass conversion and specificity. Cellulases, hemicellulases, and ligninases are the primary enzymes used for hydrolysis. Cellulases include cellobiohydrolases (EC 3.2.1.91), endoglucanases (EC 3.2.1.4), and beta-glucosidases (EC 3.2.1.21), which act synergistically to hydrolyze cellulose into glucose (Wilson, 2009). Hemicellulases are more complex and include enzymes like endo- and exo-xylanases, arabinofuranosidases, mannanases and mannosidases, acetyl xylan esterases, glucoronidases, and galactosidases (Shallom and Shoham, 2003; Yi, 2021). Several bacteria, including *Cellulomonas*, *Clostridium*, *Thermomonospora*, *Ruminococcus*, *Streptomyces*, *Acetivibrio*, etc., and fungi like *Trichoderma*, *Aspergillus*, and *Penicillium*, are currently the main sources of these enzymes. Among these, *Trichoderma reesei* RUT C30, which has excellent protein secretion and enzyme titres, is one of the most studied and industrially preferred sources of cellulases and hemicellulases (Peterson and Nevalainen, 2012).

Many factors affect the yield in the hydrolysis process, including temperature, pH, the extent of mixing, concentrations of substrate and enzyme, and surfactant addition. Developing an economical process for the production of hydrolyzing enzymes is the main research focus since it is the key parameter that influences the cost of 2GE production. Lignin acts as the primary inhibitory factor in hydrolysis, but pre-treatment processes can remove a significant fraction of it, thereby improving the accessibility of enzymes. Product and by-product inhibition are other major limiting factors for hydrolysis.

14.2.4.4 Fermentation

In this step, the sugars (C_6 and C_5 sugars) released through enzymatic hydrolysis form the carbohydrate polymers that are fermented into bioethanol by microbes. The yeast *Saccharomyces cerevisiae* is one of the most commonly used organisms for ethanol production because it gives high yields (>0.45 gg^{-1} at optimal conditions) and also exhibits very high ethanol tolerance (>100 gl^{-1} has been reported for some strains) (Casey and Ingledew, 1986; Olofsson et al., 2008b). *Zymomonas mobilis* is another organism commonly used for ethanol production, and both these organisms have been subjected to a large number of studies to impart pentose utilization capabilities (Yang et al., 2007). Fermentation can be carried out in continuous, batch, or fed-batch modes. For economically viable bioethanol production, the fermenting micro-organism should have high productivity, broad substrate range, tolerance to process temperatures, high concentrations of ethanol, and the inhibitors present in the hydrolysate. Engineered micro-organisms are often used to improve the efficiency of sugar utilization and to achieve better production benefits.

There are two approaches used for fermentation that differ in their mode of sugar supply to the system: (i) separate saccharification and fermentation (SHF) and (ii) simultaneous saccharification and fermentation (SS&F). SHF is the conventional and commonly used practice and uses two different stages and/or reactors to perform biomass hydrolysis and fermentation. Fermentation starts only after sugar generation. Pre-treated biomass is subjected to hydrolysis, where it is degraded into simple sugars, and then it is taken to the fermentation step, where the micro-organism ferments it into ethanol. End-product inhibition is the major drawback of this process, where the monomeric sugars produced in the hydrolysis step act as an inhibitor for the cellulase enzymes and the maximum sugar concentration achievable is determined by the inhibition limit of the rate-limiting enzyme (Axelsson, 2011). It also adds to the operation cost since, often, the two stages require two separate reactors operating in sequence for a longer duration.

SS&F is currently more favoured than SHF as it has been described to be more cost-effective in rice straw biorefineries (Kádár et al., 2004). SS&F removes the hydrolysis products by simultaneously converting them to ethanol/other products, thus alleviating product inhibition and avoiding the cost of separate reactors. The major limitation of this method is optimizing the process parameters since the two processes require different operation conditions. Enzymatic hydrolysis optimally proceeds at 40–50°C, and many ethanol-producing micro-organisms are unable to tolerate this temperature. This limitation could be addressed through the use of heat-tolerant ethanologenic organisms like *Candida lusitaniae*, *Kluyveromyces marxianus*, and mixed cultures like *Brettanomyces claussenii* and *Saccharomyces cerevisiae* (Spindler et al., 1988; Golias et al., 2002). Many studies have reported improved product yields when SS&F is employed instead of separate hydrolysis and fermentation. For example, SS&F of alkali-pre-treated rice straw performed using *T. reesei* and *S. cerevisiae* could yield 61.3% ethanol (Zhu et al., 2005), while Karimi et al. (2006) showed that SS&F of acid-pre-treated rice straw using three organisms – *Rhizopus oryzae*, *Mucor indicus*, and *S. cerevisiae* – could give ethanol yields of 40–74% (Karimi et al., 2006).

Butanol fermentation proceeds similarly to ethanol fermentation, except for the organisms employed and the process parameters used. The concepts of SHF and SS&F are equally applicable in this case, albeit the fermentation occurs under strictly anaerobic conditions.

14.3 CHALLENGES IN BIOFUEL PRODUCTION FROM AGRO-RESIDUES

In order to attain energy security and transition to low carbon economies, many countries are turning to biofuels as a replacement for, or more commonly, as a blend-in additive in fossil fuels. As petroleum fuels dominate the marketplace, their prices are a direct competitor for the commercialization of biofuels. Estimates have shown that biofuel production costs are still not cost-competitive against petroleum-based fuels on an energy equivalent basis (Carriquiry et al., 2011; Witcover and Williams, 2020). Biofuel production is considered viable only if the cost and energy needed for its generation are significantly lower than the energy yielded. To attain this with agro-residue-derived biofuels, several challenges need to be addressed, especially those regarding feedstock procurement and logistics, development of energy-efficient technologies for conversion, complete/near-complete utilization of byproducts, efficient strategies for waste management that may include their value-addition, distribution management, and minimization of environmental impacts (Balan, 2014). Apparently, this is complex, requiring careful planning and optimization of processes.

14.3.1 Technical Challenges

14.3.1.1 Pre-Treatment

Pre-treatment of lignocellulose is highly indispensable for efficient conversion of agro-residues/LCB as they are inherently recalcitrant with a tight, highly ordered packing of cellulose fibres within a relatively impenetrable matrix of lignin and hemicellulose. The presence of lignin protects the plant from insect attacks, excess moisture content, and the elements and prevents easy access to biomass-hydrolyzing enzymes. The role of pre-treatment is to deconstruct this structural organization by removing either the lignin or hemicellulose or both to uncover the cellulose (Sun et al., 2016). Several pre-treatment strategies have been developed based on physical, chemical, or hybrid methods and these, in addition to improving the cellulose accessibility, also change the biomass structure by changing its composition, cellulose crystallinity, the degree of polymerization, chemical bonds between cellulose, hemicellulose and lignin, etc. An array of chemicals have been tested for the pre-treatment of lignocellulose, which includes inorganic acids (e.g., H_2SO_4, HCL, H_3PO_4), organic acids (e.g., acetic acid and oxalic acid), alkalis (e.g., NaOH, KOH, NH_3), Sulfite Pretreatment to Overcome Recalcitrance of Lignocellulose (SPORL), solvents (e.g., ethanol, methanol), ionic liquids (e.g., [bmim][Cl], [amim][Cl]), deep eutectic solvents (e.g., ChCl:imidazolium, ChCl:urea), biological pre-treatments with white rot fungi as well as with lignin-degrading enzymes, etc. (Chen et al., 2017; Balan, 2014). Despite so many chemicals being tested, very few are suitable for scale-up, the rest being either not economically feasible or accompanied by serious limitations. Apparently, there is no single best method to pre-treat agro-residues, and the choice of method and its implementation have to come from experimentation with the biomass in question and the subsequent conversion strategies planned. High-temperature operations are often required, which is one of the major consumers of energy in biomass conversion, and high solids loading requires specialized designs and higher energy inputs for mixing and heat transfer (Zhang et al., 2016). This also puts a serious strain on capital expenditure as high-pressure reactors and associated infrastructure would be necessary for pre-treatment. Reactors have to be designed and optimized for the biomass considered, and often it is ideal to do modelling and simulation studies with the proposed reactor configurations, especially the impeller designs, before implementation. Most pre-treatment strategies are carried out at very low or very high pH, making it necessary to neutralize it, as the cellulase and other biomass-degrading enzymes have an optimum pH in the mild acidic range (pH 5–6). This poses challenges in the handling of the resultant salts and wastewater. One of the major roadblocks to the efficient pre-treatment of biomass is the change in the structural dynamics of lignin after pre-treatment. In certain cases, during acid pre-treatment at high temperatures, lignin is redeposited on biomass, and sometimes there is the generation of pseudo lignin, which can coat the cellulose, hindering the cellulose accessibility and increasing non-productive enzyme adsorption on biomass, thus reducing the total conversion yield (Selig et al., 2007; Li et al., 2014). Preventing lignin repolymerization and/or re-adsorption on biomass is actively being looked into, and at least for certain types of pre-treatments, pre-treatment additives that can prevent these have been proposed (Chu et al., 2021). Improving the technical- and cost-efficiencies of pre-treatment is complex and dependent on the feedstock and technology adopted for pre-treatment. Also, the pre-treatment strategies have to consider minimizing the generation of inhibitors, in addition to improving enzyme accessibility to biomass and hence the improvement in saccharification efficiency.

14.3.1.2 Enzymatic Saccharification

The efficiency of enzymatic biomass hydrolysis depends on both substrate-related and enzyme-related factors, such as biomass structure, composition, type of pre-treatment, enzyme dosage and synergism, enzyme components, etc. Enzymes may easily bind to amorphous cellulose and carry out saccharification, but it is considered difficult for the enzymes to degrade crystalline regions as it contains layers of glucose chains that are tightly packed, impeding enzyme adsorption. Lignocellulose contains cellulose and hemicellulose supported by a rigid, complex matrix of lignin, and its side groups bind non-specifically and irreversibly to the enzyme, reducing the free protein available for hydrolysis. Also, the available surface area of cellulose decreases when lignin is present. Different pre-treatment strategies result in biomass containing lignin with different hydrophobicity, aroma, molecular weight, etc., which affects enzyme adsorption differently (Leu and Zhu, 2013; Li et al., 2018). So, in the context of biomass, if the cellulose is exposed enough with minimum hindrance from lignin and hemicellulose, it can be made easily digestible. Also, biomass with lesser cellulose crystallinity is considered more amenable to digestion. This is often debated, as the measurement of biomass crystallinity is the crystallinity index (CrI), which would naturally be high after pre-treatment in biomass due to the increase in cellulose content (Park et al., 2010). So, it is often found that pre-treated biomass with high CrI is better digested than that with a lower CrI (Shi et al., 2020).

The enzyme cocktail required to digest LCB is a mixture of several enzymes like endo- and exo-glucanases, beta-glucosidases, hemicellulases like xylanase, mannase, xylosidase, and a multitude of accessory activities and proteins like swollenin, Lytic polysaccharide monooxygenases (LPMOs), galactosidase, acetyl xylan esterase, laccase, peroxidase, etc., which act in synergy or are cooperative in action (Østby et al., 2020). These enzymes are present in different proportions and quantities, and biomass with different composition require tailored enzymes for the breakdown of their complex structure. Beta glucosidase (BGl) is considered a rate-limiting enzyme as it can get inhibited by its product glucose, which in turn leads to cellobiose accumulation; this feedback inhibition can shut down the whole cellulolytic machinery (Sørensen et al., 2013). The glucose-mediated shutdown of the hydrolysis process may be overcome by the use of glucose-tolerant glucosidases (Rajasree et al., 2013). Also, a higher BGl loading in the cellulase cocktail can overcome product inhibition and help to achieve higher conversion yields in a shorter time duration. The design of commercial enzyme cocktails over the years has taken care of this, and the current preparations contain high levels of BGL (Østby et al., 2020). End-product inhibition can also be addressed effectively by SS&F, where hydrolysis is performed along with fermentation so that there is no sugar accumulation (Olofsson et al., 2008a). Enzyme cocktails currently available in the market are almost exclusively derived from fungi. Fungi are efficient degraders of cellulose, and they secrete copious amounts of enzymes, which makes them ideal candidates for the production of biomass-hydrolyzing enzymes. Major producers currently use engineered strains of *Trichoderma reesei* for cellulase production, and more often than not, the enzymes are blended with heterologous BGls, xylanases, and LPMOs (Genencor, 2011; Hu et al., 2015). Enhancing the cellulase activities of fungi used for commercial production is a route chosen by many for generating enzymes that yield more sugars per unit quantity. The method employed often includes genetic modifications to the fungi that produce the enzyme (Bischof et al., 2016; Druzhinina and Kubicek, 2017).

Enhancing hydrolysis efficiencies is also attained by improving the conditions of hydrolysis, which includes better mixing, choosing the right loading that allows maximal biomass addition without compromising the mixing efficiencies, and adding surfactants and additives to reduce/prevent non-productive enzymatic adsorption, etc.

14.3.1.3 Fermentation

The fermentation of biomass hydrolysates may be considered relatively less challenging compared to the hydrolysis and pre-treatment steps, as most of the process is already established in first-generation (1G) ethanol production technologies, which are well established commercially. The limitations here arise mainly from the production of sugar and lignin degradation products during pre-treatment, which gets carried over to the hydrolysates. Most of these compounds can inhibit the growth and metabolism of the fermenting micro-organisms (Jönsson et al., 2013). Over liming and neutralization of the biomass hydrolysates could be one of the effective means of addressing the inhibitor problem (Mohagheghi et al., 2004; 2006) and another strategy would be to detoxify the hydrolysates by passing them through charcoal or resins that can adsorb the inhibitors (Jönsson et al., 2013). Another major challenge is the fermenting organisms' sensitivity to alcohol. As fermentation progresses, the ethanol content in the fermentation liquor increases, and beyond a certain limit, growth and alcohol production by the organism ceases. The highest ethanol concentrations achievable, therefore, would depend on the tolerance of the organism to ethanol. Continuous removal of ethanol through pervaporation, gas stripping, etc. is proposed (Balan, 2014), but with classic distiller's yeast as the fermenting strain, ethanol concentrations do not become a limiting factor. A more serious concern would be the inability of typical ethanol-fermenting organisms used in industry to ferment the hemicellulose-derived pentoses. While there exist both natural and engineered ethanol-fermenting organisms that can ferment both glucose and C5 sugars, the preference for glucose often results in lower yields and productivity of ethanol (Kim et al., 2012). Co-culturing, in which two micro-organisms are concomitantly grown, is sometimes applied to augment substrate utilization and improve the fermentative abilities of the micro-organisms used. For example, *Pichia stipitis*, the C5 fermenting yeast, has been used in co-cultures with *Zymomonas mobilis* to efficiently ferment both glucose and xylose and obtain ethanol yields of up to 96% of the theoretical maximum (Fu et al., 2009). While better-engineered organisms or co-fermentation could be possible solutions, another route to address the C5 utilization challenge would be to separate the C5 stream and divert it to the production of other high-value compounds that may improve the process economies of the biorefinery.

14.3.1.4   Waste Management

A large volume of water is utilized during pre-treatment, neutralization of pre-treated slurry, washing steps for the removal of salts, and in the enzymatic saccharification, fermentation, etc., at least some of which is not reused and ends up as process wastewater. With projected water consumption of 4–5-fold or even more per unit volume of fuel produced, process wastewater is an unignorable component and recycling it is a necessity to make the process cost-effective and eco-friendly (Yuan et al., 2021). Recycling of water at certain steps like pre-treatment and washing may considerably reduce the amount of total water utilized. Solid wastes coming from all process steps like pre-treatment rejects, undigested residues from the hydrolysis step, and yeast cells and solids from fermentation liquor are the major solid wastes generated in the conversion of agro-residues to fuels, and proposed methods for their utilization include incineration to generate heat/energy that can offset production costs (Humbird et al., 2011).

14.3.2   Process Economics and Challenges in Attaining Commercial Feasibility

The economic feasibility of biofuel plants is very much dependent on the selling prices of the fuel and other byproducts. For example, 2GE plants entail very high capital costs – more than five times that of a comparably sized 1G ethanol plant – because of the larger number of unit operations associated with them. This would necessitate a higher minimum selling price, which often fails to be competitive with existing fuel options. Typically, a break-even selling price is determined at which future sales of the fuel and byproducts can balance the current capital and operating expenditures within a fixed duration allowing for profits through techno-economic analyses. This is the minimum fuel selling price (MSP), and the goal while setting up a biofuel plant is to produce the fuel below the MSP. For example, in the case of cellulosic ethanol produced from lignocellulosic feedstock, independent studies have estimated the MSP at 0.40 to 1.30 US$/l (Tao et al., 2012; Macrelli et al., 2014; Cheng et al., 2019), the variability being due to the process design, type of biomass, choice of pre-treatment, and fermentation efficiency of the organisms (Zetterholm et al., 2020).

The total cost of production is the combination of the total capital investment (calculated from equipment costs, site development costs, field expenses, etc.), operating costs (including material handling, labour, supplies, and overhead), and other indirect costs (taxes, permits, transportation equipment, etc.). According to a techno-economic analysis by National Renewable Energy Laboratory (NREL) (Humbird et al., 2011), for the conversion of corn stover to ethanol, the production cost was estimated to be 0.40 $/l. This results in an MSP of 0.57 $/l, 34.4% of which is contributed by feedstock procurement and processing charges; enzymes account for 15.8%, and the rest of the conversion processes contributes to 49.8% of the MSP. In short, the profitability of 2GE plants is mostly decided by their processing conditions, operating costs, and market prices.

14.3.2.1   Operating Costs

Operating costs include variable operating costs that are incurred only when the plant is in operation, such as raw materials costs, waste handling charges, etc., and fixed operating costs that are incurred irrespective of whether the plant is operating at full capacity or not. These include labour costs, maintenance costs, and the expenses on various overhead items. The performance of a biorefinery, i.e., the product yields and energy and mass balances, is heavily influenced by the equipment used and the chosen operating parameters.

For lignocellulose to ethanol conversion, operating costs were estimated to be between $0.35 and $0.45 per litre of ethanol (Aden et al., 2002; Lynd et al., 1996; Wooley et al., 1999), depending on the assumptions made about feedstock and enzyme costs, and the type of pre-treatment to be used. Since a typical lignocellulosic (LC) biorefinery has multiple unit operations and their associated streams, there exist many avenues for cost optimization – right from process improvements to using cheaper substrates and integrating multiple operations. Also, operating costs are driven by economies of scale, i.e., the operating costs increase more slowly as the plant size gets larger. As such, the optimal size for 2G biorefineries is estimated to be in the range of 0.9 to 1.8 billion lge (litres of gasoline equivalent) per year compared to 0.3 billion lge per year for a 1G ethanol plant (Wright and Brown, 2007).

14.3.2.2   Costs and Challenges Associated with Raw Materials

The cost of raw materials has a significant impact on the biofuel production cost. In many of the estimates, the cost of biomass feedstock alone accounts for almost a third of the total production cost (Witcover and Williams, 2020). The most common challenges in raw material cost are fluctuation and uncertainty in the market price and cost volatility, such as purchasing prices for biomass resources at farms, utility bills (e.g., electricity, stream, etc.), and expenses connected with auxiliary activities (e.g., labour, transport, etc.) (Yue et al., 2014).

14.3.2.3   Costs Associated with Logistics and the Supply Chain

Logistics and the supply chain account for a considerable amount of cost and energy for the production of 2G biofuels (Balan, 2014). The major operations involved in biomass logistics are harvesting, collecting, pre-processing (densification by compaction, pelleting, and briquetting), storing, and transporting. Both biomass collection and distribution need a significant amount of work in terms of manpower, equipment selection, shift organization, vehicle routing, and shipment scheduling. The lack of adoption of advanced machinery for these processes, especially in developing countries, can cause bottlenecks in the biomass supply chain. The cost per tonne of the feedstock is also strongly related to the yield of biomass, which is controlled by factors like soil fertility, location, and genetics (Clauser et al., 2021).

The biomass supply chain is another important aspect of large-scale biofuel production. Since each feedstock has its own complex seasonal production cycles with low or high yields (Sharma et al., 2019), efficient logistics are critical for biofuel production. Designing a biorefinery plant should therefore be based on a precise idea of the logistics on a cradle-to-grave basis to understand the technical feasibility of the process (Ulonska et al., 2018).

Transportation of LCB to biorefineries has several challenges, including variability in physical properties and texture, differences in bulk densities, variability and seasonality of availability, and dispersion across a broad geographical region. In fact, the size of the biorefinery is customized depending on the area and availability of feedstock(s) (Antonio et al., 2019), which in turn impacts the annual production cost of the biofuels. Moisture content, distance from the field to the biorefinery, infrastructure availability, on-site technology, and mode of transportation (rail, road, or waterways) will all impact the transportation costs. While various transportation networks exist, such as multi-modal transportation, combined transportation, and intermodal transportation, the overall goal is to have an integrated transportation network in which each mode of transportation is used at its best scale and operation (de Jong et al., 2017).

Storage of biomass feedstock is essential in terms of drying and pre-processing, and it also ensures regular supply, irrespective of fluctuations in feedstock availability. If optimal storage conditions are not provided, it may lead to an uncontrollable loss of biomass due to microbial action (Wendt and Zhao, 2020). To reduce the cost impact of separate storage facilities, pre-treatment depots are proposed, which would offer uniform-format feedstock(s). Such a depot would be ideally located to simplify transportation and processing and would function as a tiny satellite system that takes in diverse biomass forms from its surrounding area, converts it into uniform-format biomass, and then sends it to a network of much bigger supply terminals (Yue et al., 2014).

Mathematical optimization models are frequently used to determine the ideal, lowest cost supply chain design. Models such as the integrated biomass supply analysis and logistics model (Sokhansanj et al., 2006), the biofuel infrastructure, logistics, and transportation model (Lautala et al., 2015), the wood fuel integrated supply/demand overview mapping model (Drigo et al., 2002), and others have shown to be useful in understanding the intricate connections that exist between crop production, harvest, storage, transportation, pre-processing, and terminal distribution. A detailed study on biomass supply chain classification, optimization methods, and model development has been done by Atashbar et al. (2016), which may be referred to for a detailed understanding of the process.

14.3.2.4 Enzyme Costs

Of various factors that contribute to the high cost of production of biomass/agro-residue-based fuels, the cost of enzymes (cellulases, hemicellulases, ligninases, and other accessory proteins) remains a major challenge and the one that would often determine the process economics. Irrespective of the variability in enzyme loading, depending on the process design, studies by Klein-Marcuschamer et al. (2012; 2010), Littlewood et al. (2013), Liu et al. (2016), and others have established that, for an ethanol MSP of 0.610 $/l in a lignocellulose biorefinery, the enzyme costs impart 18–43% of the selling price, and ~44% of the raw material price (Khajeeram and Unrean, 2017), making it a focal point of targeted efforts to improve the feasibility of lignocellulosic biorefineries.

Enzyme-related costs may be attributed to two factors: the production costs of the enzymes themselves and process costs influenced by the enzyme efficiency. Irrespective of the production route, the production of biomass-degrading enzymes via microbial fermentation entails multiple unit operations. This implies that it is quite expensive to set up a production facility, even discounting all other costs. Cost optimization of the enzymatic hydrolysis step calls for the use of highly efficient enzyme mixtures that can accelerate the bioconversion process. According to modelling studies, enzyme loadings must not exceed 30 mg protein per gram cellulose (in the case of on-site produced enzyme) or 10 mg/g for purchased enzymes in order to preserve the market competitiveness of the final product (Liu et al., 2016).

Currently, different strategies are used to reduce the cost contribution of enzymes, such as increasing microbial enzyme production via strain engineering, using cheaper raw materials and more efficient inducers, process optimization, developing strategies to stabilize and recycle the enzymes, etc. (Sukumaran et al., 2017). The choice of enzyme production site is also a critical factor. An integrated production approach (in which the enzyme production is integrated via process and energy flows into the biofuel plant) reduces the enzyme cost component as compared to on-site (stand-alone enzyme plants that are located adjacent to the ethanol plant) or off-site (a central enzyme production facility that supplies numerous operations) production facilities (Johnson, 2016).

14.3.2.5 Cost of Utilities

The costs of utilities used in a biorefinery – steam, cooling water, electricity, air, and process water – play substantial roles in determining the cost-competitiveness and viability of the production process. This includes both the cost of each utility based on its input stream and the operating and capital costs of equipment associated with them, such as water treatment plants, boilers, and cooling towers. A review by Jorissen et al. (2020) on the economics of 2G biorefineries states that utility costs contribute 5–15% of the total process costs of a single output plant, and 11–33% for a plant with multiple products, with heating and cooling costs accounting for a substantial portion. Biorefineries also use copious amounts of water. It is estimated that roughly 1–4 gallons of water are needed to produce 1 gallon of biofuel (National Research Council, 2008). A study by De Fraiture et al. (2008) projected that worldwide, up to 4% of total irrigation withdrawals will be routed towards biofuel production by 2030. This fraction will be higher in regions with more established biorefineries, such as North America.

Offsetting utility costs is possible through process heat recovery, water recycling, and byproducts and wastes conversion to energy, etc. Byproducts generated in the processing of biomass feedstock, such as lignin, unutilized cellulose and hemicellulose, as well as biogas from AD and sludge from wastewater treatment, could be used to generate steam and electricity. This can potentially lower the overall fuel costs and amounts of waste generated and may even generate additional revenue through energy sales (Humbird et al., 2011). Another effective way to minimize utility costs by decreasing both the capital and operating costs associated with it is by *utility integration* i.e., recycling the utilities from one process within the biorefinery to use them in another process (Byford et al., 2009). For example, excess heat from pre-treatment may be used for operating the distillation or hydrolysis, which proceeds at significantly lower temperatures.

14.3.2.6 By-Product Generation to Offset Production Costs

In the biomass-to-biofuel conversion process, both the pre-treatment liquor and the hydrolysate contain a mixture of C6 (glucose, galactose, mannose, rhamnose) and C5 (xylose, arabinose) sugars. Additionally, a significant amount of fibrous and lignin-rich solid residue is also left over after saccharification. The amounts and qualities of the byproducts generated from these residues can improve the robustness of the fuel to market price fluctuations. A study by Ranganathan (2020 showed that the MSP of cellulosic ethanol from a rice straw biorefinery could be reduced from 0.63 to 0.25 $/l if both residual C5 sugars and lignin are converted to co-products.

Valorization may be done via the thermochemical platform or the biochemical platform. In the thermochemical platform, the biomass is gasified to a mixture of CO, CO_2, and H_2, which is then either fermented or chemically catalyzed to a range of end-products like fuels, organic acids, etc. It may also be pyrolyzed or liquefied to generate bio-oils and/or biochar. In the biochemical platform, industrial micro-organisms are used to ferment the sugars into different categories of products, of which organic acids (acetic acid, citric acid, fumaric acid, gluconic acid, succinic acid, oxalic acid, malic acid, lactic acid, and butyric acid) are the second-largest fermentation products after ethanol worldwide. These acids can be transformed into a variety of reactive compounds that have an extended range of versatile applications in industry (e.g., as platform chemicals, monomers for plastics), biomedicine, nanotechnology, and bioremediation. They are also used to derive homo-polymers with tunable material properties, like polypropylene fumarate or polylactic acid. The range of bio-based chemicals that can be sourced by biomass/biomass processing byproducts have been reviewed by the Task 42 Biorefinery of IEA Bioenergy (Ed de Jong et al., 2011). Also, there have been several studies that projected better commercial feasibility for biorefineries with by-product valorization and multiproduct output (Cai et al., 2018; Hagman et al., 2020; Philippini et al., 2020).

LCB and the sugars and fermentation products derived from it can be used as the feedstock for synthesizing a variety of polymers, either derivatized ones like methyl cellulose, carboxymethyl cellulose, and cellulose acetate, or microbially synthesized ones like dextran, xanthan, polyglutamic acid, and polyhydroxyalkanoates (PHA) (Machado et al., 2020). Lignocellulosic biomass-derived alditols also offer outstanding advantages for the preparation of biopolymers that have enhanced hydrophilicity and thus can be used in food packaging and medical devices (Isikgor and Remzi Becer, 2015).

Other platform chemicals from lignocellulose-derived sugars are ring-opened derivatives of C5 sugars, such as xylitol and arabinitol, which function as intermediates in the synthesis of various kinds of linear polycondensates. Due to their usefulness in chemical synthesis, xylitol and arabinitol have more market value than ethanol. Sorbitol, derived from glucose, is another key intermediate that finds many applications in food, pharmaceutical, and cosmetic industries and as an additive in many end-products (Kobayashi and Fukuoka, 2013). All three products of the ABE fermentation process – acetone, butanol, and ethanol – may also be considered platform chemicals in their own right because their conversion products, notably ethylene (the monomer for polyethylene), ethylene glycol, butadiene (the monomer for synthetic rubber), propene, and vinyl chloride are important polymers (Takkellapati et al., 2018).

Hydrolysates from LCB can be used to grow microbial biomass that is subsequently used as animal feed (analogous to single-cell proteins) (Fei et al., 2016). Alternatively, treated lignocellulose is also used as a natural feed supplement to increase the crude fibre content of animal food (van Kuijk et al., 2015). The fibre by-product from saccharified corn stover is sold as dried distillers grains with solubles, which are high in protein (~27%) and used as animal feed. In some cases, the lignocellulose is pre-treated with microbes to improve its protein levels (via secreted microbial proteins), digestibility, and palatability. This process requires organisms that can thrive on lignin, hemicelluloses, and celluloses like *Pleurotus eryngii*, *P. ostreatus*, and *Lentinus edodes* (Villas-Bôas et al., 2002).

Furfural and hydroxymethyl furfural (HMF) are the degradation products of pentoses and hexoses, respectively, as a result of dilute acid pre-treatment. Micro-organisms detoxify these aldehydes into less toxic alcohols – furfuryl alcohol and 5-(hydroxymethyl) furfuryl alcohol, respectively (Wierckx et al., 2011). 5-HMF and its derivatives are very promising chemical intermediates that can be used for the synthesis of different chemicals such as γ-valerolactone and succinic acidand also in preparing resins, polymers, pharmaceuticals, herbicides, and flavouring agents. Importantly, 2,5-Furandicarboxylic Acid (FDCA), which can be obtained by the bio-oxidation of HMF by bacteria like *Acinetobacter calcoaceticus* (Sheng et al., 2020), shows comparable properties to terephthalic acid used in polyesters.

Furfural is obtained via the acid-catalyzed dehydration of C5 sugars such as xylose. Reduction of the parent furan ring of furfural results in the formation of tetrahydrofuran (THF) and THF derivatives like tetrahydrofurfuryl alcohol, 2-methyltetrahydrofuran, and 2,5-bis(hydroxymethyl) tetrahydrofuran. Maleic acid and maleic anhydride are the other two industrially important furfural-derived compounds that are mostly converted to fumaric acid – a raw material for the production of polyesters and copolymers (Mathew et al., 2018).

Most of the lignocellulosic residues have significant amounts of silica-rich ash; together with lignin, its presence is detrimental to the biorefining processes. Therefore, most pre-treatment methods focus on dissolving this recalcitrant barrier. Silica has important applications as catalysts, catalyst supports, adsorbents, molecular sieves, and encapsulation agents. Lignin can be upgraded by both thermochemical and biological routes to fuels, aromatics, carbon fibres, adhesives, and resins. Biologically, bacteria that can depolymerize lignin, such as *Pseudomonas*, *Pandoraea* (Liu et al., 2019), and *Rhodococcus* sp., have been studied for the accumulation of polyhydroxyalkanoates that can be extruded as bioplastics.

Apparently, a multitude of different possibilities exists for the valorization of the lignocellulosic biorefinery wastes. There is a large body of information on such products reviewed in recent reports (Bozell and Petersen, 2010; Demirbas, 2008; Gallezot, 2012; Isikgor and Remzi Becer, 2015; Kobayashi and Fukuoka, 2013; Kumar et al., 2016; Werpy and Petersen, 2004). These and many other papers may be referred to for exhaustive information on their production routes, applications, and technical challenges to be overcome.

14.3.3 ENVIRONMENTAL CONCERNS

While 2G fuels offer undeniable advantages over 1G fuels in terms of providing energy security, reducing the carbon footprint and eliminating competition with food resources, biofuel production from biomass still entails a number of environmental concerns regarding their impact on water resources, soil quality, GHG emissions, and biodiversity. The net climatic effects of biofuel production are also hotly disputed, mainly due to the uncertainties in production routes and other indirect effects (such as alterations in land use patterns, GHG emissions from increased fertilizer use, logistics, etc.). In any case, increasing the cultivation of energy crops in order to meet the feedstock requirements for biofuel production leads to significant ecotoxicity concerns due to surface runoff and infiltration and accumulation of fertilizer compounds into groundwater reservoirs. Therefore, bio-energy production and natural resource depletion need to be carefully balanced, which can be attained by proper crop selection and optimal management of harvest rates, irrigation, and appropriate fertilization (Qin et al., 2018).

Life cycle assessment (LCA) is a tool that can be used to analyze the sustainability and environmental impacts of a particular process or product in a more comprehensive manner. Most LCA analyses on bioethanol production from LCB have reported reduced global warming potential (GWP, estimated at 0.2–1.8 kg CO_2 eq/l of ethanol), less depletion of natural resources, and decreased air acidification potential, eutrophication potential, human toxicity potential, and air odour potential (Kadam, 2002). Feedstock with lower resource (land, water, etc.) consumption has been stated to have lower GWP (Mandade et al., 2015). The contribution of by-product generation is also quite important while considering the overall GWP of a biorefinery, and operating integrated biorefineries have been proposed to reduce GHG emissions by 83% (Kumar and Verma, 2021).

The end-of-life step in LCA analysis considers carbon incorporated into a landfill or soil and can be treated as sequestered carbon, leading to negative GWP impacts. For example, compared to petrol, bioethanol with 100% biogenic carbon can offer nearly 107% savings in CO_2 eq/l release (Stichnothe and Azapagic, 2009). While GHG emissions are unavoidable during bioethanol production, it is very low when compared to the use of fossil fuels. Therefore, a better process development may be done using advanced technologies to reduce the economic and environmental impacts to make it more sustainable and feasible, which can be attained by proper management of existing crops and croplands, with due consideration for CO_2, N_2O emissions, water footprint, and coproduct allocation.

14.4 CONCLUSIONS

Agro-residues as feedstock for renewable fuels production are highly attractive as it does not compete with food/feed for land, water, and other resources, often being the by-product of food crop cultivation. Also, there are multiple options available for converting agro-residues to fuels or other value-added products: one could process it through a thermochemical route to syngas or bio-oil and from there to multiple chemicals or through the biochemical route to produce bioethanol and a multitude of products through microbial fermentation. The options are several and ever-increasing, but the level of commercial success enjoyed by the technologies for biomass conversion to biofuels is very limited, especially so with biochemical conversion-based technologies. Many reports have addressed the reasons for this, mostly in the context of technical challenges in operation, especially those in improving conversion efficiencies. While there are still limitations in biomass conversion technologies, the broader picture also needs to visualize and address a whole lot of other challenges like the supply chain of biomass and the final products, biomass storage and densification related issues, enzyme production challenges, in addition to challenges in hydrolysis efficiencies, process water and solids management (preferably involving their recycling/valorization), by-product valorization strategies that are efficient and operable at scale, reactor and process equipment related challenges, and strategies to reduce the environmental and negative socio-economic impacts. The impact of biomass conversion to biofuels on the environment and carbon emissions can be overwhelming, and LCAs and appropriate strategies to reduce the carbon footprint are essential for reducing the GWP of such plants, as the operation scales are huge. Commercial feasibility, all-inclusive sustainable development, and reduction in GWP are predicted only through integrated biorefinery operations, which can very well be assumed to be the future of agro-residue-based fuels.

REFERENCES

Abraham, A., Mathew, A.K., Sindhu, R., Pandey, A., Binod, P., 2016. Potential of rice straw for bio-refining: An overview. *Bioresour. Technol.* 215, 29–36. https://doi.org/10.1016/j.biortech.2016.04.011

Adegboye, M.F., Ojuederie, O.B., Talia, P.M., Babalola, O.O., 2021. Bioprospecting of microbial strains for biofuel production: Metabolic engineering, applications, and challenges. *Biotechnol. Biofuels.* https://doi.org/10.1186/s13068-020-01853-2

Aden, A., Ruth, M., Ibsen, K., Jechura, J., Neeves, K., Sheehan, J., Wallace, B., Montague, L., Slayton, A., Lukas, J., 2002. *Lignocellulosic Biomass to Ethanol Process Design and Economics Utilizing Co-Current Dilute Acid Prehydrolysis and Enzymatic Hydrolysis for Corn Stover.* National Renewable Energy Laboratory

Angelidaki, I., Karakashev, D., Batstone, D.J., Plugge, C.M., Stams, A.J.M., 2011. Biomethanation and its potential. In: *Methods in Enzymology.* https://doi.org/10.1016/B978-0-12-385112-3.00016-0

Antonio, F., Antunes, F., Kumar, A., Ruly, C., Hilares, T., Ingle, A.P., Rai, M., 2019. Overcoming challenges in lignocellulosic biomass pretreatment for second generation (2G) sugar production : Emerging role of nano, biotechnological and promising approaches. *3 Biotech* 9(6), 1–17. https://doi.org/10.1007/s13205-019-1761-1

Aristilde, L., Lewis, I.A., Park, J.O., Rabinowitz, J.D., 2015. Hierarchy in pentose sugar metabolism in Clostridium acetobutylicum. *Appl. Environ. Microbiol.* https://doi.org/10.1128/AEM.03199-14

Atashbar, N.Z., Labadie, N., Prins, C., 2016. Modeling and optimization of biomass supply chains: A review and a critical look. In: *IFAC-PapersOnLine.* Elsevier B.V., pp. 604–615. https://doi.org/10.1016/j.ifacol.2016.07.742

Axelsson, J., 2011. *Separate Hydrolysis and Fermentation of Pretreated Spruce.* Master Thesis, Linköping University, pp. 1–48.

Azman, S., Khadem, A.F., Van Lier, J.B., Zeeman, G., Plugge, C.M., 2015. Presence and role of anaerobic hydrolytic microbes in conversion of lignocellulosic biomass for biogas production. *Crit. Rev. Environ. Sci. Technol.* https://doi.org/10.1080/10643389.2015.1053727

Balagurumurthy, B., Singh, R., Bhaskar, T., 2015. Catalysts for thermochemical conversion of biomass. In: *Recent Advances in Thermochemical Conversion of Biomass.* https://doi.org/10.1016/B978-0-444-63289-0.00004-1

Balan, V., 2014. Current challenges in commercially producing biofuels from lignocellulosic biomass. *ISRN Biotechnol.* 2014, 1–31. https://doi.org/10.1155/2014/463074

Bharathiraja, B., Sudharsanaa, T., Bharghavi, A., Jayamuthunagai, J., Praveenkumar, R., 2016. Biohydrogen and Biogas – An overview on feedstocks and enhancement process. *Fuel* 185, 810–828. https://doi.org/10.1016/j.fuel.2016.08.030

Bischof, R.H., Ramoni, J., Seiboth, B., 2016. Cellulases and beyond: The first 70 years of the enzyme producer Trichoderma reesei. *Microb. Cell Fact.* 15(1), 1–13. https://doi.org/10.1186/s12934-016-0507-6

Boehman, A.L., Le Corre, O., 2008. Combustion of syngas in internal combustion engines. *Combust. Sci. Technol.* https://doi.org/10.1080/00102200801963417

Borand, M.N., Karaosmanoğlu, F., 2018. Effects of organosolv pretreatment conditions for lignocellulosic biomass in biorefinery applications: A review. *J. Renew. Sustain. Energy.* https://doi.org/10.1063/1.5025876

Boulton, C.A., Ratledge, C., 1981. Correlation of lipid accumulation in yeasts with possession of ATP:citrate lyase. *J. Gen. Microbiol.* 127(1), 169–176. https://doi.org/10.1099/00221287-127-1-169

Bozell, J.J., Petersen, G.R., 2010. Technology development for the production of biobased products from biorefinery carbohydrates—The US Department of Energy's "top 10" revisited. *Green Chem.* 12(4), 539–555. https://doi.org/10.1039/b922014c

Byford, M., Robbins, A., Burton, L., Fontenot, J., Bagajewicz, M., 2009. *Utility-Integrated Biorefineries.* Oklahoma. https://www.ou.edu/class/che-design/a-design/projects-2009/Biorefineries.pdf

Cai, H., Han, J., Wang, M., Davis, R., Biddy, M., Tan, E., 2018. Life-cycle analysis of integrated biorefineries with co-production of biofuels and bio-based chemicals: Co-product handling methods and implications. *Biofuels Bioprod. Biorefin.* 12(5), 815–833. https://doi.org/10.1002/bbb.1893

Capolupo, L., Faraco, V., 2016. Green methods of lignocellulose pretreatment for biorefinery development. *Appl. Microbiol. Biotechnol.* https://doi.org/10.1007/s00253-016-7884-y

Carpenter, D.L., Bain, R.L., Davis, R.E., Dutta, A., Feik, C.J., Gaston, K.R., Jablonski, W., Phillips, S.D., Nimlos, M.R., 2010. Pilot-scale gasification of corn stover, switchgrass, wheat straw, and wood: 1. Parametric study and comparison with literature. *Ind. Eng. Chem. Res.* https://doi.org/10.1021/ie900595m

Carriquiry, M.A., Du, X., Timilsina, G.R., 2011. Second generation biofuels: Economics and policies. *Energy Policy* 39(7), 4222–4234. https://doi.org/10.1016/j.enpol.2011.04.036

Carvalho, A.K.F., Rivaldi, J.D., Barbosa, J.C., de Castro, H.F., 2015. Biosynthesis, characterization and enzymatic transesterification of single cell oil of Mucor circinelloides - A sustainable pathway for biofuel production. *Bioresour. Technol.* 181, 47–53. https://doi.org/10.1016/j.biortech.2014.12.110

Casey, G.P., Ingledew, W.M.M., 1986. Ethanol tolerance in yeasts. *Crit. Rev. Microbiol.* 13(3), 219–280. https://doi.org/10.3109/10408418609108739

Chandra, R., Takeuchi, H., Hasegawa, T., 2012. Methane production from lignocellulosic agricultural crop wastes: A review in context to second generation of biofuel production. *Renew. Sustain. Energy Rev.* 16(3), 1462–1476. https://doi.org/10.1016/j.rser.2011.11.035

Chen, H., Liu, J., Chang, X., Chen, D., Xue, Y., Liu, P., Lin, H., Han, S., 2017. A review on the pretreatment of lignocellulose for high-value chemicals. *Fuel Process. Technol.* 160, 196–206. https://doi.org/10.1016/j.fuproc.2016.12.007

Chen, H.Z., Liu, Z.H., 2015. Steam explosion and its combinatorial pretreatment refining technology of plant biomass to bio-based products. *Biotechnol. J.* https://doi.org/10.1002/biot.201400705

Cheng, M.H., Wang, Z., Dien, B.S., Slininger, P.J.W., Singh, V., 2019. Economic analysis of cellulosic ethanol production from sugarcane bagasse using a sequential deacetylation, hot water and disk-refining pretreatment. *Processes* 7(10). https://doi.org/10.3390/pr7100642

Chintagunta, A.D., Zuccaro, G., Kumar, M., Kumar, S.P.J., Garlapati, V.K., Postemsky, P.D., Kumar, N.S.S., Chandel, A.K., Simal-Gandara, J., 2021. Biodiesel production From lignocellulosic biomass using oleaginous microbes: Prospects for integrated biofuel production. *Front. Microbiol.* 12, 2080. https://doi.org/10.3389/fmicb.2021.658284

Chu, Q., Tong, W., Chen, J., Wu, S., Jin, Y., Hu, J., Song, K., 2021. Organosolv pretreatment assisted by carbocation scavenger to mitigate surface barrier effect of lignin for improving biomass saccharification and utilization. *Biotechnol. Biofuels.* https://doi.org/10.1186/s13068-021-01988-w

Ciferno, J.P., Marano, J.J., 2002. *Benchmarking Biomass Gasification Technologies for Fuels, Chemicals and Hydrogen Production.* https://netl.doe.gov/sites/default/files/netl-file/BMassGasFinal_0.pdf

Clauser, N.M., González, G., Mendieta, C.M., Kruyeniski, J., Area, M.C., Vallejos, M.E., 2021. Biomass waste as sustainable raw material for energy and fuels. *Sustainability* 13(2), 1–21. https://doi.org/10.3390/su13020794

Das, D., Veziroglu, T.N., 2008. Advances in biological hydrogen production processes. *Int. J. Hydr. Energy* 33(21), 6046–6057. https://doi.org/10.1016/j.ijhydene.2008.07.098

De Fraiture, C., Giordano, M., Liao, Y., 2008. Biofuels and implications for agricultural water use: Blue impacts of green energy. *Water Policy.* https://doi.org/10.2166/wp.2008.054

de Jong, S., Hoefnagels, R., Wetterlund, E., Pettersson, K., Faaij, A., Junginger, M., 2017. Cost optimization of biofuel production – The impact of scale, integration, transport and supply chain configurations. *Appl. Energy* 195, 1055–1070. https://doi.org/10.1016/j.apenergy.2017.03.109

Demirbas, A., 2008. Products from lignocellulosic materials via degradation processes. *Energy Sources Part Recover Util. Environ. Eff.* 30, 27–37. https://doi.org/10.1080/00908310600626705

Desai, R.P., Harris, L.M., Welker, N.E., Papoutsakis, E.T., 1999. Metabolic flux analysis elucidates the importance of the acid-formation pathways in regulating solvent production by Clostridium acetobutylicum. *Metab. Eng.* 1(3), 206–213. https://doi.org/10.1006/mben.1999.0118

Dhyani, V., Bhaskar, T., 2018. A comprehensive review on the pyrolysis of lignocellulosic biomass. *Renew. Energy* 129, 695–716. https://doi.org/10.1016/j.renene.2017.04.035

Dimitriadis, A., Bezergianni, S., 2016. Hydrothermal liquefaction of various biomass and waste feedstocks for biocrude production: A state of the art review. *Renew. Sustain. Energy Rev.* 68, 113–125. https://doi.org/10.1016/j.rser.2016.09.120

Diwan, B., Parkhey, P., Gupta, P., 2018. From agro-industrial wastes to single cell oils: A step towards prospective biorefinery. *Folia Microbiol. (Praha).* 63(5), 547–568. https://doi.org/10.1007/s12223-018-0602-7

Drigo, R., Masera, O.R., Trossero, M.A., 2002. Woodfuel integrated supply/demand overview mapping - WISDOM: A geographical representation of woodfuel priority areas. *Unasylva* 53, 36–40.

Druzhinina, I.S., Kubicek, C.P., 2017. Genetic engineering of Trichoderma reesei cellulases and their production. *Microb. Biotechnol.* 10(6), 1485–1499. https://doi.org/10.1111/1751-7915.12726

Duan, P., Zhang, C., Wang, F., Fu, J., Lü, X., Xu, Y., Shi, X., 2016. Activated carbons for the hydrothermal upgrading of crude duckweed bio-oil. *Catal. Today* 274, 73–81. https://doi.org/10.1016/j.cattod.2016.01.046

Dukhnytskyi, B., 2019. World agricultural production. *Ekon. APK*, 59–65. https://doi.org/10.32317/2221-1055.201907059

Ed de Jong, Higson, A., Walsh, P., Wellisch, M., 2011. *Bio-Based Chemicals Value Added Products from Biorefineries.* IEA Bioenergy, Task 42.

Edeh, I., 2020. Bioethanol production: An overview. In: *Bioethanol* [Working Title]. https://doi.org/10.5772/intechopen.94895

Fei, Q., O'Brien, M., Nelson, R., Chen, X., Lowell, A., Dowe, N., 2016. Enhanced lipid production by Rhodosporidium toruloides using different fed-batch feeding strategies with lignocellulosic hydrolysate as the sole carbon source. *Biotechnol. Biofuels* 9. https://doi.org/10.1186/s13068-016-0542-x

Fu, N., Peiris, P., Markham, J., Bavor, J., 2009. A novel co-culture process with Zymomonas mobilis and Pichia stipitis for efficient ethanol production on glucose/xylose mixtures. *Enzyme Microb. Technol.* https://doi.org/10.1016/j.enzmictec.2009.04.006

Gabra, M., Pettersson, E., Backman, R., Kjellström, B., 2001. Evaluation of cyclone gasifier performance for gasification of sugar cane residue - Part 1: Gasification of bagasse. *Biomass Bioenergy* 21(5), 351–369. https://doi.org/10.1016/S0961-9534(01)00043-5

Gallezot, P., 2012. Conversion of biomass to selected chemical products. *Chem. Soc. Rev.* https://doi.org/10.1039/c1cs15147a

García, V., Päkkilä, J., Ojamo, H., Muurinen, E., Keiski, R.L., 2011. Challenges in biobutanol production: How to improve the efficiency? *Renew. Sustain. Energy Rev.* https://doi.org/10.1016/j.rser.2010.11.008

Genencor, 2011. *Optimized Cellulase, Hemicellulase and Beta-Glucosidase Enzyme Complex for Improved Lignocellulosic Biomass Hydrolysis.* Genencor.

Gerardi, M.H., 2003. *The Microbiology of Anaerobic Digesters, the Microbiology of Anaerobic Digesters.* John Wiley & Sons, Inc. https://doi.org/10.1002/0471468967

Golias, H., Dumsday, G.J., Stanley, G.A., Pamment, N.B., 2002. Evaluation of a recombinant Klebsiella oxytoca strain for ethanol production from cellulose by simultaneous saccharification and fermentation: Comparison with native cellobiose-utilizing yeast strains and performance in co-culture with thermotolerant yeast and Zymomonas mobilis. *J. Biotechnol.* 96(2), 155–168. https://doi.org/10.1016/S0168-1656(02)00026-3

Hagman, L., Eklund, M., Svensson, N., 2020. Assessment of by-product valorisation in a Swedish wheat-based biorefinery. *Waste Biomass Valorization* 11(7), 3567–3577. https://doi.org/10.1007/s12649-019-00667-0

Hallenbeck, P.C., Benemann, J.R., 2002. Biological hydrogen production; fundamentals and limiting processes. *Int. J. Hydr. Energy* 27(11–12), 1185–1193. https://doi.org/10.1016/S0360-3199(02)00131-3

Harvey, B.G., Meylemans, H.A., 2011. The role of butanol in the development of sustainable fuel technologies. *J. Chem. Technol. Biotechnol.* 86(1), 2–9. https://doi.org/10.1002/JCTB.2540

Hema Krishna, R., 2013. Review of research on production methods of hydrogen: Future fuel production from organic compounds, hybrid system) that are being studied for biohydrogen production. A succinct discussion of production techniques for enhancing hydrogen production from Ph. *Eur. J. Biotechnol. Bio Sci.* 1, 84–93.

Hendriks, A.T.W.M., Zeeman, G., 2009. Pretreatments to enhance the digestibility of lignocellulosic biomass. *Bioresour. Technol.* 100(1), 10–18. https://doi.org/10.1016/j.biortech.2008.05.027

Hu, J., Chandra, R., Arantes, V., Gourlay, K., Susan van Dyk, J., Saddler, J.N., 2015. The addition of accessory enzymes enhances the hydrolytic performance of cellulase enzymes at high solid loadings. *Bioresour. Technol.* https://doi.org/10.1016/j.biortech.2015.03.055

Hu, J., Li, D., Lee, D.J., Zhang, Q., Wang, W., Zhao, S., Zhang, Z., He, C., 2019. Integrated gasification and catalytic reforming syngas production from corn straw with mitigated greenhouse gas emission potential. *Bioresour. Technol.* https://doi.org/10.1016/j.biortech.2019.02.064

Huang, H., Moreau, R.A., Powell, M.J., Wang, Z., Kannan, B., Altpeter, F., Grennan, A.K., Long, S.P., Singh, V., 2017. Evaluation of the quantity and composition of sugars and lipid in the juice and bagasse of lipid producing sugarcane. *Biocatal. Agric. Biotechnol.* https://doi.org/10.1016/j.bcab.2017.03.003

Humbird, D., Davis, R., Tao, L., Kinchin, C., Hsu, D., Aden, A., Schoen, P., Lukas, J., Olthof, B., Worley, M., Sexton, D., Dudgeon, D., 2011. *Process Design and Economics for Biochemical Conversion of Lignocellulosic Biomass to Ethanol: Dilute-Acid Pretreatment and Enzymatic Hydrolysis of Corn Stover.* National Renewable Energy Laboratory. https://doi.org/10.4324/9780429290602-8

IEA BioEnergy, 2018. *Thermal Gasification of Biomass.* https://doi.org/ieatask33.org/content/thermal_gasification

Isikgor, F.H., Remzi Becer, C., 2015. Lignocellulosic biomass: A sustainable platform for production of bio-based chemicals and polymers. *Polym. Chem.* 6, 4497–4559. https://doi.org/10.1039/c3py00085k

Johnson, E., 2016. Integrated enzyme production lowers the cost of cellulosic ethanol. *Biofuels Bioprod. Biorefin.* 10(2), 164–174. https://doi.org/10.1002/bbb.1634

Jönsson, L.J., Alriksson, B., Nilvebrant, N.-O., 2013. Bioconversion of lignocellulose: Inhibitors and detoxification. *Biotechnol. Biofuels* 6(1), 16. https://doi.org/10.1186/1754-6834-6-16

Jorissen, T., Oraby, A., Recke, G., Zibek, S., 2020. A systematic analysis of economic evaluation studies of second-generation biorefineries providing chemicals by applying biotechnological processes. *Biofuels Bioprod. Biorefin.* 14(5), 1028–1045. https://doi.org/10.1002/bbb.2102

Kadam, K.L., 2002. Environmental benefits on a life cycle basis of using bagasse-derived ethanol as a gasoline oxygenate in India. *Energy Policy* 30(5), 371–384. https://doi.org/10.1016/S0301-4215(01)00104-5

Kádár, Z., Szengyel, Z., Réczey, K., 2004. Simultaneous saccharification and fermentation (SSF) of industrial wastes for the production of ethanol. In: *Industrial Crops and Products.* Elsevier, pp. 103–110. https://doi.org/10.1016/j.indcrop.2003.12.015

Kaparaju, P., Serrano, M., Thomsen, A.B.B., Kongjan, P., Angelidaki, I., 2009. Bioethanol, biohydrogen and biogas production from wheat straw in a biorefinery concept. *Bioresour. Technol.* 100(9), 2562–2568. https://doi.org/10.1016/j.biortech.2008.11.011

Karimi, K., Emtiazi, G., Taherzadeh, M.J., 2006. Ethanol production from dilute-acid pretreated rice straw by simultaneous saccharification and fermentation with Mucor indicus, Rhizopus oryzae, and Saccharomyces cerevisiae. *Enzyme Microb. Technol.* 40(1), 138–144. https://doi.org/10.1016/j.enzmictec.2005.10.046

Khajeeram, S., Unrean, P., 2017. Techno-economic assessment of high-solid simultaneous saccharification and fermentation and economic impacts of yeast consortium and on-site enzyme production technologies. *Energy.* https://doi.org/10.1016/j.energy.2017.01.090

Kim, J.S., Lee, Y.Y., Kim, T.H., 2016. A review on alkaline pretreatment technology for bioconversion of lignocellulosic biomass. *Bioresour. Technol.* 199, 42–48. https://doi.org/10.1016/J.BIORTECH.2015.08.085

Kim, S.R., Ha, S.J., Wei, N., Oh, E.J., Jin, Y.S., 2012. Simultaneous co-fermentation of mixed sugars: A promising strategy for producing cellulosic ethanol. *Trends Biotechnol.* 30(5), 274–282. https://doi.org/10.1016/j.tibtech.2012.01.005

Klein-Marcuschamer, D., Oleskowicz-Popiel, P., Simmons, B.A., Blanch, H.W., 2012. The challenge of enzyme cost in the production of lignocellulosic biofuels. *Biotechnol. Bioeng.* 109(4), 1083–1087. https://doi.org/10.1002/bit.24370

Klein-Marcuschamer, D., Oleskowicz-Popiel, P., Simmons, B.A., Blanch, H.W., 2010. Technoeconomic analysis of biofuels: A wiki-based platform for lignocellulosic biorefineries. *Biomass Bioenergy* 34(12), 1914–1921. https://doi.org/10.1016/j.biombioe.2010.07.033

Kobayashi, H., Fukuoka, A., 2013. Synthesis and utilisation of sugar compounds derived from lignocellulosic biomass. *Green Chem.* 15(7), 1740–1763. https://doi.org/10.1039/c3gc00060e

Kumar, A., Gautam, A., Dutt, D., 2016. Biotechnological transformation of lignocellulosic biomass in to industrial products: An overview. *Adv. Biosci. Biotechnol.* 07(3), 149–168. https://doi.org/10.4236/abb.2016.73014

Kumar, B., Verma, P., 2021. Life cycle assessment: Blazing a trail for bioresources management. *Energy Convers. Manag. X* 10. https://doi.org/10.1016/j.ecmx.2020.100063

Lautala, P.T., Hilliard, M.R., Webb, E., Busch, I., Richard Hess, J., Roni, M.S., Hilbert, J., Handler, R.M., Bittencourt, R., Valente, A., Laitinen, T., 2015. Opportunities and challenges in the design and analysis of biomass supply chains. *Environ. Manag.* 56(6), 1397–1415. https://doi.org/10.1007/s00267-015-0565-2

Lee, S.Y., Sankaran, R., Chew, K.W., Tan, C.H., Krishnamoorthy, R., Chu, D.-T., Show, P.-L., 2019. Waste to bioenergy: A review on the recent conversion technologies. *BMC Energy* 1(1), 1–22. https://doi.org/10.1186/s42500-019-0004-7

Leu, S.Y., Zhu, J.Y., 2013. Substrate-related factors affecting enzymatic saccharification of lignocelluloses: Our recent understanding. *Bioenergy Res.* 6(2), 405–415. https://doi.org/10.1007/s12155-012-9276-1

Li, H., Pu, Y., Kumar, R., Ragauskas, A.J., Wyman, C.E., 2014. Investigation of lignin deposition on cellulose during hydrothermal pretreatment, its effect on cellulose hydrolysis, and underlying mechanisms. *Biotechnol. Bioeng.* 111(3), 485–492. https://doi.org/10.1002/bit.25108

Li, X., Li, M., Pu, Y., Ragauskas, A.J., Klett, A.S., Thies, M., Zheng, Y., 2018. Inhibitory effects of lignin on enzymatic hydrolysis: The role of lignin chemistry and molecular weight. *Renew. Energy* 123, 664–674. https://doi.org/10.1016/j.renene.2018.02.079

Limayem, A., Ricke, S.C., 2012. Lignocellulosic biomass for bioethanol production: Current perspectives, potential issues and future prospects. *Prog. Energy Combust. Sci.* 38(4), 449–467. https://doi.org/10.1016/j.pecs.2012.03.002

Littlewood, J., Wang, L., Turnbull, C., Murphy, R.J., 2013. Techno-economic potential of bioethanol from bamboo in China. *Biotechnol. Biofuels* 6(1). https://doi.org/10.1186/1754-6834-6-173

Liu, D., Yan, X., Si, M., Deng, X., Min, X., Shi, Y., Chai, L., 2019. Bioconversion of lignin into Bioplastics by Pandoraea sp. B-6: Molecular mechanism. *Environ. Sci. Pollut. Res. Int.* 26(3), 2761–2770. https://doi.org/10.1007/s11356-018-3785-1

Liu, G., Zhang, J., Bao, J., 2016. Cost evaluation of cellulase enzyme for industrial-scale cellulosic ethanol production based on rigorous Aspen Plus modeling. *Bioprocess Biosyst. Eng.* 39(1), 133–140. https://doi.org/10.1007/s00449-015-1497-1

Liu, Z., Xu, A., Long, B., 2011. Energy from combustion of rice straw: Status and challenges to China. *Energy Power Eng.* 03(3), 325–331. https://doi.org/10.4236/epe.2011.33040

Lloyd, T.A., Wyman, C.E., 2005. Combined sugar yields for dilute sulfuric acid pretreatment of corn stover followed by enzymatic hydrolysis of the remaining solids. *Bioresour. Technol.* 96(18), 1967–1977. https://doi.org/10.1016/J.BIORTECH.2005.01.011

Lynd, L.R., Elander, R.T., Wyman, C.E., 1996. Likely features and costs of mature biomass ethanol technology. *Appl. Biochem. Biotechnol. A.* https://doi.org/10.1007/BF02941755

Machado, G., Santos, F., Lourega, R., Mattia, J., Faria, D., Eichler, P., Auler, A., 2020. Biopolymers from lignocellulosic biomass. In: *Lignocellulosic Biorefining Technologies*, pp. 125–158. https://doi.org/10.1002/9781119568858.ch7

Macrelli, S., Galbe, M., Wallberg, O., 2014. Effects of production and market factors on ethanol profitability for an integrated first and second generation ethanol plant using the whole sugarcane as feedstock. *Biotechnol. Biofuels* 71(7), 1–16. https://doi.org/10.1186/1754-6834-7-26

Mahalaxmi, S., Williford, C., 2016. Biochemical conversion of biomass to fuels. In: *Handbook of Climate Change Mitigation and Adaptation*, 2nd Edition, pp. 1777–1811. https://doi.org/10.1007/978-3-319-14409-2_26

Mandade, P., Bakshi, B.R., Yadav, G.D., 2015. Ethanol from Indian agro-industrial lignocellulosic biomass— A life cycle evaluation of energy, greenhouse gases, land and water. *Int. J. Life Cycle Assess.* 20(12), 1649–1658. https://doi.org/10.1007/s11367-015-0966-8

Mathew, A.K., Abraham, A., Mallapureddy, K.K., Sukumaran, R.K., 2018. Lignocellulosic biorefinery wastes, or resources? In: *Waste Biorefinery: Potential and Perspectives*, pp. 267–297. https://doi.org/10.1016/B978-0-444-63992-9.00009-4

Mhlongo, S.I., Ezeokoli, O.T., Roopnarain, A., Ndaba, B., Sekoai, P.T., Habimana, O., Pohl, C.H., 2021. The potential of single-cell oils derived From filamentous fungi as alternative feedstock sources for biodiesel production. *Front. Microbiol.* https://doi.org/10.3389/fmicb.2021.637381

Mikeska, M., Najser, J., Peer, V., Frantík, J., Kielar, J., 2020. Quality assessment of gas produced from different types of biomass pellets in gasification process. *Energy Explor. Exploit.* 38(2), 406–416. https://doi.org/10.1177/0144598719875272

Mohagheghi, A., Dowe, N., Schell, D., Chou, Y.-C., Eddy, C., Zhang, M., 2004. Performance of a newly developed integrant of Zymomonas mobilis for ethanol production on corn stover hydrolysate. *Biotechnol. Lett.* 264(26), 321–325. https://doi.org/10.1023/B:BILE.0000015451.96737.96

Mohagheghi, A., Ruth, M., Schell, D.J., 2006. Conditioning hemicellulose hydrolysates for fermentation: Effects of overliming pH on sugar and ethanol yields. *Process Biochem.* 41(8), 1806–1811. https://doi.org/10.1016/J.PROCBIO.2006.03.028

Mohan, D., Pittman, C.U., Steele, P.H., 2006. Pyrolysis of wood/biomass for bio-oil: A critical review. *Energy Fuels.* https://doi.org/10.1021/ef0502397

Molino, A., Chianese, S., Musmarra, D., 2016. Biomass gasification technology: The state of the art overview. *J. Energy Chem.* https://doi.org/10.1016/j.jechem.2015.11.005

Mulat, D.G., Huerta, S.G., Kalyani, D., Horn, S.J., 2018. Enhancing methane production from lignocellulosic biomass by combined steam-explosion pretreatment and bioaugmentation with cellulolytic bacterium Caldicellulosiruptor bescii. *Biotechnol. Biofuels.* https://doi.org/10.1186/s13068-018-1025-z

Mullen, C.A., Boateng, A.A., Hicks, K.B., Goldberg, N.M., Moreau, R.A., 2010. Analysis and comparison of bio-oil produced by fast pyrolysis from three barley biomass/byproduct streams. In: *Energy and Fuels.* https://doi.org/10.1021/ef900912s

Nair, R.B., Lennartsson, P.R., Taherzadeh, M.J., 2017. Bioethanol production From agricultural and municipal wastes. In: *Current Developments in Biotechnology and Bioengineering: Solid Wastes Management*, pp. 157–190. https://doi.org/10.1016/B978-0-444-63664-5.00008-3

National Research Council, 2008. *Water Implications of Biofuels Production in the United States, Water Implications of Biofuels Production in the United States.* National Academies Press. https://doi.org/10.17226/12039

Ochsenreither, K., Glück, C., Stressler, T., Fischer, L., Syldatk, C., 2016. Production strategies and applications of microbial single cell oils. *Front. Microbiol.* https://doi.org/10.3389/fmicb.2016.01539

Olofsson, K., Bertilsson, M., Lidén, G., 2008. A short review on SSF – An interesting process option for ethanol production from lignocellulosic feedstocks. *Biotechnol. Biofuels* 11(1), 1–14. https://doi.org/10.1186/1754-6834-1-7

Østby, H., Hansen, L.D., Horn, S.J., Eijsink, V.G.H., Várnai, A., 2020. Enzymatic processing of lignocellulosic biomass: Principles, recent advances and perspectives. *J. Ind. Microbiol. Biotechnol.* 47(9–10), 623–657. https://doi.org/10.1007/s10295-020-02301-8

Oudshoorn, A., van der Wielen, L.A.M., Straathof, A.J.J., 2009. Adsorption equilibria of bio-based butanol solutions using zeolite. *Biochem. Eng. J.* https://doi.org/10.1016/j.bej.2009.08.014

Park, S., Baker, J.O., Himmel, M.E., Parilla, P.A., Johnson, D.K., 2010. Cellulose crystallinity index: Measurement techniques and their impact on interpreting cellulase performance. *Biotechnol. Biofuels* 3. https://doi.org/10.1186/1754-6834-3-10

Patel, A., Karageorgou, D., Rova, E., Katapodis, P., Rova, U., Christakopoulos, P., Matsakas, L., 2020. An overview of potential oleaginous microorganisms and their role in biodiesel and omega-3 fatty acid-based industries. *Microorganisms* 8(3). https://doi.org/10.3390/microorganisms8030434

Patrizi, N., Bruno, M., Saladini, F., Parisi, M.L., Pulselli, R.M., Bjerre, A.B., Bastianoni, S., 2020. Sustainability assessment of biorefinery systems based on two food residues in Africa. *Front. Sustain. Food Syst.* 4. https://doi.org/10.3389/fsufs.2020.522614

Pattanaik, L., Pattnaik, F., Saxena, D.K., Naik, S.N., 2019. Biofuels from agricultural wastes. In: *Second and Third Generation of Feedstocks: The Evolution of Biofuels.* Elsevier, pp. 103–142. https://doi.org/10.1016/B978-0-12-815162-4.00005-7

Pavlostathis, S.G., 2011. Kinetics and modeling of anaerobic treatment and biotransformation processes. In: *Comprehensive Biotechnology*, 2nd Edition, pp. 385–397. https://doi.org/10.1016/B978-0-08-088504-9 .00385-8

Peterson, R., Nevalainen, H., 2012. Trichoderma reesei RUT-C30 - Thirty years of strain improvement. *Microbiology*. https://doi.org/10.1099/mic.0.054031-0

Philippini, R.R., Martiniano, S.E., Ingle, A.P., Franco Marcelino, P.R., Silva, G.M., Barbosa, F.G., dos Santos, J.C., da Silva, S.S., 2020. Agroindustrial byproducts for the generation of biobased products: Alternatives for sustainable biorefineries. *Front. Energy Res.* 8. https://doi.org/10.3389/fenrg.2020.00152

Pielhop, T., Amgarten, J., Von Rohr, P.R., Studer, M.H., 2016. Steam explosion pretreatment of softwood: The effect of the explosive decompression on enzymatic digestibility. *Biotechnol. Biofuels* 9, 1–13. https://doi.org/10.1186/s13068-016-0567-1

Qin, Z., Zhuang, Q., Cai, X., He, Y., Huang, Y., Jiang, D., Lin, E., Liu, Y., Tang, Y., Wang, M.Q., 2018. Biomass and biofuels in China: Toward bioenergy resource potentials and their impacts on the environment. *Renew. Sustain. Energy Rev.* 82, 2387–2400. https://doi.org/10.1016/j.rser.2017.08.073

Rahman, M.M., Liu, R., Cai, J., 2018. Catalytic fast pyrolysis of biomass over zeolites for high quality bio-oil – A review. *Fuel Process. Technol.* 180, 32–46. https://doi.org/10.1016/j.fuproc.2018.08.002

Rajasree, K.P., Mathew, G.M., Pandey, A., Sukumaran, R.K., 2013. Highly glucose tolerant β-glucosidase from Aspergillus unguis: NII 08123 for enhanced hydrolysis of biomass. *J. Ind. Microbiol. Biotechnol.* 40(9), 967–975. https://doi.org/10.1007/s10295-013-1291-5

Rajendran, K., Drielak, E., Sudarshan Varma, V., Muthusamy, S., Kumar, G., 2017. Updates on the pretreatment of lignocellulosic feedstocks for bioenergy production–a review. *Biomass Convers. Biorefin.* 82(8), 471–483. https://doi.org/10.1007/S13399-017-0269-3

Ranganathan, P., 2020. Preliminary techno-economic evaluation of 2G ethanol production with co-products from rice straw. *Biomass Convers. Biorefin.* https://doi.org/10.1007/s13399-020-01144-8

Ratledge, C., Wynn, J.P., 2002. The biochemistry and molecular biology of lipid accumulation in oleaginous microorganisms. In: *Advances in Applied Microbiology*, pp. 1–52. https://doi.org/10.1016/S0065 -2164(02)51000-5

Renewable Fuels Association, 2021. Global ethanol production by country or region. https://doi.org/afdc .energy.gov/data/10331

Rostagno, M.A., Prado, J.M., Mudhoo, A., Santos, D.T., Forster-Carneiro, T., Meireles, M.A.A., 2015. Subcritical and supercritical technology for the production of second generation bioethanol. *Crit. Rev. Biotechnol.* https://doi.org/10.3109/07388551.2013.843155

Roy, A., 2017. *Bioethanol Market Size | Global Industry Outlook & Forecast 2022*. https://doi.org/alliedmarke tresearch.com/bioethanol-market

Satlewal, A., Agrawal, R., Bhagia, S., Das, P., Ragauskas, A.J., 2018. Rice straw as a feedstock for biofuels: Availability, recalcitrance, and chemical properties. *Biofuels Bioprod. Biorefin.* 12(1), 83–107. https://doi.org/10.1002/bbb.1818

Sayara, T., Sánchez, A., 2019. A review on anaerobic digestion of lignocellulosic wastes: Pretreatments and operational conditions. *Appl. Sci.* 9(21), 4655. https://doi.org/10.3390/app9214655

Selig, M.J., Viamajala, S., Decker, S.R., Tucker, M.P., Himmel, M.E., Vinzant, T.B., 2007. Deposition of lignin droplets produced during dilute acid pretreatment of maize stems retards enzymatic hydrolysis of cellulose. *Biotechnol. Prog.* https://doi.org/10.1021/bp0702018

Shafizadeh, F., Fu, Y.L., 1973. Pyrolysis of cellulose. *Carbohydr. Res.* 29(1), 113–122. https://doi.org/10.1016 /S0008-6215(00)82074-1

Shahbandeh, M., 2021. *Wheat - Production Volume Worldwide 2011/2012-2020/1*. https://doi.org/statista.com /statistics/267268/production-of-wheat-worldwide-since-1990/

Shallom, D., Shoham, Y., 2003. Microbial hemicellulases. *Curr. Opin. Microbiol.* 6(3), 219–228. https://doi .org/10.1016/S1369-5274(03)00056-0

Sharma, B.P., Edward Yu, T., English, B.C., Boyer, C.N., Larson, J.A., 2019. Stochastic optimization of cellulosic biofuel supply chain incorporating feedstock yield uncertainty. In: *Energy Procedia*. Elsevier B.V., pp. 1009–1014. https://doi.org/10.1016/j.egypro.2019.01.245

Sheng, Y., Tan, X., Zhou, X., Xu, Y., 2020. Bioconversion of 5-hydroxymethylfurfural (HMF) to 2,5-Furandicarboxylic acid (FDCA) by a native obligate aerobic Bacterium, Acinetobacter calcoaceticus NL14. *Appl. Biochem. Biotechnol.* 192(2), 455–465. https://doi.org/10.1007/s12010-020 -03325-7

Shi, F., Wang, Y., Davaritouchaee, M., Yao, Y., Kang, K., 2020. Directional structure modification of poplar biomass-inspired high efficacy of enzymatic hydrolysis by sequential dilute acid-alkali treatment. *ACS Omega*. https://doi.org/10.1021/acsomega.0c03419

Shirkavand, E., Baroutian, S., Gapes, D.J., Young, B.R., 2016. Combination of fungal and physicochemical processes for lignocellulosic biomass pretreatment - A review. *Renew. Sustain. Energy Rev.* https://doi .org/10.1016/j.rser.2015.10.003

Sokhansanj, S., Kumar, A., Turhollow, A.F., 2006. Development and implementation of integrated biomass supply analysis and logistics model (IBSAL). *Biomass Bioenergy* 30(10), 838–847. https://doi.org/10 .1016/j.biombioe.2006.04.004

Sørensen, A., Lübeck, M., Lübeck, P.S., Ahring, B.K., 2013. Fungal beta-glucosidases: A bottleneck in indus-trial use of lignocellulosic materials. *Biomolecules* 3(3), 612–631. https://doi.org/10.3390/biom3030612

Spindler, D.D., Wyman, C.E., Mohagheghi, A., Grohmann, K., 1988. Thermotolerant yeast for simultaneous saccharification and fermentation of cellulose to ethanol. *Appl. Biochem. Biotechnol.* 17(1–3), 279–293. https://doi.org/10.1007/BF02779163

Stichnothe, H., Azapagic, A., 2009. Bioethanol from waste: Life cycle estimation of the greenhouse gas sav-ing potential. *Resour. Conserv. Recycl.* 53(11), 624–630. https://doi.org/10.1016/j.resconrec.2009.04.012

Sukumaran, R.K., Abraham, A., Mathew, A.K., 2017. Enzymes for bioenergy. In: *Bioresources and Bioprocess in Biotechnology*, pp. 3–43. https://doi.org/10.1007/978-981-10-4284-3_1

Sun, Shaoni, Sun, Shaolong, Cao, X., Sun, R., 2016. The role of pretreatment in improving the enzymatic hydrolysis of lignocellulosic materials. *Bioresour. Technol.* 199, 49–58. https://doi.org/10.1016/j.biortech .2015.08.061

Sun, Y., Cheng, J., 2002. Hydrolysis of lignocellulosic materials for ethanol production: A review. *Bioresour. Technol.* 83(1), 1–11. https://doi.org/10.1016/S0960-8524(01)00212-7

Swatloski, R.P., Spear, S.K., Holbrey, J.D., Rogers, R.D., 2002. Dissolution of cellose with ionic liquids. *J. Am. Chem. Soc.* 124(18), 4974–4975. https://doi.org/10.1021/ja025790m

Taherzadeh, M.J., Niklasson, C., 2004. Ethanol from lignocellulosic materials: Pretreatment, acid and enzy-matic hydrolyses, and fermentation, pp. 49–68. https://doi.org/10.1021/bk-2004-0889.ch003

Takkellapati, S., Li, T., Gonzalez, M.A., 2018. An overview of biorefinery-derived platform chemicals from a cellu-lose and hemicellulose biorefinery. *Clean Technol. Environ. Policy.* https://doi.org/10.1007/s10098-018-1568-5

Tao, L., Chen, X., Aden, A., Kuhn, E., Himmel, M.E., Tucker, M., Franden, M.A.A., Zhang, M., Johnson, D.K., Dowe, N., Elander, R.T., 2012. Improved ethanol yield and reduced minimum ethanol selling price (MESP) by modifying low severity dilute acid pretreatment with deacetylation and mechanical refining: 2) Techno-economic analysis. *Biotechnol. Biofuels.* https://doi.org/10.1186/1754-6834-5-69

Tekin, K., 2015. Hydrothermal conversion of Russian olive seeds into crude bio-oil using a CaO catalyst derived from waste mussel shells. *Energy Fuels* 29(7), 4382–4392. https://doi.org/10.1021/acs.energy-fuels.5b00724

Thirmal, C., Dahman, Y., 2012. Comparisons of existing pretreatment, saccharification, and fermentation processes for butanol production from agricultural residues. *Can. J. Chem. Eng.* 90(3), 745–761. https:// doi.org/10.1002/CJCE.20601

Toledo-Alarcón, J., Capson-Tojo, G., Marone, A., Paillet, F., Ferraz Júnior, A.D.N., Chatellard, L., Bernet, N., Trably, E., 2018. Basics of bio-hydrogen production by dark fermentation. In: *Green Energy and Technology*, pp. 199–220. https://doi.org/10.1007/978-981-10-7677-0_6

Ulonska, K., König, A., Klatt, M., Mitsos, A., Viell, J., 2018. Optimization of multiproduct biorefinery pro-cesses under consideration of biomass supply chain management and market developments. *Ind. Eng. Chem. Res.* 57(20), 6980–6991. https://doi.org/10.1021/acs.iecr.8b00245

Uthandi, S., Kaliyaperumal, A., Srinivasan, N., Thangavelu, K., Muniraj, I.K., Zhan, X., Gathergood, N., Gupta, V.K., 2021. Microbial biodiesel production from lignocellulosic biomass: New insights and future challenges. *Crit. Rev. Environ. Sci. Technol.* https://doi.org/10.1080/10643389.2021.1877045

van Kuijk, S.J.A., Sonnenberg, A.S.M., Baars, J.J.P., Hendriks, W.H., Cone, J.W., 2015. Fungal treated ligno-cellulosic biomass as ruminant feed ingredient: A review. *Biotechnol. Adv.* 33(1), 191–202. https://doi .org/10.1016/j.biotechadv.2014.10.014

Vasić, K., Knez, Ž., Leitgeb, M., 2021. Bioethanol production by enzymatic hydrolysis from different lignocel-lulosic sources. *Molecules.* https://doi.org/10.3390/molecules26030753

Villas-Bôas, S.G., Esposito, E., Mitchell, D.A., 2002. Microbial conversion of lignocellulosic residues for production of animal feeds. *Anim. Feed Sci. Technol.* 98(1–2), 1–12. https://doi.org/10.1016/S0377 -8401(02)00017-2

Wang, J., Yin, Y., 2017. Introduction. In: *Green Energy and Technology*. Springer, Singapore, pp. 1–17. https:// doi.org/10.1007/978-981-10-4675-9_1

Wang, Y., Ho, S.H., Yen, H.W., Nagarajan, D., Ren, N.Q., Li, S., Hu, Z., Lee, D.J., Kondo, A., Chang, J.S., 2017. Current advances on fermentative biobutanol production using third generation feedstock. *Biotechnol. Adv.* https://doi.org/10.1016/j.biotechadv.2017.06.001

Wendt, L.M., Zhao, H., 2020. Review on bioenergy storage systems for preserving and improving feedstock value. *Front. Bioeng. Biotechnol.* 8, 1–15. https://doi.org/10.3389/fbioe.2020.00370

Werpy, T., Petersen, G., 2004. Top value added chemicals from biomass. *U.S. Dep. Energy* 1, 76. https://doi.org/10.2172/926125

Wierckx, N., Koopman, F., Ruijssenaars, H.J., De Winde, J.H., 2011. Microbial degradation of furanic compounds: Biochemistry, genetics, and impact. *Appl. Microbiol. Biotechnol.* 92(6), 1095–1105. https://doi.org/10.1007/s00253-011-3632-5

Wilson, D.B., 2009. Cellulases. *Encycl. Microbiol.*, 252–258. https://doi.org/10.1016/B978-012373944-5.00138-3

Witcover, J., Williams, R.B., 2020. Comparison of "Advanced" biofuel cost estimates: Trends during rollout of low carbon fuel policies. *Transp. Res. D* 79. https://doi.org/10.1016/j.trd.2019.102211

Wooley, R., Ruth, M., Glassner, D., Sheehan, J., 1999. Process design and costing of bioethanol technology: A tool for determining the status and direction of research and development. *Biotechnol. Prog.* https://doi.org/10.1021/bp990107u

Wright, M., Brown, R.C., 2007. Establishing the optimal sizes of different kinds of biorefineries. *Biofuels Bioprod. Biorefin.* https://doi.org/10.1002/bbb.25

Xie, G., Peng, L., 2011. Genetic engineering of energy crops: A strategy for biofuel production in China. *J. Integr. Plant Biol.* https://doi.org/10.1111/j.1744-7909.2010.01022.x

Xu, Y., Hu, X., Li, W., Shi, Y., 2011. Preparation and characterization of bio-oil from biomass. In: *Progress in Biomass and Bioenergy Production.* https://doi.org/10.5772/16466

Xu, Z., Huang, F., 2014. Pretreatment methods for bioethanol production. *Appl. Biochem. Biotechnol.* 174(1), 43–62. https://doi.org/10.1007/S12010-014-1015-Y

Xue, C., Cheng, C., 2019. Butanol production by Clostridium. https://doi.org/10.1016/bs.aibe.2018.12.001

Yang, B., Tucker, M., 2013. Laboratory pretreatment systems to understand biomass deconstruction. Aqueous pretreat. *Plant Biomass Biol. Chem. Convers. Fuels Chem.*, 489–521. https://doi.org/10.1002/9780470975831.CH23

Yang, C., Wang, S., Yang, J., Xu, D., Li, Y., Li, J., Zhang, Y., 2020. Hydrothermal liquefaction and gasification of biomass and model compounds: A review. *Green Chem.* https://doi.org/10.1039/d0gc02802a

Yang, S.-T., Liu, X., Zhang, Y., 2007. Metabolic engineering – Applications, methods, and challenges BT - Bioprocessing for value-added products from renewable resources. In: *Bioprocessing for Value-Added Products from Renewable Resources: New Technologies and Applications*, pp. 73–118. Elsevier. https://www.elsevier.com/books/bioprocessing-for-value-added-products-from-renewable-resources/yang/978-0-444-52114-9

Yi, Y., 2021. Tiny bugs play big role. In: *Advances in 2nd Generation of Bioethanol Production.* Elsevier, pp. 113–136. https://doi.org/10.1016/b978-0-12-818862-0.00007-8

Yoo, C.G., Pu, Y., Ragauskas, A.J., 2017. Ionic liquids: Promising green solvents for lignocellulosic biomass utilization. *Curr. Opin. Green Sustain. Chem.* https://doi.org/10.1016/j.cogsc.2017.03.003

Yuan, H., Tan, L., Kida, K., Morimura, S., Sun, Z.Y., Tang, Y.Q., 2021. Potential for reduced water consumption in biorefining of lignocellulosic biomass to bioethanol and biogas. *J. Biosci. Bioeng.* https://doi.org/10.1016/j.jbiosc.2020.12.015

Yue, D., You, F., Snyder, S.W., 2014. Biomass-to-bioenergy and biofuel supply chain optimization : Overview, key issues and challenges. *Comput. Chem. Eng.* 66, 36–56. https://doi.org/10.1016/j.compchemeng.2013.11.016

Zetterholm, J., Bryngemark, E., Ahlström, J., Söderholm, P., Harvey, S., Wetterlund, E., 2020. Economic evaluation of large-scale biorefinery deployment: A framework integrating dynamic biomass market and techno-economic models. *Sustainability.* https://doi.org/10.3390/su12177126

Zhang, J., Hou, W., Bao, J., 2016. Reactors for high solid loading pretreatment of lignocellulosic biomass. In: *Advances in Biochemical Engineering/Biotechnology.* https://doi.org/10.1007/10_2015_307

Zhang, K., Pei, Z., Wang, D., 2016. Organic solvent pretreatment of lignocellulosic biomass for biofuels and biochemicals: A review. *Bioresour. Technol.* https://doi.org/10.1016/j.biortech.2015.08.102

Zhao, C., Ding, W., Chen, F., Cheng, C., Shao, Q., 2014. Effects of compositional changes of AFEX-treated and H-AFEX-treated corn stover on enzymatic digestibility. *Bioresour. Technol.* 155, 34–40. https://doi.org/10.1016/j.biortech.2013.12.091

Zhu, S., Wu, Y., Yu, Z., Zhang, X., Wang, C., Yu, F., Jin, S., Zhao, Y., Tu, S., Xue, Y., 2005. Simultaneous saccharification and fermentation of microwave/alkali pre-treated rice straw to ethanol. *Biosyst. Eng.* 92(2), 229–235. https://doi.org/10.1016/j.biosystemseng.2005.06.012

15 Biorefineries and the Circular Bioeconomy for the Management of Agro-Industrial Byproducts
Review and a Bibliometric Analysis

Neeraj Pandey and Akanksha Mishra

CONTENTS

15.1 INTRODUCTION

The Earth has limited resources, and humankind has a moral responsibility to have a symbiotic interface with it. Global warming, drying water aquifers, depletion of energy sources, loss of biodiversity, and desertification are signals of the careless exploitation of resources by its inhabitants that jeopardize human sustainable living goals (Bogardi et al., 2021; Opoku, 2019; Cowie et al., 2011). The adoption and use of biorefineries and circular bioeconomy practices have been a saviour and hold promise for the future. The biorefinery approach would help us sustainably use biomass for energy (Ahmad et al., 2020; Mohan et al., 2016). It uses agro-industrial byproducts and processes them to generate clean energy among other benefits (Federici et al., 2009). The circular bioeconomy emphasizes the valorization of biomass and biowaste for producing energy and new products. The circular bioeconomy will grow by US$8 trillion by 2030 (Salvador et al., 2021). It is high time that non-renewable fossil fuels are replaced by biomass raw-material-based energy substitutes

DOI: 10.1201/9781003125679-15

(Muscat et al., 2021; Stegmann et al., 2020). This would help us to reduce our carbon footprint and minimize environmental degradation. There has also been growing awareness among consumers regarding the circular bioeconomy, reflected in the demand for green products and products made from recycled materials (Pandey and Kaushik, 2012; Singh and Pandey, 2019).

The bioeconomy approach was proposed by the European Commission in 2012 along with 50 other countries (Stegmann et al., 2020; Fund et al., 2018). The definition of bioeconomy has been revisited by various scholars, including the European Commission, emphasizing circularity and sustainability (European Commission, 2018). The bioeconomy is a complementary approach to the circular economy with a focus on biomass and bioresources. Therefore, concepts like biorefinery, bioeconomy, and circular economy promote sustainable living for individuals, organizations, and society.

This study uses a bibliometric approach to analyze the literature on biorefineries and the circular bioeconomy to manage agro-industrial byproducts. There is no bibliometric study on the topic that would help develop emerging themes and research areas besides mapping the literature on the domain. The bibliometric analysis will aid in improving the understanding of pertinent areas like biorefineries and the circular bioeconomy and provide insight for academia and practitioners working in the agro-industrial sector. The review aims to explore the following research questions (RQs):

RQ1: What is the publication trend in biorefineries and the circular bioeconomy for the management of agro-industrial byproducts?

RQ2: What are the highly cited papers, authors, and collaboration networks in the field of biorefineries and the circular bioeconomy for the management of agro-industrial byproducts?

RQ3: What are the insights from the extant literature on biorefineries and the circular bioeconomy for the management of agro-industrial byproducts?

RQ4: What are the emerging areas of research for future scholars working in the field of biorefineries and the circular bioeconomy for the management of agro-industrial byproducts?

The rest of the chapter is organized as follows. The next section provides an overview of the extant literature on the given topic. This is followed by a discussion on the methodology. The next section covers the results of the bibliometric analysis. Finally, the emerging areas in biorefineries and the circular bioeconomy for the management of agro-industrial byproducts are elaborated. This will help to set an agenda for future scholars working in this domain.

15.2 LITERATURE REVIEW

The circular bioeconomy is a complementary approach supporting circular economy goals. Both concepts promote sustainable and synergistic co-living of humans and nature (Haberl et al., 2014). The health of the biomass-producing agriculture ecosystem should be sustainable to produce biomass generation after generation (Muscat et al., 2021). Awareness about sustainable practices like green product preference (Tripathi and Pandey, 2019), recycled packaging (Castle, 1994), construction material (Langston, 2008), and landfill design (Townsend et al., 2015) has also helped in espousing the cause of biorefineries and the circular bioeconomy. Biorefineries and the circular bioeconomy have also been analyzed from a regional perspective. A study conducted in northwest European countries found that a lack of clear regulations and mismatch between the investment and return on the circular bioeconomy due to the relatively small size of bio-based markets were major challenges in adopting these practices by organizations (Stegmann et al., 2020). Possas et al. (2021) examined the patent landscape in the circular bioeconomy. Many organizations like IBM and 3M used patents and licensing mechanisms as revenue generators for innovative, environmentally friendly products (Pandey and Dharni, 2014). The patents help valorize waste products by their transformation into finished products and bring revenue to the organization (Brandão et al., 2021).

Giampietro (2019) compared different narratives on the circular bioeconomy and circular economy in the extant literature and highlighted the differences and complementarities between the two approaches. D'Amato et al. (2017) proposed "green economy" as an umbrella term for the circular bioeconomy and circular economy and emphasized that it is more inclusive as it includes local biomass resource availability and its management. The business model using the principles of circular bioeconomy should have consumer insight and knowledge about the market, supply chain management of feedstock collection, biomass selection and routing, and a resilient management system (Salvador et al., 2021).

The valorization of waste into renewable feedstock can be done through various biotechnological interventions (Mohan et al., 2021); for example, food waste is being processed into biofuels, electricity, and animal feed (Dahiya et al., 2018). The waste biorefineries, which produce bioenergy products from biomass, exemplify a circular bioeconomy (Mohan et al., 2019). This kind of biorefinery approach provides sustainable living solutions. Kardung et al. (2021) found that the key driver for the success of circular bioeconomy adoption in Europe was the regulatory framework. However, it was seen that rigid regulations at times made the adoption of biorefineries and circular bioeconomy practices difficult.

15.3 METHODOLOGY

This study uses bibliometric analysis to map the intellectual structure and key themes and to derive the research agenda for scholars and practitioners working in the field of biorefineries and circular bioeconomy. The bibliometric analysis included performance analysis and science mapping. There were studies on the circular economy (Goyal et al., 2020), transfer pricing (Kumar et al., 2021), electronic word of mouth (e-WOM) (Donthu et al., 2021a), international marketing (Donthu et al., 2021b), forest bioeconomy (Paletto et al., 2020), and sustainable agriculture systems (Rocchi et al., 2020) using bibliometric analysis. However, there was no bibliometric-analysis-based study on biorefineries and the circular bioeconomy. The key search words used in the Scopus database search were "circular bioeconomy", "biorefinery", "agro-industrial", "byproducts", "circular", "agriculture", and "agri-business" using "and" and "or" combinations in the search string. The editorial, book chapters, and conference papers were excluded from the search. Only articles in the English language were considered in the dataset. The Scopus database was chosen as it is the largest research paper compilation database. Scopus is more extensive than Web of Science and other similar journal databases that index journal articles in the realm of biorefineries and the circular bioeconomy. The performance analysis in this study included author and institute affiliated publication details besides citation-related metrics. The science mapping part of the bibliometric analysis included citation, co-citation, and co-author network analysis.

15.4 RESULTS

15.4.1 PUBLICATION TRENDS

The dispersion of literature published in the last decade indicates that biorefinery and circular bioeconomy topics have gained academia's attention. The highest number of publications on biorefineries and the circular bioeconomy for the management of agro-industrial byproducts was recorded in 2021 (n = 49) and 2020 (n = 41), which reflects an increasing trend from the previous years. There were not many studies published before 2018 on this topic. However, the increasing trend in publication indicates that the concept of sustainability measures like biorefineries and the circular bioeconomy for the management of agro-industrial byproducts are becoming popular and increasingly practised by the industry (Table 15.1).

15.4.2 PUBLICATION OUTLET

The publication outlet reflects that the journal *Sustainability* is the most sought-after journal for research on biorefineries and the circular bioeconomy (n = 12 articles) (Table 15.2). This is followed

TABLE 15.1
Publication Trend

Year	Articles
2013	1
2015	1
2016	1
2017	5
2018	25
2019	23
2020	41
2021	49

TABLE 15.2
Most Relevant Sources

Sources	Articles
Sustainability (Switzerland)	12
Bioresource Technology	9
New Biotechnology	8
ACS Sustainable Chemistry and Engineering	6
Journal of Environmental Management	5
Renewable and Sustainable Energy Reviews	5
Resources Conservation and Recycling	5
Energies	4
Molecules	4
Science of the Total Environment	4

by *Bioresource Technology* and *New Biotechnology* with nine and eight articles, respectively, on biorefineries and the circular bioeconomy for the management of agro-industrial byproducts.

15.4.3 PUBLICATION PERFORMANCE

15.4.3.1 Global Citation

Global citations are the number of citations across all disciplines and languages (Baker et al., 2020). In this analysis, the highest globally cited document was the research paper entitled "Waste biorefineries: Enabling circular economies in developing countries" (Nizami et al., 2017) (n = 210) (Table 15.3). The study discussed the scope of waste biorefineries in developing countries and the potential positive impact on energy solutions as well as value-added products. The next most globally cited (n = 140) research paper was "Green bioplastics as part of a circular bioeconomy" (Karan et al., 2019). It argues the viability of bio-degradable, biorefinery, and bio-based plastics usage for the business development of the circular bioeconomy. The use of packaging made from recycled products is preferred by many consumers (Singh and Pandey, 2018).

15.4.3.2 Local Citation

Local citation is defined as citation within the identified subject domain (Baker et al., 2020). In this study, 148 documents were reviewed. The data was drawn from Scopus using relevant keywords for biorefineries and the circular bioeconomy. The research paper entitled "Agricultural waste: Review

TABLE 15.3
Highest Globally Cited Article

Document	Total citations	TC per year	Normalized TC
Nizami et al., 2017	210	42.00	3.00
Karan et al., 2019	140	31.33	4.39
Therond et al., 2017	91	18.20	1.30
Sherwood, 2020	69	34.50	4.98
Klerkx et al., 2020	64	32.00	4.62
Awasthi et al., 2019	64	21.33	2.99
Zuin et al., 2018	62	15.50	2.32
Sadhukhan et al., 2018	60	15.00	2.24
Duque-Acevedo et al., 2020	58	29.00	4.19
Hamelin et al., 2019	58	19.33	2.71
Berbel et al., 2018	54	13.50	2.02
Sarsaiya et al., 2019	38	12.67	1.78

of the evolution, approaches and perspectives on alternative uses" (Duque-Acevedo et al., 2020) was the highest locally cited article (n = 7), followed by "The significance of biomass in a circular economy" (Sherwood, 2020), which was the second-highest (n = 4) locally cited article (Table 15.4).

15.4.3.3 Top Authors and Collaborations

The publication by authors exhibits the highest number of articles published by Luis J. Belmonte-Ureña (n = 5). The top collaborators for Luis J. Belmonte-Ureña are Mónica Duque-Acevedo and Francisco Camacho-Ferre (n = 2). The other authors with the second-highest number of publications on the circular bioeconomy are Francisco Camacho-Ferre and Mónica Duque-Acevedo, with four articles each (Figure 15.1).

15.4.4 COUNTRIES

15.4.4.1 Top Countries

The literature on biorefineries and the circular bioeconomy for the management of agro-industrial byproducts has contributors from 49 countries. The most prolific literature contributing nation was Italy (n = 65 research papers), followed by Spain (n = 52), and France (n = 30). The total contributing country count is larger than the number of research papers chosen for bibliometric analysis in Table 15.5, as each paper has collaborators from multiple countries. The country-wise analysis reflects that major knowledge contribution is coming from closely connected western European countries. Country-wise analysis indicates increasing research interest in biorefineries and the circular bioeconomy in European countries and faster-growing economies like China and India.

15.4.4.2 Top Country Collaborations

The Biblioshiny visualization tool was used to demonstrate the social structure of research on biorefineries and the circular bioeconomy for the management of agro-industrial byproducts. It shows the authors with different country affiliations that collaborated in research with each other on the subject of biorefineries and the circular bioeconomy. Social network analysis revealed three major social structures. The first major social network involves Belgium, which is the primary collaborator with ten countries. These ten countries include Chile, Czech Republic, Ecuador, Germany, Indonesia, Iran, Malaysia, Mexico, Netherlands, and Saudi Arabia. The second major social network involves China's international collaborations with nine countries, which include the Czech

TABLE 15.4
Local and Global Citations

Journal	Document	Year	Local citations	Global citations
Global Ecology and Conservation	Duque-Acevedo et al., 2020	2020	7	58
Bioresource Technology	Sherwood et al., 2020	2020	5	69
Science of the Total Environment	Donner et al., 2020	2020	4	35
Agronomy	Duque-Acevedo et al., 2020	2020	4	20
Renewable and Sustainable Energy Reviews	Awasthi et al., 2019	2019	4	64
New Biotechnology	Egea et al., 2018	2018	4	28
Bioresource Technology	Nizami et al., 2017	2017	4	210
Food & Function	Jimenez-Lopez et al., 2020	2020	3	23
Biotechnology Journal	Aguilar et al., 2019	2019	3	31
Renewable and Sustainable Energy Reviews	Hamelin et al., 2019	2019	3	58
Procedia CIRP	Vea et al., 2018	2018	3	33
Trends in Plant Science	Karan et al., 2019	2019	2	94
Sustainability	Diacono et al., 2019	2019	2	39
New Biotechnology	Bell et al., 2018	2018	2	65
New Biotechnology	Lainez et al., 2018	2018	2	41
Resources, Conservation & Recycling	Harder et al., 2021	2021	1	2
Organic Agriculture	Ullmann et al., 2021	2021	1	3
Journal of Environmental Management	Teigiserova et al., 2021	2021	1	7
New Biotechnology	Woniak et al., 2021	2021	1	5
Marine Drugs	Caruso et al., 2020	2020	1	10

Republic, Belgium, Denmark, Hong Kong, Iran, Korea, Malaysia, New Zealand, and Sweden. The third major social network is France's collaborations with eight countries. It involves international collaborations with Belgium, Canada, Denmark, Germany, Netherlands, Poland, Switzerland, and the United Kingdom.

15.4.4.3   Institutions

The articles published by institutions indicate that the University of Almería, Spain is a prolific contributor to biorefinery and circular bioeconomy research (n = 7 articles). This is followed by the Institute of Agricultural Research and Training (IFAPA), Nigeria, the University of Basilicata, Italy, and the University of Bologna, Italy, each contributing five articles on biorefineries and the circular bioeconomy (Table 15.6).

15.5   EMERGING THEMES

The PageRank technique was used to explore the major emerging themes in the domain of biore-fineries and the circular bioeconomy. PageRank reflects the position of the document as its score is based on the number of links made to other research papers. In other words, the page rank score for a given document is based on the links made to other pages/documents. Also, there is a possibility that a document with a higher number of citations globally or locally may have a lower page rank or the other way around (Ding and Cronin, 2011).

Therefore, PageRank is a vital measure to distinguish the quality of a document and must be identified in bibliometric reviews, as it can bring up key articles among highly cited articles

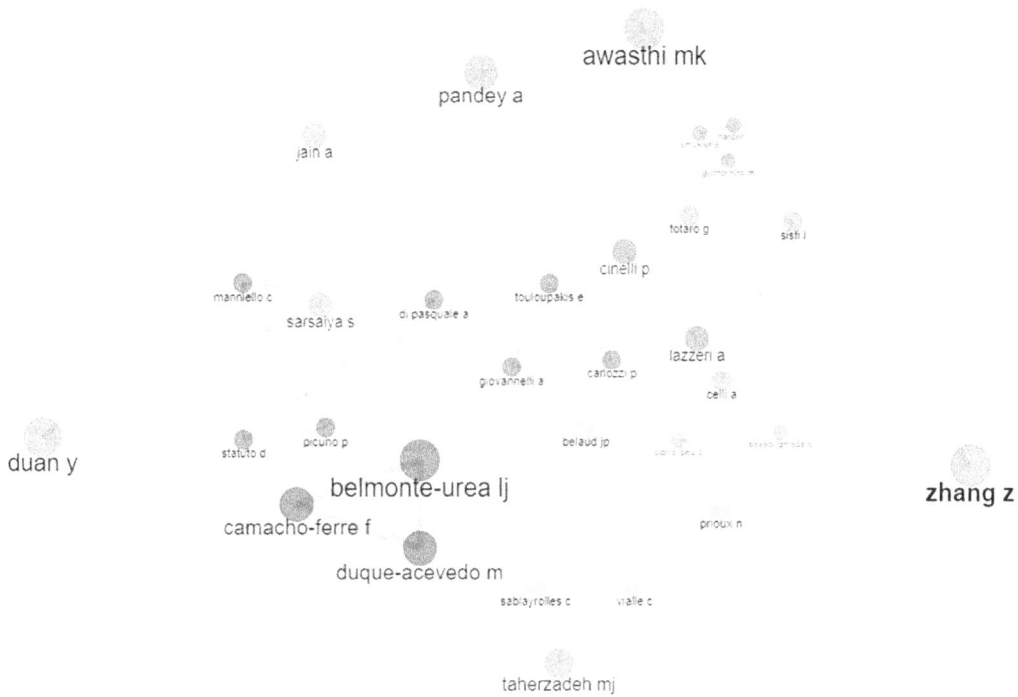

FIGURE 15.1 Co-authorship network.

TABLE 15.5
Top Contributing Countries*

Country	Documents
Italy	65
Spain	52
France	30
China	23
India	15
UK	15
Sweden	14
Greece	13
Belgium	11
Germany	11

*Each paper has collaborators from multiple countries.

(Donthu et al., 2021c). Furthermore, PageRank also generates a thematic cluster. PageRank analysis generated four major clusters from the database of this study (see Table 15.7).

Cluster 1 represents the macro-perspective of the circular bioeconomy and the roadmap of food system transition towards a circular economy. The seminal paper in this cluster was authored by Patrizia Ghisellini in 2016, entitled "A review on circular economy: The expected transition to a balanced interplay of environmental and economic systems", with the highest score in cluster 1. With a PageRank of 0.060746, this article was one of the highly cited documents on the circular economy. Cluster 2 represents the debate on guiding principles and strategies on bioeconomy along

TABLE 15.6
Most Popular Institute Affiliation

Affiliations	Articles
University of Almería, Spain	7
Institute of Agricultural Research and Training (IFAPA), Spain	5
University of Basilicata, Italy	5
University of Bologna, Italy	5
Aarhus University, Denmark	4
Northwest A and f University, China	4
Universidade De Lisboa, Portugal	4
Universit De Toulouse, France	4

with comparative analysis with other possible avenues. This theme was popular among the most cited articles in the second cluster. An article entitled "Green, circular, bioeconomy: A comparative analysis of sustainability avenues" had the highest PageRank around the above-discussed theme (PageRank = 0.051809).

Cluster 3 highlighted the theme of food waste refinery and agro-industrial complex approach with a key article entitled "An efficient agro-industrial complex in Almería (Spain): Towards an integrated and sustainable bioeconomy model". The article was authored by Francisco J. Egea (2018) with a PageRank of 0.039885. Cluster 4 focuses on the opportunities in the biorefinery and circular bioeconomy domain. Enrica Imbert's article entitled "Food waste valorization options: Opportunities from the bioeconomy" (PR = 0.064725) was the highest cited in this cluster (Table 15.7).

15.5.1 NOVEL TOPICS

Co-occurrence analysis of all keywords from 148 documents was performed in Biblioshiny software. It aided in exploring the extant research on biorefineries and the circular bioeconomy. At a fundamental level, the keywords represent the article content (Comerio and Strozzi, 2019). The co-occurrence of keywords creates the framework of intellectual structure on the subject matter (Donthu et al., 2021c). The larger the size of the nodes, the higher number of research literature is available on that particular node. Topics like circular bioeconomy, bio-based economy, biorefinery, sustainability, bioenergy, and digestive valorization have emerged as popular research areas (Figure 15.2 and 15.3).

15.6 CONCLUSIONS AND FUTURE SCOPE OF WORK

This study has mapped the extant literature and analyzed the development of domains related to biorefineries and the circular bioeconomy from different constituent perspectives. The above discussion on publication trends and publication outlets covers the first research question (RQ1). The increasing publication trend shows the growing relevance of biorefineries and the circular bioeconomy both in theory and practice. The bibliometric analysis also shows that in the case of newer research areas like biorefineries and the circular bioeconomy, open-access journals have taken the lead in terms of the number of publications and readership. The open-access journals are widely read and cited compared to pay-walled journals, which require a paid subscription. The discussion on highly cited papers, authors, and collaboration networks in the field of biorefineries and the circular bioeconomy for the management of agro-industrial byproducts covers the second research question (RQ2). The global citation of open-access journals was higher than in other category scientific journals.

The bibliometric analysis findings show that there is a higher number of cross-disciplinary global citations than local citations. It reflects the keen interest of other disciplines in the biorefinery and

TABLE 15.7
Thematic Clusters

Node	Journal	Cluster	Betweenness	Closeness	PageRank
Ghisellini P., 2016	*Journal of Cleaner Production*	1	81.13981719	0.008928571	0.06074614
Galanakis C.M., 2012	*Trends in Food Science & Technology*	1	0	0.007633588	0.034319764
Sherwood J., 2020	*Bioresource Technology*	1	17	0.005617978	0.052650957
Donner M., 2020	*Science of The Total Environment*	1	0	0.007692308	0.039484579
Jurgilevich A., 2016	*Sustainability*	1	1.99010457	0.008	0.052821491
Molina-Moreno V., 2017	*Water*	1	6.40500282	0.008264463	0.048483651
Ubando A.T., 2020	*Bioresource Technology*	1	0	0.003802281	0.032735775
Zabaniotou A., 2018	*Journal of Cleaner Production*	1	0	0.005524862	0.042077953
Duque-Acevedo M., 2020–2021	*Global Ecology and Conservation*	2	6.367206125	0.008333333	0.045050395
Stegmann P., 2020	*Resources, Conservation & Recycling*	2	0	0.0008130081	0.042366607
Damato D., 2017	*Journal of Cleaner Production*	2	14.72612379	0.008403361	0.051809418
Meyer R., 2017	*Sustainability*	2	0	0.005	0.035286614
Staffas L., 2013	*Sustainability*	2	19.81648719	0.008	0.05058426
Duque-Acevedo M., 2020–2022	*Agronomy*	2	2.542610984	0.008264463	0.044341969
Kirchherr J., 2017	*Resources, Conservation and Recycling*	3	49.84074286	0.00877193	0.059529535
Mccormick K., 2013	*Sustainability*	3	0	0.007042254	0.040192247
Dahiya S., 2018	*Bioresource Technology*	3	0	0.007751938	0.033559319
Egea F.J., 2018	*New Biotechnology*	3	0	0.007751938	0.039885877
Imbert E. 2017	*Open Agriculture*	4	1	0.002336449	0.064725541
Maina S., 2017	*Current Opinion in Green and Sustainable Chemistry*	4	0	0.002309469	0.042421469
Winans K., 2017	*Renewable and Sustainable Energy Reviews*	4	0	0.002309469	0.042421469

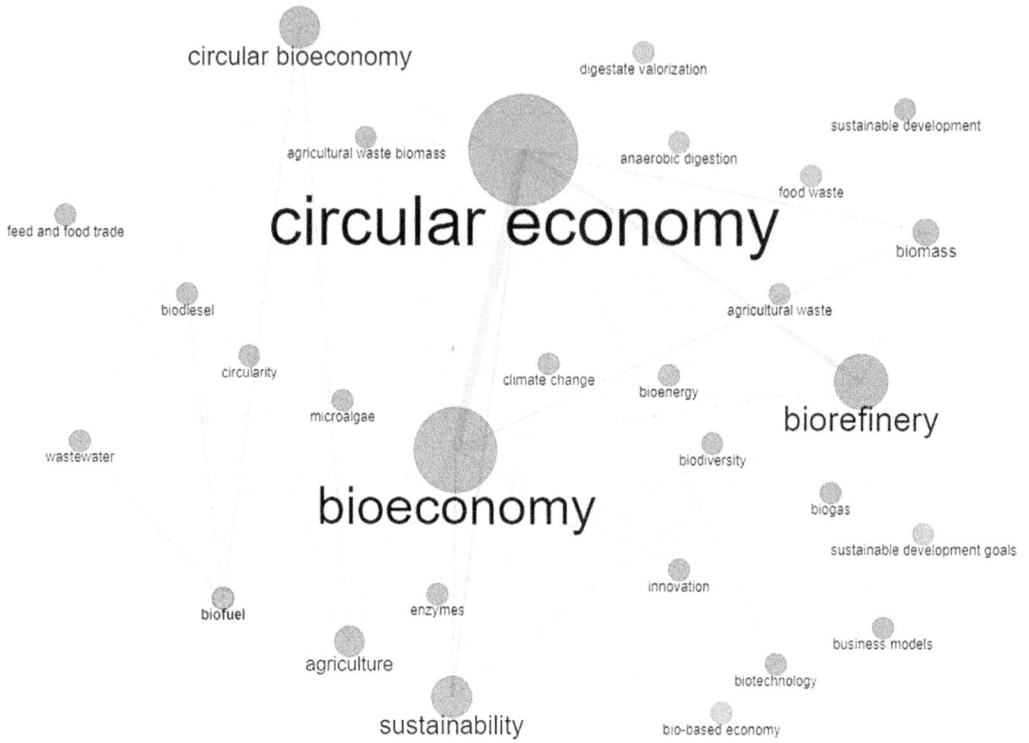

FIGURE 15.2 Co-occurrence analysis.

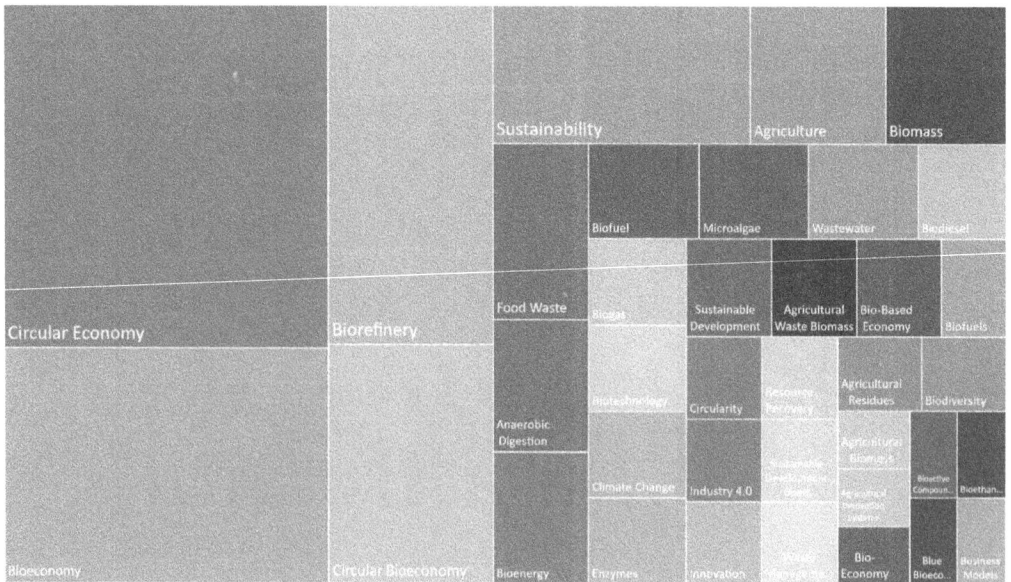

FIGURE 15.3 Treemap.

circular bioeconomy domain. These global citations include disciplines like economics, marketing, sociology, and other management sciences. The popularity of sustainable development goals, which is part of the 2030 Agenda for Sustainable Development, has also generated awareness and interest from other functional areas in topics like biorefineries and the circular bioeconomy. The collaboration network showed three strong players emerging in the area of biorefineries and the circular bioeconomy for the management of agro-industrial byproducts. These are Belgium, China, and France, which have the highest number of collaborators from other countries. These countries serve as the main node for research collaboration with other nations. Authors affiliated with Italy, Spain, and France had the highest number of contributions in the biorefinery and circular bioeconomy domain. The PageRank also helped to come up with major emerging themes like the integrative approach to economics and environment, challenges in sustaining a circular bioeconomy, and innovative waste valorization approaches. These insights on biorefineries and the circular bioeconomy covered the third research question (RQ3).

The bibliometric analysis and review highlighted that the valorization of waste products using modern scientific techniques, including machine learning and artificial intelligence, is an important area of research for future scholars. This would find applications in solid waste management, agriculture waste management, and industrial waste management. The circular bioeconomy also needs to take care of local factors, including biomass resource availability. There is scope for methodological studies, especially meta-analysis studies, on biorefineries and the circular bioeconomy. The analysis also cautions that we must consider the excessive use of chemical fertilizers on limited agricultural land to enhance farm produce. Going forward, we need to explore more marine biomass resources as it occupies two-thirds of the space of the Earth. Hence, further study on the valorization of marine biomass resources across different dimensions in the context of biorefineries and the circular bioeconomy is the need of the hour. These emerging research areas for future scholars help us cover the fourth research question (RQ4).

The review and bibliometric analysis highlighted that European countries has taken the lead in the research on biorefineries and the circular bioeconomy. It is not only in the case of the total number of publications from the region but also in terms of the total number of collaborations; European countries have been ahead of other nations. There is a growing interest from academia and practitioners in biofuels and bioenergy production. There is a need for more research studies on it, including strategies and management guidelines on agricultural waste mass valorization. The bibliometric analysis also found that there were hardly any studies covering the regulatory aspects of biorefineries and the circular bioeconomy in the context of agro-industrial products. It would be interesting to do a comparative mapping of the waste management regulations in various nations and how they can contribute to developing regulations related to biorefineries and the circular bioeconomy.

REFERENCES

Ahmad, B., Yadav, V., Yadav, A., Rahman, M. U., Yuan, W. Z., Li, Z., & Wang, X. (2020). Integrated biorefinery approach to valorize winery waste: A review from waste to energy perspectives. *Science of the Total Environment*, *719*, 137315.

Baker, H. K., Pandey, N., Kumar, S., & Haldar, A. (2020). A bibliometric analysis of board diversity: Current status, development, and future research directions. *Journal of Business Research*, *108*, 232–246.

Bogardi, J. J., Bharati, L., Foster, S., & Dhaubanjar, S. (2021). Water and its management: Dependence, linkages and challenges. In: *Handbook of Water Resources Management: Discourses, Concepts and Examples* (pp. 41–85). Springer, Cham.

Brandão, A. S., Gonçalves, A., & Santos, J. M. (2021). Circular bioeconomy strategies: From scientific research to commercially viable products. *Journal of Cleaner Production*, 295, 126407.

Castle, L. (1994). Recycled and re-used plastics for food packaging? *Packaging Technology and Science*, 7(6), 291–297.

Comerio, N., & Strozzi, F. (2019). Tourism and its economic impact: A literature review using bibliometric tools. *Tourism Economics*, *25*(1), 109–131.

Cowie, A. L., Penman, T. D., Gorissen, L., Winslow, M. D., Lehmann, J., Tyrrell, T. D., ... & Akhtar-Schuster, M. (2011). Towards sustainable land management in the drylands: Scientific connections in monitoring and assessing dryland degradation, climate change and biodiversity. *Land Degradation and Development, 22*(2), 248–260.

Dahiya, S., Kumar, A. N., Sravan, J. S., Chatterjee, S., Sarkar, O., & Mohan, S. V. (2018). Food waste biorefinery: Sustainable strategy for circular bioeconomy. *Bioresource Technology, 248*(A), 2–12.

D'Amato, D., Droste, N., Allen, B., Kettunen, M., Lähtinen, K., Korhonen, J., Toppinen, A. (2017). Green, circular, bio economy: A comparative analysis of sustainability avenues. *Journal of Cleaner Production, 168*, 716–734.

Ding, Y., & Cronin, B. (2011). Popular and/or prestigious? Measures of scholarly esteem. *Information Processing and Management, 47*(1), 80–96.

Donthu, N., Kumar, S., Mukherjee, D., Pandey, N., & Lim, W. M. (2021c). How to conduct a bibliometric analysis: An overview and guidelines. *Journal of Business Research, 133*, 285–296.

Donthu, N., Kumar, S., Pandey, N., Pandey, N., & Mishra, A. (2021a). Mapping the electronic word-of-mouth (eWOM) research: A systematic review and bibliometric analysis. *Journal of Business Research, 135*, 758–773.

Donthu, N., Kumar, S., Pattnaik, D., & Pandey, N. (2021b). A bibliometric review of International Marketing Review (IMR): Past, present, and future. *International Marketing Review, 38*(5), 840–878.

Duque-Acevedo, M., Belmonte-Urena, L. J., Cortés-García, F. J., & Camacho-Ferre, F. (2020). Agricultural waste: Review of the evolution, approaches and perspectives on alternative uses. *Global Ecology and Conservation, 22*, e00902.

European Commission. (2018). A sustainable bioeconomy for Europe: Strengthening the connection between economy, society and the environment. https://ec.europa.eu/research/bioeconomy/pdf/official-strategy _en.pdf.

Federici, F., Fava, F., Kalogerakis, N., & Mantzavinos, D. (2009). Valorisation of agro-industrial by-products, effluents and waste: Concept, opportunities and the case of olive mill wastewaters. *Journal of Chemical Technology & Biotechnology: International Research in Process, Environmental & Clean Technology, 84*(6), 895–900.

Fund, C., El-Chichakli, B., & Patermann, C. (2018). Bioeconomy policy (Part III). *Update Report of National Strategies around the World*. A Report from the German Bioeconomy Council.

Giampietro, M. (2019). On the circular bioeconomy and decoupling: Implications for sustainable growth. *Ecological Economics, 162*, 143–156.

Goyal, S., Chauhan, S., & Mishra, P. (2020). Circular economy research: A bibliometric analysis (2000–2019) and future research insights. *Journal of Cleaner Production, 287*, 125011.

Haberl, H., Erb, K. H., & Krausmann, F. (2014). Human appropriation of net primary production: Patterns, trends, and planetary boundaries. *Annual Review of Environment and Resources, 39*(1), 363–391.

Karan, H., Funk, C., Grabert, M., Oey, M., & Hankamer, B. (2019). Green bioplastics as part of a circular bioeconomy. *Trends in Plant Science, 24*(3), 237–249.

Kardung, M., Cingiz, K., Costenoble, O., Delahaye, R., Heijman, W., Lovrić, M., ... & Zhu, B. X. (2021). Development of the circular bioeconomy: Drivers and indicators. *Sustainability, 13*(1), 413.

Kumar, S., Pandey, N., Lim, W. M., Chatterjee, A. N., & Pandey, N. (2021). What do we know about transfer pricing? Insights from bibliometric analysis. *Journal of Business Research, 134*, 275–287.

Langston, C. (Ed.). (2008). *Sustainable Practices in the Built Environment*. Routledge.

Mohan, S. V., Dahiya, S., Amulya, K., Katakojwala, R., & Vanitha, T. K. (2019). Can circular bioeconomy be fueled by waste biorefineries—A closer look. *Bioresource Technology Reports, 7*, 100277.

Mohan, S. V., Nikhil, G. N., Chiranjeevi, P., Reddy, C. N., Rohit, M. V., Kumar, A. N., & Sarkar, O. (2016). Waste biorefinery models towards sustainable circular bioeconomy: Critical review and future perspectives. *Bioresource Technology, 215*, 2–12.

Muscat, A., de Olde, E. M., Ripoll-Bosch, R., Van Zanten, H. H., Metze, T. A., Termeer, C. J., ... & de Boer, I. J. (2021). Principles, drivers and opportunities of a circular bioeconomy. *Nature Food, 2*(8), 561–566.

Nizami, A. S., Rehan, M., Waqas, M., Naqvi, M., Ouda, O. K., Shahzad, K., ... & Pant, D. (2017). Waste biorefineries: Enabling circular economies in developing countries. *Bioresource Technology, 241*, 1101–1117.

Opoku, A. (2019). Biodiversity and the built environment: Implications for the Sustainable Development Goals (SDGs). *Resources, Conservation and Recycling, 141*, 1–7.

Paletto, A., Biancolillo, I., Bersier, J., Keller, M., & Romagnoli, M. (2020). A literature review on forest bioeconomy with a bibliometric network analysis. *Journal of Forest Science, 66*(7), 265–279.

Pandey, N., & Dharni, K. (2014). *Intellectual Property Rights*. PHI Learning Pvt. Ltd.

Pandey, N., & Kaushik, S. (2012). Factors affecting consumer's green product purchase decisions: An empirical approach. *International Journal of Business Competition and Growth*, 2(4), 341–356.

Possas, C., de Souza Antunes, A. M., de Oliveira, A. M., de Souza Mendes, C. D. U., Ramos, M. P., Schumacher, S. D. O. R., & Homma, A. (2021). Vaccine innovation for pandemic preparedness: patent landscape, global sustainability, and circular bioeconomy in post-COVID-19 era. *Circular Economy and Sustainability*, 1(4), 1–23.

Rocchi, L., Boggia, A., & Paolotti, L. (2020). Sustainable agricultural systems: A bibliometrics analysis of ecological modernization approach. *Sustainability*, 12(22), 9635.

Salvador, R., Puglieri, F. N., Halog, A., de Andrade, F. G., Piekarski, C. M., & Antonio, C. (2021). Key aspects for designing business models for a circular bioeconomy. *Journal of Cleaner Production*, 278, 124341.

Sherwood, J. (2020). The significance of biomass in a circular economy. *Bioresource Technology*, 300, 122755.

Singh, G., & Pandey, N. (2018). The determinants of green packaging that influence buyers' willingness to pay a price premium. *Australasian Marketing Journal (AMJ)*, 26(3), 221–230.

Singh, G., & Pandey, N. (2019). Revisiting green packaging from a cost perspective: The remanufacturing vs new manufacturing process. *Benchmarking: An International Journal*, 26(3), 1080–1104.

Stegmann, P., Londo, M., & Junginger, M. (2020). The circular bioeconomy: Its elements and role in European bioeconomy clusters. *Resources, Conservation and Recycling*, X(6), 100029.

Townsend, T. G., Powell, J., Jain, P., Xu, Q., Tolaymat, T., & Reinhart, D. (2015). *Sustainable Practices for Landfill Design and Operation*. Springer.

Tripathi, A., & Pandey, N. (2019). Voluntary pricing mechanisms for green product purchase: Altruistic versus self-enhancing consideration. *The International Review of Retail, Distribution and Consumer Research*, 29(2), 198–217.

16 Food Loss and Waste
Environmental Concerns, Life Cycle Assessment, Regulatory Framework, and Prevention Strategies for Sustainability

Sushil Koirala, Anjali Shrestha, Sarina Pradhan Thapa, and Anil Kumar Anal

CONTENTS

DOI: 10.1201/9781003125679-16

16.1   FOOD LOSS AND WASTE: GENERAL INTRODUCTION

Around one-third of the food produced for human consumption throughout the global food supply chain is estimated to be wasted. This sums to approximately 1.3 billion tonnes of edible food (Figure 16.1) (Food and Agriculture Organization (FAO), 2011). The numbers are not constant and keep growing for the food loss and food waste across countries worldwide. Europe generates about 88 million tonnes of food waste annually, costing nearly €143 billion, while it is also likely that 20% of the total food is lost or wasted along the food supply chain (EU, 2016a). The United Nations Environment Programme (UNEP) recently published food waste index report which estimates that 61% of all food waste generated in 2019 came from households, 26% came from food services, and the remaining 13% from trade and merchandizing. The report mentions the food waste generated in 2019 to be about 931 million tonnes (UNEP, 2021). These numbers are staggering considering that as many as 811 million people worldwide go hungry every day (FAO, 2020). Such food loss with immense potential, value, and resources inevitably translates into considerable environmental impacts.

Food loss and waste means reducing the amount of food in successive phases of the food supply chain (FSC) primarily projected for consumption by humans. Food waste occurs due to many reasons; it might be in response to superfluous production volume, spoilage and contamination, logistic malfunction, the incompetent aesthetic value of the commodity, improper meal planning, over-purchasing, and misinformation (Ishangulyyev, Kim, and Lee, 2019). The quantities and types of food wasted vary between and within different phases of the FSC. About 54% of food loss and waste (FLW) has been accounted for during upstream processing and post-harvest handling, whereas 46% is known to be produced during processing, supply, and consumption (Kummu et al., 2012). The problem associated with FLW is first linked with environmental and biotic factors beyond human control and, second, to the food chain contributors' behaviour, including all the operations that concern food management at all stages of retailers and consumers (Cicatiello et al., 2020). FLW is recognized as crucial for the sustainable resolution of the worldwide waste challenge. On the one hand, there are growing concerns over the intensifying greenhouse gas (GHG) emissions and other environmental, social, and economic effects connected with food waste. On the other hand, the burden on natural resources and global food security issues are also increasing. This raises questions regarding the quantities of wasted food in the global FSC that could have been channelled to provide food to hungry people (Papargyropoulou et al., 2014).

The impact of food production and consumption on the environment is amplified by food wastage. In general, food production accounts for 26% of global greenhouse emissions. It is worth noting that emissions from food that is never eaten (lost in the supply chain or consumer waste) accounts for 6% of total emissions (Cattaneo, Federighi, and Vaz, 2021). All emissions coming from the various steps of the FSC have an environmental effect. When food is wasted in a supply chain, all

" Approximately **1.3 billion tons of edible food** i.e., one-third of total food produced for human consumption, is estimated to be **wasted** throughout **global food supply chains** "

(FAO, 2011)

32%   of **global food supply by weight**

24%   of **global food supply by energy content** (calories)

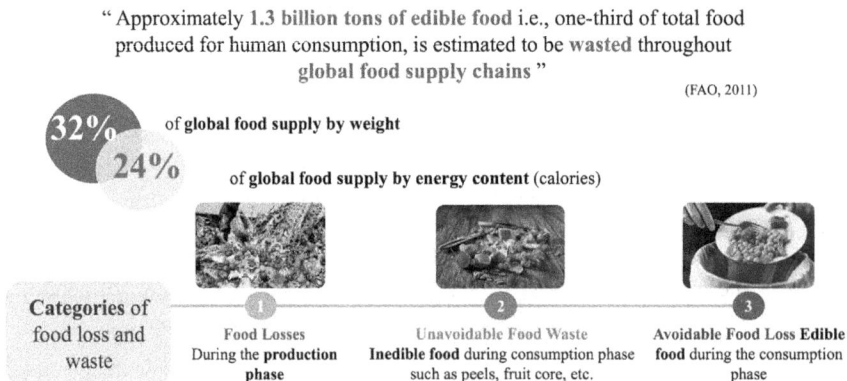

Categories of food loss and waste	① Food Losses During the **production phase**	② Unavoidable Food Waste **Inedible food** during consumption phase such as peels, fruit core, etc.	③ Avoidable Food Loss **Edible food** during the consumption phase

FIGURE 16.1   Current overview of FLW in a global scenario.

the associated activities are wasted. Hence, the longer the activity before the waste generation, the higher the impact on the environment. This is because the emission is coming out of all the activities of the supply chain (e.g., production, processing, transport). Therefore, food waste also creates a lot of energy waste in the FSC (Tonini, Albizzati, and Astrup, 2018). The environmental effect of wasted food through the supply chain and the successive waste dumping is considerable. Excessive environmental impact results from avoidable wastes in the FSC. This is mainly due to over-production. Consequently, decreasing avoidable waste could reduce food production and its overall impacts. The United Nations (UN) Sustainable Development Goal (SDG) 12.3 aims to lower food losses through the supply chain and production by the year 2030 by reducing 50% of food waste at the consumer and retail levels. All of which can potentially initiate steps to decrease environmental impacts due to food waste.

Life cycle assessment (LCA) is suggested as a useful tool for analyzing impacts on the environment due to waste prevention, valorization, and management. LCA is commonly accepted as a decision-guiding tool that can systematically assess products' and processes' environmental sustainability (Gao et al., 2018; Gentil et al., 2010; Zhao and Deng, 2014). LCA is a "cradle-to-grave" method that considers the environmental effects of the complete life cycle of a product, starting from the procurement of raw materials stage, material handling, production, shipping, consumption, disposal, and recycling (Pajula et al., 2017). The consumption of energy and resources and the emissions for the complete system are evaluated (Elginoz et al., 2020). LCA widely covers environmental and life cycle issues; it avoids shifting environmental problems between life cycle stages and impact categories (Mogensen et al., 2009). The International Organization for Standardization (ISO) has standardized the LCA approach, and its technique and measures are given in the international standards ISO 14040 and 14044. Applying this approach can also help identify possibilities for the betterment of the overall sustainability performance of food waste management while reducing the unwanted shifting of burdens among, e.g., sustainability aspects (environmental, economic, social), environmental impacts, life cycle stages, etc. It could also help define sustainable goals to prevent and recycle food waste along the supply chain and, in a broader perspective, contribute to increasing the circularity of the food sector and the overall economy.

16.1.1   UNBOXING THE BLACK BOX OF FLW DEFINITIONS

There is a significant body of literature on FLW; nevertheless, FLW definitions are not universal, and several words such as "food loss", "food waste", "bio-waste", and "kitchen trash" are interchangeable. The discrepancies in methodological design, data collecting, analysis, and interpretation of the findings may have resulted in the adoption of different definitions. Food is lost or wasted along the FSC. Waste occurs from the start of agricultural production to the end of household consumption. (FAO, 2019). The term "food loss" refers to the "reduction in the amount or quality of food". It refers to the disappearance of eatable food mass along the FSC. Food waste is a component of food loss; it refers to the discarding or non-food usage of safe and nutritious foods for human consumption. Food waste is distinct from food loss in that the processes that produce it and its remedies are distinct from those that cause food loss. Food loss refers to the diminution of edible food mass across several sectors of the FSC, including production, post-harvest handling, agro-processing, distribution, and consumption (Figure 16.2) (FAO, 2014).

The definition of food waste varies by country.

In the United States, the Environmental Protection Agency (EPA) describes food waste as

uneaten food and food preparation wastes from the residence, commercial, and institutional establishments. So, food wastes are from home, grocery stores, restaurants, bars, factory lunchrooms, and company cafeterias excluding the wastes obtained from manufacture and packaging of food i.e. (preconsumed waste).

(Lins et al., 2021)

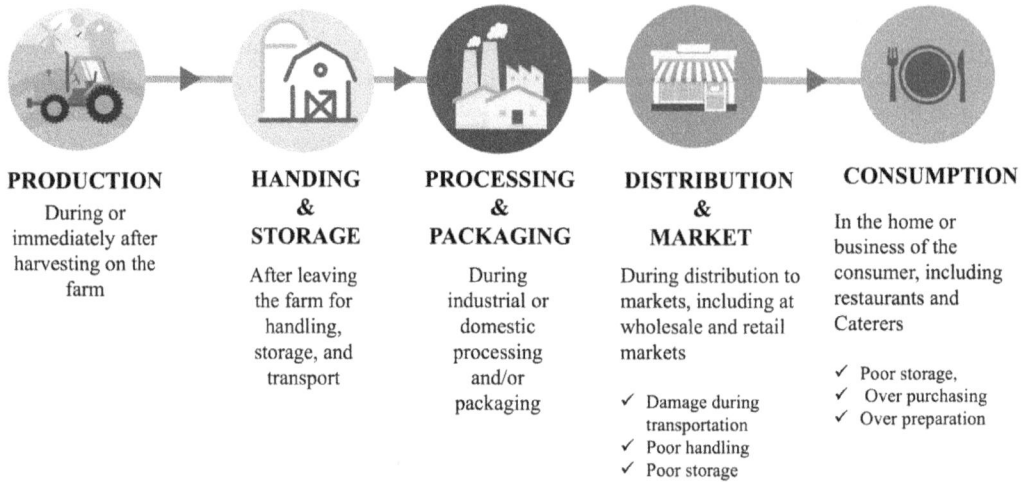

PRODUCTION	HANDING & STORAGE	PROCESSING & PACKAGING	DISTRIBUTION & MARKET	CONSUMPTION
During or immediately after harvesting on the farm	After leaving the farm for handling, storage, and transport	During industrial or domestic processing and/or packaging	During distribution to markets, including at wholesale and retail markets	In the home or business of the consumer, including restaurants and Caterers
			✓ Damage during transportation ✓ Poor handling ✓ Poor storage	✓ Poor storage, ✓ Over purchasing ✓ Over preparation

FIGURE 16.2   FLW across the FSC.

According to the United States Department of Agriculture (USDA), "Food waste is a subset of food loss, and it happens when an edible item goes unconsumed. Only food that is still edible at the time of disposal is considered waste". Three different types of losses are highlighted by the USDA along with the FSC, which include (i) losses from the primary level (e.g., farm) to retail weight; (ii) losses at the retail level; and (iii) losses at the purchaser level (Buzby et al., 2009).

In Europe, in the Agricultural and Rural Convention, food waste is defined as the

> whole of the products that are discarded from the food supply chain, for economic or esthetic reasons, or closeness to the expiry date, despite still being edible and potentially usable for human consumption, in the absence of alternative use, are rejected and disposed off. This produces negative effects from the environmental point of view, economic costs and missed revenue for companies.
>
> **(Buzby and Hyman, 2012)**

In the United Kingdom, the Waste Resources Action Programme (WRAP) classified food waste into three categories: (i) avoidable: beverages and foods that are prohibited despite being consumable (such as bread slices, fruits, and meats); (ii) possibly avoidable: beverages and foods that some individuals consume but others do not (such as bread crusts), or food that is consumable if cooked in a certain way (for example, potato skins); and (iii) unavoidable: beverages and foods that (such as meat bones, egg shells, pineapple skins, etc.) (WRAP, 2015).

16.2   IMPACT OF FOOD WASTE ON THE ENVIRONMENT

When food is wasted, all the resources used to grow it also get wasted (Prajapati, Koirala, and Anal, 2021). This also comes with a substantial amount of carbon footprint. When food is thrown away and discarded in a landfill, it decomposes, thus becoming an important contributor to GHGs such as methane. Methane is said to have 21 times the global warming potential of carbon dioxide (FAO, 2013). The amount of carbon pollution caused by growing and transporting food that is wasted is almost equal to that caused by 39 million passenger vehicles. According to the study conducted by the Consultative Group on International Agricultural Research, it is seen that up to one-third of the entire human-caused GHG emissions are caused by the global food system, making it one of the major contributors to climate change. Although energy and transportation are projected to be the primary drivers of climate change, the influence of food requires special attention (Zhao and Deng, 2014).

It is important to talk about the environmental effects of FLW combined with sustainable food systems alongside food and nutrition security. According to a study conducted, it is approximated that the total cost of global FLW is about 2.6 trillion US$. Out of this estimated cost, about 1 trillion dollars is in economic costs, 700 billion dollars in social expenditures, and 900 billion dollars in environmental expenses (Lins et al., 2021). Supporting countries and shareholders to guide investment and share choices in terms of FLW reduction measures can help lessen social, economic, and environmental pressure and maintain food and nutrition security simultaneously in the long term. Solution-oriented actions can largely help prevent, reduce, reuse, and recycle throughout the FSCs and from local to national and regional to global levels.

A more powerful GHG, methane, is produced in large amounts when food waste ends up in landfills (Tonini, Albizzati, and Astrup, 2018). The excess amount of GHGs such as carbon dioxide, methane, and chlorofluorocarbons absorb infrared radiation leading to the heating of the Earth's atmosphere, which causes global warming and climate change (Slorach et al., 2019). About 70% of the water used worldwide is utilized for agriculture. Therefore, food waste wastes a significant amount of freshwater and ground water resources (Baig et al., 2019). It is said that approximately three times the volume of water in Lake Geneva is used to harvest wasted food. By discarding one kilogram of beef, we are squandering 50,000 litres of water used to produce that meat. Similarly, about 1,000 litres of water is wasted when one glass of milk is poured down the gutter (MFH, 2021).

About 1.4 billion hectares of land, approximately one-third of the world's total farmed land, is used to grow wasted and lost food. Every year, millions of gallons of oil are also wasted to produce food that is not consumed. All this does not consider the harmful effects on biodiversity because of activities like mono-cropping and converting bare lands into agricultural areas (FAO, 2013).

The extent of the expense of natural resources used to yield food products that are eventually lost and unused has been studied worldwide regarding food losses and food waste, global water footprints of farmed products, and the FAOSTAT database. It has been seen that FLW does have significant economic effects. Studies have shown that global food waste in 2007 had an economic cost of US $750 billion. Food losses that can be avoided have a direct and undesirable effect on the returns of both farmers and consumers. People living on the margins of food insecurity are significantly affected by the reduction in FLW (UNEP, 2013). The ease of access to food can be increased for consumers by improving the efficacy of the FSC and thereby decreasing the food costs. The EPA emphasizes and highlights the monetary consequences of food waste and motivates food manufacturers, traders, and the food service sector to decrease food waste to achieve significant cost savings and reserves (Papargyropoulou et al., 2014).

The reduction in FLW aims to improve food security by preventing food loss, developing efficient use of resources, and contributing to a sustainable environment (Koester, 2014). In the yearly assessment of global hunger, FAO (2013) reported that "the world produces enough food to feed everyone"; meanwhile, one in eight people suffer from chronic undernourishment. This gap between production and consumption is occupied by FLW occurring globally in the FSC (FAO, 2013). Therefore, one major negative impact of FLW is food insecurity. In Europe and North America, food waste is projected to be as high as 280–300 kg per capita per year. In the United States, FLW at the retail and consumer level amounted to 188 kg per capita per year, worth US$165.6 billion (Garrone, Melacini, and Perego, 2014). In addition, approximately 11% of the global population is suffering from food insecurity (FAO, 2015). FLW, therefore, has a significant ethical dimension; if food resources are managed and FLW is minimized, food resources may be used to feed food-insecure people (Thyberg and Tonjes, 2016). FLW impacts three key natural resources: fresh water, farms, and fertilizers. In research by Kummu et al. (2012), about one-quarter of the produced food supply (614 kcal/cap/day) lost within the FSC is produced utilizing 24% of total freshwater resources (27 m^3/cap/yr), 23% of total cropland area (31 × 10^{-3} ha/cap/y) and 23% of total global fertilizer use (4.3 kg/cap/yr).

The amount of waste generated throughout the upstream phase of production, yield handling, and storage is believed to be greater than 50%. The remainder of the waste is generated throughout the processing, handling, distribution, and supply phases, as well as during the consumption phase or downstream phase. The FAO analysis identified a clear trend in worldwide food waste. The medium- and higher-income areas demonstrated more FLW at the downstream or consumption stage. However, poor nations were more prone to lose or waste food in the upstream phase, owing to a lack of appropriate and relevant harvest techniques and infrastructure. Thus, it is clear that food lost later in the chain is more environmentally damaging due to the energy and natural resources needed in processing, shipping, storing, and cooking. If food waste were included in a list of countries ranked by GHG emissions, it would rank third, after USA and China (FAO, 2019; 2020; 2015; 2013).

Current practices and intricate interlinkages between production, distribution, and consumption have contributed to the climate crisis due to the exhaustive and inefficient use of natural resources (Read et al., 2020). The exploitation of natural resources such as land, water, and energy and extreme usage of chemical fertilizers, pesticides, and emissions have strained the environment (Muth et al., 2019). All of this has led to the loss and destruction of habitat and the disruption of the environment.

The land is vastly used in the production of agricultural and livestock products. The transformation of land to farmland leads to a loss of biodiversity due to changes in the ecosystem brought about by land alteration. This transition affects the soil structure leading to increased soil erosion and depleted nutrients in the soil. Extreme deforestation for agricultural land can disturb wild habitats, increasing endangered species (Muth et al., 2019).

Chemicals such as fertilizers for supplementation of nitrogen, phosphorus, and carbon are used when producing food, which travels along different stages. While moving along, these elements are released in different forms under the influence of chemical transformation; not only this, but when the food is wasted, the physical presence of chemicals in food are also wasted, ultimately affecting the environment. Apart from the field, the emissions of chemicals and gases occur along with the FSC, namely during transportation, decomposition, and combustion of food waste. The transportation of food across the globe leaves greater carbon footprints. Food waste, when decomposed, releases nitrous oxide and methane (Muth et al., 2019).

The burden on the environment due to food waste can be eased with regulated intervention and the FSC. A major focus is driven to reducing FLW, which can be approached in different ways. For instance, developed nations such as the USA focus primarily on the functioning of intermediaries to diminish food waste. Initiative was taken by introducing programmes such as the US Food Loss and Waste 2030 Champions by the EPA and USDA to incentivise and acknowledge efforts towards reducing wastes and losses to minimize FLW by 50% by the year 2030.

16.3   FLW IN SUSTAINABLE DEVELOPMENT GOALS

The 17 SDGs of the 2030 agenda were adopted by the UN on 25 September 2015. The 2030 agenda includes 17 goals, 169 targets, and 230 indicators. The agenda puts forward a vision for the FAO as a crucial key to sustainable development. The strategic frameworks of the SDGs and FAO are directed to dealing with the causes of hunger and poverty and constructing a non-discriminatory society by leaving no one behind. SDG 12 aims to "ensure sustainable consumption and production patterns".

16.3.1   SDG Target 12.3

The third target of SDG 12, i.e., Target 12.3 is that "By 2030, halve per capita global food waste at the retail and consumer levels and reduce food losses along production and supply chains, including post-harvest losses".

There are two components for the SDG Target 12.3. It clearly states that two distinct indicators should measure the losses and waste, as mentioned below.

16.3.1.1   SDG Indicator 12.3.1 – Global Food Losses

16.3.1.1.1   Sub-Indicator 12.3.1.a – Food Loss Index

The Food Loss Index (FLI) focuses on the loss of food that mainly occurs from production to the selling level but does not include the process of selling and anything after that. It measures and considers the alterations in percentage for a bag of ten products by country in evaluation and comparison with a marked base period. The FLI measure will add to the quantity of the development towards SDG Target 12.3.

16.3.1.1.2   Sub-indicator 12.3.1.b – Food Waste Index

The methodology is still being developed to measure food waste at the selling, trading, and consumption levels. The UNEP is taking the lead in developing the proposal for this sub-indicator.

16.4   LCA OF FOOD WASTE

16.4.1   Quantitative Measurement Techniques

Estimating environmental impacts due to FLW can be done by developing models like LCA and environmentally extended input-output (EEIO) models considering various factors such as input and emissions along with the FSC, including cost-benefit analysis and multicriteria decision analysis. LCA models are preferred for estimating impacts due to FLW by the whole food system. In contrast, EEIO generates an estimation on a wider scale with an understanding of the comprehensive size of the environmental problem posed by FLW (Muth et al., 2019).

16.4.2   LCA

LCA is an organized procedure related to assessing possible effects on the environment through conception up until the deposition of a product. In other words, it acknowledges potential environmental impacts through the full life cycle of a product or process, from procurement and inventory management to manufacturing and final disposal (Gao et al., 2018). This methodology allows the exploration of the appropriate integration between policy and technology to limit the detrimental effect of food waste on the environment and maximize environmental benefits (Omolayo et al., 2021).

Identification and quantification of various waste management technologies by LCA provides an opportunity to recognize indicators responsible for environmental impacts. Results obtained from this model are useful information for governments and organizations for selecting appropriate technology and policies to control and minimize future food waste (Khoo, Lim, and Tan, 2010).

LCA can be structured into four stages as defined by the ISO standards (ISO, 2006) and as illustrated in Figure 16.3

  i. Goal and the scope definition
  ii. Life cycle inventory (LCI)
  iii. Life cycle impact assessment (LCIA)
  iv. Life cycle analysis (interpretation)

16.4.2.1   Goal and the Scope Definition

This is the first stage of the LCA assessment of FLW on the environment. The preliminary stage requires a thorough description of the objectives and aspirations of conducting the LCA of FLW (Dong et al., 2018). The process or product of interest is also described in detail, considering its life cycle and system boundaries. Defining the goal and scope helps analyze the necessity of conducting the LCA. In addition, it allows the generation of explicit objectives when relevant questions

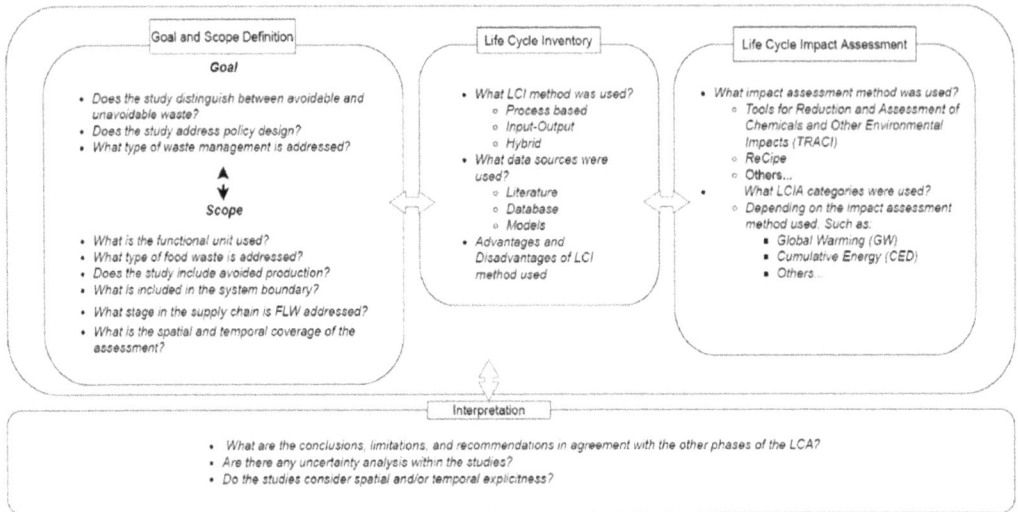

FIGURE 16.3   General principles to conduct LCA with a case study on food waste. (From Omolayo et al.,, 2020. In *Resources, Conservation and Recycling*. With permission).

are framed, directing towards the objective (Curran, 2017). Questions can be formulated to define goals that could vary depending on the studies of FLW (Figure 16.3). LCA has been perceived as a popular tool for decision support and defining long-term strategies in the food sector. Therefore, it can be used to verify study goals and objectives (Petit, Sablayrolles, and Yannou-Le Bris, 2018). Following this, scopes are defined that acknowledge all functional and technical aspects and system boundaries in use or avoided during the life cycle of the FLW. System boundaries are important aspects of LCA, which define each process depending on the study goal. Defining system boundaries is a crucial task as it is affected by several factors and is responsible for the accuracy and compatibility of studies (Heijungs, Huppes, and Guinée, 2010). Omolayo et al. (2021) elucidated farm-to-fork system boundaries in terms of FSC, such as farm production, industrial food processing, retail/distribution, consumer plate, and waste treatment in their study for the LCA of FLW in the FSC.

16.4.2.2   LCI

Necessary data collection methods are brought into practice to set up an aim and scope. LCI utilizes comprehensive information on the input and output of the system, including resources, energy, and emissions. LCI is assessed based on the LCA method used (process-based, input-output based, and hybrid LCA) (Figure 16.3). Results from LCI assist in the characterization of the environment and effects on human health based on equivalency factors for translation of inventory flows to impacts (Muth et al., 2019). Modelling, which is based on the process, input-output LCA (IO-LCA), and hybrid LCA are three approaches to LCA models. Process-based modelling considers an inventory of processes, including all inputs, outputs, and possible environmental burden of food waste, which exhibits better accuracy (Li et al., 2020). IO-LCA estimations are made regarding materials, energy, and emissions throughout the activity. A blend of process-based modelling and IO-LCA brings about a hybrid approach that values key differences between both approaches and utilizes them to benefit the database system (Onat, Kucukvar, and Tatari, 2014). The obtained data is processed as impacts by LCA, which analyzes the environmental impact of resource utilization and emissions. Several tools concerning different environmental categories can be used to translate data to environmental impact, say global warming, ozone reduction potential, human toxicity potential, the use of land and water, and more (Omolayo et al., 2021).

16.4.2.3   LCIA

LCIA translates the results from LCI into impacts. Impact scores are provided after assessing inputs and outputs from LCI against emissions and resource extractions (Figure 16.3). Impact categories are a function of the most applicable to the systems analyzed. LCIA's most commonly used tools are TRACI, ReCipe, USEtox. Global warming potential (GWP), cumulative energy demand, eutrophication, acidification, photochemical ozone creation potential, ozone depletion potential, human toxicity potential, ecotoxicity potential, land use and water use are the most common LCIA categories to determine the LCA of FLW in FSC.

16.4.2.4   Life Cycle Interpretation

Eventually, the assessment is interpreted and thoroughly studied with uncertainties and variables taken into consideration within the study and assumptions are made (Figure 16.3) (Stone et al., 2010). Uncertainty analysis fundamentally remains a key factor that needs to be addressed when conducting an LCA, with the major uncertainties being parameters, scenarios, and modelling uncertainty. Uncertainties can arise on several bases: uncertainty regarding parameters occurs when there are errors in measurement, inaccuracies, lack of reliability, and collection method (Clavreul, Guyonnet, and Christensen, 2012). Whether optimistic, pessimistic, or equiprobability criteria, different approaches can be evaluated when considering scenarios. Sensitivity analysis can be conducted to overcome different forms of uncertainties (Omolayo et al., 2021). During this stage, recommendations can be made based on the results of LCIA for future actions.

LCA exhibits a holistic approach that identifies and acknowledges various environmental impacts (Khoo, Lim, and Tan, 2010). When based on existing knowledge of policies and practices, LCA can determine the extent and methods for decreasing FLW. The LCA model has to be designed so that the basis of the problems will be identified as per their priority areas for intervention and evaluation of the impacts brought about by intervention (Corrado et al., 2017). This model bridges the knowledge gap between intentional and unintentional environmental impacts. Also, it provides cost and benefit assessment with environmental aspects in consideration. Life cycle interpretation demonstrates completion, consistency, and sensitivity checks (Omolayo et al., 2021).

16.5   SAFETY AND REGULATORY FRAMEWORK OF FOOD WASTE

Accelerating quantities of food waste have gained attention and demanded immediate measures at policy levels for sustainable solutions. Despite knowledge of food wastes occurring at various steps in the supply chain, it has been understood that the primary focus remains at the consumer level for its prevention and reuse (Nicastro and Carillo, 2021). However, for a sustainable approach, all involved in the FSC should be held responsible for reducing and eliminating food waste. The UN SDGs clarify the importance of reducing food waste and have encouraged formulating necessary policies worldwide to prevent FLW. The SDGs thus intend to reduce global food waste and reduce food losses by 50% throughout the supply chain by 2030 (Giordano et al., 2020).

Therefore, the need for legal developments to address problems related to food waste cannot be overemphasized. The legislative framework must correspond to academic literature to formulate sustainable rules and regulations regarding food safety for reusing surplus food and reducing food waste (Garske et al., 2020). Different policy approaches exist for the effective and efficient implementation of laws in different regions (Fattibene et al., 2020). Policy framework as per FUSIONS 2016 can be persuasive or regulatory. A persuasive approach will encourage behavioural change with the help of ample information and knowledge, whereas the regulatory approach relies on issuing penalties (EU, 2016b). Different instruments can further back up these approaches. For instance, market-based instruments can bring behavioural changes in the environment with taxes, charges, and subsidies. Following Mourad (2016), the framework can be based on the evaluation of the strength of prevention measures where a significant contrast between weak and strong actions is clarified to drive focus towards attaining extensive efficiency in food supply operations for weak actions and promote holistic changes in food systems when considering strong actions (Eriksson, Giovannini, and Ghosh, 2020). Even with the availability of policies, it is not easy to drive the population to abide by laws unless certain approaches are implied to alter citizens' behaviour, institutions, and businesses. Hence, an appropriate policy mix should be used during interventions (Eriksson et al., 2017).

Formulation of appropriate policies first requires a proper understanding of the effects of food waste on the environment and product life cycle. Policies thus have been primarily focused on avoiding the accumulation of food waste by either reusing or recycling it; a secondary outlook is established on lessening food waste.

16.5.1 Europe

Across the EU, 88 million tonnes of food waste are produced, accounting for a tentative €143 billion (EU, 2016a). Considering this explanation, EU guidelines have been formulated and adopted regarding food donation and the reuse of food made for human consumption as animal feed (Garske et al., 2020). EU food safety law is centred on protecting human and animal health, and therefore, guidelines are drawn as per law (Van der Meulen, 2010). Consistent food measurements and careful packaging and labelling are basic essential steps to reduce and prevent food waste. Overall, food waste prevention as per the EU Commission can be well explained and practised under three main actions: first is setting up a closely observed target and reports on its working and progress (Priefer, Jörissen, and Bräutigam, 2016). Second, cooperation and sharing of knowledge on strengthening and scaling up an action plan that exhibited the finest outcome, and third, adopting new habits by everyone in the supply chain with food prevention as a major objective, followed by the repurposing of surplus food and food wastes (Filimonau and Uddin, 2021).

In response to assessing and understanding activities undertaken to avert food waste, the European Commission established the EU Platform on Food Loss and Food Waste in the interest of supporting actors in the supply chain to prevent food waste, provide guidance on appropriate methods, and evaluate progress (Corrado et al., 2019; Vittuari et al., 2016). Addressing this issue, inclusions on strategies for reducing food waste from farm to fork for a sustainable food culture are under the European Commission as part of the European Green Deal (Priefer, Jörissen, and Bräutigam, 2016; Caldeira, De Laurentiis, and Sala, 2019). For evaluating the efficacy of different measures taken, a common framework has been established by the European Commission Joint Research Centre (EC-JRC), which gathers information on actions taken against food waste via a collection of reports in a modelled format. The frameworks developed are based on different qualitative and quantitative criteria, namely, the value of action, its usefulness, efficacy, sustainability, transferability, scalability, and intersectoral operation that will help in the assessment of actions by identifying the benefits of implementing such actions from environmental and economic aspects and staying updated with the progress made over time (Secondi, Principato, and Laureti, 2015; Garske et al., 2020; Knowles, Moody, and McEachern, 2007).

Corresponding to this, different institutions, organizations, local authorities, and businesses used different plans to reduce food waste throughout the supply chain. The European Union has countries with different propositions to combat food waste and its effects; Italy, France, and Spain approved national plans on reducing and preventing food waste. In contrast, other nations in the EU, such as Austria, the Czech Republic, and Poland, have taken up only municipal level waste management plans. Likewise, Netherlands, Sweden, and Scotland formulated action plans for food waste reduction, and the UK targeted specific food waste issues. Studies on EU policies have exposed that the EU is concentrating more on lowering resources involved in food production and its consumption to reduce adversarial effects on the environment (Grainger et al., 2018; Philippidis et al., 2019). EU Directive 2018/851 raised concerns and requested nations to take necessary actions to monitor levels of food waste and reduce it. As per Directive 2008/98/EC, 2015, a hierarchy for the prevention of waste and management policy and legislation was introduced where prevention was prioritized, followed by preparation for the reuse, reutilization, other recovery and lastly, food waste disposal. Nevertheless, avoiding surplus production in the first place and disposal of avoidable food waste is the most effective way to reduce food waste (Giordano et al., 2020).

In the EU, countries such as Italy and France passed policies in 2016 to reduce food waste with different approaches. The Italian policy is based on a persuasive approach where it urges and encourages actors throughout the supply chain to act upon reducing food waste via an extensive awareness and communication campaign. In contrast, the French policy is based on a regulatory framework with the provision of penalties if one is unable to abide by the policy. Both Italian and French policies have emphasized the prevention of waste; French policy is directly based on the food waste pyramid, whereas the Italian policy revolves around that concept (Bos-Brouwers et al., 2020).

Italian policy concentrates on the proper collection and donation of edible food waste. Food donation in Italy is a voluntary action where surplus food within food safety standards is directed for human consumption. Those that are not fit to be consumed by humans are used as fodder for animals or subjected to composting (Busetti, 2019). A step-wise procedure has been discussed in law to donate food products at the fiscal and bureaucratic levels. Communication campaigns and awareness drives on national television are effective approaches to addressing problems (Philippidis et al., 2019). Environment, agriculture, work, and health ministries work together for an awareness campaign with specific attention to using "doggy bags" at restaurants to take leftovers home.

Additionally, ministries also work on raising awareness initiatives for both food waste and inequality in food access. Besides this, active work has been conducted to develop educational programmes and policies in schools, emphasizing food sustainability and food waste. The Ministry of Health proposed to produce guidelines for reducing food waste and food donations. A market-based approach is also considered where provisions for tax exemption are available for those donating food. In aspects of improving technology, different funds are allocated and presented for works such as improvisation of food packaging to enhance shelf life, promotion, and sustainable doggy bags. Special focus is given to improving the shelf life of products (Caldeira et al., 2021; Caldeira, De Laurentiis, and Sala, 2019).

In contrast to Italian policy, French policy is directly based on the food waste pyramid and has different measures to deal with food waste designated to work on different levels. Regulatory frameworks on food donation have been at the centre of French policy, where actors in the FSC must conduct prevention, donation, reuse, composting, and energy recovery of food waste in order (Vaqué, 2017). In instances where retailers cannot abide by the policy, fines and penalties are charged. Food donation is mandatory and needs thorough regulations to be understood and considered; for instance, products segregated for donation must be donated before the expiration date and within a certain time frame (Mourad, 2016; Albizzati et al., 2019). Unpacked meat and animal products are not allowed for donation due to food safety concerns.

Most importantly, a proper traceability system for donated food must be maintained via transport documents and invoices. The French government also conducts an awareness campaign to encourage the public to make proper use of food to eliminate waste. The policy also provides prospects for collecting a firm's information on fulfilling their duties towards the environment and society to ensure the appropriate functioning of the system. French policy in market-based instruments is much stronger than Italian policy (Redlingshöfer, Barles, and Weisz, 2020).

Despite several measures and efforts made, food waste prevention remains weak with no concrete strategies to avoid generating food waste. However, both policies are dedicated to the recovery and reuse of food waste (Hebrok and Boks, 2017). Moreover, exposure to fragile food systems during COVID-19 has demanded reform in scientific findings and public awareness (Aldaco et al., 2020; Amicarelli and Bux, 2021; Brizi and Biraglia, 2021). The European Green Deal thrives in accomplishing sustainability in food systems by a farm-to-fork approach emphasizing attaining impartial, healthy, and environmentally friendly food systems (Riccaboni et al., 2021). The aspired changes can be done by formulating mitigations against climate change, neutral or beneficial impacts on the environment, reduced loss in biodiversity, and ensuring food security, nutrition, and health while maintaining food affordability. With the prevalence of COVID-19, the European Commission plans to develop a contingency plan for viable agri-food systems via trade guidelines and global cooperation instruments to ensure food security (Dupouy and Gurinovic, 2020). Payback for the prevention of FLW cannot be exaggerated. Aiming for a triple win, saving enough food for human consumption, saving resources involved in the production, and reducing environmental impact, the European Commission plans to establish and pursue effective actions on FLW (Govindan, 2018).

16.5.2 USA

A report made in 2013 by the USDA Environmental Research Service revealed that Americans generate about 417 pounds of waste per person. The data analyzed shows that over 20% of the total waste produced is from food. Wasted food is accountable for more than 5% of GHG emissions across the USA. Figures as per the USDA in 2015 show that around 42.2 million Americans live in a food-insecure household, and 11.4 million live in very low food security households. Often when food is discarded as waste, it can be redirected in many ways. Such food considered as waste might still be consumed if it is safe or can be used to nourish the soil or produce biofuels (Walia and Sanders, 2019).

If 15% of wasted food in America can be redirected for consumption, it would be suitable for 35% of the calorific requirements of food-insecure American households (Walia and Sanders, 2019). Food waste can only be redirected and repurposed after food waste has been identified, segregated, and classified as consumable or compostable. Considering food waste segregation, different laws have been formulated and brought into practice across different states and cities in the USA. Laws for separating food waste from large-scale food waste generators in New York, such as the Commercial Organic Waste Law, have existed since 2015. Following this, the City of New York Department of Sanitation began providing a distinct collection of food waste services to more than 11% of residents.

Similarly, efforts have been made, funds have been invested in anaerobic digestion, and federal subsidies have also been provided for the production of usable biofuels (Thyberg and Tonjes, 2016; 2015). Massachusetts effectively banned the commercial disposal of food waste by enacting a law that requires establishments producing more than or equal to one tonne of carbon-based waste to donate, repurpose, or turn away waste from landfills. This law allows businesses to exercise freedom and choose appropriate methods of treating food wastes. On the other hand, Seattle penalized citizens who failed to segregate food waste from other garbage. Moreover, separating food waste might encourage donating food waste. It has been observed that the aim is to divert food waste from landfills to composting. For this, efficient and effective coordination between people disposing of food and people involved in repurposing it is necessary (Gentil et al., 2010; Walia and Sanders, 2019; Raak et al., 2017).

Besides policies, several market-based solutions are also in practice led by different food rescue organizations. These organizations' supply chain management tools allow them to recognize and generate the most recipient value. Organizations such as Inter-Faith Food Shuttle (IFFS) and City Harvest have been rescuing millions of pounds of food wastes and redistributing them to the needy. IFFS emphasizes providing nutritious food by collecting rescued food and directly redistributing it to recipient homes or providing it to culinary job-training programmes where food is cooked, frozen, and distributed among soup kitchens. Different challenges have been launched with initiatives to reduce and repurpose food waste. The EPA and USDA created a Food Recovery Challenge and Food Waste Challenge, respectively; the latter aims to reduce food waste by 50% in US landfills by 2030. This initiative aims to provide education and infrastructure to businesses and organizations for practising efficient purchasing, food donations, and composting. Furthermore, programmes help build a bridge and establish coordination between organizations wasting food and ones working for its recovery. US regulations on food donation in the Bill Emerson Good Samaritan Food Donation Act (1996) and Food Donation Act (2008) ensure safe and trouble-free donation of food (Chalak et al., 2016).

16.5.3 INTERNATIONAL

It has been studied and understood that food waste occurs along different levels of the supply chain, influenced by the country's economic status, industrialization, and development levels (Parfitt, Barthel, and Macnaughton, 2010). Food waste at the post-harvest and handling stages has increased in underdeveloped and developing countries (Thi, Kumar, and Lin, 2015). This is due to the lack of proper technological and financial aspects and improper and inefficient processing and transport. On the contrary, developed countries have been identified as wasting food at consumer levels during preparation and serving with over-purchasing and poor planning (Zamri et al., 2020). The concept of "pay-as-you-throw" (PAYT) is a popular measure for combating food waste where consumers are charged for waste generated in the household. This concept has been effective in several countries, including the USA, Japan, Canada, Korea, Taiwan, Thailand, China, and Vietnam, with the expectation of lowering food waste in households and encouraging the public to recycle waste (Reichenbach, 2008; Halloran et al., 2014; Sakai et al., 2008). A study based in the Czech Republic showed that people adopting PAYT ended up recycling more, i.e., 12.1%. In contrast, those who opted for a flat fee approach for waste disposal exhibited low recycling rates, i.e., 6.9% (Chalak et al., 2016).

Laws were formulated to encourage citizens to reduce and recycle waste in Japan in 2001 that specifically targeted the promotion of recycling and appropriate treatments of cyclical food resources (Parry, Bleazard, and Okawa, 2015). Excess food or food unfit for human consumption was encouraged to be used as animal feed and fertilizers. To ensure practice efficiency, reports and surveys were conducted on the waste produced and recycled among producers (1Al-Rumaihi et al., 2020). Malaysia aimed for the ultimate segregation of food waste and aimed to recycle 20% of organic waste by 2020 under the Waste Minimization Master plan and National Strategic Plan for Food Waste Management 2005; the National Strategic Plan was accompanied by the Municipal Waste and Solid Waste Management and Public Cleaning in 2007 (Jereme et al., 2018). Similarly, South Africa works on preventing landfilling by promoting organic composting waste. A draft has also been prepared to address this concern under the Waste Classification and Management Regulations of 2010 (Dong et al., 2018).

It has been seen that most of the food that is wasted occurs at the consumer level. Therefore, it can be agreed that the knowledge of consumers on food waste and its reprocessing impacts the way food waste is controlled. Owing to this fact, several awareness campaigns and governmental and private actors informed the public on identifying and sorting food waste for reuse and useful prevention practices.

16.6 IMPACT OF COVID-19 ON FLW

The closure of restaurants and panic buying at grocery stores to stock up on food supplies in response to the risk of COVID-19 can potentially affect the quantities of household waste generated. With food insecurity still on the rise, studying this alteration in purchasing and consumption patterns is important (Pappalardo et al., 2020). Households are the primary generators of food waste. Hence, understanding the effect of the pandemic on its increase or decrease is significant since COVID-19 has had a substantial impact on hunger and food insecurity.

The decision to discard food materials is based upon many factors, including context and consumer demographic characteristics (Setti et al., 2016; Ellison and Lusk, 2018). Therefore, studying and understanding practices, patterns, and consumer habits can help us understand the nature of the food waste issue (Lazell, 2016). Besides, an alarming figure of 820 million people worldwide suffer from food scarcity or low-grade diets because of food waste, contributing to numerous micronutrient deficiencies and increased health problems (Willet et al., 2019). Thus, the focus has been driven to explore possible measures to lower food waste, along with several proposals and recommendations on respecting food and cooking skills and persuading people to keep an open mind in preparation for unexpected scenarios (Bos-Brouwers et al., 2020).

Preliminary studies on the generation of household food waste during the pandemic show that food waste is expected to decrease due to the concerns regarding disturbances in the FSC and complications in shopping because consumers tend not to waste available food (Richards and Rickard, 2020; Jribi et al., 2020). Even if many families opted for stockpiling food, given the situation, less food waste is likely to be generated with the consumption of every item in the house (Wang et al., 2020). Moreover, during crises and emergencies, people become more mindful and aware of their environment and environmental issues such as food waste (Shiller, 2020). However, it is indisputable that many people responded to the pandemic in an individualistic manner than global, which resurrected isolationism and nationalism, more so with the necessity to limit social interactions (Mair, 2020). All of this deviates attention from concerns regarding environmental protection and decreasing GHG emissions.

As of now, no recent studies exist to confirm and provide enough evidence on the reaction of households to emergency states such as pandemics when it comes to food waste production. Obtaining relevant information on critical situations is of primary importance as it impacts environmental and economic aspects. To address this situation, a nationwide survey was conducted in Italy to comprehend changes in food waste during COVID-19 and recognize factors associated with the generation of household food waste. A survey thus conducted at the prime time of strict lockdown revealed that about 5.5 million tonnes of food was wasted, approximating €15 billion (EU, 2016a). Households were found accountable for producing 80% of food waste, which significantly affected socio-economic and environmental aspects related to global GHG emissions (De Leo et al., 2015).

16.7 STRATEGIES TO REDUCE FLW FOR RESILIENT FOOD SYSTEMS

Prevention and mitigation strategies throughout the FSC for FLW can only be done with ample knowledge of the reasons behind the rejection of any food materials. A simple food waste pyramid concept (Figure 16.4) can be applied to FLW. Other strategies and interventions are described below.

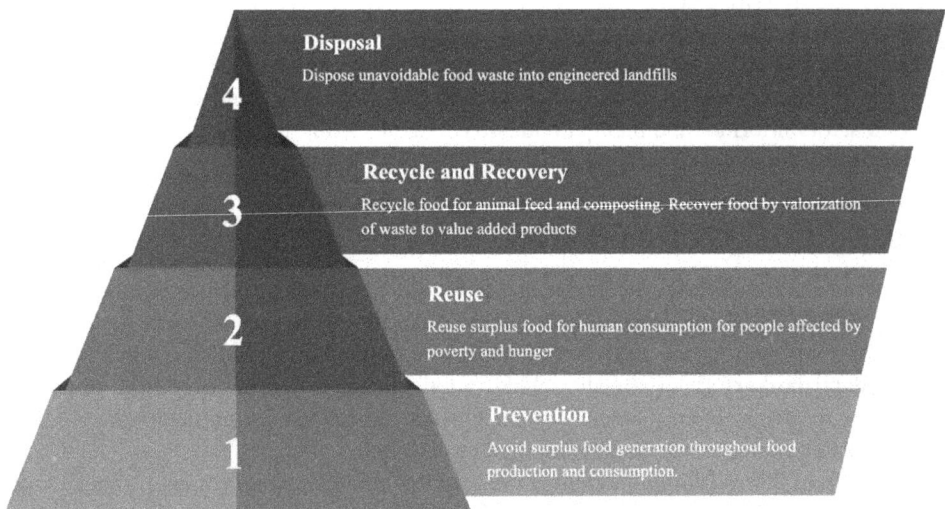

FIGURE 16.4 General strategies to reduce FLW based on food waste pyramid.

16.7.1 Behavioural Strategies

Exchange of information on FLW on a global platform with a rigorous media campaign can help raise awareness of its impact and formulate relevant and necessary solutions that can expand knowledge and modify behaviours through every actor along the food chain. Educating stakeholders and the community on the social, economic, and environmental value required for food production and the benefits of minimizing FLW can drive behavioural changes. The establishment of bins to collect organic food waste can increase awareness and encourage locals to sort domestic waste and divert waste from landfills.

Detailed guidelines and resources provided on the reduction, redirection, and disposal of food waste can benefit many aspects. Food categorized as unused and wasted can be regenerated for consumers by identifying new marketing opportunities (Rodrigues et al., 2021). Varied approaches can be undertaken to divert food waste from landfills and repurpose excess food as feed for animals, provided that the excess food poses no harm to animals both physically and nutritionally, or food waste can also be subjected to composting. Anaerobic digestion of industrial wastes such as oils and fats will provide biodiesel (Muth et al., 2019). When subjected to anaerobic digestion and composting, food waste can exhibit an achievable reduction in detrimental environmental impacts such as global warming. This can further be achieved by low energy requirements for anaerobic digestion and low carbon dioxide production (Khoo, Lim, and Tan, 2010). Moreover, the technological aspects of food products can be enhanced with better processing, transportation, and packaging systems, causing decreased food waste. In addition to this, food waste can be appreciably lowered with proper preservation and storage techniques and incentivizing the behavioural changes on the consumers' part (Read et al., 2020).

16.7.1.1 Changing Consumer Behaviours

The generation of waste and its management depends upon an individual's choice. This choice can lead to financial loss and aid in the emission of GHGs once the waste enters the landfill. Awareness of the amount of food wasted, its consequence on the household budget, and solutions on what can be done can be brought about by programmes such as the Love Food, Hate Waste Campaign (currently in the UK), which can potentially encourage the reduction of household waste. Repurposing discarded food is another way to manage food waste adopted by retailers. Along with this, providing freedom of choice in purchasing fruits and vegetables is likely to lower waste. Knowledge and understanding of the value and utilization of aesthetically imperfect fruits and vegetables can be aided with attractive financial offers such as reduced prices and discounts. In addition to this, reforming the packaging system and materials and informing consumers on methods to store their purchases for a long time can effectively reduce waste.

Most importantly, spreading awareness and education on household food waste at a national level with a focus on proper food handling practices to optimize storage life, planned purchase, and clear knowledge on packaging labels such as "use-by" and "best before" dates can positively alter consumers' behaviour towards production and handling of excess food and food wastes. Furthermore, adopting and practising repurposing and reuse of leftovers, for instance, packaging leftovers from restaurants for eating later, and composting can also help change consumer perspectives.

16.7.1.2 Engaging the Workforce to Minimize Food Waste

Capacity building of people involved in the food service business can assist in reducing food waste. Presently, many service sectors and businesses have collaborated with several bodies, including government, industries, and food rescue organizations, to reduce food waste. However, any action against food waste and its effects varies from one business to another. In the case of retail store training, proper inventory management with stock rotation, a clear understanding of labels of dates, storage instructions, quality standards, and improved management of demand- surplus produce can minimize food waste. Likewise, several other food businesses have opted to include waste auditing as part of staff training to enlighten workers on items, their disposed quantities, and the amount of financial loss that the industry has to bear. In expectation of the wider adoption of food waste reduction practices, public recognition and rewards can be arranged for companies and industries following these practices and involved in innovation to address food waste production and management problems. Given this, revolution in the perspective of food waste can be normalized and hence made acceptable.

16.7.2 Policy Interventions

In the existing literature, the primary focus is on the reduction of FLW via non-binding initiatives from governmental and non-governmental bodies, circulation of information, and implementation of policies among stakeholders in hopes of spreading awareness, financial measures, time-constricted waste reduction targets, and developing and keeping track of strategies developed for the reduction of FLW (Chalak, Abou-Daher, and Abiad, 2018). Nevertheless, arguments arise on addressing FLW as a holistic instead of a cosmetic fiscal measure. Likewise, economic concern-driven strategies to reduce FLW have become bothersome as FLW has widespread consequences. Therefore, companies' consideration of social and environmental factors can enhance the reduction of FLW (Derqui, Fayos, and Fernandez, 2016).

16.7.3 FLW Quantification

An important aspect of designing interventions and solutions to reduce FLW lies in quantifying food waste. Quantifying food waste also aids in understanding the economic aspects associated with it (Hartikainen et al., 2018). Many studies regarding the evaluation of FLW have been found in dedicated regions of the USA and Europe (Dusoruth, Peterson, and Schmitt, 2018; Griffin, Sobal, and Lyson, 2009; Johnson et al., 2018; Principato et al., 2019). SDGs have been the European Union's primary focus and have encouraged several quantification studies on FLW in Europe (UNEP and OECD, 2019).

Classifying FLW as avoidable and unavoidable helps determine appropriate strategies to mitigate and reduce wastes in an environmentally friendly manner (Dora et al., 2021). With food processing as one of the key sources of food waste, it is essential to apply and utilize the knowledge and necessary resources that can potentially reduce the chances of FLW by increasing processing efficiency. Ensuring a safe and stable power supply and maintaining uninterrupted and optimal production equipment can prevent food waste (Raak et al., 2017).

16.7.4 Investments for FLW Reduction

Commitments on a national scale regarding FLW can create opportunities to meet distinguished targets within the allocated time frame. Issues, strategies, and action plans for reducing post-harvest losses are presented in Table 16.1. Besides this, developing programmes and strategies for involvement and guidance to local and external investors to encourage investment in national investment plans can provide opportunities for the public and independent initiatives in this domain (Vilariño, Franco, and Quarrington, 2017). However, the generation of self-sufficient funds and partnerships between public and private enterprises can also reduce food waste. It is noticeable that food loss directly or indirectly affects climate change, and hence, strategies involved in the reduction of food waste can be eligible to be fuelled by multilateral funds, such as ones involved in fighting climate change such as the Green Climate Fund and Global Environment Facility (Cattaneo, Federighi, and Vaz, 2021). On the appeal of G20 chief scientists, the establishment of the Agricultural Innovation and Enterprise Facility has been formulated. It is under the guidance of partners in the Global Forum on Agricultural Research for Development (GFAR) Over US $10 billion per year can be set aside, given that there is an increase in efficiency in harvesting and post-harvest handling, packaging,

TABLE 16.1

Issues, Strategy, and Action Plan to Reduce Post-Harvest Food Loss and Waste

Operation	Issue/problem	Strategy	Action plan
Harvesting index	Absence of maturity index for vegetables for local and export markets.	Emphasis on quality, safety, and sustainability in terms of research and development.	Formulation and establishment of the maturity index.
	Price and market distance are responsible for the low adoption of available indices.	Prepare farmer-friendly harvest indices, organize extension activities.	Generate awareness via training, manuals, and posters on suitable harvest indices.
Harvest methods and time of harvesting	Rough handling; untimely harvesting.	Spread awareness on suitable procedures and time of harvest.	Circulate information, safety practices, and organize training.
	Unavailability of suitable and/or improperly designed harvesting tools, equipment, and containers.	Efficient and properly design harvesting tools and equipment-oriented research and development.	Exposure visits.
Field assembling, categorizing, grading, and packing	Lack of protocols regarding adequate sorting, grading, and packing for materials suitable for field packing.	Outline protocols for sorting, grading, and packing of vegetables; instruct farmers and stakeholders.	Upskill farmers and stakeholders and provide informative materials.
Pre-cooling	Absence of provision for pre-cooling.	Prepare a favourable policy environment to promote investment and develop alliances/commodity-based clusters for overcoming limitations regarding facilities.	Persuade subsidies from the government and investment from private enterprises.
	There is inadequate information on technologies involving pre-cooling at the commercial level; scarce information on cost benefits associated with this technology.	Emphasis on commercializing pre-cooling via research and development.	Appeal for technical assistance for gaining cost-benefit details to commercialize pre-cooling.
Grading	Absence of national standards and poor execution of standards.	Preparation of national standards.	Evaluation, circulation, execution, research and development, and maintenance via demonstration, training, etc.
	Lack of skill, awareness/financial resources.	Capacity building.	Awareness, motivation, training, and government financial support.
Procurement/packing/grading facilities	Lack of collection centres/packing houses/grading facilities.	Government support for clustering.	Identify strategic locations.

(Continued)

TABLE 16.1 (CONTINUED)

Issues, Strategy, and Action Plan to Reduce Post-Harvest Food Loss and Waste

Operation	Issue/problem	Strategy	Action plan
Packaging, labelling, and traceability	Insufficient packing technology/suitable packaging (for conveyance, storage, and consumers).	Develop and accommodate existing technologies.	Preparation of appropriate packaging which can be suitable for sites and commercialization.
	Inadequate skills and awareness on proper utilization of packaging and financial resources.	Capacity development.	Proper packaging technologies are supported by awareness campaigns, motivation training, and financial aid from the government for the stakeholders.
	Effect of packaging on the environment.	Government regulations.	Preparation of regulatory policies and regulations.
	Absence of proper labelling and traceability of produce.	Formulation of the regulatory system.	Preparation of regulatory policies and regulations.
Marketing	Insufficient information on market and absence of market strategies.	Establish a national/regional information networking system.	Devise marketing strategies and provide information on market systems.
	Inadequate market framework.	Development of market centres at different levels.	Develop market information systems and marketing strategies.
	Incapable of marketing products in domestic and international markets.	Formulation of a strategic alliance with companies and corporations at a global level.	Publicity and advertising.
Transportation	Inadequate infrastructure and appropriate transportation with refrigerated system.	Policy level support from the government and encouragement of funding from the government.	Availability of logistics and supervision to reduce cost and facilitate efficient distribution and transportation of commodities, study, and analysis on cost benefits for the appropriate transport system.
	Improper control and management of temperature, poor loading and unloading practices.	Dissemination of information on absolute transport management system.	Provide training, seminars, create and circulate information materials.
Storage	Inadequate farm storage facilities and refrigeration storage technology in the market.	Develop appropriate policy environment to encourage investment for research and development to provide definite cost and benefits of storage system.	Financial aid from the government and private enterprise, perform cost-benefit analysis on various storage systems.
	Unmanaged temperature control, unhygienic practices, and poorly facilitated storage rooms.	Provide knowledge on accurate methods of operation and storage management facilities.	Train workers and handlers involved in storage facilities.

(Continued)

TABLE 16.1 (CONTINUED)
Issues, Strategy, and Action Plan to Reduce Post-Harvest Food Loss and Waste

Operation	Issue/problem	Strategy	Action plan
	Unaware about required temperatures, sensitivity towards ethylene of mixed commodities.	Investigation and experimentation focused on relative humidity, sensitivity towards ethylene and temperature during storage.	Perform study and analysis and circulate findings.
Processing	Insubstantial provision of appropriate types of processing.	Circulate information among small processors and create suitable varieties.	Circulation of information, collection, and establishment of germ plasm for a breeding programme.
	Lack of suitable processing technologies.	Research and development.	Create innovative niche products, commercialize indigenous products.
	Insufficient commercialization of latest technologies and unavailability of basic infrastructure.	Technical, infrastructural, and policy support from government bodies.	Commencement of pilot plants, manage cost/benefits and consumer studies.
	Formation of novel/niche products.	Research and development.	Introduction of innovative niche products and commercialization of indigenous products.
	Insufficient facilities and infrastructures.	Research and development.	Create suitable facilities/infrastructure.

and monitoring of commodities. With a clarified business case for reducing FLW, powerful leadership is essential for private sectors to attain innovation and diffusion at a large scale. Several large food companies have committed to collaborating with farmers to lower emissions and fight climate change at the UN Climate Summit 2014 (Chan et al., 2015).

Acquiring accurate details on the category and quantity of food wasted in different parts of the food system allows businesses to add value to otherwise discarded products. With the repurpose, reuse, and recapture of otherwise wasted food, markets can be expanded or new markets created. This can encourage revenue for suppliers and users of raw materials from different sources and create employment opportunities. However, it is of utmost necessity that appropriate and suitable guidelines and specifications are developed to provide direction on the kinds of products that can be developed from food waste and standards by which to comply. Understanding the connection and working mechanism within the food system is important for applying the circular economy and identifying effective interventions (Principato et al., 2019; Vilariño, Franco, and Quarrington, 2017).

With insights into the actual value of food waste, several mitigation strategies and methodologies have been put forward by different businesses and organizations, which include:

- Redirection of food waste towards animal feed
- Development of organic compost and soil enrichers
- Generation of energy from food waste
- Value addition of products such as pharmaceuticals and cosmetics by extraction of nutrients from food wastes
- Repurposing of produce failing to meet specifications
- Repurposing of surplus produce by repackaging in different forms, such as marketing chopped vegetables for sale

Logical and transparent discussions on available information can generate an efficient forecast for food production, and hence, food loss due to surplus amounts can be avoided (Mena, Adenso-Diaz, and Yurt, 2011; Raak et al., 2017). Appropriate storage and transportation technology is very important in terms of decreasing food waste and loss. Improvising logistics using radio-frequency identification (RFID) technology and GPS allows effective circulation of information. Additionally, portable sensors enhance transportation and provide frequent and relevant information on shelf life. The first-in-first-out and first-expiring-first-out approaches can optimize inventory and reduce waste (Costa et al., 2013).

The prevention of food waste via intelligent packaging and enhanced shelf-life communication with consumers allows for sensible decision-making on storage, consumption, or proper disposal in instances of food waste. Cost and resources efficient utilization or repurposing of byproducts can reduce waste (Obersteiner et al., 2021).

16.7.5 THE INTERNET OF THINGS AND THE USE OF DIGITAL TECHNOLOGIES

Lately, FLW recovery has gained a new approach with the digitization of several tools (e.g., food sharing apps, data-driven farming) (Jagtap et al., 2019). Nevertheless, very little to negligible literature explains the effects of digital technologies on reducing FLW (Harvey et al., 2020; Mishra and Singh, 2018; Irani et al., 2018). Irani et al. (2018) provided an argument on the ability of these technologies to determine the impact of interventions on the reduction of FLW within the food security landscape. Traditional food movement within the FSC can be transformed with the existence of digital platforms that can create alternate food networks. This can lead to a cutback of several stages in the FSC, ultimately decreasing the probability of food wastage (Harvey et al., 2020).

Mishra and Singh (2018) explained that a representative case interprets strategies to minimize FLW utilizing data from Twitter to backtrack FSCs. Similarly, the internet of things (IoT, e.g. precision agriculture, smart farming) can be implemented to monitor the quality of food, manage food about to exceed its expiration date, and regulate the appropriate physical environment concerning temperature and humidity, which can potentially aid actors of FSC to control FLW. Moreover, technologies such as RFID can assist in lowering FLW via systematic and organized store layout and category and inventory management (Kamble et al., 2019).

16.7.6 Circular Bioeconomy Approaches to Reducing FLW

A circular bioeconomy is defined as the circularity (conservation) of materials, resources, and energy that mainly aims to reduce waste (here, food waste) by biotechnological approaches for renewability, replacing fossil fuels, and improving sustainability in the economy (Carus and Dammer, 2018). The extensive use of non-renewable resources in fossil fuels creates a strain on the linear-based bioeconomy. This prompted a more sustainable circular bioeconomy that also supports the UN's SDG 12 (sustainable consumption and production). Food waste is considered a potential source for generating biomass waste as a feedstock in a circular bioeconomy (Chiranjeevi, Dahiya, and Kumar, 2018). These feedstocks can be used to produce a diverse range of bio-based products with the help of multidisciplinary studies in science, management, and engineering (Amulya, Jukuri, and Mohan, 2015; Xiong et al., 2019). Biorefineries, biofertilizers, and bioelectricity are the three pillars of the circular bioeconomy that can utilize food waste through elimination, reduction, and reuse strategies to reduce FLW. Biorefineries can adopt biological methods such as fermentation, photosynthesis, and bioelectrogenesis as it has the potential to replace fossil fuels and encourage sustainable approaches with the least environmental impact.

The circular bioeconomy concept is developing rapidly and shaping towards an integrated vision. Recently, an innovative biorefinery was set up in Italy (August 2019) with a capacity of 750,000 tonnes of unused vegetable oils and fats to produce high-quality biofuels (Mak et al., 2020). This concept helps to reduce FLW by expanding bio-based industries and enhancing material flows within the system. LCA approaches such as LCA, LCC, and other environmental assessments of FLW in FSC help to raise awareness among policymakers to achieve a sustainable circular bioeconomy (Figure 16.5).

16.8 CONCLUSION AND FUTURE PERSPECTIVES

The increasing population needs more food, which requires more agricultural production and food processing to meet demands. While our resources remain constant, the food we produce is increasingly lost or wasted for many reasons. Increasing effects of FLW is equivalent to a country after

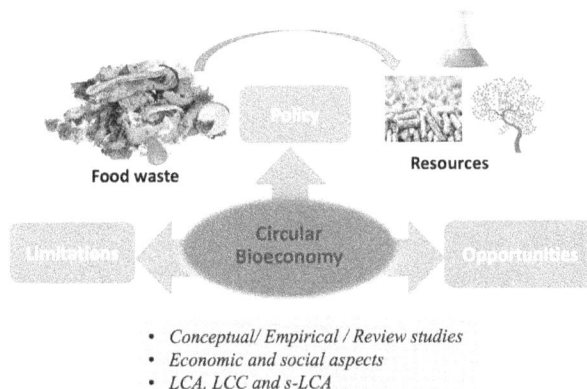

- *Conceptual/ Empirical / Review studies*
- *Economic and social aspects*
- *LCA, LCC and s-LCA*

FIGURE 16.5   Circular bioeconomy approaches to tackle FLW. (From Mak et al., 2020. With permission).

the combined GHG of US and China and their environmental impacts.. In this chapter, we have provided a brief overview of current scenarios of FLW in the FSC. Then the environmental impacts of FLW were described with case studies. A crucial factor in environmental impact assessments is LCA, which traces FLW from production to the end of its life cycle. LCA tools help to make stakeholders aware of developing policies and regulatory frameworks. Several countries have developed such tools and frameworks to deal with FLW in the future due to their massive increments in quantities of FLW. Lastly, future prevention strategies for FLW are discussed at length, such as behavioural changes, policy interventions, FLW quantifications, investments, use of IoT, and circular bioeconomy approaches that are expected to eventually reduce FLW and achieve SDGs.

REFERENCES

Al-Rumaihi, Aisha, Gordon McKay, Hamish R Mackey, and Tareq Al-Ansari. 2020. Environmental impact assessment of food waste management using two composting techniques. *Sustainability* 12(4):1595.

Albizzati, Paola Federica, Davide Tonini, Charlotte Boyer Chammard, and Thomas Fruergaard Astrup. 2019. Valorisation of surplus food in the French retail sector: Environmental and economic impacts. *Waste Management* 90:141–151.

Aldaco, R, D Hoehn, J Laso, M Margallo, J Ruiz-Salmón, J Cristobal, R Kahhat, P Villanueva-Rey, A Bala, and L Batlle-Bayer. 2020. Food waste management during the COVID-19 outbreak: A holistic climate, economic and nutritional approach. *Science of the Total Environment* 742:140524.

Amicarelli, Vera, and Christian Bux. 2021. Food waste in Italian households during the Covid-19 pandemic: A self-reporting approach. *Food Security* 13(1):25–37.

Amulya, K, Srinivas Jukuri, and S Venkata Mohan. 2015. Sustainable multistage process for enhanced productivity of bioplastics from waste remediation through aerobic dynamic feeding strategy: Process integration for up-scaling. *Bioresource Technology* 188:231–239.

Baig, Mirza B, Khodran H Al-Zahrani, Felicitas Schneider, Gary S Straquadine, and Marie Mourad. 2019. Food waste posing a serious threat to sustainability in the Kingdom of Saudi Arabia–A systematic review. *Saudi Journal of Biological Sciences* 26(7):1743–1752.

Bos-Brouwers, HEJ, Stephanie Burgos, Flavien Colin, Venice Graf, HWI van Herpen, JCM van Trijp, and EJ van Geffen. 2020. *Policy Recommendations to Improve Food Waste Prevention and Valorisation in the EU: Deliverable D 3.5. Refresh.* https://research.wur.nl/en/projects/resource-efficient-food-and-drink-for-the-entire-supply-chain

Brizi, Ambra, and Alessandro Biraglia. 2021. "Do I have enough food?" How need for cognitive closure and gender impact stockpiling and food waste during the COVID-19 pandemic: A cross-national study in India and the United States of America. *Personality and Individual Differences* 168:110396.

Busetti, Simone. 2019. A theory-based evaluation of food waste policy: Evidence from Italy. *Food Policy* 88:101749.

Buzby, Jean C, Hodan Farah Wells, Bruce Axtman, and Jana Mickey. 2009. *Supermarket Loss Estimates for Fresh Fruit, Vegetables, Meat, Poultry, and Seafood and Their Use in the ERS Loss-Adjusted Food Availability Data.* https://www.ers.usda.gov/publications/pub-details/?pubid=44309

Buzby, Jean C, and Jeffrey Hyman. 2012. Total and per capita value of food loss in the United States. *Food Policy* 37(5):561–570.

Caldeira, Carla, Valeria De Laurentiis, Agneta Ghose, Sara Corrado, and Serenella Sala. 2021. Grown and thrown: Exploring approaches to estimate food waste in EU countries. *Resources, Conservation and Recycling* 168:105426.

Caldeira, Carla, Valeria De Laurentiis, and Serenella Sala. 2019. Assessment of food waste prevention actions. Development of an evaluation framework to assess the performance of food waste prevention actions. https://publications.jrc.ec.europa.eu/repository/handle/JRC118276

Carus, Michael, and Lara Dammer. 2018. The circular bioeconomy—Concepts, opportunities, and limitations. *Industrial Biotechnology* 14(2):83–91.

Cattaneo, Andrea, Giovanni Federighi, and Sara Vaz. 2021. The environmental impact of reducing food loss and waste: A critical assessment. *Food Policy* 98:101890.

Chalak, Ali, Chaza Abou-Daher, and Mohamad G Abiad. 2018. Generation of food waste in the hospitality and food retail and wholesale sectors: Lessons from developed economies. *Food Security* 10(5):1279–1290.

Chalak, Ali, Chaza Abou-Daher, Jad Chaaban, and Mohamad G Abiad. 2016. The global economic and regulatory determinants of household food waste generation: A cross-country analysis. *Waste Management* 48:418–422.

Chan, Sander, Robert Falkner, Harro Van Asselt, and Matthew Goldberg. 2015. *Strengthening Non-State Climate Action: A Progress Assessment of Commitments Launched at the 2014 UN Climate Summit.* UN report

Chiranjeevi, P, Shikha Dahiya, and Naresh Kumar. 2018. Waste derived bioeconomy in India: A perspective. *New Biotechnology* 40(A):60–69.

Cicatiello, Clara, Emanuele Blasi, Claudia Giordano, Angelo Martella, and Silvio Franco. 2020. "If Only I Could Decide": Opinions of food category managers on in-store food waste. *Sustainability* 12(20):8592.

Clavreul, Julie, Dominique Guyonnet, and Thomas H Christensen. 2012. Quantifying uncertainty in LCA-modelling of waste management systems. *Waste Management* 32(12):2482–2495.

Corrado, Sara, Fulvio Ardente, Serenella Sala, and Erwan Saouter. 2017. Modelling of food loss within life cycle assessment: From current practice towards a systematisation. *Journal of Cleaner Production* 140:847–859.

Corrado, Sara, Carla Caldeira, Mattias Eriksson, Ole Jørgen Hanssen, Hans-Eduard Hauser, Freija van Holsteijn, Gang Liu, Karin Östergren, Andrew Parry, and Luca Secondi. 2019. Food waste accounting methodologies: Challenges, opportunities, and further advancements. *Global Food Security* 20:93–100.

Costa, Corrado, Francesca Antonucci, Federico Pallottino, Jacopo Aguzzi, David Sarriá, and Paolo Menesatti. 2013. A review on agri-food supply chain traceability by means of RFID technology. *Food and Bioprocess Technology* 6(2):353–366.

Curran, Mary Ann. 2017. Overview of goal and scope definition in life cycle assessment. In: *Goal and scope definition in life cycle assessment*, 1–62. Springer.

De Leo, Simonetta, Ines Di Paolo, Sabrina Giuca, and Francesca Giarè. 2015. *Lo spreco alimentare in Italia*, Government report. http://dspace.crea.gov.it/handle/inea/1180

Derqui, Belén, Teresa Fayos, and Vicenc Fernandez. 2016. Towards a more sustainable food supply chain: Opening up invisible waste in food service. *Sustainability* 8(7):693.

Dong, Yan, Simona Miraglia, Stefano Manzo, Stylianos Georgiadis, Hjalte Jomo Danielsen Sørup, Elena Boriani, Tine Hald, Sebastian Thöns, and Michael Z Hauschild. 2018. Environmental sustainable decision making–The need and obstacles for integration of LCA into decision analysis. *Environmental Science and Policy* 87:33–44.

Dora, Manoj, Shreyasee Biswas, Sonal Choudhary, Rakesh Nayak, and Zahir Irani. 2021. A system-wide interdisciplinary conceptual framework for food loss and waste mitigation strategies in the supply chain. *Industrial Marketing Management* 93:492–508.

Dupouy, Eleonora, and Mirjana Gurinovic. 2020. Sustainable food systems for healthy diets in Europe and Central Asia: Introduction to the special issue. *Food Policy* 96:101952.

Dusoruth, Vaneesha, Hikaru Hanawa Peterson, and Jennifer Schmitt. 2018. Estimating a local food waste baseline. *Journal of Food Products Marketing* 24(5):654–680.

Elginoz, Nilay, Kasra Khatami, Isaac Owusu-Agyeman, and Zeynep Cetecioglu. 2020. Life cycle assessment of an innovative food waste management system. *Frontiers in Sustainable Food Systems* 4:23.

Ellison, Brenna, and Jayson L Lusk. 2018. Examining household food waste decisions: A vignette approach. *Applied Economic Perspectives and Policy* 40(4):613–631.

Eriksson, Mattias, Ranjan Ghosh, Lisa Mattsson, and Alisher Ismatov. 2017. Take-back agreements in the perspective of food waste generation at the supplier-retailer interface. *Resources, Conservation and Recycling* 122:83–93.

Eriksson, Mattias, Simone Giovannini, and Ranjan Kumar Ghosh. 2020. Is there a need for greater integration and shift in policy to tackle food waste? Insights from a review of European Union legislations. *SN Applied Sciences* 2(8):1–13.

EU. 2016a. Estimates of European food waste levels. European Union, accessed 25 August. https://ec.europa.eu/food/safety/food-waste_en.

EU. 2016b. FUSIONS-food use for social innovation by optimising waste prevention strategies. European Union, accessed 28 August. http://www.eu-fusions.org/index.php/about-fusions.

FAO. 2011. What is food loss and food waste? Food and Agricultural Organization of the United Nations, accessed 10 August. http://www.fao.org/food-loss-and-food-waste/flw-data).

FAO. 2013. Food wastage footprint: Impacts on natural resources. Food and Agriculture Organization of the United Nations, accessed 28 August http://www.fao.org/3/i3347e/i3347e.pdf.

FAO. 2014. Definitional framework of food loss. Food and Agriculture Organization of the United Nations, accessed 28 August. http://www.fao.org/3/at144e/at144e.pdf.

FAO. 2015. The state of food insecurity in the world (SOFI) 2015. Food and Agriculture Organization of the United Nations, accessed August 28. http://www.fao.org/publications/card/en/c/c2cda20d-ebeb-4467-8a94-038087fe0f6e/.

FAO. 2019. The state of food and agriculture: Moving forward on food loss and waste reduction. Accessed 28 August. http://www.fao.org/3/ca6030en/ca6030en.pdf.

FAO. 2020. Food security and nutrition around the world in 2020. Food and Agriculture Organization of the United Nations, accessed 28 August.

Fattibene, Daniele, Francesca Recanati, Katarzyna Dembska, and Marta Antonelli. 2020. Urban food waste: A framework to analyse policies and initiatives. *Resources* 9(9):99.

Filimonau, Viachaslau, and Riaz Uddin. 2021. Food waste management in chain-affiliated and independent consumers' places: A preliminary and exploratory study. *Journal of Cleaner Production*319: 128721.

Gao, Si, Jingling Bao, Xiaojie Liu, and Asa Stenmarck. 2018. Life cycle assessment on food waste and its application in China. (Vol. 108, No. 4, p. 042037) *IOP Conference Series: Earth and Environmental Science.*

Garrone, Paola, Marco Melacini, and Alessandro Perego. 2014. Opening the black box of food waste reduction. *Food Policy* 46:129–139.

Garske, Beatrice, Katharine Heyl, Felix Ekardt, Lea Moana Weber, and Wiktoria Gradzka. 2020. Challenges of food waste governance: An assessment of European legislation on food waste and recommendations for improvement by economic instruments. *Land* 9(7):231.

Gentil, Emmanuel C, Anders Damgaard, Michael Hauschild, Göran Finnveden, Ola Eriksson, Susan Thorneloe, Pervin Ozge Kaplan, Morton Barlaz, Olivier Muller, and Yasuhiro Matsui. 2010. Models for waste life cycle assessment: Review of technical assumptions. *Waste Management* 30(12):2636–2648.

Giordano, Claudia, Luca Falasconi, Clara Cicatiello, and Barbara Pancino. 2020. The role of food waste hierarchy in addressing policy and research: A comparative analysis. *Journal of Cleaner Production* 252:119617.

Govindan, Kannan. 2018. Sustainable consumption and production in the food supply chain: A conceptual framework. *International Journal of Production Economics* 195:419–431.

Grainger, Matthew James, Lusine Aramyan, Katja Logatcheva, Simone Piras, Simone Righi, Marco Setti, Matteo Vittuari, and Gavin Bruce Stewart. 2018. The use of systems models to identify food waste drivers. *Global Food Security* 16:1–8.

Griffin, Mary, Jeffery Sobal, and Thomas A Lyson. 2009. An analysis of a community food waste stream. *Agriculture and Human Values* 26(1):67–81.

Halloran, Afton, Jesper Clement, Niels Kornum, Camelia Bucatariu, and Jakob Magid. 2014. Addressing food waste reduction in Denmark. *Food Policy* 49:294–301.

Hartikainen, Hanna, Lisbeth Mogensen, Erik Svanes, and Ulrika Franke. 2018. Food waste quantification in primary production–the Nordic countries as a case study. *Waste Management* 71:502–511.

Harvey, John, Andrew Smith, James Goulding, and Ines Branco Illodo. 2020. Food sharing, redistribution, and waste reduction via mobile applications: A social network analysis. *Industrial Marketing Management* 88:437–448.

Hebrok, Marie, and Casper Boks. 2017. Household food waste: Drivers and potential intervention points for design–An extensive review. *Journal of Cleaner Production* 151:380–392.

Heijungs, Reinout, Gjalt Huppes, and Jeroen B Guinée. 2010. Life cycle assessment and sustainability analysis of products, materials and technologies. Toward a scientific framework for sustainability life cycle analysis. *Polymer Degradation and Stability* 95(3):422–428.

Irani, Zahir, Amir M Sharif, Habin Lee, Emel Aktas, Zeynep Topaloğlu, Tamara van't Wout, and Samsul Huda. 2018. Managing food security through food waste and loss: Small data to big data. *Computers and Operations Research* 98:367–383.

Ishangulyyev, Rovshen, Sanghyo Kim, and Sang Hyeon Lee. 2019. Understanding food loss and waste—Why are we losing and wasting food? *Foods* 8(8):297.

ISO. 2006. Environmental management–life cycle assessment–Principles and framework–Requirements and guidelines. Accessed 8th September. https://www.iso.org/standard/37456.html.

Jagtap, Sandeep, Chintan Bhatt, Jaydeep Thik, and Shahin Rahimifard. 2019. Monitoring potato waste in food manufacturing using image processing and internet of things approach. *Sustainability* 11(11):3173.

Jereme, Innocent A, Chamhuri Siwar, Rawshan Ara Begum, Basri Abdul Talib, and Er Ah Choy. 2018. Analysis of household food waste reduction towards sustainable food waste management in Malaysia. *The Journal of Solid Waste Technology and Management* 44(1):86–96.

Johnson, Lisa K, Rebecca D Dunning, J Dara Bloom, Chris C Gunter, Michael D Boyette, and Nancy G Creamer. 2018. Estimating on-farm food loss at the field level: A methodology and applied case study on a North Carolina farm. *Resources, Conservation and Recycling* 137:243–250.

Jribi, Sarra, Hanen Ben Ismail, Darine Doggui, and Hajer Debbabi. 2020. COVID-19 virus outbreak lockdown: What impacts on household food wastage? *Environment, Development and Sustainability* 22(5):3939–3955.

Kamble, Sachin S, Angappa Gunasekaran, Harsh Parekh, and Sudhanshu Joshi. 2019. Modeling the internet of things adoption barriers in food retail supply chains. *Journal of Retailing and Consumer Services* 48:154–168.

Khoo, Hsien H, Teik Z Lim, and Reginald BH Tan. 2010. Food waste conversion options in Singapore: Environmental impacts based on an LCA perspective. *Science of the Total Environment* 408(6): 1367–1373.

Knowles, Tim, Richard Moody, and Morven G McEachern. 2007. European food scares and their impact on EU food policy. *British Food Journal*.109(1): 43–67.

Koester, Ulrich. 2014. Food loss and waste as an economic and policy problem. *Intereconomics* 49(6):348–354.

Kummu, Matti, Hans De Moel, Miina Porkka, Stefan Siebert, Olli Varis, and Philip J Ward. 2012. Lost food, wasted resources: Global food supply chain losses and their impacts on freshwater, cropland, and fertiliser use. *Science of the Total Environment* 438:477–489.

Lazell, Jordon. 2016. Consumer food waste behaviour in universities: Sharing as a means of prevention. *Journal of Consumer Behaviour* 15(5):430–439.

Li, Shaobin, Yuwei Qin, Jeyamkondan Subbiah, and Bruce Dvorak. 2020. Life cycle assessment of the US beef processing through integrated hybrid approach. *Journal of Cleaner Production* 265:121813.

Lins, Maísa, Renata Puppin Zandonadi, António Raposo, and Veronica Cortez Ginani. 2021. Food waste on foodservice: An overview through the perspective of sustainable dimensions. *Foods* 10(6):1175.

Mair, Simon. 2020. How will coronavirus change the world? *BBC Future* 31. https://www.bbc.com/future/article/20200331-covid-19-how-will-the-coronavirus-change-the-world

Mak, Tiffany MW, Xinni Xiong, Daniel CW Tsang, KM Iris, and Chi Sun Poon. 2020. Sustainable food waste management towards circular bioeconomy: Policy review, limitations and opportunities. *Bioresource Technology* 297:122497.

Mena, Carlos, Belarmino Adenso-Diaz, and Oznur Yurt. 2011. The causes of food waste in the supplier–retailer interface: Evidences from the UK and Spain. *Resources, Conservation and Recycling* 55(6):648–658.

MFH. 2021. The environmental impact of food waste. Move for Hunger, accessed 28 August. https://moveforhunger.org/the-environmental-impact-of-food-waste.

Mishra, Nishikant, and Akshit Singh. 2018. Use of twitter data for waste minimisation in beef supply chain. *Annals of Operations Research* 270(1):337–359.

Mogensen, Lisbeth, John E Hermansen, Niels Halberg, Randi Dalgaard, JC Vis, and B Gail Smith. 2009. Life cycle assessment across the food supply chain. *Sustainability in the Food Industry* 35:115.

Mourad, Marie. 2016. Recycling, recovering and preventing "food waste": Competing solutions for food systems sustainability in the United States and France. *Journal of Cleaner Production* 126:461–477.

Muth, Mary K, Catherine Birney, Amanda Cuéllar, Steven M Finn, Mark Freeman, James N Galloway, Isabella Gee, Jessica Gephart, Kristal Jones, and Linda Low. 2019. A systems approach to assessing environmental and economic effects of food loss and waste interventions in the United States. *Science of the Total Environment* 685:1240–1254.

Nicastro, Rosalinda, and Petronia Carillo. 2021. Food loss and waste prevention strategies from farm to fork. *Sustainability* 13(10):5443.

Obersteiner, Gudrun, Marta Cociancig, Sandra Luck, and Johannes Mayerhofer. 2021. Impact of optimized packaging on food waste prevention potential among consumers. *Sustainability* 13(8):4209.

Omolayo, Yetunde, Beth J Feingold, Roni A Neff, and Xiaobo Xue Romeiko. 2021. Life cycle assessment of food loss and waste in the food supply chain. *Resources, Conservation and Recycling* 164:105119.

Onat, Nuri Cihat, Murat Kucukvar, and Omer Tatari. 2014. Scope-based carbon footprint analysis of US residential and commercial buildings: An input–output hybrid life cycle assessment approach. *Building and Environment* 72:53–62.

Pajula, Tiina, Katri Behm, Saija Vatanen, and Elina Saarivuori. 2017. Managing the life cycle to reduce environmental impacts. In: *Dynamics of Long-Life Assets*, 93–113. Springer, Cham.

Papargyropoulou, Effie, Rodrigo Lozano, Julia K Steinberger, Nigel Wright, and Zaini bin Ujang. 2014. The food waste hierarchy as a framework for the management of food surplus and food waste. *Journal of Cleaner Production* 76:106–115.

Pappalardo, Gioacchino, Simone Cerroni, Rodolfo M Nayga Jr, and Wei Yang. 2020. Impact of Covid-19 on household food waste: The case of Italy. *Frontiers in Nutrition* 7:291.

Parfitt, Julian, Mark Barthel, and Sarah Macnaughton. 2010. Food waste within food supply chains: Quantification and potential for change to 2050. *Philosophical Transactions of the Royal Society of London Series B Biological Sciences* 365(1554):3065–3081.

Parry, Andrew, Paul Bleazard, and Koki Okawa. 2015. *Preventing Food Waste: Case Studies of Japan and the United Kingdom*, OECD Food, Agriculture and Fisheries Papers 76, OECD Publishing

Petit, Gaëlle, Caroline Sablayrolles, and Gwenola Yannou-Le Bris. 2018. Combining eco-social and environmental indicators to assess the sustainability performance of a food value chain: A case study. *Journal of Cleaner Production* 191:135–143.

Philippidis, George, Martina Sartori, Emanuele Ferrari, and Robert M'Barek. 2019. Waste not, want not: A bio-economic impact assessment of household food waste reductions in the EU. *Resources, Conservation and Recycling* 146:514–522.

Prajapati, Saugat, Sushil Koirala, and Anil Kumar Anal. 2021. Bioutilization of chicken feather waste by newly isolated keratinolytic bacteria and conversion into protein hydrolysates with improved functionalities. *Applied Biochemistry and Biotechnology*, 193: 1–19.

Priefer, Carmen, Juliane Jörissen, and Klaus-Rainer Bräutigam. 2016. Food waste prevention in Europe–A cause-driven approach to identify the most relevant leverage points for action. *Resources, Conservation and Recycling* 109:155–165.

Principato, Ludovica, Luca Ruini, Matteo Guidi, and Luca Secondi. 2019. Adopting the circular economy approach on food loss and waste: The case of Italian pasta production. *Resources, Conservation and Recycling* 144:82–89.

Raak, Norbert, Claudia Symmank, Susann Zahn, Jessica Aschemann-Witzel, and Harald Rohm. 2017. Processing-and product-related causes for food waste and implications for the food supply chain. *Waste Management* 61:461–472.

Read, Quentin D, Samuel Brown, Amanda D Cuéllar, Steven M Finn, Jessica A Gephart, Landon T Marston, Ellen Meyer, Keith A Weitz, and Mary K Muth. 2020. Assessing the environmental impacts of halving food loss and waste along the food supply chain. *Science of the Total Environment* 712:136255.

Redlingshöfer, Barbara, Sabine Barles, and Helga Weisz. 2020. Are waste hierarchies effective in reducing environmental impacts from food waste? A systematic review for OECD countries. *Resources, Conservation and Recycling* 156:104723.

Reichenbach, Jan. 2008. Status and prospects of pay-as-you-throw in Europe–A review of pilot research and implementation studies. *Waste Management* 28(12):2809–2814.

Riccaboni, Angelo, Elena Neri, Francesca Trovarelli, and Riccardo M Pulselli. 2021. Sustainability-oriented research and innovation in "farm to fork" value chains. *Current Opinion in Food Science*, 68(2): 189–194.

Richards, Timothy J, and Bradley Rickard. 2020. COVID-19 impact on fruit and vegetable markets. *Canadian Journal of Agricultural Economics/Revue Canadienne d'Agroeconomie* 68(2):189–194.

Rodrigues, Vasco Sanchez, Emrah Demir, Xun Wang, and Joseph Sarkis. 2021. Measurement, mitigation and prevention of food waste in supply chains: An online shopping perspective. *Industrial Marketing Management* 93:545–562.

Sakai, Shin-ichi, Tatsuhito Ikematsu, Yasuhiro Hirai, and Hideto Yoshida. 2008. Unit-charging programs for municipal solid waste in Japan. *Waste Management* 28(12):2815–2825.

Secondi, Luca, Ludovica Principato, and Tiziana Laureti. 2015. Household food waste behaviour in EU-27 countries: A multilevel analysis. *Food Policy* 56:25–40.

Setti, Marco, Luca Falasconi, Andrea Segrè, Ilaria Cusano, and Matteo Vittuari. 2016. Italian consumers' income and food waste behavior. *British Food Journal*, 118(7): 1731–1746

Shiller, Robert J. 2020. *Narrative Economics: How Stories Go Viral and Drive Major Economic Events.* Princeton University Press.

Slorach, Peter C, Harish K Jeswani, Rosa Cuéllar-Franca, and Adisa Azapagic. 2019. Environmental sustainability of anaerobic digestion of household food waste. *Journal of Environmental Management* 236:798–814.

Stone, Vicki, Bernd Nowack, Anders Baun, Nico van den Brink, Frank von der Kammer, Maria Dusinska, Richard Handy, Steven Hankin, Martin Hassellöv, and Erik Joner. 2010. Nanomaterials for environmental studies: Classification, reference material issues, and strategies for physico-chemical characterisation. *Science of the Total Environment* 408(7):1745–1754.

Thi, Ngoc Bao Dung, Gopalakrishnan Kumar, and Chiu-Yue Lin. 2015. An overview of food waste management in developing countries: Current status and future perspective. *Journal of Environmental Management* 157:220–229.

Thyberg, Krista L, and David J Tonjes. 2015. A management framework for municipal solid waste systems and its application to food waste prevention. *Systems* 3(3):133–151.

Thyberg, Krista L, and David J Tonjes. 2016. Drivers of food waste and their implications for sustainable policy development. *Resources, Conservation and Recycling* 106:110–123.

Tonini, Davide, Paola Federica Albizzati, and Thomas Fruergaard Astrup. 2018. Environmental impacts of food waste: Learnings and challenges from a case study on UK. *Waste Management* 76:744–766.

UNEP. 2013. UN report: one–third of world's food wasted annually, at great economic, environmental cost. United Nations Environment Programme, accessed 28 August. https://news.un.org/en/story/2013/09/448652.

UNEP. 2021. UNEP food waste index report 2021. United Nations Environment Programme, accessed 25 August. https://www.unep.org/resources/report/unep-food-waste-index-report-2021.

UNEP, and OECD. 2019. IISD (2019) masuring fossil fuel subsidies in the context of the sustainable development goals. *UN Environment.*

Van der Meulen, BMJ. 2010. Development of food legislation around the world. In: *Ensuring Global Food Safety Exploring Global Harmonization*, 5–69. Academic Press.

Vaqué, Luis González. 2017. French and Italian food waste legislation: An example for other EU member states to follow? *European Food and Feed Law Review* 12(3):224–233.

Vilariño, Maria Virginia, Carol Franco, and Caitlin Quarrington. 2017. Food loss and waste reduction as an integral part of a circular economy. *Frontiers in Environmental Science* 5:21.

Vittuari, Matteo, Paolo Azzurro, Sivia Gaiani, Manuela Gheoldus, Stephanie Burgos, Lusine Aramyan, Natalia Valeeva, David Rogers, Karin Östergren, and Toine Timmermans. 2016. Recommendations and guidelines for a common European food waste policy framework. *Fusions.*

Walia, Bhavneet, and Shane Sanders. 2019. Curbing food waste: A review of recent policy and action in the USA. *Renewable Agriculture and Food Systems* 34(2):169–177.

Wang, Erpeng, Ning An, Zhifeng Gao, Emmanuel Kiprop, and Xianhui Geng. 2020. Consumer food stockpiling behavior and willingness to pay for food reserves in COVID-19. *Food Security* 12(4):739–747.

Willet, W, J Rockström, B Loken, M Springmann, T Lang, S Vermeulen, T Garnett, D Tilman, F DeClerck, and A Wood. 2019. Food in the anthropocene: The EAT–Lancet Commission on healthy diets from sustainable food systems. *EAT Lancet* 393(10170):447–492.

WRAP. 2015. Household food waste in the UK. The Waste and Resources Action Programme, accessed August 28.

Xiong, Xinni, KM Iris, Daniel CW Tsang, Nanthi S Bolan, Yong Sik Ok, Avanthi D Igalavithana, MB Kirkham, Ki-Hyun Kim, and Kumar Vikrant. 2019. Value-added chemicals from food supply chain wastes: State-of-the-art review and future prospects. *Chemical Engineering Journal* 375:121983.

Zamri, Gesyeana Bazlyn, Nur Khaiyum Abizal Azizal, Shohei Nakamura, Koji Okada, Norul Hajar Nordin, NorÁzizi Othman, Fazrena Nadia MD Akhir, Azrina Sobian, Naoko Kaida, and Hirofumi Hara. 2020. Delivery, impact and approach of household food waste reduction campaigns. *Journal of Cleaner Production* 246:118969.

Zhao, Yan, and Wenjing Deng. 2014. Environmental impacts of different food waste resource technologies and the effects of energy mix. *Resources, Conservation and Recycling* 92:214–221.

Index

For Product Safety Concerns and Information please contact our EU
representative GPSR@taylorandfrancis.com
Taylor & Francis Verlag GmbH, Kaufingerstraße 24, 80331 München, Germany

www.ingramcontent.com/pod-product-compliance
Lightning Source LLC
Chambersburg PA
CBHW080712220326
41598CB00033B/5389